Interior Design Towards the Sustainable Environment: People, Environment, Design, Technology

Interior Design Towards the Sustainable Environment: People, Environment, Design, Technology

Topic Editors

Oleg Kapliński
Agata Bonenberg
Jan Słyk
Wojciech Bonenberg

MDPI

Basel • Beijing • Wuhan • Barcelona • Belgrade • Novi Sad • Cluj • Manchester

Topic Editors
Oleg Kapliński
Faculty of Architecture
Poznań University of Technology
Poznań
Poland

Agata Bonenberg
Faculty of Architecture
Poznan University of Technology
Poznan
Poland

Jan Słyk
Faculty of Architecture
Warsaw University of Technology
Warszawa
Poland

Wojciech Bonenberg
Faculty of Architecture
Poznań University of Technology
Poznań
Poland

Editorial Office
MDPI AG
Grosspeteranlage 5
4052 Basel, Switzerland

This is a reprint of the Topic, published open access by the journals *Buildings* (ISSN 2075-5309) and *Sustainability* (ISSN 2071-1050), freely accessible at: https://www.mdpi.com/topics/OU59ZRX6C0.

For citation purposes, cite each article independently as indicated on the article page online and as indicated below:

Lastname, A.A.; Lastname, B.B. Article Title. *Journal Name* **Year**, *Volume Number*, Page Range.

ISBN 978-3-7258-6886-5 (Hbk)
ISBN 978-3-7258-6887-2 (PDF)
https://doi.org/10.3390/books978-3-7258-6887-2

Contents

About the Editors

Oleg Kapliński

Oleg Kapliński is a professor of civil engineering and collaborates with the Faculty of Architecture, Poznań University of Technology, in Poznań, Poland. His academic achievements cover the theory of decision making, including multicriteria decision aiding; construction processes organization and modeling, including analysis of the phenomena of waiting, the phenomena of equilibrium, balancing of the construction processes in conditions of uncertainty, risk in management, network planning, and reliability of production systems; research on resentment and predilection to risk in the light of utility theory; work ethos; integrated design and management; and sustainable development. Currently, his research includes interactions between architects and engineers. He is a Doctor honoris causa of VGTU, Lithuania; an honorary member of the EURO Working Group: Operations Research in Sustainable Development and Civil Engineering (EWG-ORSDCE); and was recognized for his work by being awarded the Integral Humanism Medal.

Agata Bonenberg

Agata Bonenberg is a professor at the Faculty of Architecture at Poznan University of Technology and Director of the Institute of Interior Design and Industrial Design. Agata Bonenberg is also an elected member of the Committee of Architecture and Urban Planning of the Polish Academy of Sciences. Throughout her professional career, she has worked for renowned architectural firms, including the Renzo Piano Building Workshop in Genoa, Italy, and has contributed to a variety of architectural projects in Germany, Australia, and Scotland. From 2014 to 2021, she taught at the Scuola di Architettura e Società and Architettura Urbanistica Ingegneria delle Costruzioni at Politecnico di Milano. Her research focuses on sustainability in architecture and interior design, with an emphasis on the use of innovative materials such as mycelium-based composites for growing furniture and surface finishes. She also leads studies on the perception of interior spaces, supported by eye-tracking methods.

Jan Słyk

Jan Słyk is a professor of architecture at Warsaw University of Technology and was Vice-Dean and Dean of the Faculty of Architecture from 2012 to 2020. Currently, he is the Vice-Rector of Warsaw University of Technology and has been a member of the Steering Committee of ENHANCE European University Alliance since 2020. His research and didactic activity are focused on the methodology of design and IT implementation in architecture. He has authored international workshops exploring augmented architectural environments and has supervised doctoral and master's students writing their theses on topics related to IT techniques in architecture. As well as being the author of over seventy publications—two books, articles, and chapters—and the co-author/editor of collective monographs, he has also participated in international conferences and sat on scientific councils and editorial committees. He has led and participated in numerous international and national research projects. In addition to this, he is an active, practicing architect with a full design license, having authored over 70 design projects (housing, commercial, public buildings, mainly in Warsaw) and been the author or co-author of multiple award-winning architectural competition projects.

Wojciech Bonenberg

Wojciech Bonenberg is a professor at the Faculty of Architecture, Poznań University of Technology, where he served two consecutive terms as Dean. His scholarly work engages contemporary research methodologies, including data-driven design, evidence-based design, and AI-assisted generative design. His research interests encompass inclusivity in architectural practice, sustainable design, and energy efficiency in the built environment. He has authored over 200 architectural projects, notably trans-European border terminals within major transportation corridors linking Europe and Asia: the Swiecko Terminal in Frankfurt (Oder), the Koroszczyn Terminal, and the Olszyna Terminal in Forst. These works represent critical infrastructural interventions at continental interfaces. In recognition of his contributions, he was conferred a Doctor honoris causa by the Cracow University of Technology in 2023.

Preface

With the increased interest in sustainable development, the issue of "Interior Design Towards the Sustainable Environment" has gained greater significance. New paradigms in the field of the sustainable environment set new challenges for designers and define new areas and methods of research. This reprint includes 18 articles, presenting a collection of diverse scientific studies, and is an attempt to respond to the above-mentioned challenges. The subject matter is multifaceted and multidimensional. In their work, the authors introduce concepts that have already become permanently established in research on interior architecture. Examples include terms derived from neurology (sensory design, synaesthesia) as well as such notions as salutogenic design, spatial perception, quality of work life, well-being, and biophilia. A characteristic combination of three words – safety, satisfaction, and contentment – often appears in these articles. The research presented inspires actions aimed at inclusive design and sustainable well-being. Readers will have the chance to become acquainted with new achievements. We welcome you to read further.

Oleg Kapliński, Agata Bonenberg, Jan Słyk, and Wojciech Bonenberg
Topic Editors

buildings

MDPI

Editorial

Interior Design Towards the Sustainable Environment: People, Environment, Design, Technology—Editor's Comment

Oleg Kapliński

Faculty of Architecture, Poznan University of Technology, 60-965 Poznań, Poland; oleg.kaplinski@put.poznan.pl

1. Introduction

The present Topic concerns interior design with a focus towards the sustainable environment. It is precisely this branch of design that is distinctive, linking well-being with the technical solutions shaping our everyday living conditions, while simultaneously contributing to the development of a sustainable society.

The borderland between engineering and art, which interior architecture represents, constitutes a rewarding field for research and modelling, yet it is also entrusted with expectations of supporting the implementation and achievement of the Sustainable Development Goals (SDGs). Their implementation requires the consolidation of academic and economic entities. Research and innovative projects are indispensable: this Topic poses a genuine challenge to specialists. Research in this field is being conducted intensively. The present Topic addresses the interpenetration of such studies and their prestige within interior architecture, and within the technical and social infrastructure. It offers an opportunity to acquaint oneself with new achievements and to appreciate the significance of knowledge, the symbiosis of scientific disciplines, and the need for cooperation among specialists, as well as the identification and examination of phenomena together with their implications.

From the manuscripts submitted, following rigorous substantive selection by editors and reviewers, 17 articles were accepted and published. The Topic attracted considerable attention within the academic community; ultimately, 42 authors from 14 countries took part, with the largest groups representing Australia, India and Poland. A broader perspective on the distribution of authors by country is presented in Figure 1.

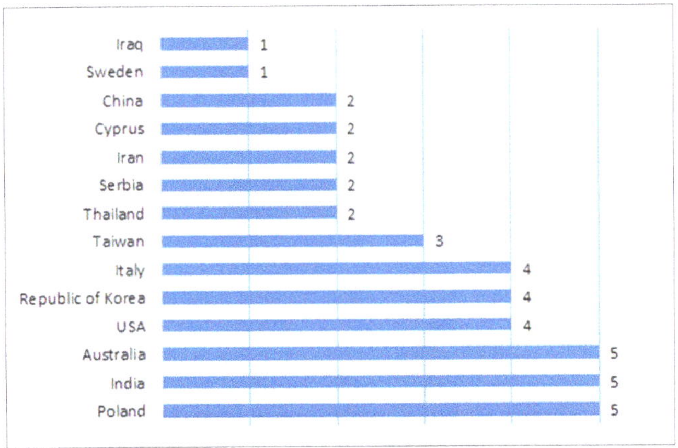

Figure 1. Distribution of authors by country.

2. Overview of the Contributions

The subject matter of the accepted articles is multifaceted, and has resulted in the application of diverse research techniques. Despite their originality, the contents of these articles intersect to some extent, and may be classified into four thematic groups. The criterion for assigning an article to one of these groups is its dominant substantive feature (the principal issue or disciplinary focus). Naturally, such classification remains subjective. The groups selected are as follows: (A) pro-life technical aspects; (B) human beings and spatial satisfaction; (C) pro-health actions; (D) a sustainable society.

A list of the contributions to the present Topic, arranged according to these groups, is presented in Table 1. To facilitate readers in locating issues of particular interest, Table 1 includes three additional columns answering three guiding questions: What is the subject of the research? What constitutes the research problem? What research techniques (or instrumentality) have been applied? The following paragraphs provide a synthetic collation.

Table 1. The content of the Topic "Interior Design Towards the Sustainable Environment: People, Environment, Design, Technology", and synthetic collation.

Author	Subject of the Research	Research Problem	Research Technique Instrumentality
Group A: Pro-life technical aspects			
Wang et al.	Wayfinding in interior environments: signage systems Experimental site: the National Taiwan University Hospital	Understanding wayfinding behaviour within healthcare facilities	Interactive experiments with virtual wayfinding Tracking the Wayfinding Index Blender and Unity Software
Cho and Suh	Spatial perception ability Integration of Virtual Reality (VR) technology into spatial design processes	Investigation of spatial comprehension ability	The Architecture and Interior Design Domain—specific spatial ability test The object–spatial imagery questionnaire Interactive VR AISAT mode and OSIQ mode
Park et al.	Quality of indoor lighting Daylight-induced glare Workspaces, Hongik University offices	Indoor light environment Location and direction of glare	Fisheye field of view Simulated and measured glare levels The human field of view (FOV)
Kubiak	Interior lighting design Lighting technologies	Sustainable design Discrepancies between sustainable design principles and actual user practices	Design Thinking Methodology Integrated Design Process
Ma and Yang	Scenarios of different spatial locations for Guidance Signage System Passenger flow routes	Enhancement of user navigation efficiency Maximising visual attention and ensuring smooth passenger flow Pedestrian movement	Depthmap10 and Massmotion11.5 software combined with eye tracking

Table 1. *Cont.*

Author	Subject of the Research	Research Problem	Research Technique Instrumentality
Group B: Human beings and spatial satisfaction			
Selanon and Chuangchai	Residential interior design for people with special needs Identifying relationships between home mobility and physical abilities Physical and psychological well-being of residents	Ergonomics in design Examination of physical abilities	Design recommendations regarding layout, furniture, flooring and lighting Level of mobility within interior spaces Statistical analysis (SPSS Statistics)
Dincer et al.	Interior design: bedrooms and sleep hygiene	Sleep environment research	A survey: multiple-choice and open-ended questionnaire items Likert scale Kruskal–Wallis H Test
Emami et al.	Waiting areas in dental clinics Arrangement styles	Shaping clinical environments that reduce patient stress	Discrete choice experiment Eight interior design attributes
Wlazły and Bonenberg	Visual perception of interior architecture Décor and spatial perception (living room scene) First impression effect	Spatial perception analysis using eye-tracking technology	Mobile eye-tracking device Décor
Krizmanić and Milovanović	The relationship between architectural attributes of workspaces and employee satisfaction Offices of three corporations in Belgrade	Territoriality in organisational environments Quality of Work Life (QWL)	The Workspace Characteristics Scale (WCS) The Workspace Satisfaction Scale (WSS) Multiple regression analyses Questionnaire-based methodology
Group C: Pro-Health Actions			
Mohammed et al.	Determining biophilic content in interiors: preschool facilities in Duhok (Iraq) Indoor Environment Regeneration	Relationship: architecture–human–nature Study of environmental characteristics	The Biophilic Design Matrix (BDM) Biophilic attributes
Rek-Lipczyńska	Health-oriented architecture Enhancing the health performance of the built environment in Poland	Participatory design Analysis of Polish legal regulations in the context of international standards: WELL Building Standard	Identifying key salutogenic factors supporting users' health and well-being Well-being standards analysis
Rek-Lipczyńska and Kowalska	Perception of space through various senses (sight, hearing, smell, touch and thermal sensations) Inclusiveness in interior design	The phenomenon of synaesthesia The concept of Design for All (the elderly and visually impaired people)	Comparative study on the perception of red and blue spaces
Group D: Sustainable Society			
Jason et al.	Housing instability, women's employment, and care homes within the Oxford House network (USA)	Maintaining ecological balance in communities Access to employment and reintegration of people recovering from addiction Community integration	Employment indicators Promotion of social and environmental policy

Table 1. *Cont.*

Author	Subject of the Research	Research Problem	Research Technique Instrumentality
Shetty et al.	Critical factors shaping the healthcare work environment Healthcare facilities (hospitals)	Physical environment: impact on hospital staff	Structured interviews Databases: PubMed and Scopus
Song and Dincer	Student dormitories Interaction: space use, space layout, user performance Shanghai universities	Spatial layout versus privacy, personalisation and satisfaction Mitigation of comfort deficit	PRISMA reporting. Cross-analysis. Chinese social platform Case study
Bellini et al.	Study of waiting spaces of local socio-healthcare facilities for primary care (Case della Comunità) New model of territorial primary care facility in Italy	Relationship between the architectural design of waiting areas in primary care centres and health promotion Age-it project	Perceptual analysis through interviews and questionnaires Sensory design

Group (A): Pro-life technical aspects. This thematic group embraces studies on workspaces, indoor lighting, signage systems in rooms and internal routes, pedestrian movement transit, the Design Thinking methodology, and spatial ability. Wang et al. [1] discuss the assessment of Wayfinding Behaviours on the basis of a case study conducted at the National Taiwan University Hospital (NTUH). The navigation system in healthcare buildings constitutes the subject of their analysis. The investigation of wayfinding was preceded by a preliminary test, after which a virtual-environment model of part of the hospital campus was developed for experimental purposes. The findings indicated appropriate indoor signage arrangements that improved the legibility and consistency of information. At the same time, the results elucidated behavioural patterns during wayfinding in complex spatial environments.

Research concerning spatial ability was undertaken by Cho and Suh [2]. Their article examines a significant topic within the fields of interior design and architecture, namely the assessment and comprehension of spatial abilities in professional education and practice. The study compares traditional static modes (a paper–desktop mode) with interactive virtual reality (VR) modes for assessing spatial abilities, employing instruments such as VR AISAT and VR OSIQ. The participants were students of the Architecture and Interior Design Domain, and the classrooms at one of the universities in Seoul served as the experimental setting.

Two subsequent articles address a closely related subject. They concern studies on the lighting (in various aspects) of workplaces, more specifically office environments. The quality of interior lighting represents an important element of interior architecture, owing to its impact on well-being. Both articles highlight the need to incorporate lighting design into the overall architectural design process. The first, written by Park et al. [3], presents actual measurements and simulations concerning daylight-induced glare. In the study of glare perception, a fisheye field of view (FOV) was applied. The measurements took into account the location and direction of glare and were conducted in office rooms at Hongik University in Seoul. The second article (Kubiak [4]) introduces the Design Thinking (DT) methodology in the lighting design of office rooms. Lighting devices should be designed and installed to efficiently and comfortably illuminate the work task and work environment. The article presents a holistic approach to lighting design, and the priority of requirements is oriented to users. The primary users of the office are administrative staff working at computer stations. The background of the analysis lies in the discrepancies (and their minimisation) between sustainable design and the use of the modernised interior. The author recommends applying the DT methodology during lighting modernisation works.

The article by Ma and Yang [5] concludes this technical group. It addresses the integration of various functional spaces and passenger flow routes within railway stations, underground areas, and commercial zones. The essence of the research is to determine spatial locations for the Guidance Signage System. The experiments presented highlight the areas that connect different spaces and track passengers' visual attention, providing scientific insights for the design and implementation of the national Transit-Oriented Development (TOD) model in China. The Tobii Glasses 2 eye-tracking device was employed in these studies.

Group (B) (Human and spatial satisfaction) is characterised by the following keywords: residents' physical abilities (in-home mobility), sleep hygiene, patient stress, first impression effect, quality of work life, ergonomics, and inclusive design. The subject matter of the articles in Group (B) is introduced by the paper authored by Selanon and Chuangchai [6]. The article presents evidence-based findings on ergonomic design that promote mobility, safety, and healthy living by embedding these considerations within interior design. The creation of residential spaces in the face of knowledge, integrative design, and even the ergonomics of ageing is complemented by inclusive design. The adjustment of interior decoration for people with disabilities, as well as the physical and mental well-being of the residents, the increase in self-esteem, and the reduction in dependence on caregivers, enrich the content of the article.

Dincer et al. [7] introduce the world of the sleep environment, focusing on the use of the bedroom, on sleep research, on user preferences and their needs. The majority of the surveyed Australians declared the desire for physical and architectural changes and the need for functional changes as well as changes in aesthetics and comfort. It was comfort that was indicated as the primary necessity. Of course, owners and tenants defined different preferences. The designer must pay attention to the relationship between the quality of sleep and the need for privacy.

Waiting rooms in a clinic/dental surgery, a visit, stress, a certain nervousness, time standing still—who doesn't know that pain? Help may come from the interior architect: it is possible to shape clinical/consulting environments in such a way as to alleviate patients' stress. Experiments and proposals for interior arrangements were carried out in three Iranian clinics. The respondents made use of computer-generated images of waiting rooms. Eight attributes of the interior design of the waiting rooms were taken into account; 'ambient lighting' proved to be the most important attribute in increasing patient comfort. The article by Emami et al. [8] harmonises with the themes of health, arrangement styles, and well-being.

Wlazły and Bonenberg [9] introduce the reader to visual perception in interior architecture. The research investigates a fundamental question of how decorative elements influence spatial perception. Indirectly, it also explores how different individuals—architects, users, and developers—visually engage with their interior surroundings. The living room served as the study environment, and the research instrument was a mobile eye-tracking device. A clear impact of decorative styling on the design process was identified. This is where we encounter the effect of first impressions.

When designing workplaces, attention should be paid to both employee and spatial satisfaction. We enter the sensory and sociological dimensions of design: Krizmanić and Milovanović [10] shared the results of a study identifying predictors of workplace satisfaction, namely aesthetics, flexibility, and a sense of belonging. The empirical setting comprised the offices of three corporations in Belgrade. An assessment of Quality of Work

Life (QWL) was accompanied by the Workspace Satisfaction Scale (WSS). The title of the article is telling: From Space to Satisfaction.

Group (C) is Pro-Health Activities, including biophilic design, health quality of the environment, legal regulations, standards, including the WELL Building Standard, salutogenic design, synaesthesia, and Design for All. From the notion of well-being, we move to biophilic design; that is, to a direction in interior design. This represents a path towards better well-being through interiors inspired by nature (a healthy space for people). Mohammed et al. [11] present a study of biophilic content in preschool buildings in a town in Iraq. The existence of biophilic features and attributes was examined, and the Biophilic Interior Design Matrix was used for their identification. Interest in the relationship between architecture, human beings, and nature is becoming increasingly common. This is an interesting aspect of sustainability.

Another article [12] offers a contextual literature review of salutogenic design, sustainable development, and health-focused architecture. It discusses the tension between the pathogenic model that has permeated Polish regulations and the salutogenic model. The author identified key salutogenic factors and compared Polish regulations with international standards, primarily with the WELL Building Standard. The discussion as a whole is conducted in the spirit of sustainable, human-centred design.

Research on the role of colour perception extending beyond the visual modality is presented in [13]. The article emphasises how multisensory colour coding can support accessibility and well-being in interior spaces. In this case, colours can be coded and perceived through senses other than sight. Particular attention was paid to the needs of older adults and individuals with visual impairments. Three concepts—the phenomenon of synaesthesia, inclusivity in interior design, and Design for All—are central to the present discussion.

Group (D) (A Sustainable Society) is formed by strong values, that is, recovery homes in the Oxford House network, addiction recovery homes, student dormitories, community homes (Case della Comunità), healthcare work environments, social and environmental policy (case study of treatment through work), and factors shaping the work environment. A significant social problem, situated within the network of recovery homes, is presented by Jason et al. [14]. The subject of their research is women with addiction and the sustainable society, while the objects are workplaces, the employment of women in communities, and housing problems. The exploration of the relationship between women's employment at the community level and internal variables in care homes within the Oxford House network (USA) is the essence of this article. The background in care homes, including the importance of community integration for sustainable development, is also articulated.

The main issue addressed by the following article [15] is a comprehensive understanding of the factors of the physical environment affecting personnel in healthcare institutions, which has been analysed through a systematic review of the existing literature. The physical environment is considered in three dimensions, which are architectural design, interior design, and environmental elements. The key element of the study is the search for determinants of staff satisfaction and the building of community in healthcare facilities.

The use of space, its layout, and the efficiency of users is the domain of the research presented in [16]. The subject of consideration is student rooms, and the background is room occupancy. Privacy, personalisation, and satisfaction are discussed. In turn, the article [17] focuses on the design of waiting rooms in the Italian "Casa della Comunità," aiming to promote health, active ageing, and social interaction. The theoretical framework rests on three pillars: territorial care, waiting space, and sensory design. Eight spatial typologies are discussed within the context of Italy's new Casa della Comunità healthcare model. The article draws attention to the spatial requirements of waiting areas, and in particular to how sensory design and environmental configurations enhance psychological comfort, social interaction, physical activity, and the adoption of healthy behaviours.

3. Discussion and Comments

A review of the contributions and achievements of the present Topic makes it possible to formulate several findings.

A technical theme can be traced across each of the four thematic groups discussed in the contributions. It will always exist in interior architecture, to a greater or lesser extent, regardless of time. However, in the present issue, this aspect has not been sufficiently emphasised: there are no articles devoted entirely to the technical equipment of the building, HVAC, or domotics (home automation system). These and related issues are and will remain indispensable in design, and technological progress continues to make a permanently astonishing advancement. The research instruments and methods applied are adequate for the problems analysed. Their diversity is evident. In further studies, attention should be given, among other areas, to the Internet of Things (IoT) and to AI + BIM, which is the combination of Artificial Intelligence (AI) with Building Information Modelling (BIM) technology. We already know that AI is revolutionising the construction industry (for example [18,19]). It is not yet known to what extent it will strengthen design within the field of interior architecture.

The synthesis of the articles assembled in the present Topic indicates high cognitive value. The very identification and subsequent tackling of a refined phenomenon, and the kinds of approaches to design, ennoble the scientific worth of the collection. A juxtaposition of these notions is presented in Figure 2. All those listed can be included among the new trends in environmental and social design, and more precisely Human-Centred Design and Sustainable Design. The overview of instrumentality is presented in Table 1. All this confirms that the research issues in the domain of interior architecture are multi-layered and multi-parametric. Even the WELL Building Standard [20] included in this collection contains 10 health-related parameters (10 concept areas and 120 features). In a sphere with an abundance of subjectivity and artistry, it is still difficult today to measure qualitative factors by quantitative methods. What is important is that all the papers submitted have a research character or deal with research.

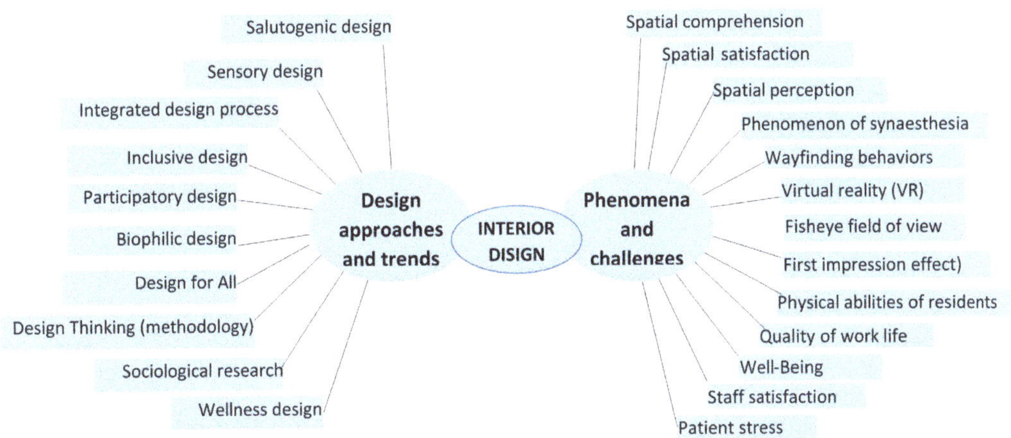

Figure 2. The cognitive value of a collection of articles. Trends used in research and in environmental design.

The participation of authors from fourteen countries and the range of topics raise the question of whether, in the era of globalisation, research results can be transferred to other regions. Chinese problems concerning the solution of communication (pedestrian signage [5]), or the Italian experiences of creating new community care homes [17], the outcomes of Design for All [13], and similar examples, may be utilised by other designers, though perhaps with a degree of caution. It is easier to transfer experiences from the implementation of biophilic solutions [11]; it is somewhat harder to replicate the results of the previously mentioned organic design, which will certainly differ. In all four groups of articles, themes concerning styles (within interior architecture and arrangement) feature subtly. Architects are familiar with the emergence, interpenetration, and disappearance of styles; they recognise changes along the axis of time and their evolution, for example in the context of population migration. Migrants most often carried with them elements of their own culture, including architectural and interior styles. For instance, nineteenth-century emigrants from Pomerania (from Germany and Poland) brought to Brazil half-timbered structures, the layout of rural settlements, traditions of shared space, and the rustic style [21], whereas the wave of settlers from Bamberg (Bavaria, Germany) to Poland in the eighteenth century introduced farmsteads with an enclosed courtyard, characteristic wooden furniture—dowry chests, benches, and cupboards—and embroidery and textiles in the Bavarian style [22]. The remaining aspects are connected with the process of cultural assimilation. However, the persistent standardisation of styles (of interiors) is premature, as confirmed by Lefaivre and Tzonis [23].

There is an important feature characterising all articles submitted to this Topic. Regardless of whether the subject is housing or workplaces, or whether the facilities are social, industrial, or involve pedestrian areas, their foremost qualities are safety, satisfaction, and contentment. The subject is the human being. This accords with the global trend in interior architectural design [24,25], and the attributes emphasised in papers on wellness design have become a permanent component of sustainability.

Health and well-being, discussed in the articles of Group (C) and partly in Group (B), relate not only to the role of the healthcare environment, care homes, or student residences, but also to the enhancement of psychological comfort, social interaction, and the health quality of the interior environment. This is manifested especially in Biophilic Design and Organic Design (see [11]). The interior architect distinguishes between these two concepts: biophilic design is the creation of spaces integrating people with nature, whereas organic design focuses on forms inspired by nature. If, following Coulthard (paraphrasing the title of his book [26]), biophilia in interior architecture can be defined as YOU + NATURE + HOME, then it represents an exceptionally expressive component of sustainability.

4. Conclusions

The subject matter of interior architecture develops at the intersection of architecture and arrangement; it is an attractive field and offers a rewarding area for scientific research. However, it is also multidimensional, even multidisciplinary (for instance, combining neuroscience, psychology, and design). It requires refined research methods and becomes a collective endeavour. Although it is impossible to encompass the entire complexity of the field, every step (whether by the designer, developer, or researcher) is both intriguing and responsible. The processes, studies, and projects presented in this Topic are not randomised but purposeful.

The triad of values mentioned, namely safety, satisfaction, and contentment, is supported by the four thematic groups of articles: pro-life technical aspects, human beings and spatial satisfaction, pro-health actions, and a sustainable society. What is clearly visible is the manifestation of the primacy of the human being in various aspects of public life—in design, in science, and in research on interior architecture.

It is worth reviewing the research findings.

Funding: This research received no external funding.

Conflicts of Interest: The author declares no conflicts of interest.

References

1. Wang, C.-Y.; Chen, C.-I.; Zheng, M.-C. Exploring Sign System Design for a Medical Facility: A Virtual Environment Study on Wayfinding Behaviors. *Buildings* **2023**, *13*, 1366. [CrossRef]
2. Cho, J.Y.; Suh, J. Spatial Ability Performance in Interior Design and Architecture: Comparison of Static and Virtual Reality Modes. *Buildings* **2023**, *13*, 3128. [CrossRef]
3. Park, J.; Lee, K.; Kim, K. Comparative Validation of Light Environment Simulation with Actual Measurements. *Buildings* **2023**, *13*, 2742. [CrossRef]
4. Kubiak, K. Design Thinking in Lighting Design to Meet User Needs. *Sustainability* **2024**, *16*, 3561. [CrossRef]
5. Ma, X.; Yang, Q. Research on the Spatial Location Design of Guidance Signage Systems to Connect the Space of Transit-Orientated Development Sites Based on Multi-Software Analysis. *Buildings* **2025**, *15*, 683. [CrossRef]
6. Selanon, P.; Chuangchai, W. Residential Interior Design for People with Special Needs in Thailand Based on Physical Abilities: Age, Gender, and Living Environment Considerations. *Buildings* **2023**, *13*, 2170. [CrossRef]
7. Dincer, D.; Tietz, C.; Dalci, K. Beyond Sleep: Investigating User Needs in Today's Bedrooms. *Buildings* **2024**, *14*, 1061. [CrossRef]
8. Emami, M.; Pazhouhanfar, M.; Stoltz, J. Evaluating Patients' Preferences for Dental Clinic Waiting Area Design and the Impact on Perceived Stress. *Buildings* **2024**, *14*, 3160. [CrossRef]
9. Wlazły, W.; Bonenberg, A. Modulating Perception in Interior Architecture Through Décor: An Eye-Tracking Study of a Living Room Scene. *Buildings* **2025**, *15*, 48. [CrossRef]
10. Krizmanić, V.; Milovanović, A. From Space to Satisfaction: Investigating Architectural Interior Determinants of Quality Work Environments. *Buildings* **2025**, *15*, 2256. [CrossRef]
11. Mohammed, I.; Onur, Z.; Çağnan, Ç. An Exploration of Biophilic Design Features within Preschool Interiors. *Sustainability* **2023**, *15*, 11913. [CrossRef]
12. Rek-Lipczyńska, A. Salutogenic Factors and Sustainable Development Criteria in Architectural and Interior Design: Analysis of Polish and EU Standards and Recommendations. *Sustainability* **2025**, *17*, 9661. [CrossRef]
13. Rek-Lipczyńska, A.; Kowalska, A. Multisensory Coding of Red and Blue in Interior Design for Older Adults and Visually Impaired Users: An Inclusive Design Perspective. *Sustainability* **2025**, *17*, 9381. [CrossRef]
14. Jason, L.A.; Bell, J.S.; Bobak, T.J. Positive Associations between Women's Employment at the Community-Level and Opportunities and Interpersonal Sharing within Recovery Homes. *Sustainability* **2024**, *16*, 6551. [CrossRef]
15. Shetty, R.S.; Kamath, G.B.; Rodrigues, L.L.R.; Nandineni, R.D.; Shetty, S.R. The Impact of the Physical Environment on Staff in Healthcare Facilities: A Systematic Literature Review. *Buildings* **2024**, *14*, 2773. [CrossRef]
16. Song, X.; Dincer, D. Spatial Constraints and User Adaptation: A Mixed-Methods Study of High-Density Students Dormitories. *Buildings* **2025**, *15*, 4330. [CrossRef]
17. Bellini, E.; Setola, N.; Rossi, L.; Stara, V. Active and Healthy Case Della Comunità: Model Research for Spatial Requirements of Waiting Spaces. *Sustainability* **2025**, *17*, 9467. [CrossRef]
18. Li, J.; Liu, Z.; Han, G.; Demian, P.; Osmani, M. The Relationship Between Artificial Intelligence (AI) and Building Information Modeling (BIM) Technologies for Sustainable Building in the Context of Smart Cities. *Sustainability* **2024**, *16*, 10848. [CrossRef]
19. Alnaser, A.A.; Elmousalami, H. Exploring Critical Success Factors of AI-Integrated Digital Twins on Saudi Construction Project Deliverables: A PLS-SEM Approach. *Buildings* **2025**, *15*, 3543. [CrossRef]
20. *WELL v2: Dynamic, Resilient, Validated: The Next Version of the WELL Building Standard™*; International WELL Building Institute: New York, NY, USA, 2020; Available online: https://kb-well-support.s3.amazonaws.com/space-5/mediaLibrary/2526/well-building-standard-v2-q4-2020-wellapv2-02-03-23.pdf (accessed on 20 September 2025).
21. Angoneze-Grela, E. Tożsamość migrantów a tożsamość miejsca: Architektura wiejska Pomorza Zachodniego w świetle masowych migracji ludności w XIX i XX wieku. *Zesz. Nauk. Politech. Poznańskiej Archit. Urban. Archit. Wnętrz* **2021**, *7*, 39–53. [CrossRef]
22. Kapliński, O. The Creeping Gentrification of a Street: The Case of the Historic Oval-Shaped Kościelna Street in Poznań. *Architecture* **2025**, *5*, 13. [CrossRef]
23. Lefaivre, L.; Tzonis, A. *Architecture of Regionalism in the Age of Globalization: Peaks and Valleys in the Flat World*; Routledge: London, UK, 2021.
24. McGee, S. *The Art of Home*; Harper Horizon: Nashville, TN, USA, 2023.

25. Fiell, C.; Fiell, P. *Design of the 20th Century*; Taschen: Cologne, Germany, 2022.
26. Coulthard, S. *Biophilia: You + Nature + Home. A Handbook for Bringing the Natural World into Your Life*; Octopus Publishing Group: London, UK, 2020.

buildings

MDPI

Article

Exploring Sign System Design for a Medical Facility: A Virtual Environment Study on Wayfinding Behaviors

Ching-Yuan Wang, Ching-I Chen and Meng-Cong Zheng *

Department of Industrial Design, National Taipei University of Technology, Taipei 10608, Taiwan; t109588008@ntut.org.tw (C.-Y.W.); zoeychen116@gmail.com (C.-I.C.)
* Correspondence: zmcdesign@gmail.com; Tel.: +886-2-27712171 (ext. 2827)

Abstract: The National Taiwan University Hospital (NTUH) is a significant institution in modern medicine in Taiwan. Its West Campus, the oldest medical center in the country, has been recognized as a municipal historic site due to its architectural and historical importance. However, visitors have faced navigation difficulties for a long time. To address this issue, a study was conducted to find navigation solutions within the constraints of architectural structure, hardware, and busy crowds. Blender and Unity were used to recreate the environment, and interactive virtual wayfinding experiments were conducted with 64 participants divided into two groups. Each group completed 12 tasks in two scenarios, and their task performance, wayfinding behavior, and questionnaire responses were collected to evaluate the overall environment. The pretest helped identify problems in the existing signage system and weak areas, leading to redesigning of a new signage system. The main objective of the posttest was to evaluate the new design's effectiveness. Our research contributes to future signage system layout design references, enhancing readability and information coherence while recommending locations within medical facilities. We set specific design standards to facilitate wayfinding signage systems in complex environments.

Keywords: wayfinding; sign system design; virtual environment; medical facility; indoor navigation; signage design

1. Introduction

Finding one's way in a building is a prerequisite for successfully achieving goals [1]. However, locating a destination in a medical institution is also one of the first steps for a patient to seek treatment. Wayfinding problems can lead to confusion, frustration, anger, stress, increased blood pressure, headaches, and fatigue [2,3]. Staff also constantly interrupt work to provide directions, wasting staff time and reducing concentration [4]. Therefore, improving the indoor wayfinding system is crucial to promote the user's wayfinding experience and spatial awareness.

National Taiwan University Hospital (NTUH) is the core institution of modern medicine in Taiwan. The NTUH Original Building, also known as West Campus, is the oldest medical center in Taiwan and has been registered as a municipal monument due to its architectural and historical significance. The existing interior layout is configured with a central corridor throughout the building. Still, the West Campus is steeped in complexity and sign systems due to its long history, new services, and surrounding expansion. Visitors quickly get lost in the hospital, which leads to the need for some volunteers to help guide the way. However, as the West Campus is a municipality-designated monument, its historical value must be retained, so it is difficult to arbitrarily change its space and interior decoration. So, the sign system is more critical as environmental information can be added additionally and is less expensive to change.

In summary, providing a clear and effective sign system for visitors to find their way into medical institutions is essential. Still, since NTUH West Campus is a monument, the optimization of the signage system is compared to changing the structure of the building or interior design easier to achieve. Our study considered the busyness of the medical environment and employed experiments in virtual environments.

1.1. Wayfinding

Wayfinding is an indispensable behavior in daily life, a concrete socialization activity that occurs almost daily in our lives [5]. The term wayfinding was first coined by Kevin Lynch [6] in his book *The Image of the City*, in which he defined it as one of the elements of urban intentions as "the process of using spatial and environmental information to navigate to a destination". The term was not widely used until the 1970s, when it largely replaced the term "spatial orientation" used in academic writing [7]. Allen [8] believes that wayfinding is a motivated, purposeful process. He divides wayfinding tasks into three categories according to functional goals, the most challenging of which is that when people move in unfamiliar environments, they are more likely to have a low level of awareness of the environment and cause a temporary loss of the situation [9–12].

Down and Stea [13] also define the wayfinding process as the following four stages: positioning orientation, path decision-making, path monitoring, and target confirmation. The most crucial factor in Down and Stea's definition is that the third stage measures the repeated recursive behavior of all decisions based on continuous environmental cognition. So, Conroy-Dalton [14] has a new definition of wayfinding: wayfinding refers to reaching a destination through a constantly recursive route selection process while assessing spatial decisions before the continuous perception of the environment.

1.2. Sign Systems in Indoor Space

Wayfinding is challenging for many, especially in complex buildings such as hospitals, airports, and office buildings [7]. Many environmental factors affect wayfinding behavior in indoor spaces. The layout configuration of buildings, the degree of difference between areas, visibility, and signs are essential factors in predicting wayfinding behavior [1]. Jamshidi et al. [15], based on the four factors proposed by Weisman [1] and Lynch's [6] study of the outdoor environment, proposed several environmental factors: area, edge, path, node, landmark, plan configuration, sign, map, etc. This paper mainly focuses on the research results of wayfinding related to medical institutions, which are divided into three aspects: layout configuration, regional differences, and signs and landmarks.

A study pointed out that the layout's complexity negatively impacts the wayfinding behavior, and a space with high integrity can improve wayfinding efficiency [16]. In addition to layout configuration, support for regional differences is required. Some studies found that in polyclinics, symmetrical building structures and path differences are less repetitive and monotonic, which makes it easier for more people to feel lost, and more people need the uniqueness of the area to help find their way [17–20]. However, the actual outpatient space cannot be changed arbitrarily, so it is even more important to add additional environmental information that can be added and is less expensive to change, such as signs and landmarks. If signs are appropriately set and positioned, they can help reduce wayfinding difficulties, while the presence of landmarks can also reduce cognitive workload and promote higher wayfinding success rates [21].

The sign system represents clear information in the overall configuration and structure of the environment. As the [14] functionality of the medical facility increases, the expansion leads to an increasingly complex building layout, and the user's wayfinding in the building is also negatively affected [3,17]. When people are unfamiliar with the environment or the environment is too complex for people to fully familiarize themselves, the demand for signs also increases [19,22–25]. Each site in a medical institution has information that must be conveyed to users. The information may be too complex and invalidate the signage if not integrated and managed. So, providing clear and unambiguous signs in medical institutions is vital to help users quickly obtain the correct information and arrive at their destination without a burden [26,27].

Color coding is also a consideration for studies that process complex information quickly. Making the most of color-coded communication helps readers decipher the encoded data and maximizes the effectiveness of our communication [28–31]. Some studies show that color coding plays a role in classifying complex information [32–34] and even helps achieve better performance in mazes [34].

1.3. Wayfinding Research in the Virtual Environment

Since NTUH West Campus is a municipal monument, its historical value must be preserved, and it is difficult to modify or change it arbitrarily regarding space and interior decoration. The sign system is relatively easy to implement at a low cost and can be added and changed in the building. However, among the facilities of established medical institutions, it is difficult to make a rigorous comparison through any form of wayfinding function research. We created virtual environments to mirror the physical world as a "digital twin" for physical assets, processes, or systems—the term derived from Michael Grieves' concept [35], creating "cheap information bits" that virtualize "expensive physical atoms" to improve the performance benefits of design and testing [36]. Advances in the technology of "digital twins" (DTs) have allowed us to understand the nature of reality and its virtual forms better [37]. This concept has also been tested in practice [38]. The diversity of DTs in different application scenarios, such as cultural heritage and medical treatment, shows that the DT concept has been widely accepted by academia and industry [39].

In recent years, the rapid development of virtual environment technology, from computer desktops to fully immersive virtual environment settings, has provided new experimental methods for studying wayfinding behavior and spatial knowledge acquisition [14,20,40–44]. More and more of the fielding of wayfinding research is carried out in a virtual environment. With this technology, all changes in the experimental setting can be accomplished with a lower financial cost and less time [45,46]. Many researchers have also compared the wayfinding cognitive process and spatial knowledge acquisition of the virtual environment with prototypes of their natural world. They found that the way people find their way in the virtual environment and the natural world is similar, and there is no noticeable difference [14,47].

Therefore, many scholars have begun using the virtual environment as an experimental method to explore the influence of different signs and internal design conditions on hospital wayfinding behavior. Saleh Kalantari et al. [48] built a virtual reality test platform to evaluate the impact of signage and hospital internal design conditions on wayfinding tasks. Al-Sharaa et al. [49] also investigated the existing wayfinding system in Malaysian public hospitals through two virtual environment experiments, conducted the experiment again with a new proposal, and cross-compared the two results. The results show that the time spent by participants in the task in the new proposal is reduced, and the walking distance is shortened, which proves that the new proposal can effectively improve the wayfinding efficiency. These studies demonstrate that virtual environments can be an effective tool for participatory and evidence-based design, avoiding the need for costly changes after physical construction.

Based on the previous background and motivation, the aim of this study was to use the virtual environment to discover the design of sign systems in NTUH West Campus. We wanted to observe participants' wayfinding behavior in medical institutions and their ways of viewing signs through experiments. We also wanted to summarize the problems and areas for improvement in the existing sign system. The basic design principles of the signage system are formulated, a new design proposal is proposed, and then the same experimental process is used to verify whether the new design proposal can effectively improve the wayfinding performance of the participants. Finally, the design focus and suggestions for the future logo system are further summarized. We also propose the advantages and disadvantages of using a virtual environment for research.

2. Method

The advantage of this study is that the experimental process did not affect the hospital's operation, and it can control the variation and scope of the experimental field and save a lot of time and physical strength for testing. Participants conducted the task in the first-person view, giving researchers a clearer idea of what they were looking at. However, there are still some limitations in this experimental method, that is, the influence of the "crowd" on feelings and wayfinding behavior.

2.1. Situation Investigation and Material Design

2.1.1. Field Exploration

We visited the location and recorded each sign in NTUH West Campus. Wall signs are used as a spatial identification system in addition to the hanging type sign system of the circulation corridor (Figure 1) at the entrance hall, around the corner, around the elevator and the stairs, and outside each clinic or examination room (Figure 2). We can find how different indications are distributed in the figure.

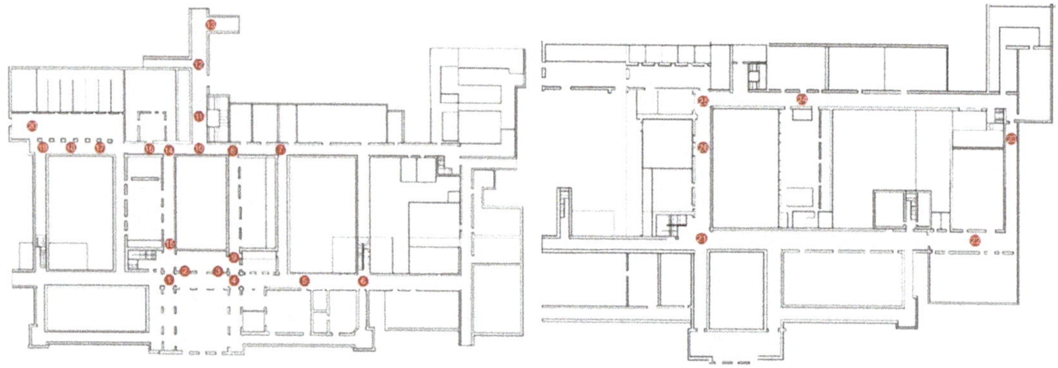

Figure 1. Locations and distribution of the walkway's hanging-style sign. The left graph is the first floor; the right is the second floor. (The numbers in the picture are for researchers' records only).

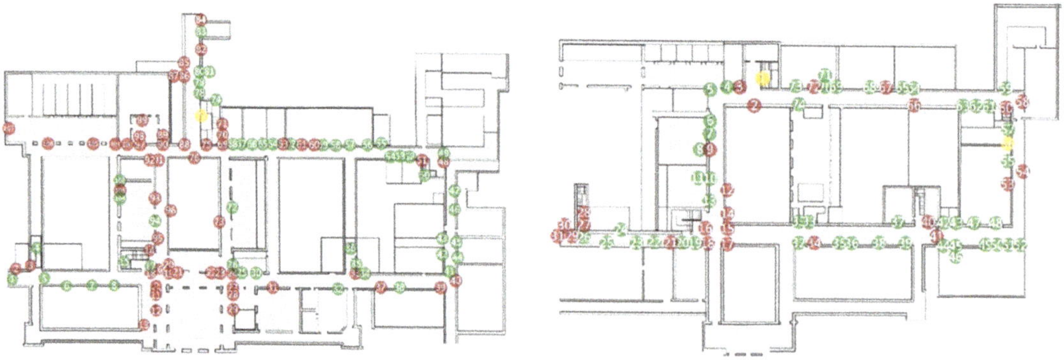

Figure 2. The location of the walkway's wall signs (the left graphic is the first floor; the right is the second floor). (The numbers in the picture are for researchers' records only).

We also found that some signs were pasted to add or remove information, resulting in a visual crowding of layout information (Figure 3). The various wall signs also have inconsistencies in visual design (Figure 4).

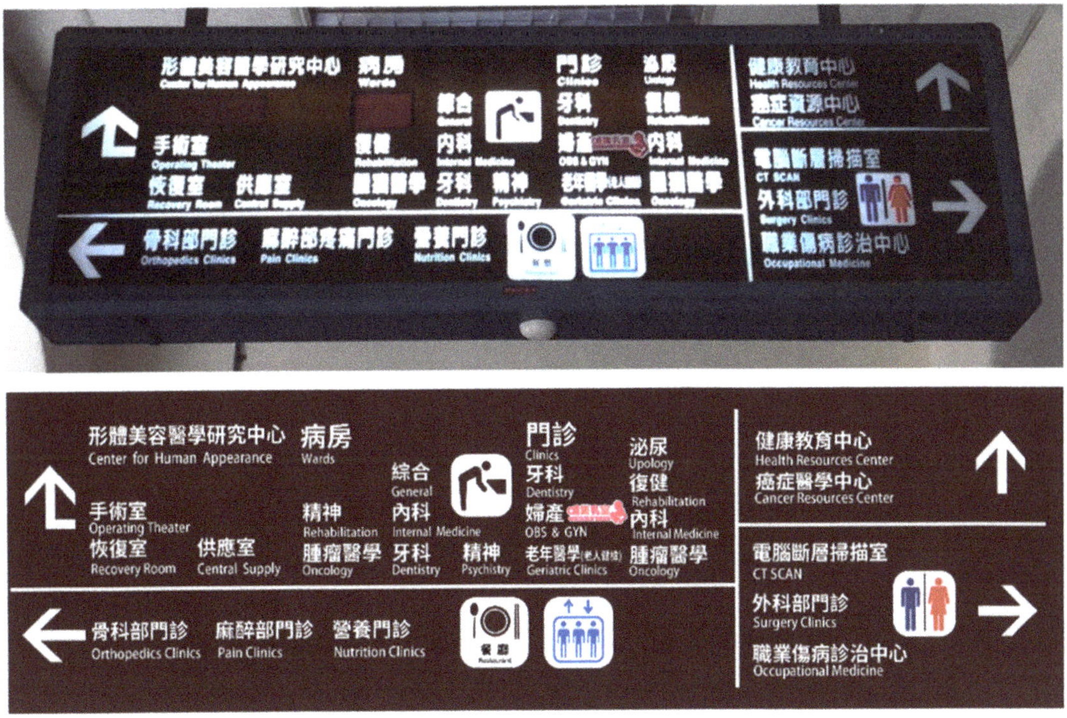

Figure 3. The presentation of information on existing hanging-style signs; the picture above is a photo from the scene, and the figure below is a graph drawn on an as-is basis for use in a virtual environment. (Please refer to Appendix A. Figure A1 for the English content of the signs).

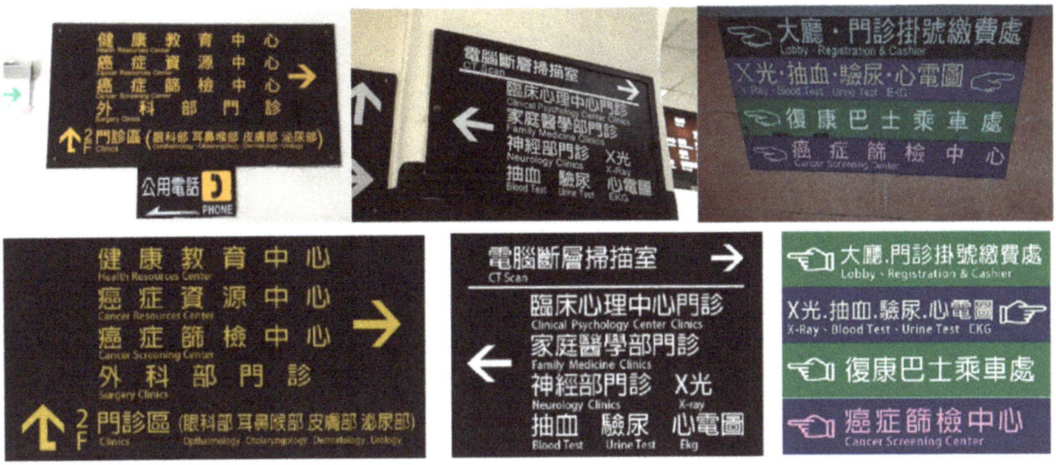

Figure 4. The presentation of information on existing wall signs; the pictures above are photos from the scene, and the figure below is a graph drawn on an as-is basis for use in a virtual environment. (Please refer to Appendix A. Figure A2 for the English content of the signs).

2.1.2. Material Design

According to the current situation, we drew the sign system, size, font, typeface, typography, etc., and built a virtual environment. Two models were built using the 3D modeling software "Blender" (Figure 5). One was a practice room model, to familiarize participants with the operation of the entire experiment before the formal test. Another model was for the West Campus, which contained a virtual model of the circulation corridor on the two floors of the entrance area and the main outpatient area. It covered all signs except temporarily posted signs and illuminated all environments evenly. After that, we imported the models into Unity 3D to simulate a scene that allows participants to move freely around the building at an equal speed (Figure 6). Participants watched the computer display screen, proceeded with the keyboard's arrow keys, and used the mouse to rotate the perspective.

Figure 5. Screenshot of the 3D modeling software Blender (the left picture is the practice room model, and the right is the NTUH West Campus model).

Figure 6. The screen displayed while walking.

In Unity, we added a way to set the character as a Rigibody of the physics engine and bind the camera to simulate a scene. It allowed the test to move freely inside the building at a consistent speed from a first-person view (FPV). After finishing the work, we uploaded this file to the Unity Play website to make an online game. It is convenient for all test participants to click on the URL to conduct virtual environment experiments.

2.2. Participants

In the pilot test, we found that participants aged 40–60 who were unfamiliar with the relevant operations in the past had more difficulty in control and were more likely to feel intense dizziness during the process. Therefore, formal experiments limited this study's age group to 20–39. Participants must not have been to the National Taiwan University Hospital Original Building, and participants were required to have no experience with "visually induced motion sickness" (VIMS) [50–52] but be familiar with computer operation.

We recruited 64 participants, with 32 people in the pretest and 32 people in the posttest. There was no significant difference in the analysis after age and wayfinding performance.

2.3. Experiment Design

The experimental design of this study was to use the "existing sign system" for wayfinding, to understand the wayfinding performance and behavior of visitors under NTUH's existing sign system, and to unify the problem points. After the design conditions, we proposed an optimized version of the sign system. In the posttest, the same procedures were conducted to test the newly designed sign system. In addition, "color coding" was added to this primary sign as a variable, and two virtual environments were set up. One is the conventional format, and the other is the color-coded format; participants were randomized into two groups for the experiment. Under the same virtual experimental scene, experimental process, and situational tasks as the pretest, we compared the pre- and posttests in terms of evaluation of task performance, wayfinding behavior, anxiety, and readability of experimental participants. On the other hand, we wanted to know if the color coding on the sign system influenced the users' wayfinding behavior (Figure 7).

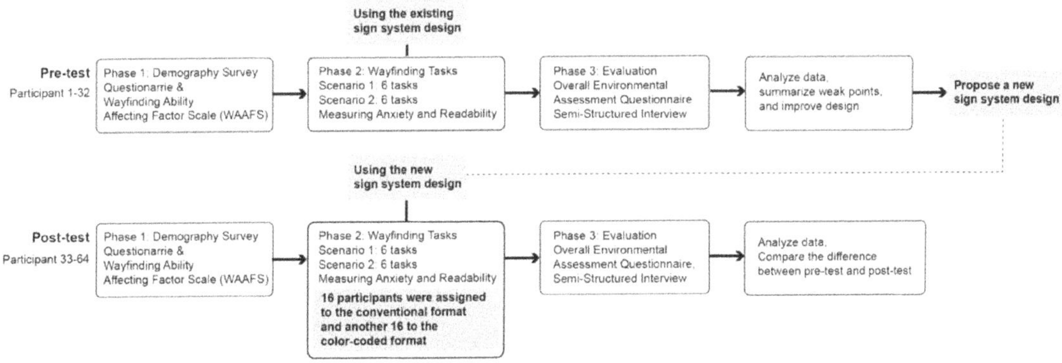

Figure 7. Experiment flowchart.

The experimental processes of the pretest and posttest were the same. Each experimental process was divided into three stages: (1) Participants were asked to complete the pretest questionnaire, including basic information and experience playing 3D games. The Wayfinding Ability Affecting Factors Scale (WAAFS) [53] was used to understand participants' essential wayfinding ability and confidence. (2) A simulated situational task wayfinding experiment was conducted; it began in the practice room to let participants learn how to move in the virtual space, and then two sets of scenarios with twelve wayfinding tasks were carried out. Each set of scenarios began with a different entrance and provided a text scenario description that asked participants to imagine hurried wayfinding in a healthcare facility. Each group's six tasks required upper and lower floors (Figure 8) and a return to the original entrance in the last task. In this process, in addition to recording the task completion time, we also counted the number of signs participants saw in the task and determined whether the signs they saw provided target information to calculate the efficiency of viewing the signs. We asked them to think aloud at the same time during the task. In addition, participants answered two questions after each task to report their anxiety level and the mark evaluation of the route. (3) At the end of all tasks, we asked participants to complete the overall environmental assessment questionnaire, which included an evaluation of the overall spatial layout and signage. A semi-structured interview was conducted to assess the participants' wayfinding behavior and the perception of signs. For the execution of the experiment, please refer to the environment and operation described in Section 2.1.2.

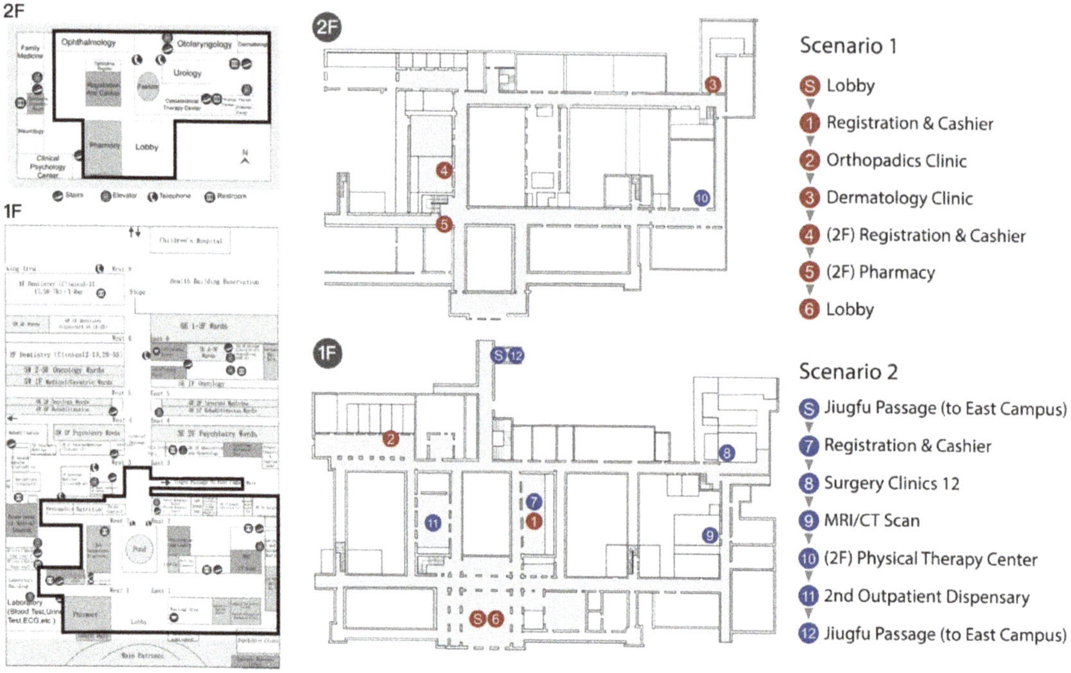

Figure 8. The left map is an original map from NTUH (https://www.ntuh.gov.tw/ntuh/Fpage.action?muid=3484&fid=3257 (accessed on 4 February 2021). The gray part of the middle floor plan is the area where participants can walk in the virtual environment. The number marked above is the task number in the scenario description, and S is the starting point of the task. After that, the task numbers are referred to simply as T1 through T12.

The overall environmental assessment questionnaire questions are shown in Table 1, designed with a four-point Likert scale, from 1 to 4 points representing strong disagreement to strong agreement.

Table 1. Contents of the overall environmental assessment questionnaire.

Overall, I think the spatial layout of this hospital can be well-oriented.
Overall, I think the hospital sign system is helpful for wayfinding.
Overall, I think the hospital sign system is easy to understand.
Overall, I think the hospital's hanging-style sign system from the ceiling is straightforward to understand.
Overall, I think the hospital's hanging-style sign system from the ceiling has helped me a lot.
Overall, I think the signs on this hospital's walls are straightforward to understand.
Overall, I think the wall signs of this hospital have helped me a lot.

2.4. New Sign System Design

Through the forward test experiment, we summarized the participants' problem points and interview results in the wayfinding process and derived their design conditions. This resulted in a reformulation of the design specifications for the new marking system and the readjustment of all signs for subsequent verification using this rule. The optimization content is as follows.

2.4.1. Hanging-Style Sign System

The hanging-style signs were uniformly changed to more recognizable white text on a black background (Figure 9), the leading guide for visitors to find their way to the hospital. In the central corridor, complete information was presented, including all the hospital's critical administrative services and outpatient departments. The services' functions were divided into outpatient, medical, laboratory, and ward. We focused on information consistency and the principle of legibility. Services in the same building were prioritized in order.

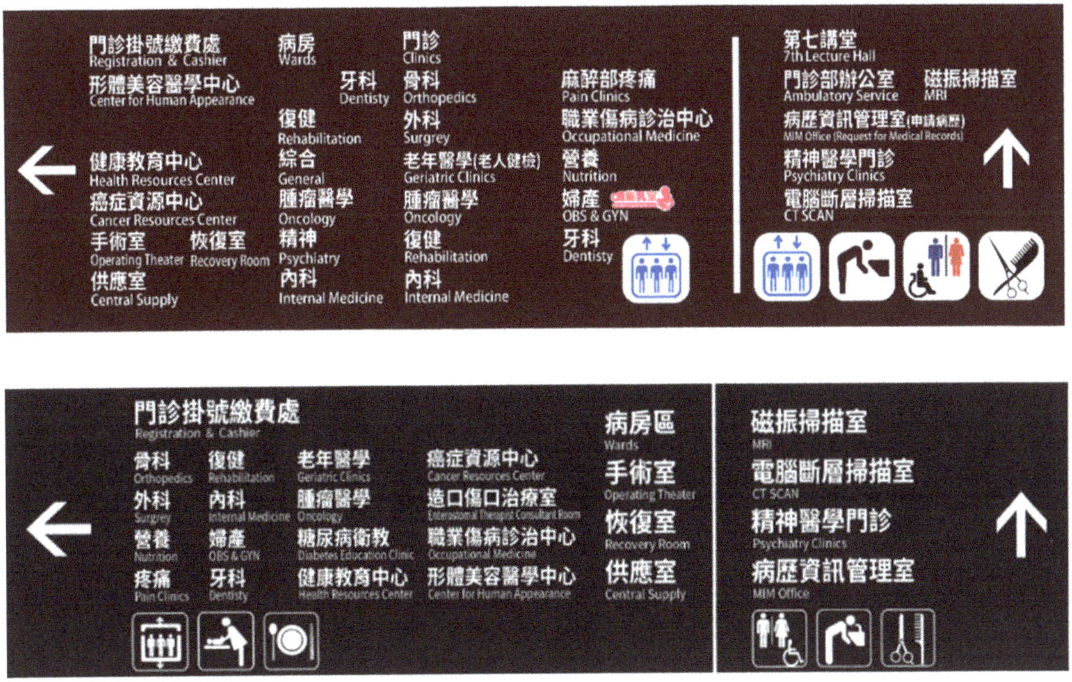

Figure 9. Comparison of hanging-style signs (The image above is the existing sign, and the lower image is the redesigned sign. Please refer to Appendix A. Figure A3 for the English content of the signs).

2.4.2. Wall Sign System

First, we removed any duplicate information signs on the wall, set up a corridor code and direction guide in each corridor, and added floor space information signs to the stairs and elevators. In addition, the auxiliary signs initially posted on the lower position on the wall in the hospital were changed to uniform white characters on a gray background. Although it was distinguished from other signs in the background color, it maintained the same form as the overall sign system in the font and directional arrows. All wall signs followed the specifications of the comprehensive sign system (Figures 10 and 11).

Figure 10. New design for wall sign system (The left is the sign of the corridor passage on the left, and the right is the sign near the stairs on the right. Please refer to Appendix A. Figure A4 for the English content of the signs).

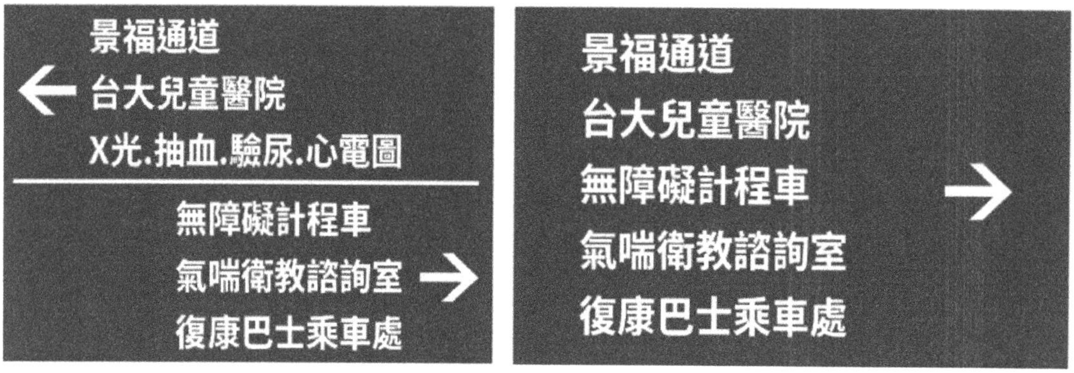

Figure 11. Auxiliary wall signs in the new sign system (Please refer to Appendix A. Figure A5 for the English content of the signs).

2.4.3. Color-Coded Design Proposals

Since the hospital provides many services, the amount of information presented on the sign is still hefty, so we hoped to add other ways to help viewers quickly filter information. Many participants also mentioned in interviews that adding color coding to the sign system is recommended to help find the way. So, based on the above design, color coding was added to the second version of the design proposal (Figures 12–14), which adopted a unified background color and text color. We used the presentation of the color bar on the elevator sign of the existing sign system as a form of color coding. We integrated it into all signs except auxiliary signs. The color selection was divided into four colors according to the active type of service provided, red for outpatient services, blue for laboratory services, green for forward services, and yellow for other medical administrative services. The color block preceded all sign information in the sign system except for auxiliary signs.

Figure 12. Color-coded format hanging-style sign system (Please refer to the appendix for the English content of the signs; the content is the same as Figure 9).

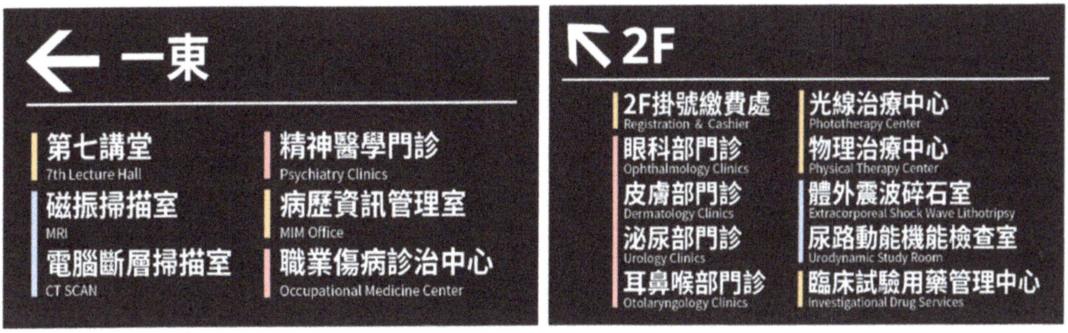

Figure 13. Color-coded format wall sign system (on the left is the corridor passage sign, and on the right is the staircase sign).

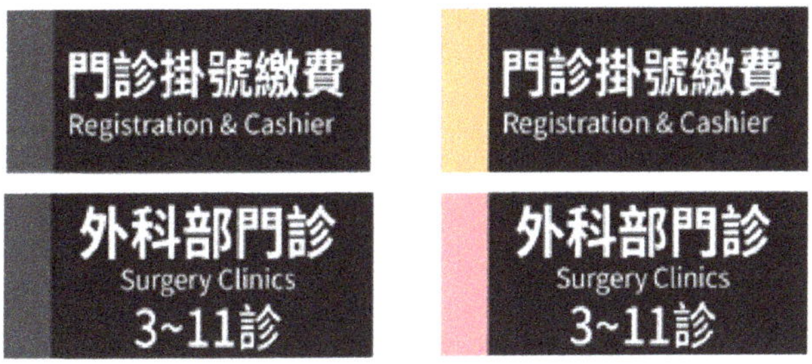

Figure 14. Spatial identification signs (on the left is the conventional format, and on the right is the color-coded format).

3. Results

3.1. Wayfinding Performance

Wayfinding performance is judged on two metrics; one is the time spent on the task, and the other is the valid rate of the signs.

3.1.1. Task Completion Time

For calculating task time spent, except for the starting point to T1 in Scenario 1 and the starting point to T7 in Scenario 2, observers told participants before the experiment to say "I found it" upon arrival at each task location. This speaking time point was used to calculate their time to the next task until all tasks were over. After comparing the task performance of the pretest experiment and the posttest, we found that the posttest is almost always better than the pretest experiment in task performance. Eleven tasks are shorter than the pretest in the average task completion time, and nine of the independent sample T-tests are significant (Table 2).

Table 2. Descriptive analysis of task completion time for pretest and posttest.

		M	SD	t-Value	p-Value
T1	pretest	51.00	36.11	2.506 ***	0.000
	posttest	30.56	28.73		
T2	pretest	42.97	20.23	1.439	0.078
	posttest	36.53	15.21		
T3	pretest	143.53	69.65	1.279	0.103
	posttest	122.72	60.19		
T4	pretest	50.97	25.67	2.508 **	0.008
	posttest	37.19	17.53		
T5	pretest	29.69	38.80	2.476 **	0.009
	posttest	12.22	9.35		
T6	pretest	69.06	47.10	3.371 ***	0.000
	posttest	36.50	27.71		
T7	pretest	62.19	31.86	3.536 ***	0.000
	posttest	36.50	25.95		
T8	pretest	56.47	47.57	2.481 **	0.009
	posttest	34.84	12.97		
T9	pretest	71.72	46.77	3.43 ***	0.000
	posttest	40.69	20.79		
T10	pretest	259.16	85.29	13.696 ***	0.000
	posttest	45.75	22.24		
T11	pretest	74.50	33.62	−0.135	0.447
	posttest	75.47	22.87		
T12	pretest	160.06	116.05	3.648 ***	0.000
	posttest	73.94	66.10		

Note: There were 32 participants each in the pretest and posttest; ** $p < 0.01$, *** $p < 0.001$.

Analysis of T3 and T10 in the pretest

The results of the forward test show that the average time spent on all tasks is 89.3 s. The tasks that took the longest in both groups were T3 (to find the first consultation with Dermatology) and T10 (to find the Physical Therapy Center). Both tasks needed to be found from the stairs on the first floor to the second floor. In T3, there were a total of six signs on the first floor, with information from the Dermatology Clinics on the second floor, placed in the lobby, the central corridor, and near the stairs and elevators (Figure 15). So, although participants looked back and forth on the first floor, they still were able to find these specific signs and successfully go upstairs. However, in T10, there were only two signs with information about the Physical Therapy Center on the second floor, a sign next to the elevator with all floor information and a sign next to the staircase nearby. As a result, participants spent a lot of time searching for the Physical Therapy Center on the first floor, and some did not even find the sign before going upstairs. We found from the think-aloud log that some participants in these two tasks could not find clear information on the sign. They began speculating that it was classified under other relevant departments and went to those departments to look for it. This leads to further deviations from the target location. For example, in Task 3, the majority of people speculated that the Dermatology Clinics were affiliated with Surgery Clinics. In T10, most people assumed that the Physical Therapy Center was affiliated with Rehabilitation Clinics.

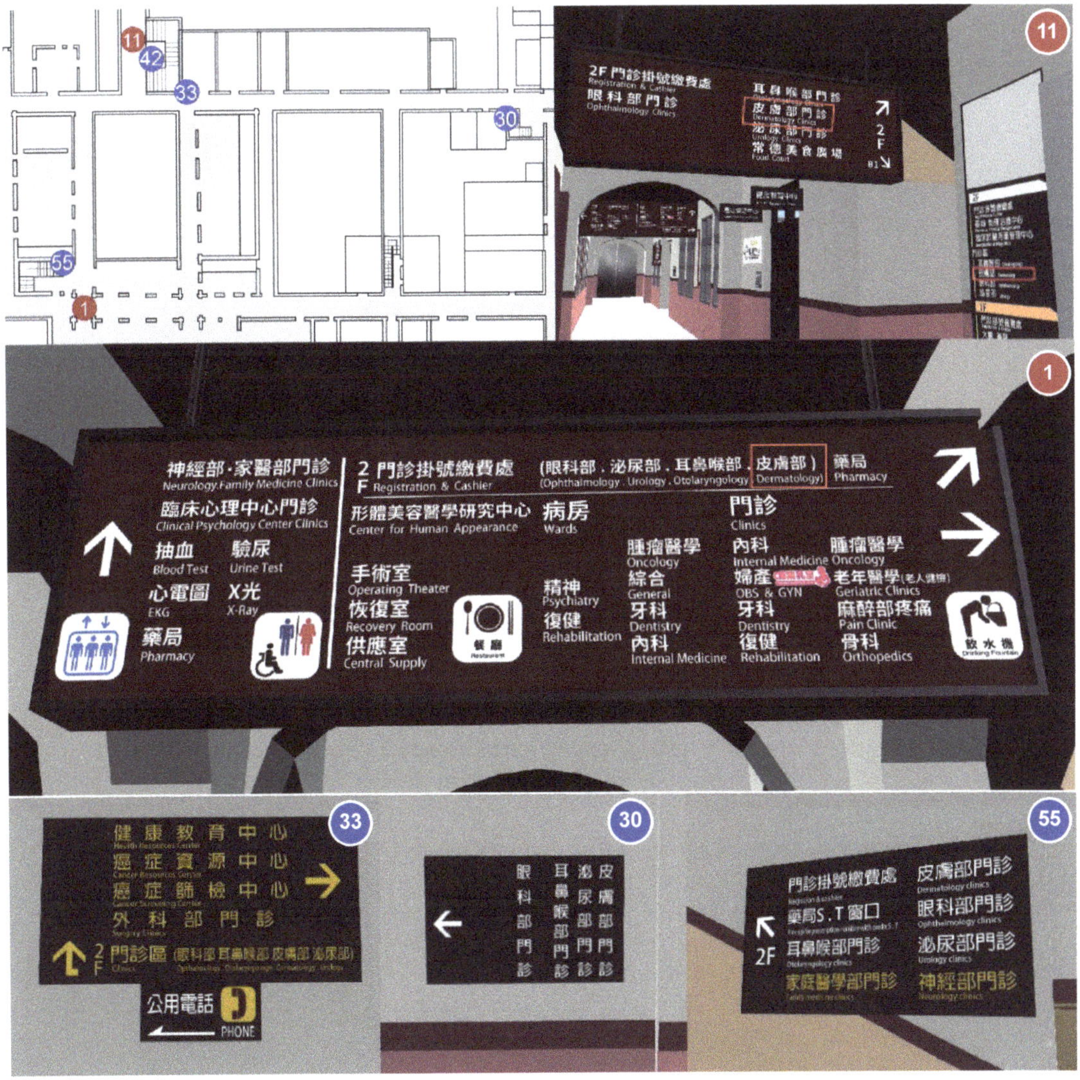

Figure 15. T3-related signs in the pretest. The numbers on the floor plan correspond to the five pictures.

Analysis of T3 and T11 in the posttest

The posttest results show that the average time spent on all tasks is 48.6 s. The tasks that took the longest to complete in both groups of situations were T3 (look for the first diagnosis of Dermatology) and T11 (look for the second Outpatient Dispensary). In the posttest, the newly designed sign system in T3 placed the relevant signs of floor guidance next to the stairs and elevators (Figure 16), which differed from the existing sign system that mixed information on the wall and the hanging-style signs. While the information is presented more consistently, we also observed that this change causes participants to go through stairs or near elevators to perceive this information, which is why T3 took longer to perform than other tasks. However, the same task to go upstairs for which the longest average completion time was found for the pretest participants was T11 (look for the Physical Therapy Center). Since participants noticed signs next to the stairs and looked for them after leaving the previous task location, their task performance was significant even if they did not go through the elevator. This also indirectly caused almost all participants to be guided up and down the stairs farthest from the second Outpatient Dispensary, so T11 (look for second Outpatient Dispensary) also has a longer average time to complete.

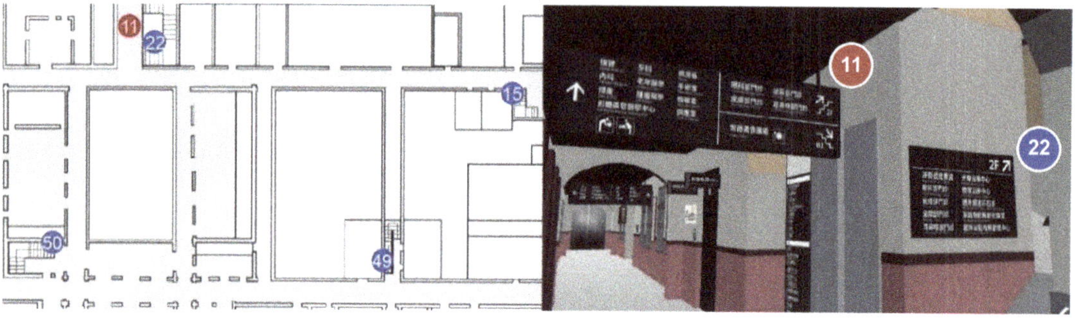

Figure 16. Schematic diagram of signs and locations related to T3 in the posttest. The numbers on the floor plan map to the five pictures. The positions of No. 11 and No. 22 in the floor plan correspond to the signs on the right figure.

3.1.2. Valid Rate of Signs

By counting the number of signs viewed by participants in the task and whether the signs they considered provided task target information divided into valid and invalid signs, we can know the effective rate of the signs viewed by participants in each task. The average viewing efficiency of the posttest was higher than that of the pretest in nine tasks, of which seven were significant (Table 3). We found from the participants' movement trajectory that the participants' movement path in the posttest is more uniform. Posttest experiments also exhibited significantly reduced behaviors such as losses, reentries, or misses that would have occurred in the task.

Table 3. Descriptive analysis of valid rate for pre- and posttest.

		M	SD	*t*-Value	*p*-Value
T1	pretest	0.63	0.31	−3.170 **	0.001
	posttest	0.85	0.24		
T2	pretest	0.84	0.21	−6.35	0.264
	posttest	0.87	0.17		
T3	pretest	0.36	0.17	2.160 *	0.018
	posttest	0.28	0.11		
T4	pretest	0.84	0.19	0.514	0.305
	posttest	0.82	0.17		
T5	pretest	0.63	0.36	−4.159 ***	0.000
	posttest	0.92	0.32		

Table 3. *Cont.*

		M	SD	*t*-Value	*p*-Value
T6	pretest	0.30	0.33	0.099	0.461
	posttest	0.29	0.32		
T7	pretest	0.49	0.13	−8.003 ***	0.000
	posttest	0.73	0.11		
T8	pretest	0.84	0.22	1.225	0.113
	posttest	0.78	0.15		
T9	pretest	0.56	0.22	−2.777 **	0.004
	posttest	0.70	0.17		
T10	pretest	0.17	0.07	−23.270 ***	0.000
	posttest	0.89	0.16		
T11	pretest	0.61	0.27	−2.270 **	0.014
	posttest	0.74	0.18		
T12	pretest	0.12	0.13	−7.208 ***	0.000
	posttest	0.42	0.19		

Note: There were 32 participants each in the pretest and posttest; * $p < 0.05$, ** $p < 0.01$, *** $p < 0.001$.

The results show that the new sign system can play a higher guiding role for users who are finding their way in hospitals, and visitors can make correct wayfinding decisions through the guidance of the signs. It helped to find the target faster, rather than wandering around the corridor to find the target department. In addition, during the pretest, some users began to question the name of the task location on the sign system after not finding the target information on the sign many times and speculated about going to the department that is considered to find it, thus completely deviating from the target direction. This phenomenon does not appear again in the posttest experiment.

Analysis of T6 and T12 in the pretest

In the forward test results, the tasks with the lowest average efficiency in the two sets of scenarios were T6 (look for the entrance of the lobby; valid rate = 0.30) and T12 (look for the entrance of Jingfu Passage; valid rate = 0.12). Both T6 and T12 require participants to retrieve the entry task at the beginning of the situational task. Since there are few signs in hospitals that provide information about the location of exits, but the lobby is a space with high ceilings and is quite different from other areas, most people in T6 can find the hall without relying on signs. However, in T12, most participants had to go back and forth on the ground floor to find information about various corridors and signs, and only a very few remembered the location of the Jingfu Passage. There were only three signs about Jingfu Passage in the hospital, scattered and incoherent guidance, which also led to participants being unable to find Jingfu Passage through a single sign even if they noticed the sign information.

Analysis of T6 and T12 in the posttest

The tasks with the lowest valid rates of the average viewing markers in the posttest are still T6 (look for Dermatology Clinic) and T12 (look for Jingfu Passage). However, in the new sign system, the information on the Jingfu Passage was changed to an auxiliary sign placed lower on the wall and placed at the turning point of the central corridor (Figure 17). Compared with the results of the pretest, the performance of this task has increased significantly in terms of completion time and valid rate. However, due to the low position of the auxiliary sign, it was more difficult for participants to notice the sign the first time, which also led to a lower valid rate of viewing the sign than other tasks. Refer to the movement track diagram of T12 (Figure 18).

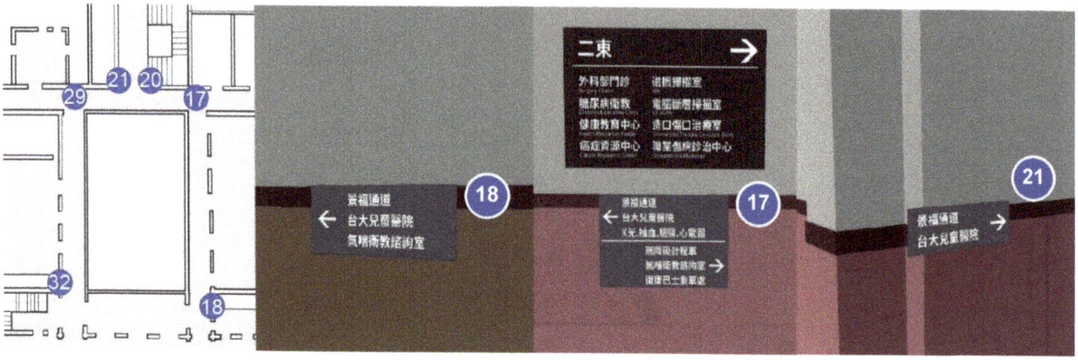

Figure 17. Schematic diagram of the relevant signs and locations of T12 in the posttest. The positions of No. 17, No. 18, and No. 21 in the floor plan correspond to the signs on the right figure.

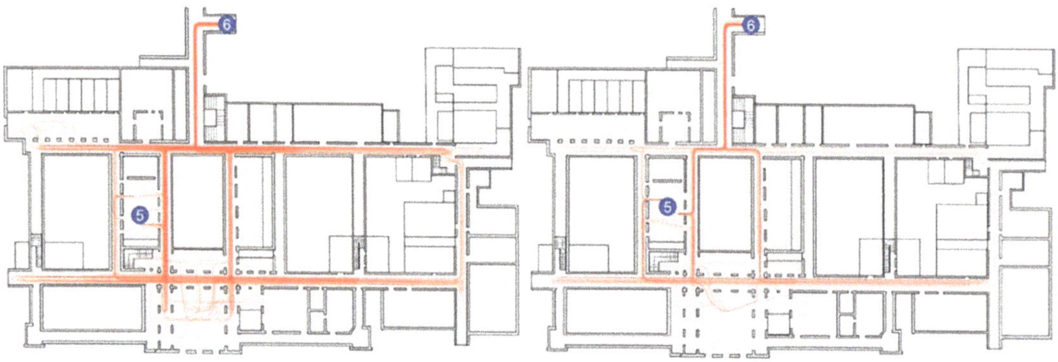

Figure 18. The movement track diagram of T12 (the left is the pretest, and the right is the posttest).

3.2. Participant Feedback

After each task in the pretest and posttest, participants answered two questions. One recorded their anxiety level (this task makes me feel very anxious). The other was the readability evaluation of signs on the task route (the sign setting in this task is easy to understand). Both questions were answered on a five-point Likert scale.

Participants in the posttest reported less anxiety in each task than in the pretest, and seven tasks were significant (Table 4). All scores were also higher than those in the pretest on task scoring, and 11 items were significant (Table 5). The result shows that the new sign system can significantly reduce the negative emotions generated during wayfinding while improving the efficiency of wayfinding in the hospital. Whether the participants gave the readability score of the route in each task or the overall sign system after the test, the score of the new sign system was higher than that of the existing sign system in the hospital.

The results showed that participants reported lower readability in T3 (readability: M = 2.72; anxiety value: M = 3.00) and T10 (readability: M = 1.69; anxiety value: M = 3.84) and felt more significant anxiety (overall task readability M = 2.94; anxiety: M = 2.5). These two tasks also took the longest to complete, which showed that participants experienced increased anxiety due to delays in finding related signs. The Physical Therapy Center in the first-floor sign system had fewer signs than the Dermatology Clinic, so participants felt higher anxiety levels. This also lowered the evaluation score for readability in this route.

Table 4. Descriptive statistics of anxiety values for each task in the pre- and posttests.

		M	SD	*t*-Value	*p*-Value
T1	pretest	2.44	1.29	2.662	0.050
	posttest	1.69	0.93		
T2	pretest	1.94	0.95	1.603	0.057
	posttest	1.59	0.76		
T3	pretest	3.00	1.46	0.582	0.282
	posttest	2.81	1.09		
T4	pretest	1.91	0.96	1.659	0.051
	posttest	1.53	0.84		
T5	pretest	1.69	0.96	3.001 **	0.002
	posttest	1.53	0.84		
T6	pretest	2.59	1.48	2.818 **	0.004
	posttest	1.69	1.06		
T7	pretest	2.50	1.34	3.083 **	0.002
	posttest	1.59	0.98		
T8	pretest	2.06	0.98	1.865 *	0.034
	posttest	1.66	0.75		
T9	pretest	2.50	1.19	2.502 **	0.008
	posttest	1.84	0.88		
T10	pretest	3.84	1.30	6.253 ***	0.000
	posttest	1.97	1.09		
T11	pretest	2.28	1.17	0.316	0.377
	posttest	2.19	1.20		
T12	pretest	3.28	1.49	2.984 **	0.002
	posttest	2.25	1.27		

Note: There were 32 participants each in the pretest and posttest; * $p < 0.05$, ** $p < 0.01$, *** $p < 0.001$.

Table 5. Descriptive statistics of readability for each task in the pre- and posttests.

		M	SD	*t*-Value	*p*-Value
T1	pretest	3.00	1.37	−3.878 ***	0.000
	posttest	4.13	0.91		
T2	pretest	3.56	1.19	−3.518 ***	0.000
	posttest	4.47	0.84		
T3	pretest	2.72	1.33	−1.545	0.064
	posttest	3.19	1.09		
T4	pretest	3.56	1.08	−3.031 **	0.002
	posttest	4.31	0.90		
T5	pretest	3.53	1.08	−5.107 ***	0.000
	posttest	4.69	0.69		
T6	pretest	2.78	1.16	−3.898 ***	0.000
	posttest	3.84	1.02		

Table 5. *Cont.*

		M	SD	*t*-Value	*p*-Value
T7	pretest	3.22	1.04	−3.597 ***	0.000
	posttest	4.13	0.98		
T8	pretest	3.53	1.16	−2.763 **	0.004
	posttest	4.22	0.79		
T9	pretest	3.00	1.27	−3.177 **	0.001
	posttest	3.91	1.00		
T10	pretest	1.69	1.03	−8.833 ***	0.000
	posttest	4.03	1.09		
T11	pretest	2.72	1.20	−3.084 **	0.002
	posttest	3.66	1.23		
T12	pretest	1.91	1.00	−5.412 ***	0.000
	posttest	3.47	1.30		

Note: There were 32 participants each in the pretest and posttest; ** $p < 0.01$, *** $p < 0.001$.

3.2.1. Overall Environmental Assessment

From the evaluation results, we found that the tasks in which participants felt the most anxious and had the lowest average readability evaluation scores in the two scenarios; these tasks were T3 (look for Dermatology Clinic; readability: M = 3.19, anxiety value: M = 2.81) and T12 (look for Jingfu Passage; readability: M = 3.47, anxiety value: M = 2.25). These two tasks were also given a lower valid rate by participants in the two scenarios (all tasks readability: M = 4.00, anxiety value: M = 1.83). Even if they looked at many signs, participants in this task may not have been able to find information on the direction to the target. This illustrates that when participants cannot find the information they need on the sign for the first time, it will increase anxiety and result in a low evaluation of the medical facility's signs.

The results of the overall environmental assessment questionnaire after the test (Table 6) showed that the average score of the participants for the new sign system was much higher than that for the original sign system, and the readability of the spatial layout was also improved. They rated the entire hospital as medium or above, agreeing that the new sign system could help with wayfinding and easy understanding. In addition, as in the pretest, the participants gave the hanging-style sign a higher score than the wall sign, indicating that users thought hanging style from the ceiling was more helpful in finding a department in the hospital.

The results of the overall environmental assessment questionnaire filled out after all tasks in the pretest showed that from the first three questions, the participants did not have a high average evaluation of the spatial layout and signs of the overall hospital. However, they still agreed that the signs were helpful for wayfinding. In addition, based on the results of the last four questions, we found that after the pre- and posttest, participants gave higher scores to signs hanging from the ceiling than signs posted on the wall after experiencing the sign system. This illustrates that they believe the signs hanging from the hospital ceiling are more straightforward and easier to understand and help them find their way.

Table 6. Descriptive analysis of the pre- and posttest overall environmental assessment questionnaires.

OVERALL, I THINK ...		M	SD	*t*-Value	*p*-Value
the spatial layout of this hospital can be well-oriented.	pretest	2.53	1.05	−3.574 ***	0.0005
	posttest	3.59	1.32		
the hospital sign system is helpful for wayfinding.	pretest	3.56	1.01	−4.506 ***	0.000
	posttest	4.53	0.67		
the hospital sign system is easy to understand.	pretest	2.75	1.02	−6.307 ***	0.000
	posttest	4.28	0.92		
the hospital's hanging-style sign system from the ceiling is straightforward to understand.	pretest	3.00	1.22	−4.981 ***	0.000
	posttest	4.31	0.86		
the hospital's hanging-style sign system from the ceiling has helped me a lot.	pretest	3.75	0.80	−3.735 ***	0.000
	posttest	4.50	0.80		
the signs on this hospital's walls are straightforward to understand.	pretest	2.81	0.97	−5.424 ***	0.000
	posttest	4.19	1.06		
the wall signs of this hospital have helped me a lot.	pretest	2.66	1.00	−4.481 ***	0.000
	posttest	3.94	1.27		

Note: There were 32 participants each in the pretest and posttest; *** $p < 0.001$.

3.2.2. Results of Semi-Structured Interviews

In the pretest interviews, most participants indicated that they would prioritize looking at the signs hanging from the ceiling to find directions during the wayfinding process, citing the belief that the hanging-style signs would provide complete information. Since the hanging-style signs seemed more important from a height place, the information provided by the signs posted on the wall seemed more fragmented, and those only served as auxiliary signs in the wayfinding process. Participants suggested that the hospital signs provided too much information, and the typography and interspersed icons would distract the reading order. As a result, they struggle to find the information they need and miss it. The current way of classifying information in wards and clinics is very misleading, so a more clearly categorized and easier-to-read layout is desired. The arrows diagonally above the direction indicators are confusing by sometimes indicating going upstairs and sometimes indicating going straight to turn. Participants were troubled by the frequent incoherence of information between signs, and if the target was to go straight and turn left, they expected the next sign to tell them whether to continue straight or turn left, and if they did not have this message, they may turn back to look at it (Figure 19).

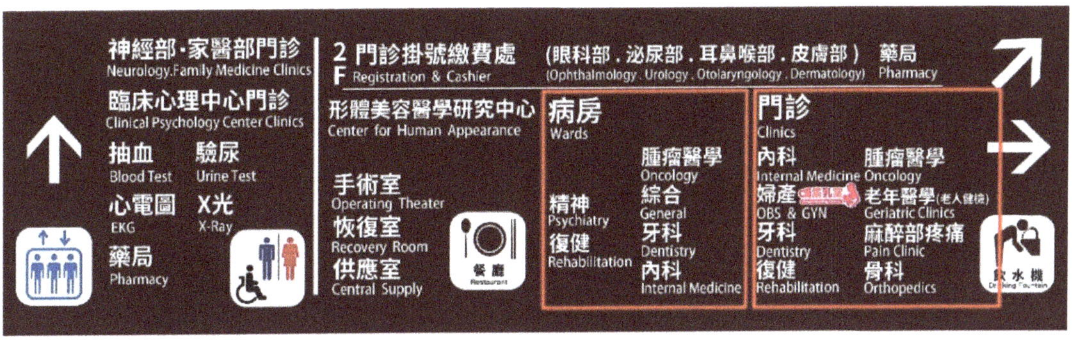

Figure 19. The information classification method of existing hanging-style signs.

Most participants in the posttest still said in interviews that they would preferentially rely on hanging signs to find directions. For suggestions that the verification sign system could be improved, they mentioned adding more information or signs, such as instructions on stairs, exits, or parking lots. In interviews with color coding, only 7 of the 16 people in the color-coded group said they used color coding to assist in wayfinding. After understanding the differences between the two versions, 22 people in the two groups had a higher preference for a color-coded sign system, and the rest indicated no preference. The rationale for preferring color coding is that there is an additional way to help identify and classify information.

3.3. Individual Differences in Participants

We took the participants' basic wayfinding ability and game experience as individual differences and analyzed whether both affect task performance. Through the WAAFS filled in by the participants before the test, the participants were divided into high, medium, and low abilities according to the three aspects of wayfinding anxiety, direction perception ability, and path memory ability, and then analyzed with one-way ANOVA with task performance. No relevant significance was found in the aspect of path memory ability. Participants with low wayfinding anxiety or high directional awareness in a particular task were more likely to reach mission objectives faster. Highly directionally aware people may also have less anxiety during tasks. In addition, all participants move at an equal rate during the simulated wayfinding. However, in some cases, it will still take more time for those without game experience to reach the mission site than those with game experience. The result is similar to the findings of the pretest, that even though all participants moved freely in the environment at an equal rate, participants without gaming experience still need to spend more time reaching the target location in certain situations and might be more likely to feel anxious. The reason is that they were less familiar with this type of operation and therefore found it more challenging to navigate in a virtual environment.

3.4. The Effect of Color Coding on Wayfinding Performance

The results show that compared with the existing sign system in the pretest, the difference between the conventional and color-coded formats is less noticeable. Regarding task performance, eight tasks had a lower average completion time in the color-coded format group than in the conventional format group. However, no significant result was found. However, two tasks are significant in the T-test: T7, which is to find the first floor of Registration and Cashier ($p = 0.046 < 0.05$), and T12, which is to look for Jingfu Passage ($p = 0.042 < 0.05$). However, we observed the movement tracks and wayfinding behaviors of T3 and found the cause of the difference is that the participants of the conventional format group arrived at Registration and Cashier almost without any problems. Some "color-coded" group members missed certain signs and returned to the lobby.

However, this condition does not seem to be related to the color-coded variation on the logo. Color coding is not provided on the auxiliary logo identifying Jingfu Passage in T12. Overall, no color-coded variation on the signs has been found to lead to significant task performance change. In the interview, it was also found that only 44% (seven people) of the participants in the color-coded group used color coding to assist in wayfinding. In addition, very few participants could clearly articulate the meaning of the color coding (distinguished by service function). Therefore, the main factors that affected the wayfinding performance were the position adjustment, information presentation, and the typesetting method of the overall signs.

On the other hand, from the results of the anxiety and readability assessments, 10 of the tasks scored slightly higher than the conventional format. The same results were found in the overall environmental assessment questionnaire (Table 7), especially in the hanging-style sign for wayfinding ($p = 0.015 < 0.05$). Most participants who learned about the differences between the two formats in the interview had a higher preference for a sign system with a color-coded format. Therefore, although classifying information by color coding could not directly affect participants' wayfinding performance in this study, it still improved visitors' impressions.

Table 7. Overall environmental assessment questionnaire with different color-coded descriptive analyses.

OVERALL, I THINK ...		M	SD	*t*-Value	*p*-Value
the spatial layout of this hospital can be well-oriented.	Conventional format	3.56	1.15	−0.132	0.448
	Color-coded format	3.63	1.50		
the hospital sign system is helpful for wayfinding.	Conventional format	4.38	0.72	−1.333	0.097
	Color-coded format	4.69	0.60		
the hospital sign system is easy to understand.	Conventional format	4.19	0.98	−0.568	0.288
	Color-coded format	4.81	0.40		
the hospital's hanging-style sign system from the ceiling is straightforward to understand.	Conventional format	4.19	0.91	−0.819	0.210
	Color-coded format	4.44	0.81		
the hospital's hanging-style sign system from the ceiling has helped me a lot.	Conventional format	4.19	0.98	−2.357 *	0.015
	Color-coded format	4.81	0.40		
the signs on this hospital's walls are straightforward to understand.	Conventional format	3.94	1.24	−1.351	0.094
	Color-coded format	4.44	0.81		
the wall signs of this hospital have helped me a lot.	Conventional format	3.94	1.24	0.000	1.000
	Color-coded format	3.94	1.34		

Note: N = 32; * $p < 0.05$.

4. Discussion

This study explored the influence of the sign systems in medical institutions on wayfinding behavior using the virtual environment as an experimental method. NTUH West Campus was used as the research site. For this purpose, we conducted a field study of NTUH, a pretest, and a posttest. We discussed the results of the two tests and further proposed medical facility sign system design methods and follow-up research.

4.1. Wayfinding for Moving between Floors

In past studies, wayfinding between different floors was a common challenge for visitors [18,19,54]. From the experimental results, we found that most participants would preferentially rely on hanging-style signs on the ceiling to find directions and search for and read more informative signs. It was more difficult to find tasks that needed to move between floors. There needed to be a clear sign informing the target of the location of the floor. Otherwise, visitors would not easily go upstairs to find it. During the wayfinding process, if the target information was not seen on the sign, the participant would wander the corridor repeatedly or return to the lobby. Some participants might reason independently to look for other related subjects, resulting in a complete deviation from the correct route. Therefore, the hanging signs in the lobby and the central corridor should play the main role of guiding visitors. The signs on the walls could complement and compensate for the shortcomings of the hanging signs. Participants looked at the sign and expected the next sign with the information they needed, and incoherent information led to confusion or return.

Due to the hospital's lack of signs for floor instructions, participants generally found it challenging to go from the first to the second floor to find the target. Unless the sign with the target location was on the second floor, they did not quickly go upstairs but repeatedly wandered on the first floor to search, increasing their negative emotions when finding the way.

4.2. Weak Points in Information Architecture

The main purpose of information architecture is findability [55]. However, the term is designed with its origins in the rise of the Internet and extended to the user experience. However, it largely corresponds to the paths, edges, districts, nodes, and landmarks in

Kevin Lynch's *The Image of the City* [6]. Since the NTUH West Campus is a Baroque building and a completely symmetrical structure, the concept of "landmark" is not applicable in this indoor environment. However, the lobby is very different from the rest of the building, and most visitors can find it without relying on signs. However, the number of signs with information about the location of other exits in the hospital is small, such as the Jingfu Passage linking the West Campus to the East Campus, which is not easy to find. However, to exit to the East Campus, most participants must go back and forth to find various corridors and signs for information. Therefore, the sign system in the hospital still needs to add signs and smaller specific exits to assist with wayfinding.

Currently, the classification method of information on the hanging-style sign is difficult for unfamiliar visitors to understand. They were even misled by the two classification titles of ward and outpatient. This is not in line with the "labeling system" in the information architecture that should consider viewers' manner or mental model [55]. The auxiliary icon was interspersed with text information without rules, which interfered with the user's reading flow. It is necessary to arrange the information content neatly and regularly; otherwise, it is easy to ignore the information. The meanings of directional arrows must also be unified to avoid misunderstandings. Currently, the walls of the circulation corridor of the hospital are full of signs of various styles and sizes and too many kinds of presentation methods. Too much repetitive information and the lack of consistency will make the visitors question and be confused about finding their way, reducing trust in this sign system [56]. Therefore, it is necessary to unify the design specifications to make and decide the location of the installation.

4.3. The Impact of Basic Wayfinding with/without 3D Gaming Experience on Wayfinding Performance in a Virtual Environment

The experimental results show that the essential wayfinding ability of the participants in the pretest affected the task performance in some tasks. High-ability people are more likely to reach task goals faster and may generate less anxiety during the task. Still, similar results were not found in the posttest. We speculate that this may be due to the optimization of the sign system so that people of both high and low ability can better reach their destination in the hospital. In terms of experience with FPV 3D games, it was found in both experiments that although all participants moved at an equal rate during the simulated wayfinding, in some cases, those with no game experience still took more time to reach the task targets than those with game experience. Therefore, in a virtual environment, the participant's "experience with/without 3D games" is more likely to impact experimental performance than "essential wayfinding ability".

4.4. Pros and Cons of Virtual Environments for Wayfinding Research in Medical Facilities

The advantage of this study is that the experimental process did not affect the hospital's operation, and it can control the variation and scope of the experimental field and save a lot of time and physical strength for testing. Participants conducted the task in the first-person view, giving researchers a clearer idea of what they were looking at. However, there are still some limitations in this experimental method, that is, the influence of the "crowd" on feelings and wayfinding behavior. In addition, this study was conducted on a desktop screen display, which is relatively narrow compared to walking in the real field, and it is easy to ignore the scenery on both sides; in particular, the corridor of NTUH West Campus has many pillars and arches, which easily cause blind spots. In addition, due to the long experimental time during the pilot test, a small number of people reported dizziness during the process of going up and down the stairs, and the way of climbing the stairs was changed to touching the specified icon and then being teleported upstairs so that the number of participants who reported dizziness was greatly reduced. If participants are inexperienced with 3D games, they can reduce the mouse sensitivity beforehand to avoid dizziness caused by uncomfortable operations. Although there are many parts to pay attention to when using a virtual environment as an experimental method, it still has many advantages for related wayfinding research in medical institutions because it can easily change the environment and verify the design effectiveness without affecting the hospital's operation. This presents the benefit of the "digital twin" concept [35,36,38].

4.5. Design Priorities and Recommendations for Future Sign Systems

Based on the pretest results, this study formulates the basic sign system design principles. We readjust the information presentation, classification, and layout of all hanging-style signs and various types of wall signs in the experimental environment and propose a new sign system. The information on the hanging sign is changed to outpatient, medical, laboratory, and ward services according to the type of function provided. Based on the Gestalt principles [57] and synthesizing the study results, we make three recommendations: (1) Important hospital administrative services and outpatient departments must be presented. (2) The information on the sign is processed consistently to avoid interruptions in reading, and the layout is based on the principle of not affecting the columnar format. (3) Services on the same floor should be prioritized in order.

Repetitive information is removed from the wall, and each corridor is marked with its corridor code and directions. Stairs and elevators are also equipped with a sign system with columnar format floor space information, which follows the exact specifications of the overall sign system. At the same time, adding a color classification system to the above design can positively impact the wayfinding behavior of visitors.

The results of the posttest show that the new sign system is better than the existing sign system in the performance of many tasks, which shows that the basic design principles set are conducive to visitors finding destination information in the hospital sign system. People can make correct wayfinding decisions through the guidance of the sign, which helps to find the target faster and significantly reduces the negative emotions generated during wayfinding. In addition, we have also found that when there is perfect sign guidance in a medical facility, people of high or low ability can smoothly follow the sign guidance in the hospital to reach their destination, so the wayfinding behavior tends to be consistent.

Through the design verification of the posttest, the following design points and suggestions are summarized in this study for reference in future medical institution marking systems:

Hanging-style sign design

The information on the signs on important circulation corridors should include essential administrative services and outpatient departments in the hospital. The information between the signs should be presented consistently, and the direction arrows should be used uniformly and standardized because of visual consistency; the viewer's accessibility in reading information will be improved [58,59]. The diagonal upper arrows can be emphasized by adding floor or staircase icons if they mean upper and lower floors. In addition, in the posttest, we also found that some participants could not find their direction for a while after going upstairs, so a hanging sign could be added to indicate the staircase on the second floor. The auxiliary icons on the sign should not be interspersed with principal text information, and stairs, exits, or parking lots can be added as needed to assist visitors in finding their way.

How information is arranged

Information can be classified according to the type of function provided, such as outpatient, medical, laboratory, and ward services. According to proximity and common region in Gestalt principles [57], the same type of information can be placed in the same block, and the order is not affected by the principle of column reading. The layout of the sign system is uniformly arranged in this order for easy access so that the user can quickly find the target information.

Wall sign design

For the wall, the use of many repetitive signs should be avoided, and all signs should be designed with the exact specifications of the overall sign system. Each staircase and elevator should provide complete and precise floor information regarding floor guidance. In the posttest, the task performance of going upstairs to find the Dermatology Clinic and the Jingfu Passage is still low. So, the wall near the entrance can be modified to contain the floor map and spatial system information of the entire hospital to improve the wayfinding efficiency of visitors.

Other recommendations

Since medical facilities will gradually add new services and facilities over time, or the location of departments will be adjusted, it is possible to pay more attention to the use of types that can be easily changed in the sign system so that flexible adjustments can be made at any time. In addition, an overall wayfinding sign and a color-coded four-color description were added at the lobby entrance and used in conjunction with the West Campus map colors. Madson and Goodwin's [60] research has demonstrated that color coding systems positively impact wayfinding, which can amplify the impact of color coding and improve wayfinding performance.

5. Conclusions

In this study, we conducted a field study and created a virtual environment that could map the physical NTUH West Campus to address the difficulties of conducting experiments in a physical environment. We conducted the study using task-oriented controlled experiments in which we identified current marker system issues from task performance, participant evaluation, and interview feedback. The design solution was then proposed, and the posttest assessed the same tasks to validate the design. We observed people's behavior and habits when observing indicator wayfinding and pointed out weaknesses in the current indicator system by analyzing task performance. In addition, the impact of basic wayfinding with/without 3D gaming experience on wayfinding performance in virtual environments was explored through the questionnaire results. Further, we proposed the pros and cons of using virtual medical facilities for wayfinding research. Posttest experiments verified that the signage system outperforms existing signage systems in multiple tasks, showing that the basic design principles are beneficial for visitors to find destination information in hospital signs. People can use signs to guide them to make the right wayfinding decisions, help them find their goals faster, and significantly reduce the negative emotions generated during the wayfinding process.

Study Limitations and Recommendations for Follow-Up Studies

The experimental site was the virtual model of the NTUH West Campus. However, when the experimental conditions change, there may be different findings. Because of the severe global COVID-19 epidemic during the study period, to avoid congregation, the experimental process of this study was conducted online, and the participants were requested to share their screen pictures in the online conference room. The different sizes of computer screens used by each participant may also affect the speed at which signs can be viewed during wayfinding. In the future, it is suggested that the test subjects be invited to a unified computer device to conduct experiments to reduce the variables that affect the performance of the task. In the experimental field, due to the very large scope of the entire NTUH West Campus, this study only built a virtual environment model of part of the NTUH West Campus for wayfinding experiments, and there is still insufficient understanding of how visitors will find their way between departments over longer distances or between each building. Therefore, we suggest that the scope of the virtual environment can be expanded for further research.

In addition, the visitors of NTUH are diverse, but the experimental participants in this study ranged in age from 20 to 39 years old and were mainly young people skilled in various computer operations. For the insufficient sample of older adults, the individual differences of other age groups and different groups should be explored to expand the understanding of wayfinding behavior in medical institutions.

Author Contributions: Conceptualization, software, formal analysis, investigation, writing—original draft preparation, C.-Y.W., methodology, C.-Y.W. and M.-C.Z.; validation, resources, supervision, project administration, M.-C.Z.; data curation, writing—review and editing, visualization, C.-I.C. All authors have read and agreed to the published version of the manuscript.

Funding: This research received no external funding.

Institutional Review Board Statement: Not applicable.

Informed Consent Statement: Informed consent was obtained from all subjects involved in the study.

Data Availability Statement: Due to confidentiality agreements, supporting data can only be made available to bona fide researchers subject to a non-disclosure agreement. Details of the data and how to request access are available from Meng-Cong Zheng (zmcdesign@gmail.com) at the National Taipei University of Technology.

Acknowledgments: The authors are grateful for the helpful comments of Kiminobu Sato, Design Research Institute, Environmental Design, Chiba University, whose advice was invaluable. The authors would like to thank all the anonymous reviewers and editors who contributed their time and knowledge to this study.

Conflicts of Interest: The authors declare no conflict of interest.

Appendix A

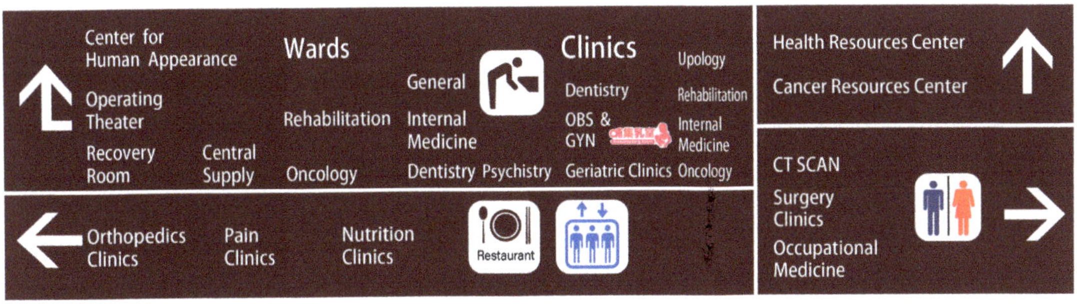

Figure A1. It's the English translation of Figure 3.

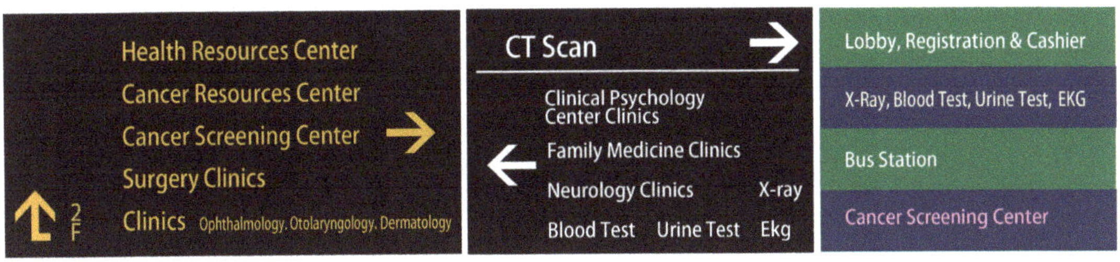

Figure A2. It's the English translation of Figure 4.

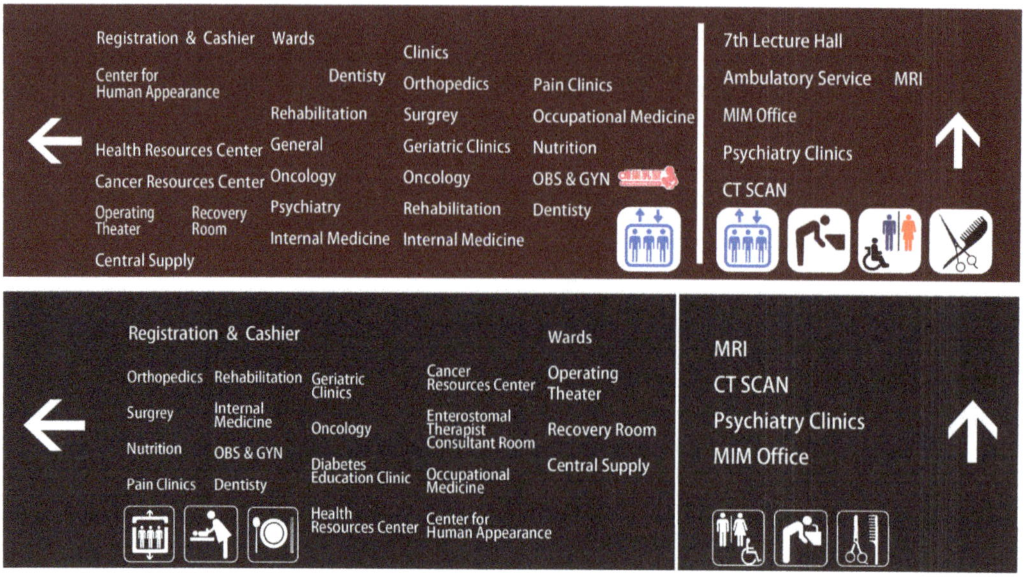

Figure A3. It's the English translation of Figure 9.

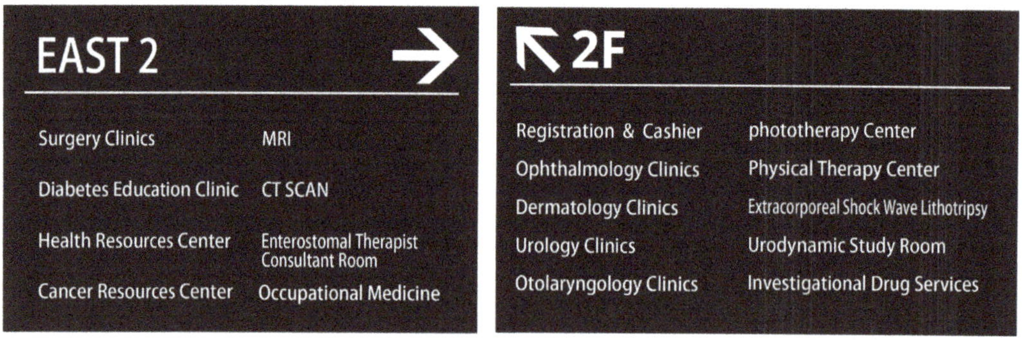

Figure A4. It's the English translation of Figure 10.

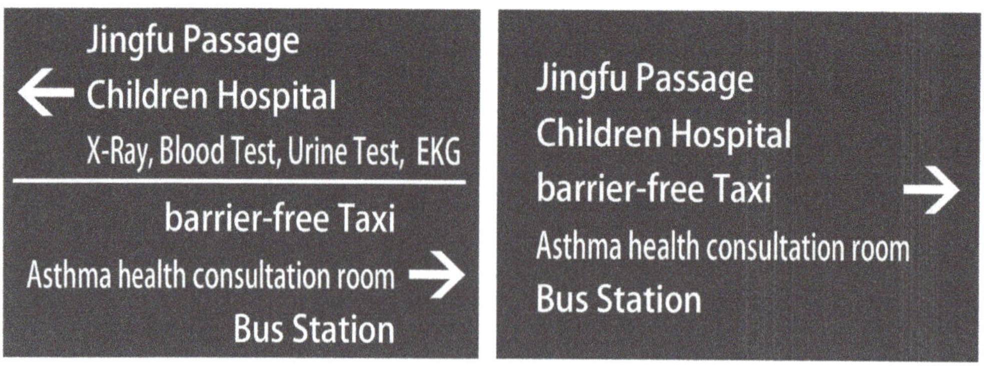

Figure A5. It's the English translation of Figure 11.

References

1. Weisman, J. Evaluating Architectural Legibility. *Environ. Behav.* **1981**, *13*, 189–204. [CrossRef]
2. Carpman, J.R.; Grant, M.A.; Simmons, D.A. Hospital Design and Wayfinding. *Environ. Behav.* **1985**, *17*, 296–314. [CrossRef]
3. O'Neill, M.J. Evaluation of a Conceptual Model of Architectural Legibility. *Environ. Behav.* **1991**, *23*, 259–284. [CrossRef]
4. Water, T.; Wrapson, J.; Reay, S.; Ford, K. Making Space Work: Staff Socio-Spatial Practices in a Paediatric Outpatient Department. *Health Place* **2018**, *50*, 146–153. [CrossRef]
5. Symonds, P.; Brown, D.H.K.; Lo Iacono, V. Exploring an Absent Presence: Wayfinding as an Embodied Sociocultural Experience. *Sociol. Res. Online* **2017**, *22*, 48–67. [CrossRef]
6. Lynch, K. *The Image of the City*; The MIT Press: Cambridge, MA, USA, 1964.
7. Arthur, P.; Passini, R. *Wayfinding: People, Sings, and Architecture*; Focus Strategic Communications Incorporated: Brantford, ON, Canada, 2002; ISBN 0973182202.
8. Allen, G.L. Cognitive Abilities in the Service of Wayfinding: A Functional Approach. *Prof. Geogr.* **1999**, *51*, 555–561. [CrossRef]
9. Golledge, R.G. *Wayfinding Behavior: Cognitive Mapping and Other Spatial Processes*; Golledge, R.G., Ed.; Johns Hopkins University Press: Baltimore, MD, USA, 1999; ISBN 978-0801859939.
10. Carpman, J.R.; Grant, M.A.; Simmons, D.A. No More Mazes. *Prog. Arch.* **1985**, *66*, 156–157.
11. Carling, T.; Böök, A.; Lindberg, E. Spatial Orientation and Wayfinding in the Designed Environment: A Conceptual Analysis and Some Suggestions for Postoccupancy Evaluation. *J. Archit. Plan. Res.* **1986**, *3*, 55–64.
12. Evans, G.W.; McCoy, J.M. When Buildings Don't Work: The Role of Architecture in Human Health. *J. Environ. Psychol.* **1998**, *18*, 85–94. [CrossRef]
13. Stea, D. *Image and Environment*; Routledge: Oxfordshire, UK, 2017; ISBN 9780203789155.
14. Conroy-Dalton, R.A. *Spatial Navigation in Immersive Virtual Environments*; University of London, University College London: London, UK, 2001.
15. Jamshidi, S.; Ensafi, M.; Pati, D. Wayfinding in Interior Environments: An Integrative Review. *Front. Psychol.* **2020**, *11*, 2930. [CrossRef] [PubMed]
16. Aksoy, E.; Aydın, D.; İskifoğlu, G. Hastanelerde Plan Şeması ve Yön Bulma Kararları Arasındaki Ilişkinin Analizi. *MEGARON Yıldız Tech. Univ. Fac. Archit. E-J.* **2020**, *15*, 509–520. [CrossRef]
17. Baskaya, A.; Wilson, C.; Özcan, Y.Z. Wayfinding in an Unfamiliar Environment. *Environ. Behav.* **2004**, *36*, 839–867. [CrossRef]
18. Gath-Morad, M.; Thrash, T.; Schicker, J.; Hölscher, C.; Helbing, D.; Aguilar Melgar, L.E. Visibility Matters during Wayfinding in the Vertical. *Sci. Rep.* **2021**, *11*, 18980. [CrossRef]
19. Hölscher, C.; Meilinger, T.; Vrachliotis, G.; Brösamle, M.; Knauff, M. Up the down Staircase: Wayfinding Strategies in Multi-Level Buildings. *J. Environ. Psychol.* **2006**, *26*, 284–299. [CrossRef]
20. Kuliga, S.F.; Thrash, T.; Dalton, R.C.; Hölscher, C. Virtual Reality as an Empirical Research Tool—Exploring User Experience in a Real Building and a Corresponding Virtual Model. *Comput. Environ. Urban Syst.* **2015**, *54*, 363–375. [CrossRef]
21. Nothegger, C.; Winter, S.; Raubal, M. Selection of Salient Features for Route Directions. *Spat. Cogn. Comput.* **2004**, *4*, 113–136. [CrossRef]
22. Smitshuijzen, E. *Signage Design Manual*; Lars Müller Publishers: Baden, Switzerland, 2007; ISBN 978-3-03778-096-1.
23. Heft, H. The Ecological Approach to Navigation: A Gibsonian Perspective. *Constr. Cogn. Maps* **1996**, *32*, 105–132. [CrossRef]
24. Norman, D.A. Cognition in the Head and in the World: An Introduction to the Special Issue on Situated Action. *Cogn. Sci.* **1993**, *17*, 1–6. [CrossRef]
25. Dalton, R.C. The Secret Is To Follow Your Nose. *Environ. Behav.* **2016**, *35*, 107–131. [CrossRef]
26. Jiang, S.; Allison, D.; Duchowski, A.T. Hospital Greenspaces and the Impacts on Wayfinding and Spatial Experience: An Explorative Experiment Through Immersive Virtual Environment (IVE) Techniques. *HERD Health Environ. Res. Des. J.* **2022**, *15*, 206–228. [CrossRef]
27. Min, Y.H.; Ha, M. Contribution of Colour-Zoning Differentiation to Multidimensional Spatial Knowledge Acquisition in Symmetrical Hospital Wards. *Indoor Built Environ.* **2021**, *30*, 787–800. [CrossRef]
28. Wong, B. Color Coding. *Nat. Methods* **2010**, *7*, 573. [CrossRef] [PubMed]
29. Ozcelik, E.; Karakus, T.; Kursun, E.; Cagiltay, K. An Eye-Tracking Study of How Color Coding Affects Multimedia Learning. *Comput. Educ.* **2009**, *53*, 445–453. [CrossRef]
30. Keller, T.; Gerjets, P.; Scheiter, K.; Garsoffky, B. Information Visualizations for Knowledge Acquisition: The Impact of Dimensionality and Color Coding. *Comput. Hum. Behav.* **2006**, *22*, 43–65. [CrossRef]
31. Nurdiansyah, D.M.R.; Asyid, S.A.; Parmawati, A. Using color coding to improve students' english vocabulary ability. *PROJECT Prof. J. Engl. Educ.* **2019**, *2*, 358. [CrossRef]
32. McDonald, J.E.; Molander, M.E.; Noel, R.W. Color-Coding Categories in Menus. In *Proceedings of the SIGCHI Conference on Human Factors in Computing Systems–CHI '88*; ACM Press: New York, NY, USA, 1 May 1988; Volume Part F130202, pp. 101–106.
33. Lamberski, R.J.; Dwyer, F.M. The Instructional Effect of Coding (Color and Black and White) on Information Acquisition and Retrieval. *Educ. Commun. Technol. J.* **1983**, *31*, 9–21. [CrossRef]
34. Wang, Y.; Punithakumar, K.; Boulanger, P. The Impact of Color Coding in Virtual Reality Navigation Tasks. In *Proceedings of the Optical Architectures for Displays and Sensing in Augmented Virtual, and Mixed Reality (AR, VR, MR)*; Kress, B.C., Peroz, C., Eds.; SPIE: Bellingham, WA, USA, 19 February 2020; Volume 11310, p. 40.

35. Grieves, M. Digital Twin: Manufacturing Excellence through Virtual Factory Replication. *White Paper* **2014**, *1*, 1–7.
36. Githens, G. Product Lifecycle Management: Driving the Next Generation of Lean Thinking by Michael Grieves. *J. Prod. Innov. Manag.* **2007**, *24*, 278–280. [CrossRef]
37. Batty, M. Digital Twins. *Environ. Plan B Urban Anal. City Sci.* **2018**, *45*, 817–820. [CrossRef]
38. Haag, S.; Anderl, R. Digital Twin–Proof of Concept. *Manuf. Lett.* **2018**, *15*, 64–66. [CrossRef]
39. Minerva, R.; Lee, G.M.; Crespi, N. Digital Twin in the IoT Context: A Survey on Technical Features, Scenarios, and Architectural Models. *Proc. IEEE* **2020**, *108*, 1785–1824. [CrossRef]
40. Larsen, C.R.; Oestergaard, J.; Ottesen, B.S.; Soerensen, J.L. The Efficacy of Virtual Reality Simulation Training in Laparoscopy: A Systematic Review of Randomized Trials. *Acta Obstet. Gynecol. Scand.* **2012**, *91*, 1015–1028. [CrossRef] [PubMed]
41. Darken, R.P.; Sibert, J.L. Wayfinding Strategies and Behaviors in Large Virtual Worlds. In *Proceedings of the SIGCHI Conference on Human Factors in Computing Systems Common Ground–CHI '96*; ACM Press: New York, NY, USA, 1996; pp. 142–149.
42. Tanja-Dijkstra, K.; Pahl, S.; White, M.P.; Andrade, J.; Qian, C.; Bruce, M.; May, J.; Moles, D.R. Improving Dental Experiences by Using Virtual Reality Distraction: A Simulation Study. *PLoS ONE* **2014**, *9*, e91276. [CrossRef] [PubMed]
43. Loomis, J.M.; Blascovich, J.J.; Beall, A.C. Immersive Virtual Environment Technology as a Basic Research Tool in Psychology. *Behav. Res. Methods Instrum. Comput.* **1999**, *31*, 557–564. [CrossRef]
44. Cliburn, D.; Winlock, T.; Rilea, S.; Van Donsel, M. Dynamic Landmark Placement as a Navigation Aid in Virtual Worlds. In *Proceedings of the 2007 ACM Symposium on Virtual Reality Software and Technology*; ACM: New York, NY, USA, 5 November 2007; pp. 211–214.
45. Morganti, F.; Carassa, A.; Geminiani, G. Planning Optimal Paths: A Simple Assessment of Survey Spatial Knowledge in Virtual Environments. *Comput. Hum. Behav.* **2007**, *23*, 1982–1996. [CrossRef]
46. Ulrich, R.S.; Zimring, C.; Zhu, X.; DuBose, J.; Seo, H.B.; Choi, Y.S.; Quan, X.; Joseph, A. A Review of the Research Literature on Evidence-Based Healthcare Design. *HERD* **2008**, *1*, 61–125. [CrossRef]
47. Dong, W.; Qin, T.; Yang, T.; Liao, H.; Liu, B.; Meng, L.; Liu, Y. Wayfinding Behavior and Spatial Knowledge Acquisition: Are They the Same in Virtual Reality and in Real-World Environments? *Ann. Am. Assoc. Geogr.* **2022**, *112*, 226–246. [CrossRef]
48. Kalantari, S.; Tripathi, V.; Rounds, J.D.; Mostafavi, A.; Snell, R.; Cruz-Garza, J.G. Evaluating Wayfinding Designs in Healthcare Settings through EEG Data and Virtual Response Testing. *bioRxiv* **2021**. bioRxiv:2021.02.10.430638. [CrossRef]
49. Al-Sharaa, A.; Adam, M.; Amer Nordin, A.S.; Alhasan, A.; Mundher, R.; Zaid, O. Enhancing Wayfinding Performance in Existing Healthcare Facilities Using Virtual Reality Environments to Revise the Distribution of Way-Showing Devices. *Buildings* **2022**, *12*, 790. [CrossRef]
50. Bos, J.E.; Bles, W.; Groen, E.L. A Theory on Visually Induced Motion Sickness. *Displays* **2008**, *29*, 47–57. [CrossRef]
51. Kennedy, R.S.; Drexler, J.; Kennedy, R.C. Research in Visually Induced Motion Sickness. *Appl. Ergon.* **2010**, *41*, 494–503. [CrossRef] [PubMed]
52. Keshavarz, B.; Golding, J.F. Motion Sickness: Current Concepts and Management. *Curr. Opin. Neurol.* **2022**, *35*, 107–112. [CrossRef] [PubMed]
53. Ma, S.-F. *Exploring the Way-Finding Ability of College Students and Its Affecting Factors*; Chiao Tung University: Hsinchu, Taiwan, 2008.
54. Ishikawa, T.; Murasawa, K.; Okabe, A. Wayfinding and Art Viewing by Users of a Mobile System and a Guidebook. *J. Locat. Based Serv.* **2009**, *3*, 277–293. [CrossRef]
55. Rosenfeld, L.; Morville, P.; Arango, J. *Information Architecture: For the Web and Beyond EBook: Rosenfeld, Louis, Morville, Peter, Arango, Jorge: Kindle Store*, 4th ed.; O'Reilly Media: Sebastopol, CA, USA, 2015.
56. Huang, Z.; Benyoucef, M. Usability and Credibility of E-Government Websites. *Gov. Inf. Q.* **2014**, *31*, 584–595. [CrossRef]
57. Koffka, K. *Principles of Gestalt Psychology*; Routledge: Abingdon, UK, 2013; ISBN 9781136306815.
58. Wakimoto, D.K. Librarians and Graphic Design: Preparation, Roles, and Desired Support. *Public Serv. Q.* **2015**, *11*, 171–182. [CrossRef]
59. Watzman, S.; Re, M. Visual Design Principles For Usable Interfaces: Everything Is Designed: Why We Should Think Before Doing. In *Human-Computer Interaction: Design Issues, Solutions, and Applications*; CRC Press: Boca Raton, FL, USA, 2009; pp. 19–44.
60. Madson, M.; Goodwin, K. Color Coding the "Labyrinth": How Staff Perceived a Two-Part Intervention to Improve Wayfinding in an Adult Emergency Department. *HERD* **2021**, *14*, 429–441. [CrossRef]

buildings

MDPI

Article

Spatial Ability Performance in Interior Design and Architecture: Comparison of Static and Virtual Reality Modes

Ji Young Cho [1,*] and Joori Suh [2]

1 Department of Housing and Interior Design, Kyung Hee University, Seoul 02447, Republic of Korea
2 School of Architecture and Interior Design, University of Cincinnati, Cincinnati, OH 45221-0016, USA; joori.suh@uc.edu
* Correspondence: jcho@khu.ac.kr

Abstract: Recent advancements in virtual reality (VR) technology have enabled its integration into learning diverse aspects of spatial components and relationships in the field of spatial design, as well as designing, communicating, collaborating, and managing complex building projects. With the growing interest in incorporating VR technology in spatial design, examining whether people understand, perceive, and perform spatial tasks in the same way in VR as they do in static modes is essential. Thus, the purpose of this study was to compare spatial ability performance in a conventional static paper–desktop mode and an interactive VR mode. Thirty students completed the Architecture and Interior Design Domain–Specific Spatial Ability Test in both modes. Their visual cognitive style was measured with the Object–Spatial Imagery Questionnaire, and their responses to the usability of the VR mode were analyzed. The results revealed: (a) significant difference in performance between static and VR modes, including better performance in three spatial visualization subconstructs in static mode than in VR; (b) no gender difference in VR mode; (c) a tendency of spatial visualizers to benefit from VR mode; and (d) a tendency of people with high spatial ability to be more susceptible to test mode. Overall, the results contribute to expanding our understanding of spatial ability performance in different test modes and provide insights concerning the integration of VR into the development of spatial ability tools and education.

Keywords: design education; interactive; interior design; spatial ability; test mode; virtual reality

1. Introduction

Spatial ability is one of the key cognitive abilities required to perform design tasks in the field of interior design and architecture. In creating new spaces and buildings, designers exercise their imaginations, visualizing the composition and volumetric relationships among spatial components. In addition, to communicate spatial properties and their relationships, visual representation methods involving two-dimensional (2D) and three-dimensional (3D) drawings are used among designers, engineers, and constructors. Thus, two of the primary skills imbedded in education in interior design and architecture are spatial awareness and versatility in communicating spatial information.

With the development of virtual reality (VR) technology, tremendous changes have taken place in communicating spatial objects and components, especially in education in interior design and architecture. The high sense of presence and the sense of immersion with the VR technology allow students to learn content more vividly in a virtual environment (VE). Furthermore, various spatial interactions, such as teleporting and manipulating objects in VE, are now possible. Real-time modification and iterations of design can be made within VE, and collaboration among designers and multiple stakeholders from multiple locations is also possible in VR. Researchers have shown the benefit of using VR technology in design studios, including facilitating creativity in the student design process [1], enhanced perception of solid–void characteristics of a model [2], and motivation [1].

With growing interest in incorporating VR technology in various aspects of the spatial design process, the following question must be asked: Do people understand, perceive, and perform spatial tasks in the same way in VR as they do in static mode? Since the measurement of spatial ability innately involves 3D space, the potential of VR technology in spatial ability has been investigated by researchers [3–5]. Specifically, Lochhead et al. (2022) investigated differences in mental rotation (MR) performance in 2D images versus 3D stimuli, both in VR mode, and found better participant performance in 3D stimuli [3]. In measuring spatial ability, however, examining what type of spatial ability is measured, what test mode must be used, and how the spatial ability questions should be presented is important since they may impact spatial ability performance depending on test mode. Moreover, recent neuroscience studies revealed that responses to the same task in virtual reality and desktop mode vary considerably [6]. Different parts of the brain are activated, and since the networks of the brain regions are dissimilar, deep understanding of spatial ability performance in interactive VR versus conventional static mode is necessary. Whether performance in VR mode resembles that in the original test modes has rarely been reported; consequently, researchers have pointed out the strong need to understand spatial ability performance in various test modes [3].

Thus, the purpose of this study was to identify the characteristics of spatial ability performance in interactive VR mode in comparison with that in a static paper–desktop mode, using the Architecture and Interior Design Spatial Ability Test (AISAT), developed to measure domain-specific spatial ability in the field of architecture and interior design. Test data and self-reported information were analyzed to examine potential factors related to the test mode effect on the spatial ability test. Performance in the two modes was compared using a statistical program. Two modes were examined; therefore, any impact that might have occurred as a result of the order of performing the tasks was examined. Performance in relation to individual differences, such as gender and one's visual cognitive style were also examined. The Object–Spatial Imagery Questionnaire (OSIQ) [7] was administered to identify participants' tendency toward a particular visual cognitive style and its relationship with performance in the two test modes of the AISAT. Performance in VR mode by high vs. low spatial ability score in static mode was examined to identify who benefits from using the VR mode. Moreover, a usability survey was also included to evaluate the tool used for the VR version of the AISAT.

The main research questions were:

- Is spatial ability performance on the AISAT similar or different in the static mode and VR mode?
- Does the order of taking the tests in a particular mode (VR mode first vs. static mode first) have an impact on performance?
- Does gender difference exist in spatial ability performance in the two modes?
- Does visual cognitive style matter in spatial ability performance?
- Do individuals with low spatial ability benefit from the VR test mode?
- How do participants evaluate the usability of the VR AISAT?

2. Background

2.1. *Spatial Ability Measurement*

Spatial ability is the proficiency needed to understand spatial conditions among objects and environment and problem-solving capacity regarding spatial information. It is defined as "skill in representing, transforming, generating, and recalling symbolic, nonlinguistic information" ([8], p. 1482) and "the ability to visualize, manipulate and interrelate real or imaginary configurations in space" ([9], p. 3). It is a significant cognitive capacity required in everyday problem-solving, including wayfinding, design, technical drawings, and graphic visualization [10]. Existing instruments for measuring SA were developed mainly in the 1970s by researchers in the fields of cognitive and developmental psychology with the primary purpose of predicting successful academic or professional performance. Spatial ability can be categorized as ability in mental rotation, spatial visualization (SV), and spatial perception [8]. Some of the well-known and validated tools include mental rotation, developed in 1978 [11]; paper folding, introduced in 1976 [12]; and a test for perspective-taking and spatial orientation, introduced in 2004 [13]. These tools were originally developed in static mode in a paper-and-pencil version; later some of them were converted from static to an interactive computer version or interactive VR version by other researchers [3]. In the domain of architecture and interior design, examples of tools include the 3D Ability Test (3DAT), developed in 2010 [14]; and the Urban Layout Test, Indoor Perspective Test, and Packing Test, introduced in 2021 [15].

In the domain of interior design and architecture, spatial ability can be understood as "the mental manipulative skills required to perform mental processes such as the rotation of objects, the understanding of how objects appear in different positions, and the conceptualization of how objects relate to each other in space" ([16], p. 2). Unlike in other domains, practitioners of interior design and architecture deal with spatial relationships of human-scale or large environment-scale components as well as the translation of 2D drawings into 3D representation of spatial components within habitable spaces. In architecture and interior design, designers handle spaces that are more specific and concrete as well as abstract. The AISAT was developed with the intention to measure domain-specific spatial ability in architecture and interior design, different from general spatial ability; its reliability and validity were investigated and reported in a prior study [17]. In the AISAT, instead of using small abstract objects typical in general spatial ability tools, large-scale spatial forms relevant to environmental design were used to test mental rotation and spatial visualization. The predictive potential of the developed AISAT for design performance and creativity has been investigated and reported [18,19].

2.2. Nature of VR and Its Application for Spatial Ability Measurement

VR technology is a medium, comprising computer simulations, that allows the viewer to experience full immersion within the simulations [20]. In a VE, spatial representation as well as human–object interaction becomes more realistic, allowing the user to experience the sense of the scale of the environment. VR simulation also makes spatial objects appear closer to their real form with an increased sense of depth and three-dimensionality. With the provision of constantly changing perspective views, the viewer can experience more extensive perception of the spatial forms from many different angles, which is impossible with a static mode.

VR has the effect of creating a sense of immersion and presence as users interact with visual stimuli in a 3D space. Sense of presence and fidelity are key characteristics of VR, the former being the degree to which users feel that they are present in the VE—or that virtual objects are present—without being aware of their virtual presence. Fidelity refers to the degree to which users experience interactions within the VR as if they were interacting in the real world [21]. A review of previous studies has shown that VR experiences help to expand cognitive abilities and improve spatial perception along with increasing immersion, confidence, and motivation [22]. Since the development of VR technology, the design process, modifications of design, and the way designers collaborate have been influenced by VR technology across domains. Along with more extended application of the immersive experience encompassing the virtual and physical worlds, many new approaches to design and management methods using the eXtended Realty (XR) platform have been introduced. For example, Banfi at al. (2019) proposed informative models for architectural heritage using the scan-to-building information modeling (BIM) process for VR and augmented reality (AR) that allows an immersive experience of built heritage to discover hidden historical values [23]. Other examples include a platform for immersive and interactive experiences for cultural heritage [24] and a platform for reducing the rework of design changes in BIM using VR and AR [25]. Alizadehsalehi and Yitmen (2021) proposed a digital twin-based automated construction progress monitoring system that allows a user to collaborate to create, analyze, manage, and visualize construction progress using BIM and XR technologies [26]. With the potential of integrating artificial intelligence and machine learning within the VR-based educational process, including design exploration and communication, deeply understanding how designers' spatial skills differ in the virtual and physical interactions has become more crucial.

One example of the VR version of spatial ability-measuring tools is the Immersive Mental Rotations Test by Lochhead et al. (2022), who fit the mental rotation test originally developed by Vandenberg and Kuse in 1978 to VR head mounted display (HMD)–Oculus Quest [3]. They did not, however, compare performance in VR and static modes. Instead, they compared performance of 2D images versus 3D objects, both in VR mode and found that participants performed better in terms of accuracy and speed in "stereo 3D stimuli than with 2D images of those stimuli" (p. 1). In addition, they pointed out that conventional paper-and-pencil MR measures two things—one is the ability to visualize 2D to 3D, and the other is the ability to rotate that 3D—and the immersive VR can alleviate the cognitive load of the first stage by providing 3D objects in VR. Finally, they argued that with the emergence of new technologies such as VR, new questions on utility, interface, application, and users in VE must be asked.

2.3. The Development and Test Modes of AISAT

The original version of the AISAT was available only in static mode—on paper or a computer screen. The VR version of the AISAT was developed to explore the potential of implementing VR technology in measuring spatial ability in the domain of interior design and architecture. Implementation of VR technologies using large-scale simulated environments may have benefit in terms of providing a better sense of scale and enhancing the understanding of spatial relationships among the components presented at an environmental scale, but each mode has its own strengths and weaknesses (See Table 1).

Table 1. Strengths and weaknesses of two modes of the AISAT.

	Static Paper—Computer Mode AISAT	Interactive VR Mode AISAT
Strengths	Display of the entire space is possible. Convenience of preparing for the test	Realistic simulated view of the space Dynamically coordinated perspective views according to the viewer's positions High resolution images Interaction within the space (e.g., teleporting, changing direction of views, and rotating objects using controller)
Weakness	Printing of each page required. Low resolution Limited interaction Lack of realistic experience	Cumbersome and heavy HMD Requires longer preparation time and training for familiarity Discomfort (e.g., dizziness)

The paper-and-pencil version has benefits in ease of use and distribution. It is also economical since no specific equipment is needed. The structure of each question entails one question and four multiple choices. One can look at all options and choose one answer. Its weakness, however, lies in the resolution of question images and details due to the difficulty in conveying realistic and high-resolution images on a small piece of paper. In addition, it is not sustainable in the long run since questions must be printed each time the test is given. As a result of the low resolution and limitation in detail on the paper version, the spatial visualization part of the AISAT was prepared for the computer. Questions were prepared in an online survey format (qualtrics.com, accessed on 1 June 2021), and the link was shared with participants. In the paper–desktop mode, all questions and answer options were provided in static mode. In contrast, in the VR version of the AISAT, all questions and answer options appear in interactive mode.

The VR version of the AISAT was developed using the questions from the paper–desktop test mode. Three-D modelling was created using SketchUp 2020.2 and Rhino 6 software. The 3D modelling was converted to a fbx file, an extension that can be used on the Unity game engine. To test the AISAT in VR, user interface and user experience design of the virtual platform were required to induce appropriate behaviors and experiences for users. In the VR version of the AISAT, test takers can look around the environment in question, rotate the objects in the 3D test space (on the MR test), look at 3D simulations from diverse angles by shifting their bodies and by using the controller, and move within the 3D simulation space using the controller and teleport function. Unity, a game engine, was used for the software development and HTC VIVE was used as the HMD for the VR experience.

2.4. Spatial Ability, Gender, and VR

Gender difference has been frequently reported, particularly men's outperformance in MR (e.g., [8,27–29]). In our prior studies, in static mode AISAT, no gender difference was observed in any of the SV subconstructs [18], but gender difference was found in MR [17]. Reporting no gender difference in VR can be found in literature. For example, Park (2009) found that in VR, gender did not affect spatial ability in MR and SV [30]. Samsudin et al. (2011) also found that in MR, no significant gender difference occurred in the VR version [31]. Thus, we assumed that no gender difference would be observed in the VR mode of the AISAT.

2.5. Spatial Ability and Visual Cognitive Style

"Visual cognitive style" is a term that refers to "an individual's cognitive capacity to process information in terms of object and spatial images" ([32], p. 198). The intent of the OSIQ, developed by Blajenkova et al. (2006), was to measure tendencies in visual cognitive style. One's visual cognitive style can be divided into object visualization style and spatial visualization style; those processing styles were called object visualizers and spatial visualizers, respectively [7]. Object visualizers tend to be good at processing object information, such as colorful and pictorial images of objects, and spatial visualizers tend to excel in processing the spatial relationships of objects. The OSIQ was used to identify its relationship with performance in the two modes of the AISAT and any differences in advantaging in a certain mode. The literature has shown that spatial visualizers who have high spatial scores tend to have higher spatial ability than object visualizers [32,33].

2.6. Spatial Ability Level and Benefit of VR

A review of prior studies showed the potential benefit of VR in people with different levels of spatial ability. For example, in a comparison of learning performance in PowerPoint slides versus desktop VR in people with high spatial ability (HSA) and low spatial ability (LSA) [34], the performance of people with LSA increased in VR but that of people with HSA did not, indicating that VR benefits people with LSA. An electroencephalography (EEG) study by Sun et al. (2019) showed that VR helped reduce the cognitive load for people with LSA [35]. In fact, some researchers have argued for a compensator hypothesis [36,37], meaning that those with LSA benefit since VR can alleviate cognitive load, but others have argued that those with HSA benefit under the enhancer hypothesis [37]. Investigating those who benefit in VR mode according to one's spatial ability is worthwhile.

3. Research Method

The primary focus of this research was to understand performance in a static mode and the VR mode of the AISAT. Based on key findings from our previous research and literature review, we hypothesized the following for the current research:

H1: Performance in the static mode of AISAT and the VR mode of AISAT will be similar.

H2: The order of taking the tests (VR mode first vs. static mode first) will not influence spatial ability performance.

H3: No gender difference will be observed in the performance of the VR AISAT.

H4: Spatial visualizers in visual cognitive style will perform better in VR mode than object visualizers.

H5: Those with LSA will benefit from VR mode more than those with HSA.

Figure 1 shows a graphic summary of the research framework and hypothesis of the current study.

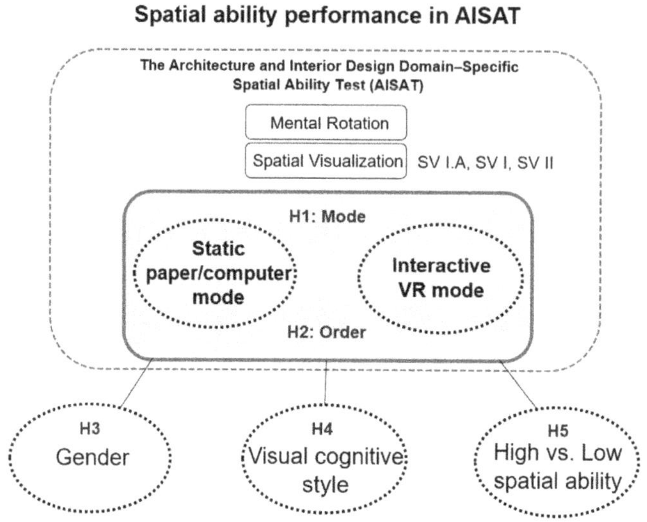

Figure 1. Research framework and hypothesis.

3.1. Procedure

Data were collected from June 2021 to June 2022. A total of 46 college students (7 men and 39 women) enrolled in an interior design program at one university in Seoul, Republic of Korea signed up to participate. The dominance of women in interior design programs is commonly observed in diverse countries, such as in the US [38] and UK [39]. An official report in 2020 indicated that the proportion of female interior design students was 83% between 2014 to 2018 [39]. Although no official report on the proportion of gender in interior design programs is available in Korea, one may reasonably say the dominance of females in the current research represents the tendency in the discipline. Participants were randomly assigned to two groups, and each group performed the two modes of tests in a different order in a one-week period to ensure no impact from the sequence of solving the tasks.

For the static mode, participants sat in one classroom together and completed the task individually. One researcher administered the AISAT. Fourteen minutes were given for the static mode; including the tutorial and sample questions, about twenty minutes were used. For the VR mode, each participant performed the VR AISAT individually. Participants went to another room, where they stood while wearing HMD and completed the VR version. The questions themselves required 23 min. However, including the tutorial and sample questions, about 40 min were needed. The rationale behind the time allotted derived from the results of the pilot study, in which the 10 participants taking the test in VR mode needed more time to attain a score similar to what they earned in static mode. Thus, we determined the time allotted would be what participants needed to earn similar scores in both modes. Once finished, participants returned to a waiting area and completed the usability test questionnaire.

3.2. Instrument

Participants completed two modes of the AISAT, the OSIQ, a demographic information questionnaire, and the usability test questionnaire for the VR version.

3.2.1. AISAT

The constructs of the static mode and the VR mode of the AISAT and time given for each construct of questions appear in Table 2.

Table 2. AISAT constructs in two modes.

Subconstruct of Spatial Ability	Description	Static Mode		Interactive VR	
		Number of Questions	Time	Number of Questions	Time
Mental rotation	MR The ability "to mentally rotate 3D spatial forms and visualize them rapidly"	7	4 min	7	2 min
Spatial visualization	SV I.A (2D → 3D in abstract information) The ability "to read 2D information and convert it to 3D and find correct location for the correct viewpoint with respect to the orientation of test takers' own body"	7	2.5 min	7	4 min
	SV I (2D → 3D) The ability "to read 2D spatial information (e.g., floor plan drawing), expand it into volumetric forms, mentally proceeding through various possibilities to locate the correct 3D exterior or interior shapes of the building"	10	5 min	10	10 min
	SV II (3D → 2D) The ability "to read 3D volumetric information, compress complex volumetric information, convert it to 2D information, and find the correct 2D floor plan"	5	2.5 min	5	7 min
Total		29	14 min	29	23 min

Sample questions for each construct in the static mode and the VR mode of the AISAT appear in Table 3.

The AISAT comprises of two main constructs—MR and SV—and three subconstructs within SV: SV I.A, SV I, and SV II. MR measures the ability to mentally rotate 3D spatial forms and visualize them rapidly. The questions in MR comprise "abstract environmental information, such as vertical dividers and horizontal elements familiar from an isometric architectural perspective and furniture drawings" ([17], p. 16). SV I.A and SV I measure the ability to translate 2D information into 3D information. The difference between SV I.A and SV I is that SV I.A uses "abstract environmental information with appropriate human-eye level" ([17], p. 16) while SV I uses more concrete information, such as a 2D floor plan and 3D interior/architectural perspectives. SV II measures the ability to translate 3D information into 2D information with concrete spatial elements, such as 3D interior/architectural perspectives and 2D floor plan.

The AISAT was prepared as two sets (Set A and Set B). Each set has the same number of questions and participants were randomly assigned one of the sets. Within each set, the order of the questions was randomized to counter for possible learning and carry-over effects. The VR mode provides a tutorial on how to use the button in the system and a few sample example questions for users, allowing them to familiarize themselves with the system. Onscreen text as well as verbal instructions were delivered by the research moderator. After the introduction and tutorial sessions, an overview of the entire procedure was introduced before they began the test. In addition, the moderator asked participants whether they understood how to use the equipment, had any issue regarding the HMD, or had any questions so that all participants could begin with sufficient understanding of the test tool and feel comfortable.

Table 3. Sample questions from the AISAT static mode and VR mode.

Subconstruct of Spatial Ability		Static Mode (Paper–Computer)	Interactive Mode (VR)
MR		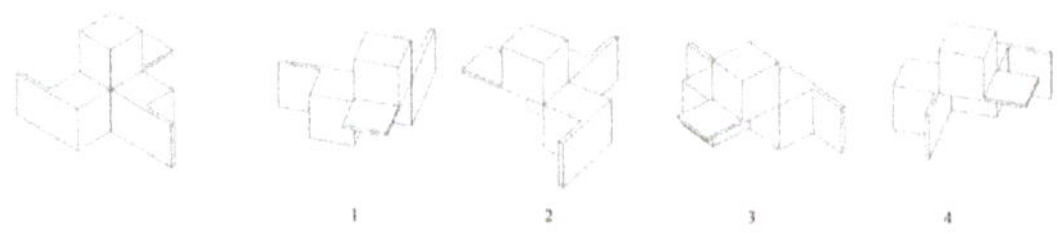	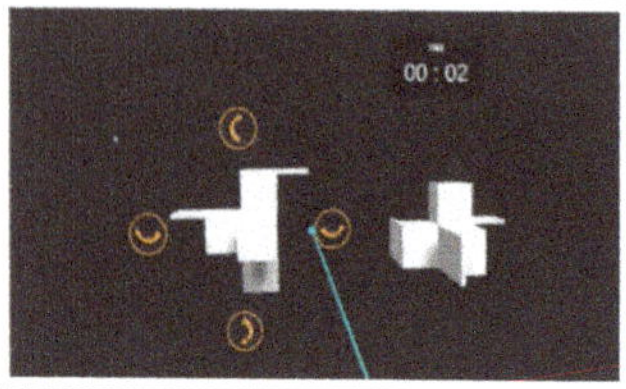
		Paper version	
		Users are asked to imagine rotating the objects along a vertical axis, and find two images that are the same as the model aside from its orientation.	Users are asked to rotate an object to match the provided object in question.
SV	SV I.A	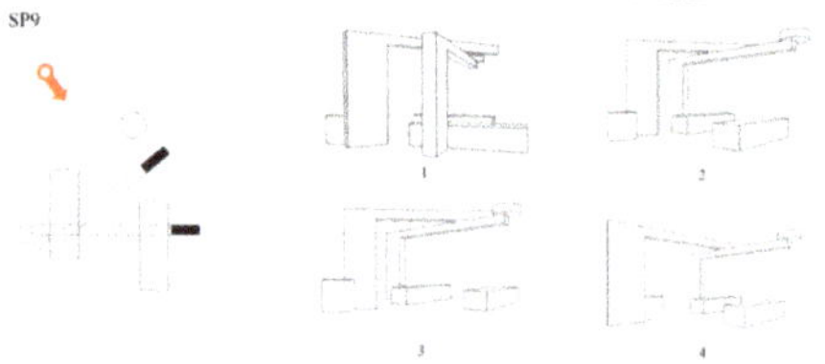	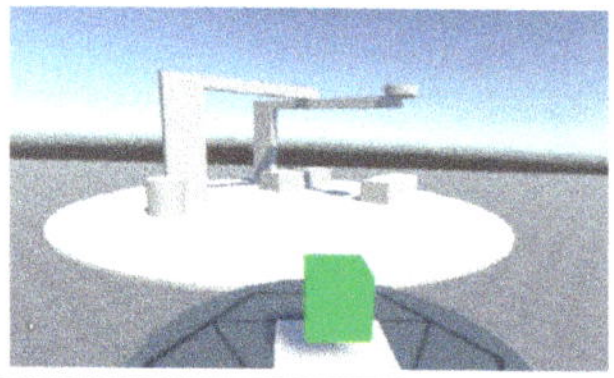
		Paper version	
		Users are asked to look at the 2D plan, imagine that they are standing inside the circle looking in the direction of the arrow, and find the correct perspective among options.	Users are asked to look at 2D drawing and find the correct 3D abstract models of spatial information
	SV I	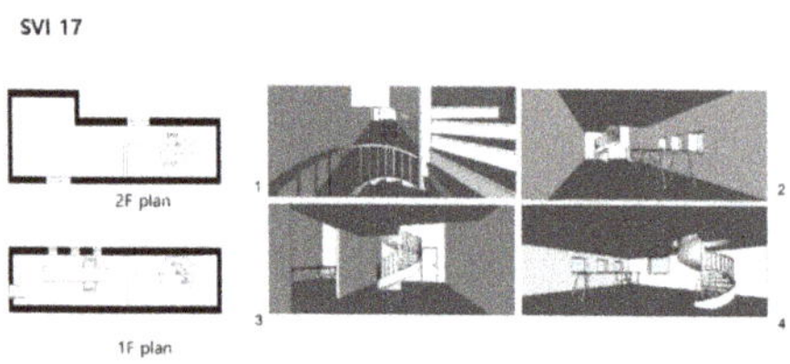	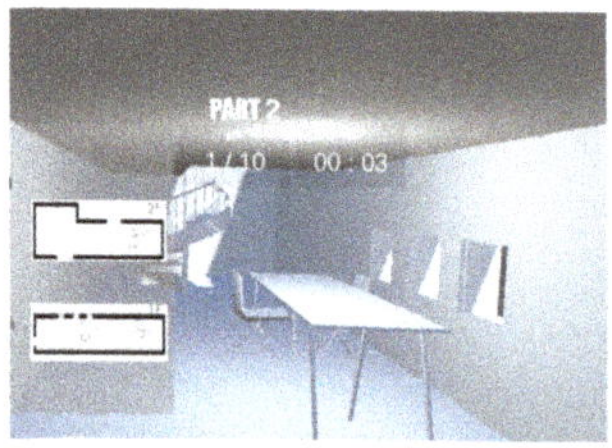
		Computer version	
		Users are asked to look at the floor plan, imagine its three-dimensional condition, and select the correct 3D perspective among the options.	Users are asked to look at a 2D drawing and select the correct 3D architectural model among options.
	SV II	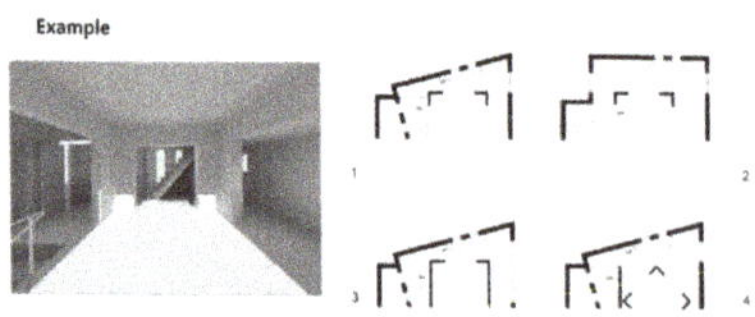	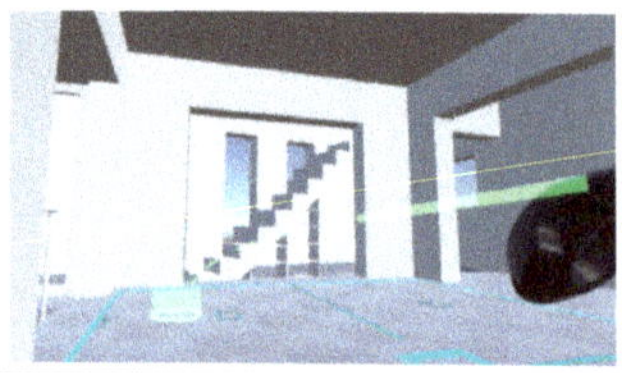
		Computer version	
		Users are asked to look at the perspective image, imagine its two-dimensional condition, and select the correct 2D drawing among options.	Users are asked to explore 3D space and select correct 2D drawing among options.

3.2.2. OSIQ

After completion of the VR mode of AISAT, participants' visual cognitive style was measured using the OSIQ. The original OSIQ is a self-reported measurement consisting of 30 statements with responses recorded on a 7-point Likert scale, ranging from 1 (totally disagree) to 7 (totally agree). In this study a simple version comprising 10 statements was used (five questions for spatial visualizer and five for object visualizer).

3.2.3. Demographic Information Questionnaire

The questionnaire on demographic information included gender, age, year in design major, VR experience, and experience with VR games.

3.2.4. Usability Test Questionnaire

The usability test comprised 34 statements with responses recorded on a 5-point Likert scale, ranging from 1 (totally disagree) to 5 (totally agree). Table 4 shows the categories and sample statements appearing on questionnaire for the usability of the VR AISAT.

Table 4. Questionnaire categories and sample statements.

Item	Sample Statements	Number of Questions
Usability	I'm more comfortable with a traditional paper test. I am more comfortable with the virtual reality method. The controls in the program are easy to use.	7
Easy to understand tutorial	The tutorial for the first type, MR, helped me understand how to solve the problem.	4
Task performance	I understood exactly what the goal of the task was.	4
Exploration and navigation	The menus of the interface were easy to navigate.	4
Satisfaction	Doing the virtual AISAT is a valuable experience for me.	5
Immersion	It was immersive while solving the virtual AISAT.	3
Discomfort	I felt dizzy while using Virtual AISAT.	7
Open-ended questions	Please feel free to share your thoughts on solving the problems. Tell us how we can improve AISAT-VR	2

3.3. *Participants*

Participants were recruited from students enrolled in an interior design program at one university in Seoul, Republic of Korea. Students voluntarily participated in the study and received a gift card valued at $20 as compensation for their time. Data were collected ethically, following approved Institutional Review Board protocol at the university.

Originally, a total of 46 students signed up for the research. However, as a result of COVID-19, only 30 participants (12 participants who performed the VR mode first and 18 participants who performed the static mode first) completed both modes, which were used for the comparison analysis.

4. Results

In the static mode, MR and SV I.A questions were in pencil-and-paper format, and SV I and SV II questions were on the computer screen display format. Scores for the paper version were recorded by the researchers. In the computer screen display version, their responses were recorded in Qualtrics. On the VR AISAT, participants' selection of answers and response time were recorded on a desktop, and correctness was automatically measured. In this research, only the accuracy score was measured and analyzed since each test had a time limit, and thus the response time was not analyzed.

4.1. Demographic Characteristics Analysis

Participants included 6 men and 24 women (average age = 22.36). A total of 3 of them were sophomores, 21 were juniors, and 6 were seniors. A total of 46.7% of them had no prior experience in VR, 26.7% had one, 20% had two, and 6.7% had four VR experiences. Regarding VR game experience, 53.3% answered no experience; 30%, one; 6.7%, two, and 10%, three times. Approximately half the participants had no prior experience in either VR or VR games. When analyzing any difference in performance due to participants' past VR experience and VR game experience, no statistical difference was observed. So, both modes of the AISAT could be used as a measuring tool without past VR-related experiences.

4.2. Performance Comparison by Test Mode (Test of H1)

For comparison purposes, scores have been converted based on a maximum score of 100. When comparing participants' performance in static mode and VR mode, overall, they performed better on SV in static mode but better on MR in VR mode. When examining the performance score in each subconstruct, the pattern between static mode and VR showed differences; in static mode, scores were high in SV I.A, MR, SV I, and SV II in that order; scores in VR mode were high in MR, SV I.A, SV I, and SV II in that order. Overall, participants performed better in SV I.A and MR but worse in SV I and SV II. The average score of each mode was 66.78 (SD = 14.59) in static mode and 56.09 (SD = 12.80) in VR mode out of 100. See Figure 2.

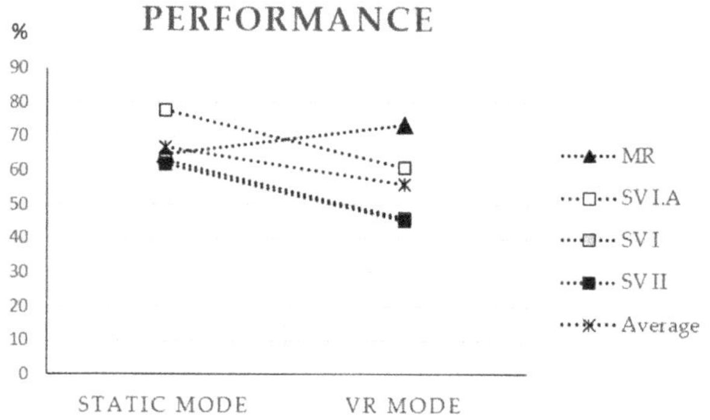

Figure 2. Observed scores (% out of 100) by static and VR mode.

Examination of statistical differences using a paired sample t-test revealed significant differences in three SV items with better performance in static mode. A statistically significant difference was found in SV 1.A performance between static mode (M = 77.62, SD = 14.86) and VR mode (M = 60.95, SD = 26.51) with t(29) = 3.22, p = 0.003, d = 0.78; SV I performance between static mode (M = 63.00, SD = 20.03) and VR mode (M = 46.00, SD = 15.67) with t(29) = 4.39, p = 0.000, d = 0.95; SV II performance between static mode (M = 62.00, SD = 28.45) and VR mode (M = 45.33, SD = 28.74), with t(29) = 4.09, p = 0.000, d = 0.58; and the average of all subconstruct score between static mode (M = 66.78, SD = 14.59) and VR (M = 56.09, SD = 12.80), with t(29) = 4.61, p = 0.000, d = 0.78. Results indicated that participants performed significantly better in the three SV tests in static mode. The three SV subconstructs measure the following spatial visualization abilities: SV I.A (2D to 3D in abstract information), SV I (2D to 3D in concrete information), and SV II (3D to 2D in concrete information). The dimensionality crossing between 2D and 3D in SV seemed to impact performance in VR more. The SD of the average score of static mode was larger than that in VR, indicating larger individual differences in static mode performance than in VR.

4.3. Performance Comparison by the Order of Mode (Test of H2)

Of 30 participants, 12 performed VR mode first, and 18 performed static mode first. When comparing any difference by the order, the independent *t*-test result shows no statistically significant difference in the performance by the order except SV II in VR, meaning the order of test mode did not bring effect on performance in most cases but it did on performance of VR SV II. Those who completed VR first performed better in VR SV II than those who did static mode first. See Figure 3.

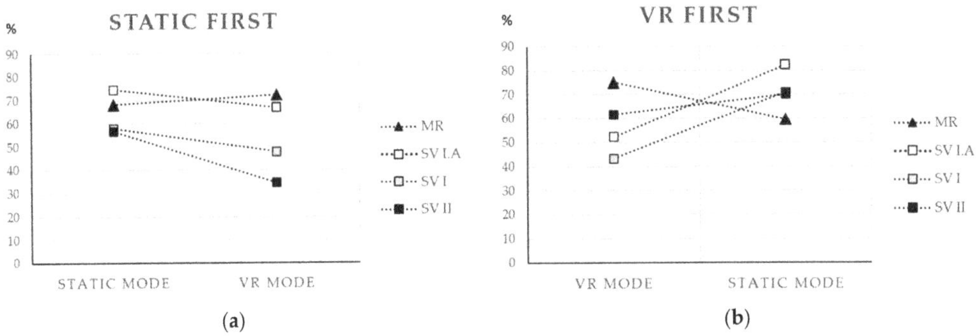

Figure 3. (**a**) Observed scores by static first group; (**b**) Observed scores by VR first group.

When calculating the difference between the score in static mode and that in VR mode and then conducting an independent *t*-test using that score, significant differences were found in SV I.A and SV I. A significant difference occurred in the SV I.A scores between static first group (M = 7.94, SD = 25.09) and VR first group (M = 29.76, SD = 28.87) with t(28) = 2.20, p = 0.036 and in the SV I scores between static first group (M = 10.00, SD = 19.40) and VR first group (M = 27.50, SD = 20.01); t(28) = 2.39, p = 0.024. In the two subconstructs, those who completed VR first had larger difference in that score, meaning the VR first group performed better in static mode later compared to static first group, indicating the potential VR training effect on performance.

4.4. Performance Comparison by Gender (Test of H3)

The participants included 6 men and 24 women. Although the proportion is somewhat unbalanced for comparison, any difference according to gender was examined to confirm that VR AISAT does not measure general SA but domain-specific SA. Overall, men performed better than women in all items except MR in VR. When examining any difference using an independent *t*-test, significant difference was observed only in MR in static mode between men (M = 88.10, SD = 14.05) and women (M = 58.93, SD = 23.95) with t(28) = −2.84, p = 0.008 and in the average score of static mode between men (M = 79.31, SD = 9.75) and women (M = 63.65, SD = 14.03) with t(28) = −2.57, p = 0.016. This result indicates that performance on most components of the AISAT in either static or interactive VR modes is not influenced by gender.

Figure 4 shows that compared with men, for whom all scores in VR mode decreased, women had higher MR scores in VR mode, indicating for women VR mode was much easier than the static mode. The reason for better performance in interactive MR seemed to be the reduced cognitive load in VR mode since participants were able to see the rotated view of the shape quickly by clicking the button until it matched the shape suggested in the question. In addition, for women, the slopes of the graph of three SV subconstructs tended to be much gentler than for men, meaning that women did not receive much impact in performance in VR compared to men.

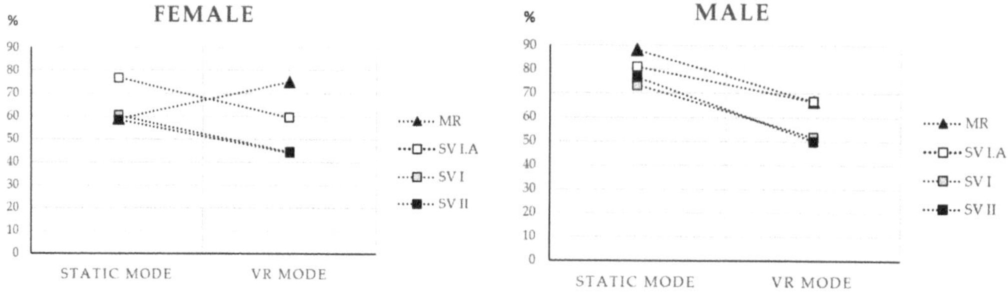

(**a**) (**b**)

Figure 4. (**a**) Observed scores by female; (**b**) Observed scores by male.

4.5. Performance Comparison by the Visual Cognitive Style (Test of H4)

Based on the OSIQ score, participants were categorized into two groups: a spatial visualizer group (N = 13), who had high spatial visualization scores, and an object visualizer group (N = 13), who had high object visualization score. Overall, the spatial visualizer group performed better than the object visualizer group. The independent *t*-test result showed that significant difference was observed in static SV II, VR SV I, and VR SV II. Significant difference was observed in SV II between the spatial visualizer group (M = 72.31, SD = 23.86) and the object visualizer group (M = 46.15, SD = 28.73) with t(24) = 2.53, $p = 0.019$; in VR SV I between the spatial visualizer group (M = 53.08, SD = 14.94) and the object visualizer group (M = 36.92, SD = 14.37) with t(24) = 2.81, $p = 0.01$; and in VR SV II between the spatial visualizer group (M = 56.92, SD = 25.62) and the object visualizer group (M = 30.77, SD = 29.00) with t(24) = 2.44, $p = 0.023$. In the three items, the spatial visualizer group performed significantly better than the object visualizer group. The better performance in two of SV subconstructs in VR indicated the potential of spatial visualizers' benefit in VR. See Figure 5.

4.6. Performance Comparison by High vs. Low Spatial Ability (Test of H5)

Participants were divided into two groups based on the sums of their scores on performance in static mode: high spatial ability group (HSA, N = 10, static sum is 22 or more out of 29) and low spatial ability group (LSA, N = 8, static sum is 17 or less out of 29); their performance was compared using an independent *t*-test. In the VR mode, HSA performed better than the LSA in SV I and SV II, meaning that people with high spatial ability perform well in SV I and SV II in VR but performance in MR and SV I.A in VR mode was not influenced by spatial ability. See Figure 6.

Examination of the statistical differences of performance in static mode and VR mode using a paired sample *t*-test within each group revealed different patterns in people with HSA and LSA. A significant difference was observed in three SV items for HSA with better performance in static mode. This indicated that those with HSA tended to be more influenced by and sensitive to the test mode than those with LSA. However, for LSA most differences disappeared, but only difference in SV I was observed; and the MR score of LSA was even higher in VR mode than that in static mode although statistically not significant. In addition, the degree of decline from SV scores in static to that in VR mode was smaller in those with LSA. This indicated the potential benefit of VR for LSA. For HSA, statistically significant difference was found in SV 1.A between static mode (M = 87.14, SD = 8.11) and VR mode (M = 57.14, SD = 25.20) with t(9) = 3.37, $p = 0.008$, d = 1.60; SV I performance between static mode (M = 75.00, SD = 17.16) and VR mode (M = 55.00, SD = 7.07), with t(9) = 3.35, $p = 0.008$, d = 1.52; and SV II performance between static mode (M = 86.00, SD = 16.47) and VR mode (M = 62.00, SD = 23.94), with t(9) = 3.67, $p = 0.005$, d = 1.17. For LSA, statistically significant difference was found in SV II between static mode (M = 37.50, SD = 22.52) and VR mode (M = 22.50, SD = 24.93) with t(7) = 2.39, $p = 0.048$, d = 0.63.

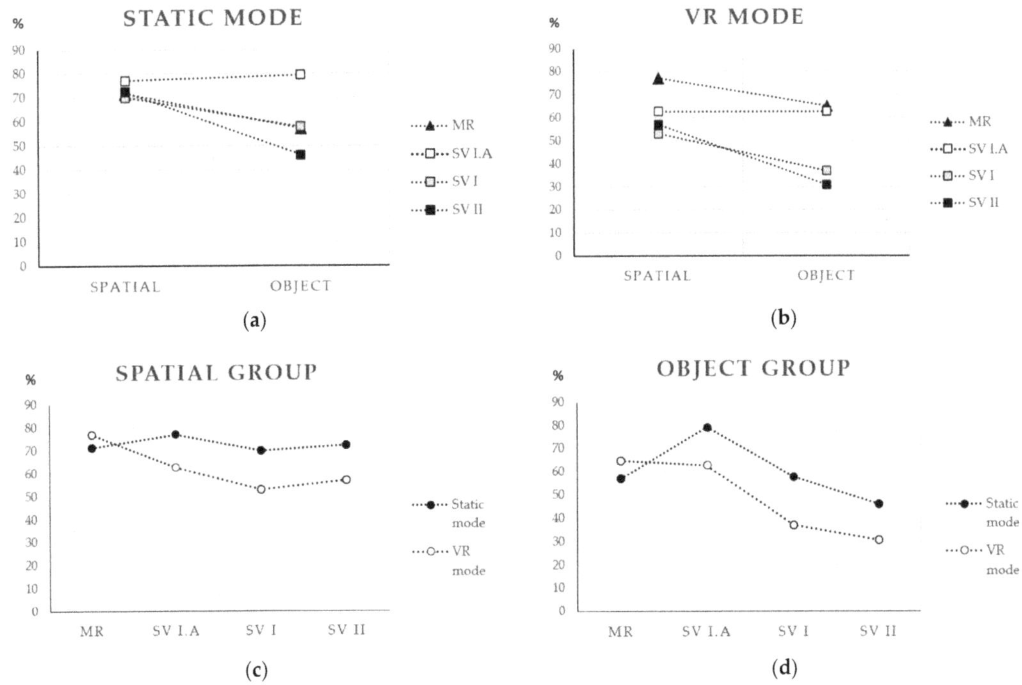

Figure 5. (**a**) Observed scores by OSIQ score in static mode; (**b**) Observed scores by OSIQ score in VR mode; (**c**) Observed scores by spatial group; (**d**) Observed scores by object group.

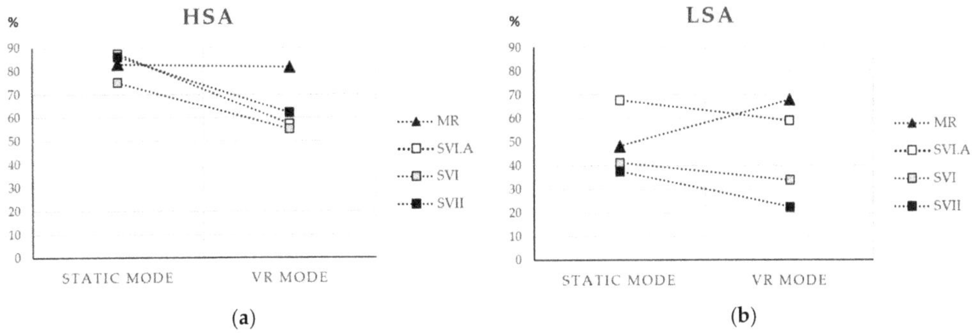

Figure 6. (**a**) Observed scores by HSA; (**b**) Observed scores by LSA.

4.7. Correlation between Static and VR Mode Performance

A correlation among static and VR mode performance and visual cognitive style scores was conducted to understand their relationship. The Pearson correlation analysis results showed that the average of all subconstructs in static mode correlated with that in VR ($r = 0.577$). However, only the performance in SV II in static mode shows statistically significant correlation with that in VR mode. The result implies that the overall performance in each of the two modes is related. However, when measuring the detailed subconstruct of spatial ability, differences in performance may be brought about due to the different test modes. As for the visual cognitive style, spatial scores on the OSIQ correlated with VR SV I ($r = 0.463$) and VR average score ($r = 0.447$) while object scores negatively correlated with the paper version SV I.A (r = −0.395). The spatial visualizer group seems to have benefited from VR mode. See Table 5.

Table 5. Correlation between paper and VR AISAT performance.

	MR	SV I.A	SV I	SV II	Static_ Average	VR MR	VR SV I.A	VR SV I	VR SV II	VR Average	Spatial Score	Object Score
MR	1	0.096	0.267	0.323	**0.674** **	0.054	0.283	**0.481** **	−0.031	0.358	0.246	−0.046
SV I.A		1	0.250	0.179	**0.464** **	−0.182	0.152	0.216	**0.428** *	0.246	−0.112	**−0.395** *
SV I			1	**0.401** *	**0.780** **	0.250	−0.208	0.314	**0.367** *	0.290	0.256	0.143
SV II				1	**0.704** **	**0.499** **	−0.076	**0.514** **	**0.695** **	**0.687** **	0.262	−0.292
Static_average					1	0.264	0.031	**0.574** **	**0.500** **	**0.577** **	0.284	−0.147
VR MR						1	**−0.388** *	0.280	0.229	**0.491** **	0.295	−0.074
VR SV I.A							1	0.180	−0.105	0.350	0.010	−0.119
VR SV I								1	**0.447** *	**0.819** **	**0.463** **	−0.120
VR SV II									1	**0.633** **	0.273	−0.259
VR average										1	**0.447** *	−0.245
Spatial score											1	**0.369** *
Object score												1

Note. * Significant at the 0.05 level; ** Significant at the 0.01 level (2-tailed).

4.8. Usability Test of VR AISAT

Participants' opinions on the usability of VR AISAT were analyzed. Table 6 summarizes the mean score of each question, reliability with Cronbach's alpha, and item average. To check the internal consistency of the questions in each category, Cronbach's alpha was computed. The alpha for tutorial helpfulness, task performance, exploration and navigation, satisfaction, and discomfort formed a scale that has reasonable internal consistency reliability with over 0.66.

Overall, participants perceived static mode as easier to handle than VR. They perceived the traditional mode as 'more comfortable' (M = 3.67) than VR mode (M = 2.73). Regarding tutorial helpfulness, participants indicated that understanding the tutorials on the four types of VR spatial ability tasks was 'helpful in completing the test' (M > 4.10). Regarding task performance, their answers indicated that 'they knew task goals and procedures' (M > 3.90), but since task results were not provided to users in the system, 'seeing the results' was rated relatively low (M = 3.37). Regarding exploration and navigation, although participants perceived the 'interface menus as easy to navigate' (M = 4.17), 'free exploration', 'knowing the exact location in the simulated environment', and 'free manipulation of the virtual objects' were not rated particularly high (M < 3.40). Overall, participants were satisfied with the VR AISAT experience (M > 4.00). They said 'doing the VR AISAT was fun and a good experience' (M = 4.23). Participants reported 'high immersion' (M = 4.03). In terms of discomfort, reports on 'feeling dizzy while wearing HMD' were higher than average (M = 3.44), and 'heavy headset' also received high marks (M = 4.13).

In addition to the structured usability questionnaire, participants were asked two open-ended questions: "Freely talk about your feelings while solving problems on the VR AISAT" and "Please suggest anything to improve the VR AISAT". Table 7 is a summary of participants' responses.

The main feelings that participants experienced while solving the VR AISAT were quite diverse: negative comments highlighted dizziness and uncomfortable HMD. Positive comments included 'helpful' and 'fun experience.' For improvement, participants suggested improving graphics for clarity and allowing more freedom of movement.

Table 6. Usability of VR AISAT item and mean score.

Subcategory	Statements	Mean of Each Question	Cronbach's Alpha	Item Average
Usability	I am more comfortable with a traditional paper test.	3.67	-	3.67
	I am more comfortable with the virtual reality method.	2.73	-	2.73
	The controls in the program are easy to use.	3.83	(0.371)	3.86
	The eye-level position in the program is appropriate.	3.67		
	The brightness of the lighting in the program is adequate.	4.07		
	The level of graphics in the program is appropriate.	3.67		
	The size of the objects, such as models and buttons, is appropriate.	4.07		
Tutorial helpfulness	The tutorial for the first type, MR, helped me understand how to solve the problem.	4.10	0.891	4.14
	The tutorial on the second type, SV I. A, helped me understand how to solve the problem.	4.14		
	The tutorial on the third type, SV I, helped me understand how to solve the problem.	4.13		
	The tutorial on the third type, SV I, helped me understand how to solve the problem.	4.17		
Task performance	I understood exactly what the goal of the task was.	4.20	0.710	3.88
	I was able to decide what I wanted to do in the task.	4.00		
	I could easily see the results of my tasks.	3.37		
	I could see exactly what the next task was.	3.93		
Exploration and Navigation	I could freely explore the experimental spaces.	3.37	0.661	3.49
	I knew exactly where I was	3.03		
	The menus of the interface were easy to navigate.	4.17		
	I was able to manipulate the virtual objects freely.	3.40		

Table 6. *Cont.*

Subcategory	Statements	Mean of Each Question	Cronbach's Alpha	Item Average
Satisfaction	I think I can measure my spatial ability through the virtual AISAT.	4.07	0.760	4.08
	I think the information and knowledge I gained from the virtual AISAT will help me improve my spatial ability in real life.	4.03		
	The virtual AISAT was fun and a good experience.	4.23		
	Doing the virtual AISAT is a valuable experience for me.	4.27		
	I would highly recommend the Virtual AISAT to others.	3.80		
Immersion	It was immersive while solving the virtual AISAT.	4.03	-	4.03
	I wish there were more environmental devices for immersion. I wish I could see my body for immersion.	3.93 3.17	-	3.55
	I wish I could see my body for immersion.	3.17		
Discomfort	I felt dizzy while using Virtual AISAT.	3.43	0.788	2.86
	I felt nauseous while using Virtual AISAT.	2.93		
	I felt motion sickness while repositioning.	2.63		
	I felt motion sickness while rotating my head.	2.80		
	The headset hurts where it touches my skin.	2.37		
	The headset is heavy.	4.13		
	The controller is difficult to use.	1.73		
Open-ended questions	Please feel free to share your thoughts on solving the problems. Tell us how we can improve the AISAT-VR			

Note. Grey highlight means reverse item; that is, a high score indicates high negativity.

Table 7. Summary of open-ended responses.

Question	Answers	Freq. (Multiple Answers Possible)	Question	Answers	Freq. (Multiple Answers Possible)
Freely talk about your feelings while solving problems on the VR AISAT.	Dizziness	9	Please suggest anything to improve the VR AISAT.	Graphics	6
	Helpful, fun	9		Freedom of movement	6
	Uncomfortable HMD	6		Dizziness	2
	Need graphics improvement	4		Accidental click on button resulting in moving forward to next question	2
	Difficult questions	4		Focus is not clear	2
	Problems are intuitive and easily understood.	4		Uncomfortable with VR devices	2
	Unfamiliarity	2		Adding background music	1
	Lack of freedom of movement	1		How to check your view is uncomfortable	1
	3D implementation is well done	1		Changing the way tutorials work	1
	Immersive experience	1		Adding one's location	1
				Show remaining time	1
				Hard to see the bottom of the field of view	1
				Difficult to see the point of view in the SV I view	1
				Change MR model and arrow button layers	1

In summary, the questions concerning how research results supported our hypotheses are reported below:

H1: Performance in the static mode of AISAT and the VR mode of AISAT will be similar. → Mostly rejected. A significant difference was revealed in the performance in three SV constructs with better performance in static mode. Although no statistical difference was found in MR, MR performance was better in VR mode.

H2: The order of taking the tests (VR mode first vs. static mode first) will not influence spatial ability performance. → Mostly supported. Difference was observed only in SV II.

H3: No gender difference will be observed in the performance of the VR AISAT. → Supported. No gender difference was observed in any of the constructs in the VR mode.

H4: Spatial visualizers in visual cognitive style will perform better in VR mode than object visualizers. → Partially supported. Spatial visualizers performed better in VR SV I and VR SV II out of four constructs.

H5: Those with LSA will benefit from VR mode more than those with HSA. → Partially supported. The difference in performance between static mode and VR mode was observed in only one construct with those with LSA but three with HSA. The degree of decline from scores in static to that in VR mode was smaller in LSA, and HSA seems to be more susceptible to test mode.

5. Discussion

5.1. Differences in Spatial Ability Performance in Static and VR Modes

The results of this study showed a difference in performance on the AISAT in static mode and VR mode. Participants generally performed better in static mode than interactive VR except in MR. Possible reasons behind such differences in the test takers' performance could be (a) challenges caused by the VR system itself and (b) potential issues caused by slightly different procedures designed for the AISAT VR software. One of the challenges caused by the VR system itself was cybersickness. General discomfort, including dizziness, uncomfortable HMD, and lack of movement as indicated in the open-ended responses may relate to the test takers' low performance in VR. Researchers have theorized the potential causes of cybersickness. Among them, sensory conflict theory explains that cybersickness occurs with a mismatch between visual senses and the vestibular senses [40]. The teleporting option added to SV II was intended to provide options to navigate to view the space from many different positions, allowing the viewer to have a holistic understanding of the given space. However, the sudden positional change of the users might have caused them to experience cybersickness.

Another possible reason for performance difference in the two modes might be the slightly different procedures designed for the AISAT VR software: First, an additional action was required in taking the test of each subconstruct of spatial ability in the VR mode. For example, compared to the static mode, in VR more steps in the thinking procedure were required to solve problems. In VR SV I.A, the user's position in relation with the 3D objects matters, so the four answer options were in four directions in a radial organization, requiring users to rotate their bodies 90 degrees to see each option. The test takers had to use their hands to manipulate the controller. Second, differences existed in the field of vision in the two test modes. The ease of grasping the entirety of the space in question in the static mode compared with the VR mode allowed the user to perform better, especially in the spatial visualization test. Encoding the geometry of a room in VR was reported difficult [41] perhaps as a result of the field of view. To take advantage of the benefit of immersive quality of VR technology, real-world scale was applied to the simulated models in question in AISAT VR mode. Therefore, compared to the static mode, where test-takers could see the spatial objects in question and multiple choices together, VR users needed to explore the space and the answer choices to fully understand the question, thereby causing more cognitive load. Last, although we provided a training tutorial with a few sample questions before they took the test in VR mode, more exercises might have been needed for some of the test takers. Providing enough exercises and a substantial training period is critical in VR [21].

In fact, prior studies investigating the effect of VR on spatial ability revealed inconsistency in findings regarding the benefit of VR. For example, better SV performance was observed in a 2D floor plan-based task than in VE [42]. In other studies, however, better SV performances were reported in VR than non-VR [43] and in printed materials [31]. In contrast, previous research comparing MR performance showed more consistent results, reporting better performance in VR than non-VR mode [5,44], with printed materials [31], and in 2D [4]. Although no statistical difference was found, the results of the current study support the previous study on MR performance regarding better performance in VR than that in static mode. In MR in VR mode, users could strategically rotate the option until they could equate it with the shape suggested in the question, reducing the test takers' cognitive load. Especially with the function that allows the test taker to rotate the object in real time, the participants had the advantage of quickly grasping the rotated forms to solve the problem.

The factors noted above did not diminish the benefit of using VR for developing spatial ability measuring tools. In VR mode, users could experience the sense of scale that is part of spatial skills not formally recognized in previous spatial ability studies. The ease of viewing the entirety of the space in question in the static mode benefited users allowing them to perform better in the static mode. However, the static mode did not provide a sense of presence and immersion. Based on the results of this study, we believe the perceptual experience of spatial dimensions and solving spatially relevant problems cannot be considered the same cognitive activities. Therefore, this research highlights the necessity of careful consideration in designing and developing VR test modes for spatial ability measurement.

5.2. Less Gender Difference in VR

The results of this study regarding the gender difference in VR align with prior research. Gender difference was found only in MR in the static mode. However, no gender difference was found in any of the subconstructs in VR mode. Lochhead et al. (2022) found that less gender difference in 3D stimuli MR in VR than 2D image stimuli in VR may have occurred since the 3D stimuli do not require the 2D to 3D transformation step, "alleviating the cognitive burden of dimensionality crossing" ([3], p. 15), imposed by 2D representation of a 3D object. The difference in strategies used by participants of each gender in solving MR problems may be another possible reason behind less gender difference in performance in VR mode. Functional brain imaging in a study by Jordan et al. (2002) showed that activation patterns in men and women differed when they solved MR problems [45]: The brain areas known to be involved in the cognitive processing of object–part identification, object categorization, and spatial analysis were activated more in women than in men; whereas brain areas involving more concrete imagined rotation similar to hands-on approach were activated in men (p. 2406). The function in the VR version AISAT that allowed the test takers to view the rotated forms of the given object in real time by one click may have advantaged female test takers to quickly grasp the rotated forms and expedited their cognitive processing of object-part identification and decision making based on spatial analysis (i.e., matching between rotated form and the image in question). For male test takers the way MR test is designed in VR may have forced them to shift the way to solve mental rotation task using different types of spatial analysis focused more on matching skills, disadvantaging their skills of imagined mental rotation using primary motor cortex. Nonetheless, previous research findings also showed the lack of gender difference in MR tested in VR [30,31]. Less gender difference in VR modes indicates the potential of VR to benefit female users in spatial learning and thinking using VR.

5.3. Who Benefits from VR

In the current study those with HSA generally received higher scores than those with LSA regardless of test mode. When comparing their performance in each subconstruct of spatial ability, however, the spatial visualization scores of those with HSA were lower when the test was conducted in the VR mode, but for LSA most of the differences were diminished. In fact, only a difference in SV I was observed. In addition, the MR scores of those with LSA were higher in VR mode compared to that in static mode. This suggests that individuals with high spatial ability as assessed in the static mode do not consistently benefit from the VR test mode in spatial visualization, and those with HSA tend to be more susceptible to test mode than those with LSA. In addition, this study revealed spatial visualizers with visual cognitive style benefit from the VR mode. This suggests the potential challenges for those with an object-oriented cognitive style when using VR technology.

5.4. Implications

This study shows the potential of innovative and practical application and provides following implications to the community of spatial design researchers, practitioners, and educators. First, providing a means of holistic understanding of the spatial information in question in VR mode is necessary for better performance. Although the same questions were used for both test modes, when the static image of a space is converted into a simulated volumetric spatial form in a VE, a perceptual alteration may occur. In static mode, test takers can see the entire space in question, allowing them to grasp a holistic view of the space. When the real scale is applied in the VE and a viewer is put in front of the simulated building or inside the space, however, due to the limited field of vision in the VE, the viewer may need more time to reach a holistic understanding of the given space.

Second, making the procedure in solving problems in VR as simple as possible is essential. The complexity of the VEs and wearing HMD itself could bring about a sense of overwhelming and discomfort. In the AISAT, compared to the simple procedure and instruction in static mode, in VR a more complex procedure and learning curve occurred with each construct of SV. In addition, per each construct slightly different actions were needed, such as clicking the option button, looking at options in a larger scale, and teleporting. Such a difference in procedure in each construct might require more working memory, cognitive load, and effort to adapt to each type of question. This might have been the reason that the participants performed better in the static mode irrespective of the benefit of experiential quality becoming more realistic in the VE. Whether the static version or the VR version of the spatial ability test is closer to the spatial ability required to solve spatial problems in the real world is unknown.

Third, using VR mode for training purposes and static mode for measuring purposes is encouraged. Although findings were limited, one result shows the VR first group performed better in SV I.A and SV I static mode later compared to static first group, indicating the potential benefit for developing a VR version for training of spatial ability. Previous research [16,46] also has shown the benefit of virtual simulation technology in aiding spatial learning. For example, after using extended reality technology in a design project, participants' spatial ability improved [16]. Although measuring spatial ability in static mode and in VR mode is not perfectly compatible, the use of the VR mode of the AISAT as a training tool may have a potential benefit in enhancing at least spatial visualization skills.

Fourth, educators must be cautious and avoid the hasty development of VR tools for spatial learning and education without a thorough consideration of the complex nature of VR technology and its interaction with individual differences. Despite the advantages of VR technology, the user's perceptual understanding of spaces in VEs may differ from that in static test modes.

5.5. Limitations

The first limitation is the sample size (30 participants) and the imbalance between the number of female and male students. The participants were interior design students in Korea; whether the findings in this research are consistent in other cultures and more skilled interior designers is unknown. Even though we planned a total of 46 participants, some completed only one mode due to COVID-19 and could not be included for analysis. In addition, due to the time-consuming nature of the AISAT designed to measure domain-specific spatial ability and the limited number of VR HMDs, voluntary participation was low. However, the gender proportion in interior design programs is female dominant at 85% in UK, not much different in Korea. Thus, the disparity of the gender proportion in this study is likely to represent the status quo of the population. The second limitation regards the VR HMD used in the study. Since the VR AISAT was developed for HTC VIVE specifically, participants reported its heaviness as part of the inconvenience of the test mode. Since the up-to-date lighter devices have been introduced, applying them to different types of HMD seems necessary. The third is the experimental design limitations. In terms of time allotted for VR mode, ours was based on a pilot study of 10 students, necessitating more investigation into the time needed with a greater number of participants.

6. Conclusions and Future Works

In this study we explored the influence of test modes on domain-specific spatial ability performance in the field of architecture and interior design. We examined two distinct test modes: a static paper–desktop mode and an interactive VR mode. Conclusions are as follows:

1. Our findings reveal that participants demonstrated better performance in the spatial visualization test when using the static mode as opposed to the VR mode. Conversely, mental rotation performance was better in the VR mode.
2. The order of test mode minimally influenced performance.
3. The gender difference typically observed in mental rotation tests was not evident in the VR mode, indicating potential advantages for female users in employing VR for spatial learning related to mental rotation.
4. Individuals with a strong spatial visualizer cognitive style excelled, particularly in the SV I and SV II conducted in VR mode.
5. People with high spatial ability tended to be more influenced by test mode than those with low spatial ability.
6. Spatial visualizers and low spatial ability groups tended to benefit from VR mode.
7. Regarding usability, participants expressed positive reactions to VR AISAT, but challenges due to the nature of virtual reality occurred, including dizziness and participants' perception that solving spatial problems in the static mode was easier than in the VR mode.

These findings highlight the intricate nature of spatial ability across test modes, further influenced by individual differences.

In future, the following activities should be carried out by researchers: First, reexamining the time allotted for each construct in the VR mode of the AISAT is necessary with more participants and can result in higher a level of correlation in each construct of the static and VR modes. Second, improving graphic quality and the usability of the AISAT based on participants' responses, especially using lighter HMD for ease of use and comfort would be of considerable value. Third, making a greater number of questions in each construct with the aid of rule-based question generation will be helpful in the development of the AISAT. Artificial Intelligence technology can also be used to customize the question types and number of questions in real time to measure more detailed spatial ability specific to an individual. Fourth, improving the VR mode of AISAT and recruiting a larger number of participants from different cultures and with diverse levels of experience will facilitate an understanding of how and what areas the VR mode benefits; generalizability will follow. In addition, conducting a study with bio and brain sensors, such as an eye-tracker and an EEG, to understand the strategies of solving space-related problems involving high vs. low spatial ability and the cognitive and neural mechanism of spatial thinking and visualization would be highly valuable.

Author Contributions: Conceptualization, J.Y.C. and J.S.; methodology, J.Y.C. and J.S.; formal analysis, J.Y.C.; validation, J.Y.C. and J.S.; investigation, J.Y.C. and J.S.; resources, J.Y.C.; data curation, J.Y.C.; writing—original draft preparation, J.Y.C. and J.S.; writing—review and editing, J.Y.C. and J.S.; visualization, J.Y.C.; funding acquisition, J.Y.C. All authors have read and agreed to the published version of the manuscript.

Funding: This research was funded the BK21 plus program "AgeTech-Service Convergence Major" through the National Research Foundation (NRF) funded by the Ministry of Education of Korea [5120200313836] and by the NRF grant funded by the Korea government (MSIT) [2020R1A2C1009689].

Institutional Review Board Statement: The study was approved by the Institutional Review Board of Kyung Hee University IRB (protocol code: KHSIRB-21-146 (NA) and date of approval: 6 April 2021 and 6 April 2022).

Informed Consent Statement: Informed consent was obtained from all subjects involved in the study.

Data Availability Statement: The data presented in this study are available on request from the corresponding author. The data are not publicly available due to [privacy].

Acknowledgments: We acknowledge all participants in this study.

Conflicts of Interest: The authors declare no conflict of interest.

References

1. Obeid, S.; Demirkan, H. The influence of virtual reality on design process creativity in basic design studios. *Interact. Learn. Environ.* **2023**, *31*, 1841–1859. [CrossRef]
2. Ceylan, S. Using virtual reality to improve visual recognition skills of first year architecture students: A comparative study. In Proceedings of the 12th International Conference on Computer Supported Education (CSEDU), Prague, Czech Republic, 2–4 May 2020; Volume 2, pp. 54–63.
3. Lochhead, I.; Hedley, N.; Çöltekin, A.; Fisher, B. The immersive mental rotations test: Evaluating spatial ability in virtual reality. *Front. Virtual Real.* **2022**, *3*, 820237. [CrossRef]
4. Passig, D.; Eden, S. Virtual reality as a tool for improving spatial rotation among deaf and hard-of-hearing children. *CyberPsychol. Behav.* **2001**, *4*, 681–686. [CrossRef] [PubMed]
5. Yurt, E.; Sünbül, A.M. Effect of modeling-based activities developed using virtual environments and concrete objects on spatial thinking and mental rotation skills. *Educ. Sci. Theory Pract.* **2012**, *12*, 1987–1992.
6. Kim, N.; Gero, J.S. Neurophysiological responses to biophilic design: A pilot experiment using VR and EEG. In *Design Computing and Cognition*; Gero, J.S., Ed.; Springer International Publishing: Cham, Switzerland, 2022; pp. 235–253. [CrossRef]
7. Blajenkova, O.; Kozhevnikov, M.; Motes, M.A. Object-spatial imagery: A new self-report imagery questionnaire. *Appl. Cogn. Psychol.* **2006**, *20*, 239–263. [CrossRef]
8. Linn, M.C.; Petersen, A.C. Emergence and characterization of sex differences in spatial ability: A meta-analysis. *Child Dev.* **1985**, *56*, 1479–1498. [CrossRef] [PubMed]
9. Gaughran, W. Cognitive modelling for engineers. In Proceedings of the 2002 American Society for Engineering Education Annual Conference and Exposition, Montreal, QC, Canada, 16–19 June 2002; American Society for Engineering Education: Washington, DC, USA, 2002. [CrossRef]
10. Hegarty, M.; Waller, D. Individual differences in spatial abilities. In *The Cambridge Handbook of Visuospatial Thinking*; Shah, P., Miyake, A., Eds.; Cambridge University Press: Cambridge, UK, 2005; pp. 121–169. [CrossRef]
11. Vandenberg, S.G.; Kuse, A.R. Mental rotations, a group test of 3-dimensional spatial visualization. *Percept. Mot. Ski.* **1987**, *47*, 599–604. [CrossRef] [PubMed]
12. Ekstrom, R.B.; French, J.W.; Harman, H.H.; Dermen, D. *Manual for Kit of Factor Referenced Cognitive Tests*; Educational Testing Service: Princeton, NJ, USA, 1976.
13. Hegarty, M.; Waller, D. A dissociation between mental rotation and perspective-taking spatial abilities. *Intelligence* **2004**, *32*, 175–191. [CrossRef]
14. Sutton, K.; Williams, A. Implications of spatial abilities on design thinking. In *Proceedings of the Design and Complexity: DRS International Conference 2010, Montreal, QC, Canada, 7–9 July 2010*; Durling, D., Bousbaci, R., Chen, L., Gauthier, P., Poldma, T., Roworth-Stokes, S., Stolterman, E., Eds.; Design Research Society: London, UK, 2010; Available online: http://www.drs2010.umontreal.ca/data/PDF/115.pdf (accessed on 10 September 2023).
15. Berkowitz, M.; Gerber, A.; Thurn, C.M.; Emo, B.; Hoelscher, C.; Stern, E. Spatial abilities for architecture: Cross sectional and longitudinal assessment with novel and existing spatial ability tests. *Front. Psychol.* **2021**, *11*, 4096. [CrossRef]

16. Darwish, M.; Kamel, S.; Assem, A. Extended reality for enhancing spatial ability in architecture design education. *Ain Shams Eng. J.* **2023**, *14*, 102104. [CrossRef]
17. Cho, J.Y.; Suh, J. The architecture and interior design domain–specific spatial ability test (AISAT): Its validity and reliability. *J. Inter. Des.* **2022**, *47*, 11–30. [CrossRef]
18. Cho, J.Y.; Suh, J. Understanding spatial ability in interior design education: 2D–to–3D visualization proficiency as a predictor of design performance. *J. Inter. Des.* **2019**, *44*, 141–159. [CrossRef]
19. Suh, J.; Cho, J.Y. Linking spatial ability, spatial strategies, and spatial creativity: A step to clarify the fuzzy relationship between spatial ability and creativity. *Think. Ski. Creat.* **2020**, *35*, 100628. [CrossRef]
20. Sherman, W.R.; Craig, A.B. *Understanding Virtual Reality*; Morgan Kauffman: San Francisco, CA, USA, 2003.
21. Waller, D.; Hunt, E.; Knapp, D. The transfer of spatial knowledge in virtual environment training. *Presence* **1998**, *7*, 129–143. [CrossRef]
22. Ahn, J.M.; Cho, J.Y. The potential of VR and AR in improving designers' spatial ability. *Conf. Korean Inst. Inter. Des.* **2018**, *20*, 47–49.
23. Banfi, F.; Brumana, R.; Stanga, C. Extended reality and informative models for the architectural heritage: From scan-to-BIM process to virtual and augmented reality. *Virtual Archaeol. Rev.* **2019**, *10*, 14–30. [CrossRef]
24. Silva, M.; Teixeira, L. Developing an extended reality platform for immersive and interactive experiences for cultural heritage: Serralves museum and coa archeologic park. In Proceedings of the 2020 IEEE International Symposium on Mixed and Augmented Reality Adjunct (ISMAR-Adjunct), Recife, Brazil, 9–13 November 2020; IEEE: Piscataway, NJ, USA, 2020; pp. 300–302.
25. Panya, D.S.; Kim, T.; Choo, S. An interactive design change methodology using a BIM-based Virtual Reality and Augmented Reality. *J. Build. Eng.* **2023**, *68*, 106030. [CrossRef]
26. Alizadehsalehi, S.; Yitmen, I. Digital twin-based progress monitoring management model through reality capture to extended reality technologies (DRX). *Smart Sustain. Built Environ.* **2023**, *12*, 200–236. [CrossRef]
27. Ju, J.E.; Kim, A.Y. Gender differences in spatial ability and using strategies for solving spatial problems. *Korean J. Woman Psychol.* **2010**, *15*, 829–851.
28. Maeda, Y.; Yoon, S.Y. A meta-analysis on gender differences in mental rotation ability measured by the Purdue spatial visualization tests: Visualization of rotations (PSVT: R). *Educ. Psychol. Rev.* **2013**, *25*, 69–94. [CrossRef]
29. Reilly, D.; Neumann, D.L. Gender-role differences in spatial ability: A meta-analytic review. *Sex Roles* **2013**, *68*, 521–535. [CrossRef]
30. Park, J.-R. The Analysis of Spatial Ability Difference According to Learning Style and Gender in Virtual Reality Based Learning. Master's Thesis, Korea National University of Education, Cheongju, Republic of Korea, 2009. Available online: https://www-riss-kr.openlink.khu.ac.kr/link?id=T11590539 (accessed on 10 September 2023).
31. Samsudin, K.; Rafi, A.; Hanif, A.S. Training in mental rotation and spatial visualization and its impact on orthographic drawing performance. *Educ. Technol. Soc.* **2011**, *14*, 179–186.
32. Cho, J.Y. An investigation of design studio performance in relation to creativity, spatial ability, and visual cognitive style. *Think. Ski. Creat.* **2017**, *23*, 67–78. [CrossRef]
33. Kozhevnikov, M.; Kozhevnikov, M.; Yu, C.J.; Blazhenkova, O. Creativity, visualization abilities, and visual cognitive style. *Br. J. Educ. Psychol.* **2013**, *83 Pt 2*, 196–209. [CrossRef] [PubMed]
34. Lee, E.A.L.; Wong, K.W. Learning with desktop virtual reality: Low spatial ability learners are more positively affected. *Comput. Educ.* **2014**, *79*, 49–58. [CrossRef]
35. Sun, R.; Wu, Y.J.; Cai, Q. The effect of a virtual reality learning environment on learners' spatial ability. *Virtual Real.* **2019**, *23*, 385–398. [CrossRef]
36. Hays, T.A. Spatial abilities and the effects of computer animation on short-term and long-term comprehension. *J. Educ. Comput. Res.* **1996**, *14*, 139–155. [CrossRef]
37. Huk, T. Who benefits from learning with 3D models? The case of spatial ability. *J. Comput. Assist. Learn.* **2006**, *22*, 392–404. [CrossRef]
38. Meneely, J.; Portillo, M. The adaptable mind in design: Relating personality, cognitive style, and creative performance. *Creat. Res. J.* **2005**, *17*, 155–166.
39. British Institute of Interior Design. Ground-Breaking Diversity Analysis of Interior Design Students. 10 December 2020. Available online: https://biid.org.uk/resources/ground-breaking-diversity-analysis-interior-design-students (accessed on 20 November 2023).
40. LaViola, J.J., Jr. A discussion of cybersickness in virtual environments. *ACM Sigchi Bull.* **2000**, *32*, 47–56. [CrossRef]
41. Kimura, K.; Reichert, J.F.; Olson, A.; Pouya, O.R.; Wang, X.; Moussavi, Z.; Kelly, D.M. Orientation in virtual reality does not fully measure up to the real-world. *Sci. Rep.* **2017**, *7*, 18109. [CrossRef] [PubMed]
42. Schnabel, M.A.; Kvan, T. Spatial understanding in immersive virtual environments. *Int. J. Archit. Comput.* **2003**, *1*, 435–448. [CrossRef]
43. Chen, C.J. Are spatial visualization abilities relevant to virtual reality? *E-J. Instr. Sci. Technol.* **2006**, *9*, n2.
44. Molina-Carmona, R.; Pertegal-Felices, M.; Jimeno-Morenilla, A.; Mora-Mora, H. Virtual reality learning activities for multimedia students to enhance spatial ability. *Sustainability* **2018**, *10*, 1074. [CrossRef]

45. Jordan, K.; Wüstenberg, T.; Heinze, H.J.; Peters, M.; Jäncke, L. Women and men exhibit different cortical activation patterns during mental rotation tasks. *Neuropsychologia* **2002**, *40*, 2397–2408. [CrossRef]
46. Jiang, E.; Laidlaw, D.H. Practicing in Virtual Reality Improves Mental Rotation Ability: Lower Scorers Benefit More. 2019. Available online: https://cs.brown.edu/media/filer_public/26/02/2602a1b3-b630-4e19-9cf6-5f077f5f3271/jiangelaine.pdf (accessed on 10 September 2023).

buildings

MDPI

Article

Comparative Validation of Light Environment Simulation with Actual Measurements

Juhyang Park, Kyungsun Lee * and Kirim Kim

School of Architecture, Hongik University, Seoul 04066, Republic of Korea; c2111113@g.hongik.ac.kr (J.P.); kkl881008@g.hongik.ac.kr (K.K.)
* Correspondence: ksunlee@hongik.ac.kr; Tel.: +82-10-3731-2170

Abstract: The quality of indoor lighting significantly influences human well-being, emphasizing the need to integrate lighting planning into the architectural design process. To optimize indoor lighting conditions, light environment simulations are commonly employed. While much of the relevant literature clearly shows that simulations are widely used to predict lighting environments, there is limited active research validating these simulations. Therefore, this study aimed to assess the alignment between actual measurements and simulations, specifically focusing on daylight-induced glare. To achieve this, a comparative analysis and verification of glare levels between simulations and actual measurements were conducted that accounted for glare location and direction. Disparities between the simulated and measured glare levels were revealed contingent on the glare location and direction. These variations primarily arose from the simulation's utilization of a fisheye field of view (FOV) for glare measurement. To improve the accuracy of glare analysis in simulations, it is advisable to follow the standards related to the human perception of glare, such as the human field of view (FOV), instead of solely depending on a fisheye FOV. The study's limitations include challenges in environmental replication, minor measurement errors, and tree branch shading interference. Despite the potential for simulations to not replicate temporary glare effects, consistent differences with actual measurements indicate that the fisheye FOV was a key contributing factor.

Keywords: indoor light environment; luminance; glare; actual measurement; simulation; workspace

1. Introduction

In modern society, light wields a profound influence on human physiology and psychology, making it a pivotal environmental factor for enhancing health and well-being [1–5]. As the majority of individuals now predominantly remain indoors, the need arises to meticulously craft lighting environments tailored to specific indoor spaces [6]. These spaces encompass residential, educational, and work-oriented spaces, each of which entails extended indoor stays, thus necessitating comprehensive lighting environment planning. Notably, individuals in workspaces are acutely sensitive to prolonged exposure to the lighting environment, which significantly affects work efficiency, worker health, and overall satisfaction [7].

The strategic development of an optimal lighting environment relies on precise measurement and evaluation techniques. Two primary methods are employed for light environment measurement: utilizing measuring equipment in actual spaces and conducting computer-based 3D modeling simulations [8]. While measurement devices offer precise assessments of luminance, color rendering, and illuminance, their application is confined to spaces that are already constructed. In contrast, simulations enable the modeling of various scenarios in virtual environments, proving invaluable during the initial design phases [9]. However, it should be recognized that simulations can exhibit deviations when compared to actual measurements [8].

The evaluation of the light environment yields crucial insights into its impact on psychological comfort and physical health, based on factors such as luminance, color rendering, and illuminance obtained through measurement [10]. In the context of an evolving emphasis on sustainable eco-friendly design, considerations extend beyond energy and environmental aspects to encompass human well-being [11]. Numerous green building certification systems have acknowledged this shift, incorporating evaluations of the light environment within indoor spaces as a certification criterion. Among these systems are LEED (Leadership in Energy and Environmental Design, USA), BREEAM (Building Research Establishment Environmental Assessment Method, UK), and DGNB (German Sustainable Building Council, Germany). Recent certifications, such as WELL and Fitwel, which underscore health and wellness, have also identified the lighting environment as a pivotal component. Notably, Korea's G-SEED (Green Standard for Energy and Environmental Design, Republic of Korea) evaluates the lighting environment in educational spaces but does not extend this evaluation to workspaces. The various certification systems that integrate daylighting evaluations have set specific criteria to foster suitable lighting conditions and minimize glare, which is a common issue for occupants of indoor spaces. These glare standards predominantly evaluate the integration of glare prevention mechanisms in artificial lighting or glare prevention systems based on daylight. To effectively design a glare prevention system influenced by natural lighting, it is crucial to anticipate how glare manifests within indoor spaces. During the design phase, such predictions are feasibly achievable primarily through simulations [9].

The primary objective of this study was to conduct a comparative analysis distinguishing between predicted glare levels derived from light environment simulations and actual measurements. The effective implementation of a system to mitigate glare caused by natural lighting necessitates a proactive approach to predicting how glare may manifest before constructing the space.

2. Materials and Methods

This study consisted of a comparative analysis between actual measurements and simulation results, specifically targeting glare. The methodology and process of this study were as follows (Figure 1):

1. A comprehensive review of previous studies was performed to clarify the characteristics and effects of artificial lighting and natural daylight, which collectively contribute to the composition of the light environment. Additionally, the underlying causes of natural daylight deficiency within office environments were identified, and the specific challenges that must be addressed to facilitate the integration of natural daylight indoors were highlighted.
2. In relation to the glare effect, which makes it difficult to introduce natural daylight indoors, the terminology and measurement methods related to glare were organized. A comparison was then made to assess how luminance (glare) was evaluated in green building certification systems, such as LEED, BREEAM, DGNB, WELL Certification, Fitwel Certification, and G-SEED. By reviewing these evaluation methods, the necessity of simulations to effectively prevent glare was examined. The research also verified the effectiveness of commonly used light environment simulations in optimizing lighting conditions in planning within actual construction environments. Based on the findings of this verification research, the need for further studies on simulation validation was confirmed.
3. To validate the simulations, actual measurements in real spaces were essential. Therefore, existing methods used for measuring space glare from previous studies were analyzed and compared with the measurement method employed in this study to identify its unique features and differentiating factors.
4. The research methodology for this study involved conducting actual measurements and simulations of luminance in three selected workspaces based on specific criteria. Actual measurements were carried out during two separate time slots, 2:00 PM and 4:00 PM, corresponding to the times when sunlight most significantly enters indoor spaces. Daylight glare probability (DGP) was chosen as the measurement metric

for comparison with the actual measurements. This choice was influenced by the widespread use and validation of the DGP for evaluating the probability of daylight-induced glare in spaces, following its proposal by Wienold and Christoffersen in 2006. To perform these simulations, ClimateStudio (Version 1.9) [12] was selected as the simulation program for the study due to its ability to measure glare based on DGP and its effective integration with Rhino, a popular architectural modeling tool.

5. Comparisons were made between the previously obtained actual measurement values and those generated from the simulation of the light environment. The results of these comparisons were then analyzed. This analysis aimed to identify the factors required to achieve more accurate predictions by simulations concerning light environments.

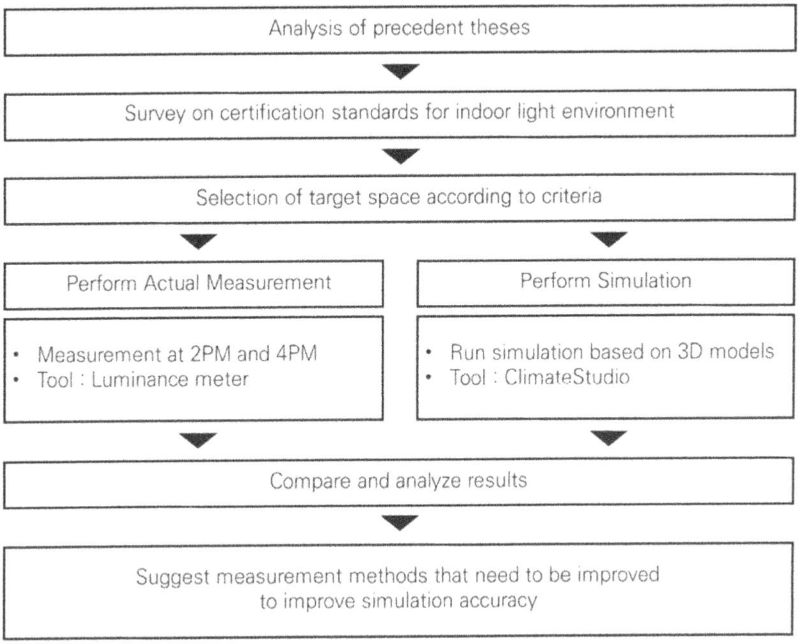

Figure 1. Research plan.

3. Previous Reviews

3.1. Characteristics and Effects of Daylight and Artificial Lighting

The light environment of an indoor space is composed of daylight and artificial lighting. These two sources have different characteristics and can affect occupants differently. First, artificial lighting has the advantage of ease in maintaining a uniform lighting environment, and its lighting properties, such as illuminance, color temperature, and color rendering, can be set in various ways. These features of artificial lighting enable a great deal of control over the light environment according to user needs, which can enhance efficiency by creating a suitable light environment for specific work activities. In contrast, in the case of daylight, it is difficult to maintain a constant light environment, as with artificial lighting, but it provides different advantages. The effective use of daylight can reduce dependence on artificial lighting, thereby reducing electric consumption. This leads to cost savings in building operations and enables environmentally friendly and sustainable building management [13]. Moreover, daylight improves the quality of the indoor environment and positively affects human health and comfort [14–17]. Illumination originating from sunlight can help regulate circadian rhythms and strengthen the immune system; it also allows for better psychological satisfaction and visual comfort in work or living environments [18–20].

Despite the numerous benefits of access to daylight in indoor spaces, many existing workplaces do not actively use it [21], because daylight's dynamic nature makes it difficult to control; furthermore, it could cause glare or hinder thermal comfort [22,23]. To minimize these negative effects while providing daylight in indoor spaces, it is necessary to consider how it will enter the space during the building design stage.

3.2. Glare-Related Terms and Metrics

Various methods are available for measuring the indoor lighting environment, with each green building certification system applying distinct metrics for its evaluation. Among the many light environment evaluation factors, emphasis is placed on defining and summarizing luminance and glare, which form the central point of this discussion.

First, luminance is a physical concept that represents the amount of light emitted or reflected from a specific surface or light source. The SI unit for luminance is candelas per square meter (cd/m^2). Luminance appears differently, depending on the angle between the glare source and the observer's line of sight. There are thresholds for perceived glare based on absolute luminance values and contrast thresholds based on differences in brightness between the glare source and background.

Regarding the luminance threshold according to absolute luminance values, Weinold and Christofferson [24] suggested that, when an occupant is positioned parallel to a window, glare is perceived as below 2000 cd/m^2, and visual discomfort is experienced as above 4000 cd/m^2. When an occupant looks at a window at a 45 degree angle, glare is perceived as below 1500 cd/m^2, and visual discomfort is experienced as above 4500 cd/m^2. Suk et al. [25] stated that, when looking vertically at a window, occupants feel visually comfortable with a luminance of 1920 cd/m^2 or below and experience visual discomfort with values above 5014 cd/m^2.

Regarding contrast thresholds based on brightness differences, Weinold and Christofferson [24] and Veitch and Newsham [26] argued that the upper limit of tolerable glare occurs when the contrast ratio between a glaring light source and its background reaches 1:10. Schiler [27] suggested that this threshold occurs when the ratio reaches 1:3 with respect to average ambient background levels, and Egan [28] proposed it happens with ratios of darkest-to-brightest sources reaching up to 1:40. These differing views arise because individual perceptions of glare vary by person and environment.

Second, glare is a visual impairment phenomenon that occurs when a high luminance source or large luminance contrast is presented in the field of view. It can occur when the iris reaction, which responds to changes in light, is delayed and cannot focus or when visual sensitivity is disrupted due to a scattering of light entering the eyeball. As previously explained, even with the same intensity and contrast of luminance, the degree of glare an individual perceives can vary. Although glare is not entirely quantifiable, typical evaluations are conducted in comfortable and uncomfortable light environments for specific environmental conditions.

Glare can be classified into disability glare and discomfort glare. Disability glare impairs the ability to visually perceive objects and often occurs to nighttime drivers. Discomfort glare does not impair visual perception but causes psychological discomfort and displeasure, such as eye fatigue. The way to measure and evaluate glare may vary according to the certification system; most evaluate glare caused by indoor artificial lighting, with the UGR (unified glare rating) being commonly used for evaluation. The UGR is a system for evaluating glare proposed by the CIE (International Commission on Illumination), mainly used for predicting indoor lighting systems. However, the current study conducted an evaluation of glare caused by daylight rather than artificial lighting. Thus, the DGP (daylight glare probability) was utilized as a system for evaluating glare, as it is especially applicable in scenarios with a wide range of light sources, including daylight.

The DGP has been widely used and validated for evaluating the probability of daylight-induced glares since it was proposed by Wienold and Christoffersen in 2006 [29]. ClimatеStudio, a lighting environment analysis tool, incorporates DGP evaluations. It considers multiple parameters, such as vertical illuminance at the observer's eye level (E_v), source luminance (L_s), the solid angle subtended by the source (ω_s), and the Guth position index (P) [30]. The value related to the probability occurrence of daylight-induced glares can be derived through calculation with the formula shown in Figure 2, which results in a value in the range of 0–100%. DGP values are divided into four bands according to their magnitude; higher values imply higher probabilities of occurrence.

$$DGP = c_1 \cdot E_v + c_2 \cdot \log\left(1 + \sum \frac{L_{s,i}^2 \cdot \omega_{s,i}}{E_v^{a_1} \cdot P_i^2}\right) + c_3$$

E_v : Vertical eye illuminance in lx L_s : Source luminance in cd/m²

ω_s : Solid angle of source P : Position index

c_1 : $5.87 \cdot 10^{-5}$ c_2 : $9.18 \cdot 10^{-2}$ c_3 : 0.16 a_1 : 1.87

Glare rating	DGP
Imperceptible	DGP ≤ 0.34%
Perceptible	34% < DGP ≤ 38%
Disturbing	38% < DGP ≤ 45%
Intolerable	45% < DGP

Figure 2. Calculation formula and glare rating of the DGP.

3.3. Glare Evaluation Method by Certification System

The evaluation of the visual environment in LEED is divided into Interior Lighting, Daylight, and Quality Views. Among them, glare is evaluated in the interior lighting area, along with color rendering, lighting control, and surface reflectance. In this area, one point is given if one out of four items is satisfied, and two points are awarded if three items are met. To meet the glare control item requirement, the luminance of indoor lighting fixtures must be less than 7000 cd/m^2 or have a UGR rating of less than 19.

BREEAM evaluates the visual environment based on Control of Glare, Daylighting, View Out, and Lighting Levels/Zoning/Control. The evaluation of glare control assesses potential glare occurrence in indoor spaces and whether measures such as shading have been established for control. These mitigation strategies should aim to maximize daylight use without increasing energy consumption from lighting under all weather conditions.

DGNB evaluates the visual environment based on Daylight, Visual Contact, Antiglare, Artificial Light, and Color Rendering. Among these categories, a glare prevention assessment depends on whether a system has been installed to prevent glare caused by daylight and grades the system accordingly. When a building incorporates both solar protection systems and glare prevention systems, the evaluation is conducted based on the higher-rated system of the two.

LEED, BREEAM, and DGNB are early certification systems that emerged alongside discussions about eco-friendly architecture and continue to be revised, reflecting the current needs. While initial certification systems began with evaluations from an energy and environmental perspective, they now also incorporate assessments from an occupant well-being perspective. However, the tendency toward evaluation from an energy and environmental perspective remains stronger. Compared to the three aforementioned certifications, WELL and Fitwel are relatively recently established and place stronger emphasis on the environment's impact on occupant well-being and health in their evaluations. Although there are differences in overall trends between certification systems, WELL is similar to LEED and Fitwell is similar to BREEAM and DGNB in terms of evaluating glare.

In WELL certification, the visual environment evaluation consists of nine categories: (1) Light Exposure, (2) Visual Lighting Design, (3) Circadian Lighting Design, (4) Electric Light Glare Control, (5) Daylight Design Strategies, (6) Daylight Simulation, (7) Visual Balance, (8) Electric Light Quality, and (9) Lighting Control. The glare control category is intended to prevent glare caused by indoor lighting, and it assesses the glare index associated with indoor lighting. To satisfy antiglare control requirements, indoor luminosity fixtures must have luminance less than 6000 cd/m^2 or an UGR rating less than 16.

Fitwel certification evaluates Daylight, View of Nature, and Operable Shading in regard to the light environment within workspaces. Operable shading is an evaluation factor for glare prevention. For spaces designated as general occupancy areas, all windows should be equipped with shades that can be controlled by the occupants.

G-SEED evaluates the presence of shading as a factor in glare assessment. However, this evaluation is conducted only for educational facilities and not office buildings.

The aforementioned certification systems adopt UGR in their glare evaluation systems. While this is a suitable criterion for assessing indoor lighting, it is not an appropriate system for evaluating glare caused by natural daylight. In cases where there are measures to control glare induced by natural daylight, most evaluations primarily focus on verifying the presence of glare prevention systems. The evaluation schemes for all the certification systems described are summarized in Table 1.

Table 1. Glare evaluations of the certification schemes.

Certification	Workspace Evaluation	Glare Evaluation Method			Score
LEED	O	Interior Lighting	Glare Control	<7000 cd/m^2 or UGR < 19	1~2 *
			Color Rendering	RA ≥ 90	
			Lighting Control	Brightness Control of Lighting (90% of exclusive area)	
			Surface Reflectance	Ceiling $\geq 80\%$	
				Wall $\geq 55\%$	
BREEAM	O	Control of Glare		Glare Prevention Strategy (There should be no increase in lighting energy usage)	1
DGNB	O	Antiglare		Installation of Glare Prevention System	8
			Glare Prevention System Grade	=Grade 1	12
				$\geq$Grade 2	16
WELL	O	Electric Light Glare Control		<6000 cd/m^2 or UGR ≤ 16	2
Fitwel	O	Operable Shading		Installation of Operable Shading	0.28
G-SEED	X	Antiglare		Installation of Shading (Educational Facilities)	2

* Satisfying one of the detailed evaluation items: 1 point. Satisfying three of the detailed evaluation items: 2 points.

3.4. Distinction from Previous Studies

From the literature retrieved from SCOPUS using the keywords "simulation" and "glare", Diva and ClimateStudio (formerly referred to as DIVA-for-Rhino) were found to be the major analysis tools. Additionally, it was confirmed that the DGP method was frequently used. Moreover, most of the studies using DIVA or ClimateStudio for glare analysis appeared to utilize simulations to predict and optimize lighting conditions [31–59].

However, none of these studies addressed ways to improve the potential limitations of these simulations.

This study investigated the lighting environment in workspaces subject to glare caused by daylight using both actual measurements and simulations. Through a comparison of these results, we sought to identify and discuss elements that deserve further consideration in lighting environment simulations.

The high dynamic range (HDR) image creation method is commonly used for measuring a lighting environment's luminance and illuminance [60,61]. This method involves capturing a scene in space with a fisheye lens and digital camera, measuring the luminance at one point in that scene with a luminance meter and applying the obtained value as a correction to the image. When luminance measurements are conducted using the HDR method, the luminance of all images captured within the fisheye lens is measured.

However, according to Safranek [62], the HDR method has errors due to lens vignetting, lens flare, sensor blooming/luminous overflow, and the spectral responsivity of the sensors. Furthermore, in the current study, because only glare caused by daylight is considered, it was deemed inappropriate to use the HDR method for actual occupied workspaces, as they could be influenced by the luminance of artificial lighting and digital devices.

To mitigate potential errors associated with the HDR method and to exclude the influence of artificial lighting, this study utilized a luminance meter (CS-150, Konica Minolta) that measured the luminance at localized points (measuring angle 1°) (see Figure 3). Luminance measurements were conducted in the direction of the windows, where luminance from natural daylight was anticipated. As depicted in Figure 3, the luminance measurement was carried out in an environment illuminated by both natural and artificial light sources. However, since the measurement specifically targeted a window receiving natural light, the impact of artificial lighting on the recorded luminance value was minimal.

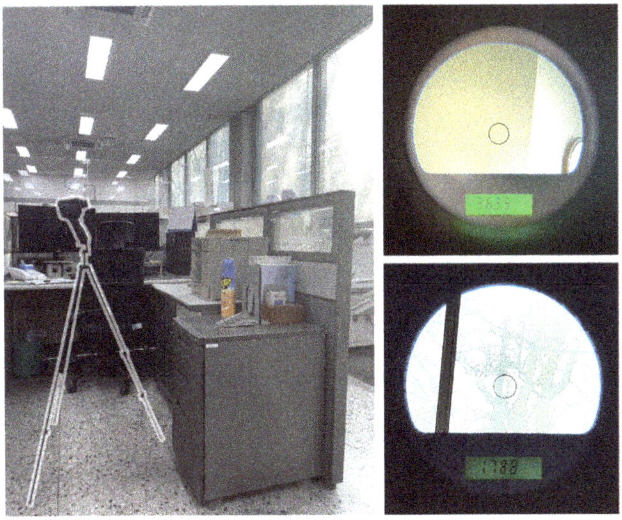

Figure 3. Luminance meter installation and viewport.

4. Experimental Space and Methodology

4.1. Space Selection and Criteria

In comparing the actual measurements to the simulation results, the Offices of Academic Affairs, Administration, and Industry-Academic Cooperation at Hongik University (Seoul, Republic of Korea, 37°33′02.6″ N 126°55′33.5″ E) were designated as the focal spaces. All three offices were situated in the MH Building on campus. The criteria for choosing buildings for the light environment assessment were accessibility for research measurements and unobstructed natural light ingress devoid of interference from nearby structures or installations. Given these criteria, the MH Building was selected from among the accessible campus buildings; due to its elevated location, it ensured unhindered natural light during the study. The building's substantial size, spanning 16 stories, further influenced our choice given the diversity of spaces it offered. Within the MH Building, our chosen spaces took into account a mix of attributes, such as size, floor level, geographical position, and altitude, ensuring a variety of characteristics (Table 2).

Table 2. Characteristics of each target space.

Space	Characteristics				
	Floor	**Altitude**	**Area**	**Floor Height**	**Location**
A	1F	1.05 m	129 m^2	4.08 m and 2.3 m	North of MH Building
B	8F	26.6 m	102 m^2	2.8 m	Center of MH Building
C	1F	1.05 m	153 m^2	4.08 m and 2.3 m	South of MH Building

The Office of Academic Affairs (Space A) was located on the first floor to the north, with windows facing west (Figure 4). The Industry-Academic Cooperation Foundation Office (Space B) was located on the 8th floor, with windows on three sides (none to the east). The Office of General Affairs (Space C) was located on the first floor to the south, with windows facing west. Measurement points were arranged at 2 m × 2 m intervals in each space, depending on each space's unit area. There were 40 measurement points in Space A, 27 in Space B, and 40 in Space C (Figure 5).

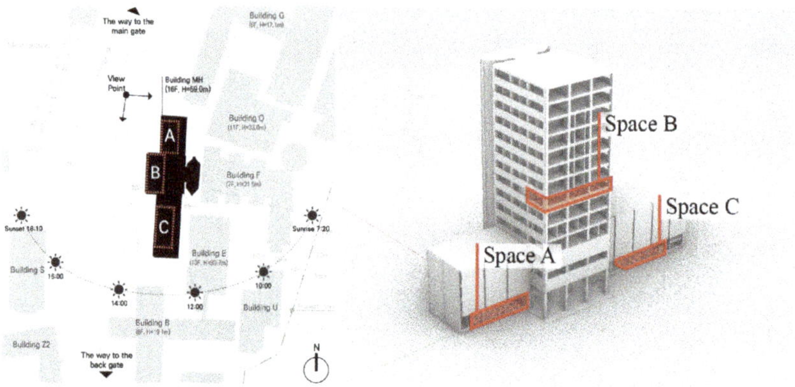

Figure 4. Site plan and positions of the target spaces.

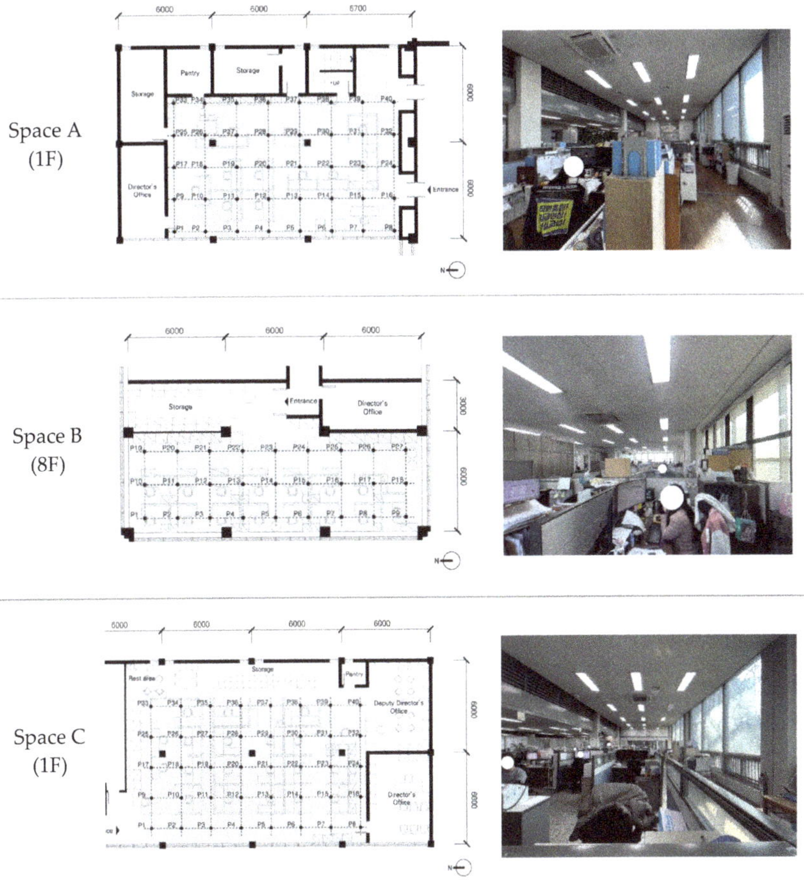

Figure 5. Floor plans and measurement points (**left**), and photographs of the target spaces (**right**).

4.2. Method for Luminance Measurement Using a Luminance Meter

In this study, we aimed to measure the occurrence of glare caused by daylight. Consequently, we planned to measure the luminance in the direction of the windows, where most of the glare was expected due to daylight. For the target spaces, as visual openness through the windows primarily faced west, measurements were conducted in three directions: northwest, west, and southwest. The interval between each measurement point was set at 2 m × 2 m (Figure 6).

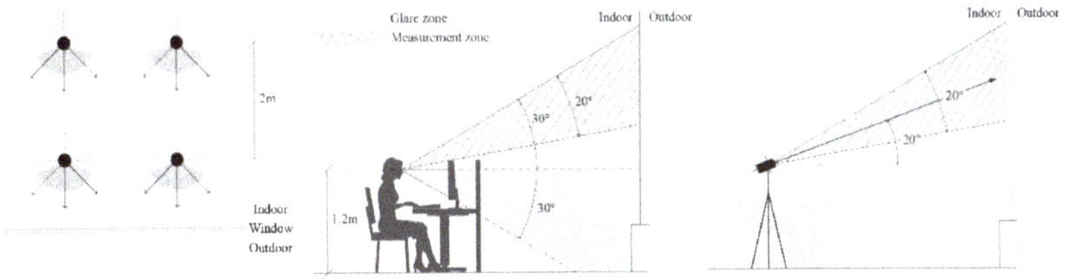

Figure 6. Measurement points and directions (**left**), glare zone (**center**), and measurement settings (**right**).

To enhance the accuracy of the luminance measurements, specific points were pre-marked on the floor of the measurement space. Additionally, to ensure consistent vertical and horizontal angles during the luminance readings, corresponding angles were indicated on the tripod that held the luminance meter. This process aimed to mitigate potential discrepancies during each measurement. However, minor errors might have been introduced due to the physical process of initiating each measurement, a common challenge with manual experimental readings.

When scheduling the actual measurements, shading effects from the exterior landscaping on Spaces A and C had to be considered, given their locations on the lower floors. As a solution, readings were conducted during the winter months, specifically from 21 to 23 February 2023. The measurements for Space A (1F) took place on 21 February, Space B (8F) on 22 February, and Space C (1F) on 23 February. While this timing presented a limitation in observing luminance variations that might occur in other seasons, it provided a more controlled environment. This control facilitated a focus on differences stemming solely from natural lighting rather than other environmental factors.

Regarding the specific times for measurements, the readings were segmented into two slots: 2:00 PM, when the onset of direct natural light into the west-facing areas was anticipated, and 4:00 PM, when a more profound penetration of natural light into these spaces was expected. A broader time range for the measurements was not feasible, because the spaces were operational work areas.

In the case of the actual physical measurements using a luminance meter, there were multiple measurement points and directions, so there were limitations in measuring the light environment at exactly the same time for each measurement point. Therefore, in this experiment, luminance measurements were carried out over a period of time that included the exact simulation time: from 2:00 to 2:50 PM and from 4:00 to 4:50 PM. A CS-150 luminance meter (Figure 7), manufactured by Konica Minolta (Tokyo, Japan) [63], was employed for these measurements. This device is widely utilized to assess luminance in diverse applications, including displays, LEDs, outdoor and indoor lighting fixtures, diverse light sources, and various other lighting setups. In this experiment, it was employed to gauge the luminance generated by daylight in a workspace. On all three measurement days, the weather conditions in Seoul were clear, with cloud coverage between zero and two-tenths [64]. The external illuminance measured using a Konica Minolta CL-500a spectroradiometer (Figure 7) during each time slot is shown in Table 3 [65].

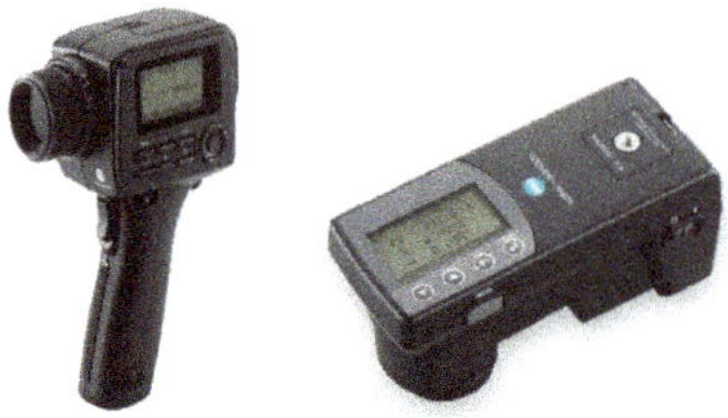

Figure 7. Luminance meter (**left**) and illuminance meter (**right**).

Table 3. Measurements of external illuminance (Unit: lx).

Time	Day		
	21 February 2023	**22 February 2023**	**23 February 2023**
2:00 PM	70,310	68,006	60,840
4:00 PM	33,147	24,886	27,247

Traditional methods of measuring glare typically involve the use of HDR imaging with a fisheye lens to set the measurement range at 180 degrees. However, according to the research by Suk et al. [66], the field of view (FOV) of the human vision system is smaller than that of a fisheye FOV. In their study, it was found that, during typing tasks, the human vision system's FOV accurately detected 97% of glare sources, while the fisheye FOV accurately detected only 55% of glare sources in a scene. Consequently, they argued for using the human vision system's FOV when evaluating scenes for glare.

The human eye contains photoreceptor cells, described as cones and rods. Cones function well under high illumination levels during daylight and are responsible for color discrimination; rods function well at low illumination levels at night and distinguish shades of gray. These photoreceptor cells are differentially distributed across the retina: cones exist at a low density across most of the retina but are denser toward the fovea (the center), while rods are densely packed throughout most parts of the retina but dramatically decrease in density near the fovea [67]. This distribution corresponds with suggestions that humans perceive light and color better within approximately 30 degrees centered on their line of sight.

In this research, we therefore defined an angle within 30 degrees from one's gaze as our "glare zone". Given that our target space was used as an office space with furniture arrangements potentially obscuring some areas from view, we excluded these obscured regions from our measurements. We assumed that any glare caused by daylight would occur 10–30 degrees above one's line of sight (Figure 6). Accordingly, we set up our luminance meter to measure luminance centered on an angle 20 degrees above a typical individual's line of sight.

4.3. Method for Glare Measurement Using Simulation

For the simulations, this study used ClimateStudio (Version 1.9), a plug-in program developed by Solemma that can be used in conjunction with Rhino, a 3D modeling program developed by Robert McNeel. Therefore, Rhino 7 (Version 7.13) was used to create 3D models for the simulation. The floor plans and elevations of Spaces A, B, and C were used to create the 3D models. For the actual spaces, the ceiling was made of white textile with a reflectance of 50–70%, the walls were covered by a light cream color paint with about 50–60% reflectance, and the floor was composed of materials with a reflectance of 20–30%. The windows were double-glazed and had a very slight tint, possessing performance values similar to the visible transmittance (VTIS) applied in the simulation. The VTIS is a value that affects the amount of light entering indoors. A similar VTIS value in the actual space and the simulation space implies that a similar amount of light entered both indoor environments. Because the materials applied in the actual space had properties similar to those recommended by the LEED standard, all spatial elements in the 3D model were applied with materials recommended by the LEED standard (Figure 8). After assigning materials according to the architectural elements, we set up location-specific climate data. The climate data utilized were an energy plus weather format (EPW) file provided by Energy Plus, an energy simulation program developed by the U.S. Department of Energy (DoE). Among the EPW files available, we used KOR_SO_Seoul.471080_TRY_PHIKO.epw, which was based on weather data measured by the Korea Meteorological Administration in Seoul. The sky settings were configured within ClimateStudio to match the sky conditions on the days when the actual measurements were taken—the CIE Standard Clear Sky setting was used in this case. Based on this, an annual glare analysis was conducted in ClimateStudio, targeting the created 3D model. From the results obtained through the simulation analysis, only the data from the same date as the actual measurements were extracted for comparison. Because the simulation provided glare values at 30 min intervals, the DGP values extracted at 2:30 PM and 4:30 PM were used for comparison. Furthermore, measurement points were spaced at intervals of 2 m × 2 m and were aligned to correspond with the actual measurement locations. The direction of measurement in the simulation was also aligned in the same direction as the actual measurements.

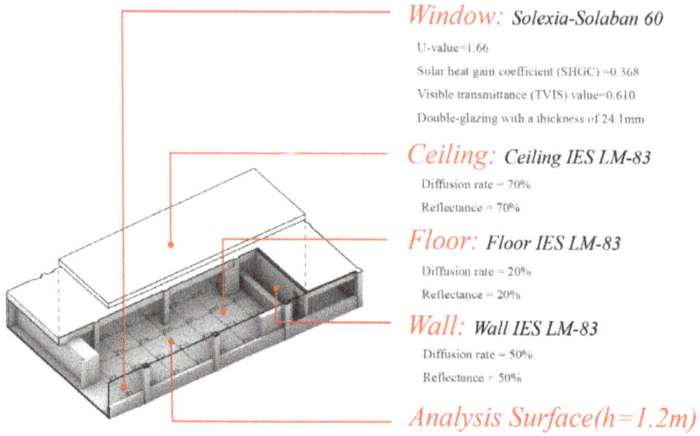

Figure 8. Three-dimensional model for simulation.

5. Results: Comparison of Simulation and Actual Measurements

5.1. Space A

Space A (Figure 9; additional details in Appendix A) had windows located on the west side, with the floor height of the space where the measurement points P1–P24 were located being 4.05 m, and the space where P25–P40 were located had a lower floor height of 2.3 m. The windows started at 0.85 m above the floor and reached the ceiling, with a total height of 3.2 m. At approximately 2:00 PM, sunlight did not directly enter the field of view in Space A. On 21 February, at 2:00 PM, the sun's position was at an azimuth of 203 degrees and an elevation angle of 38 degrees, passing the meridian. Consequently, in the absence of any external obstructions, sunlight should have been visible in west-facing Space A. However, Space A did not receive direct sunlight, because it was blocked by a midsection of the MH Building protruding about six meters to the west.

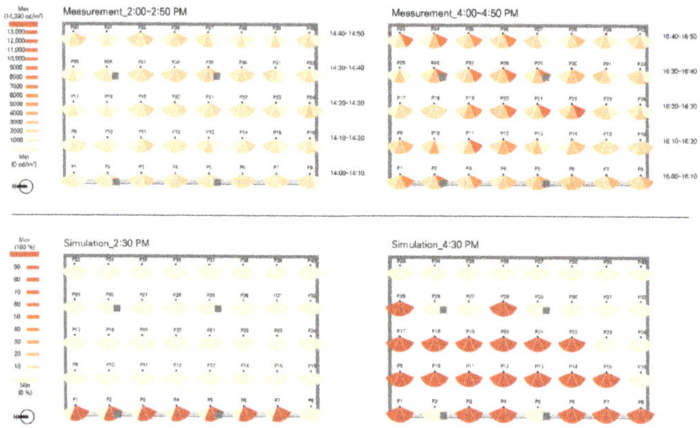

Figure 9. The measurement and simulation results of Space A (The gray box area in this Figure is the same as the gray area in Figure 5 above, which indicates the measurement area. Detailed data are included in Appendix A).

The results of the luminance measurements conducted on Space A showed that, at 2:00 PM, there was no influx of direct sunlight; hence, no points within the space exhibited especially high values. However, it was observed that, generally, higher luminance values were recorded when measurements were taken in the southwest direction. This trend became more pronounced around 4:00 PM as direct sunlight began to enter the space. The luminance values measured in the southwest direction showed significantly higher figures, and with each 45 degree rotation of the measurement direction toward the north, there was a marked decrease in these values.

The simulation results for Space A revealed that, at 2:30 PM, only the positions P1~P7 located near the window exhibited a DGP value of 100% in the southwest and west directions. Other measurement points showed almost negligible DGP values at this time, registering at 0.3%. At 4:30 PM, as the sun moved farther west and its altitude decreased, there was an increase in locations within the space experiencing glare. At these glare occurrence points, the DGP values were recorded as 80% in the northwest direction and consistently 100% in both the southward and westward directions. In other areas during this time frame, the DGP was virtually nonexistent, with a value of 0.3%.

5.2. Space B

Space B (Figure 10; additional details in Appendix B) was the only space among the target areas that had windows facing the south, west, and north. The floor height in Space B was uniformly set at 2.8 m throughout, with no variation. The windows began 1.1 m above the floor and reached the ceiling, with a total window height of 1.7 m. In the case of Space B, due to its lower ceiling height compared to Spaces A and C, there were occasions where the luminance of the ceiling was mistakenly measured instead of the window luminance when measurements were made 20 degrees upward for points P19–P27, which were farthest from the window. However, according to the actual measurement data at 4:00 PM in this experiment, there was still a spot showing strong luminance at that location. This may have been due to the refraction of light by the window and the reflection of light by the ceiling, as these measurement points were close to the window edge.

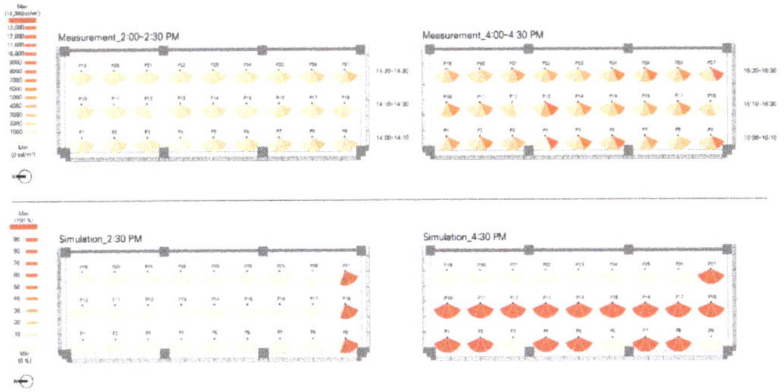

Figure 10. The measurement and simulation results of Space B (The gray box area in this Figure is the same as the gray area in Figure 5 above, which indicates the measurement area. Detailed data are included in Appendix B).

Unlike Spaces A and C, Space B had windows facing south, west, and north. Therefore, it was anticipated that Space B would exhibit stronger luminance compared to the other two spaces.

However, the actual luminance measurements in Space B did not identify any points with higher luminance values compared to the other spaces. This could be due to the fact that the measurement points for luminance were taken pointing upwards at a 20 degree angle from a typical observer's eye level (1.2 m), and at around 2:00 PM on 22 February, the sun's altitude was 39 degrees, located higher than this viewing angle. By around 4:00 PM, the sun's altitude dropped to about 24 degrees, which aligned closely with the luminance meter's measurement angle. At this time, the sun's azimuth angle was approximately 233 degrees, which was close to directly southwest. Therefore, high luminance values were observed when measurements were taken in the southwest direction.

The simulation results for Space B revealed that, at 2:30 PM, glare with a DGP of 100% was predicted to occur in the west and southwest directions from positions P9, P18, and P27 located near the south-facing windows. In the other locations, the DGP value was observed to be nearly negligible at 0.3%. By 4:30 PM, glare was detected not only at the south-facing window positions but also at those facing west. The DGP values for both the southwest and west directions were found to be 100% at the points where glare occurred, while it was in the range of 79–82% only in the northwest direction. The innermost locations within the space, specifically P19–P26, exhibited almost no occurrence of glare, with a DGP of 0.3%.

5.3. Space C

Like Space A, Space C (Figure 11; additional details in Appendix C) had windows facing west. However, it distinguished itself from Space A in that there were no buildings obstructing the influx of sunlight during the afternoon hours. The ceiling height of the space where the measurement points P1–P24 were located was 4.05 m, identical to that of Space A, and the area where P25–P40 were located also had a ceiling height of 2.3 m. The windows began 0.85 m above the floor and reached the ceiling, with a total height of 3.2 m. On 23 February, when the measurements for Space C were conducted, at 2:00 PM, the sun's azimuth was 203 degrees, and its altitude was 39 degrees; at 4:00 PM, its azimuth was 234 degrees, with an altitude also at 39 degrees.

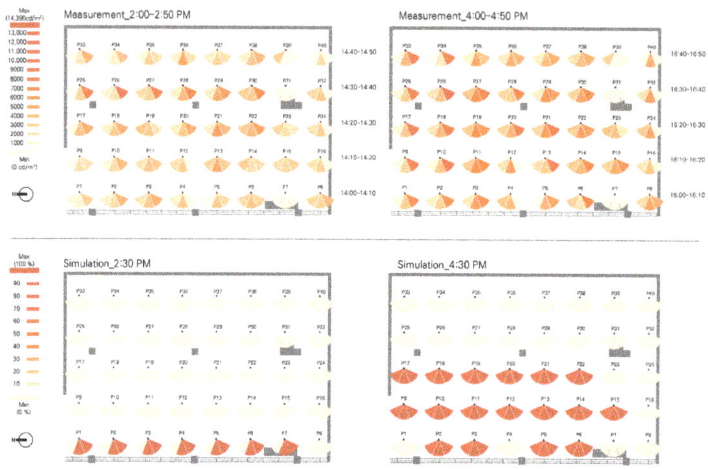

Figure 11. Measurement and simulation results of Space C (The gray box area in this Figure is the same as the gray area in Figure 5 above, which indicates the measurement area. Detailed data are included in Appendix C).

Similar to the other areas, the luminance data measured around 2:00 PM did not show significant variations among the different points and directions. However, around 4:00 PM, as more light entered the space due to the lower solar altitude, noticeable differences in the luminance values were observed among the various points and directions.

The simulation results for Space C indicated that, at 2:30 PM, only the positions P1–P7 located near the west-facing window exhibited a DGP value of 100% in the southwest and west directions. In other locations, the DGP value was approximately 0.3%, indicating almost no occurrence of glare. At 4:30 PM, as the sun's altitude decreased, allowing more sunlight to enter the interior, there was an increase in the locations experiencing glare. However, measurement points P23–P40, located in areas with lower ceilings, showed almost no occurrence of glare (DGP values about 0.3%). Measurement points located at a higher floor height generally showed a high probability of glare occurrence. At the point where glare occurred, the DGP values were 100% in both the southwest and west directions and 83% in the northwest direction.

6. Discussion

Upon comparing the results of the field measurements and the simulations for Spaces A, B, and C, it was observed that both the actual measurement and simulation confirmed more glare occurrence between 4:00 PM and 4:50 PM than between 2:00 PM and 2:50 PM. Although the external illuminance measured earlier (Table 3) showed stronger illuminance at 2:00 PM than at 4:00 PM, the glare inside the space appeared stronger around 4:00 PM when the sun's altitude was lower. This suggests that glare was more influenced by the presence or absence of direct sunlight entering a space due to changes in the solar altitude and azimuth angle rather than by external illuminance.

Between 2:00 PM and 2:50 PM, although luminance changes occurred according to the direction of the field measurements, all values (except those from P2, P3, P10, P13, P17, P18, P20, P21, P22, P28, P29, P33, and P38) in Space C measured below 4000 cd/m^2. Therefore, according to the standards suggested by Weinold and Christofferson [19], almost no locations in Spaces A and B would cause visual discomfort for individuals due to glare. Nevertheless, the simulation predicted that, in all the spaces (Spaces A, B, and C), the DGP values would reach 100 at measurement points near the windows, signifying the occurrence of glare. This highlights an inconsistency between the actual measurements and the simulation results. For the more effective use of simulations in light environment planning, understanding and addressing the sources of this discrepancy in the simulation is essential.

For the period between 4:00 PM and 4:50 PM, the field measurements showed stronger variations in luminance depending on the direction and a significant increase in points where the luminance was high enough to cause visual discomfort. During the same time period, the simulation also predicted an increase in locations likely to experience glare. Notably, in the field measurements, there were no instances of strong luminance causing discomfort in the northwest direction. However, the stimulations identified numerous points with DGP values of approximately 80% in this direction. In addition, concerning the actual measurements, elevated luminance values capable of causing discomfort were observed, even in areas relatively far from the windows. Conversely, the simulations showed that, in the innermost parts of the space, the DGP values were almost negligible at 0.3%.

The prediction of strong luminance occurrence in the northwest direction through simulations can be attributed to differences arising due to the light collection method used for the DGP calculation. This method employed fisheye rendering techniques to collect light sources within a 180 degree angle, which may have resulted in disparities based on light information collection and associated computation methods. The glare analysis using fisheye rendering in the simulation is based on the same principle as the HDR method with a fisheye lens, a technique commonly employed in actual glare measurements. However, when the results of luminance measurements and simulations from this research were compared, there were instances where the simulation predicted high glare from directions not evident in the actual measurements. This discrepancy suggests that the simulation may have used an overly broad light capture range. Given that the human field of view (FOV) is more limited than that of a fisheye lens, using fisheye rendering to predict glare might lead to overestimations of glare compared to human perceptions. This could introduce inaccuracies in the simulation's predictions due to its light capture and analysis methods. Therefore, to enhance the accuracy of the simulation, further investigation is needed on glare analysis methods using a human field of view (FOV) that considers the subjective feedback of occupants.

Concerning obstructions within spaces, for Spaces A and C, columns present within the measurement areas were accounted for in both the actual measurements and the simulation. For Spaces A, B, and C, the studies were designed to ensure that indoor furniture would not influence either the real measurements or simulation outcomes, resulting in negligible deviations due to furniture placement. Conversely, the effects of building columns were factored in, leading to instances in both the actual measurements and simulations in which the luminance readings were lower due to column obstructions. A summary of the comparisons is presented in Table 4.

Table 4. Comparisons of the actual measurements and simulation.

	Actual Measurement	Simulation
2:00 PM to 2:50 PM	- Except for some points in Space C, there were no areas exceeding 4000 cd/m^2.	- Glare was predicted to occur at window-side locations in Spaces A, B, and C.
4:00 PM to 4:50 PM	- Numerous points with luminance exceeding 4000 cd/m^2 were detected in Spaces A, B, and C. - Points with luminance exceeding 4000 cd/m^2 were also detected in the innermost part of the space. - There was a tendency for a significant difference in luminance to occur between the NW, W, and SW.	- It was predicted that glare would not occur in the innermost part of the space. - There was no significant glare difference between the NW, W, and SW.

7. Conclusions

While natural lighting offers numerous benefits, many workspaces resort to covering windows with blinds to mitigate glare. This approach not only obstructs external views but also reduces the influx of natural light. Therefore, architectural designs and placements that balance external views, natural light, and glare prevention are crucial. Simulations can be instrumental in light environment design. However, this study unveiled discrepancies between existing simulation results and actual measurements, possibly because the simulations employed fisheye rendering. Suk et al. [66] highlighted that a human's field of view (FOV) is narrower than that of a fisheye lens, leading to potential inaccuracies in glare capture. To enhance indoor glare assessments, simulations should incorporate a human FOV to discern glare variations by direction. Furthermore, beyond just the ClimateStudio program used in this research, a comparative analysis of various simulation tools is essential for refining the lighting environment.

The limitations of this study include the inability to perfectly replicate identical environments, minor measurement errors (such as those arising from the physical act of taking measurements) during data collection, and the incomplete removal of the shading effects of tree branches (despite conducting the experiments in winter to reduce these influences). Additionally, the simulations might have missed some temporary glare effects, because they predicted only specific time points. However, even with these challenges, consistent discrepancies were observed between the actual measurements and the simulation results, particularly in terms of the measurement direction and points. These differences suggest that the primary cause may not be factors such as external landscaping or unpredicted temporary glare effects in the simulations but, rather, a more significant factor, such as the light capture method in the simulations using the fisheye FOV.

For future research involving luminance measurements, it is imperative to develop a glare measurement method tailored to the specific environmental conditions of the chosen study space, such as partitions, occupants' eye level, and spatial depth. The glare zone defined in this research was specifically designed for the selected space. Moreover, to grasp the variability of the data across time and space, further research should be conducted in environments and at times different from those in this experiment.

Author Contributions: K.L. conceived and designed the experiments; J.P. and K.K. performed the experiments; J.P. analyzed the data; and J.P. and K.L. wrote the paper. All authors have read and agreed to the published version of the manuscript.

Funding: This work was funded by the National Research Foundation of Korea (NRF) grant issued by the Korean government (MSIT) (no. NRF-2021R1A2C2011849) and by the 2023 Hongik University Research Fund.

Data Availability Statement: The data presented in this study are available in article.

Conflicts of Interest: The authors declare no conflict of interest. The funding sponsors had no role in the design of the study; in the collection, analyses, or interpretation of the data; in the writing of the manuscript; or in the decision to publish the results.

Appendix A

Table A1. Comparison of the luminance measurements and DGP simulations in Space A. Unit: cd/m_2 (measurement) and DGP (%) (simulation).

	Point	Measurement		Simulation		Measurement		Simulation
2:00 ~ 2:30 PM	P1	NW	914	0.3	P21	NW	840	0.3
		W	876	* 100.0		W	1321	0.3
		SW	44	* 100.0		SW	1778	0.3
	P2	NW	847	0.3	P22	NW	1097	0.3
		W	686	* 100.0		W	1244	0.3
		SW	1120	* 100.0		SW	1594	0.3
	P3	NW	870	0.3	P23	NW	1310	0.3
		W	561	* 100.0		W	1571	0.3
		SW	1761	* 100.0		SW	355	0.3
	P4	NW	698	0.3	P24	NW	974	0.3
		W	1367	* 100.0		W	2275	0.3
		SW	1736	* 100.0		SW	71	0.3
	P5	NW	968	0.3	P25	NW	145	0.3
		W	1183	* 100.0		W	1319	0.3
		SW	1288	* 100.0		SW	116	0.3
	P6	NW	1200	0.3	P26	NW	1203	0.3
		W	1380	* 100.0		W	1467	0.3
		SW	1350	* 100.0		SW	118	0.3
	P7	NW	896	0.3	P27	NW	1266	0.3
		W	850	* 100.0		W	1839	0.3
		SW	1376	* 100.0		SW	2021	0.3
	P8	NW	1192	0.3	P28	NW	1257	0.3
		W	1294	0.3		W	1767	0.3
		SW	35	0.3		SW	2094	0.3

	Point	Measurement		Simulation	Point	Measurement		Simulation
4:00 ~ 4:30 PM	P1	NW	1100	* 81.8	P21	NW	1518	* 81.8
		W	2180	* 100.0		W	1900	* 100.0
		SW	5576	* 100.0		SW	* 11,150	* 100.0
	P2	NW	1200	0.3	P22	NW	1420	* 81.8
		W	1900	0.3		W	3117	* 100.0
		SW	7365	0.3		SW	* 14,390	* 100.0
	P3	NW	1164	* 81.8	P23	NW	2001	0.3
		W	1427	* 100.0		W	1909	0.3
		SW	5023	* 100.0		SW	565	0.3
	P4	NW	846	* 81.8	P24	NW	1114	0.3
		W	2349	* 100.0		W	2309	0.3
		SW	2908	* 100.0		SW	93	0.3
	P5	NW	1410	0.3	P25	NW	264	* 81.8
		W	2994	0.3		W	2659	* 100.0
		SW	3933	0.3		SW	172	* 100.0
	P6	NW	1605	* 81.8	P26	NW	1147	0.3
		W	3115	* 100.0		W	3221	0.3
		SW	2954	* 100.0		SW	* 9727	0.3
	P7	NW	1490	* 81.8	P27	NW	1354	0.3
		W	1270	* 100.0		W	2357	0.3
		SW	3590	* 100.0		SW	* 7486	0.3
	P8	NW	1546	* 81.8	P28	NW	2464	* 81.8
		W	1672	* 100.0		W	2442	* 100.0
		SW	282	* 100.0		SW	* 9290	* 100.0

Table A1. *Cont.*

	Point	Measurement		Simulation	Point	Measurement		Simulation		Point	Measurement		Simulation	Point	Measurement		Simulation
2:00 ~ 2:30 PM	P9	NW	201	0.3	P29	NW	1367	0.3	4:00 ~ 4:30 PM	P9	NW	1405	* 81.8	P29	NW	1051	0.3
		W	1009	0.3		W	2245	0.3			W	2605	* 100.0		W	2589	0.3
		SW	1047	0.3		SW	2336	0.3			SW	3513	* 100.0		SW	* 8576	0.3
	P10	NW	785	0.3	P30	NW	902	0.3		P10	NW	1339	* 81.8	P30	NW	966	0.3
		W	1172	0.3		W	1764	0.3			W	* 4035	* 100.0		W	2886	0.3
		SW	102	0.3		SW	1778	0.3			SW	228	* 100.0		SW	3753	0.3
	P11	NW	114	0.3	P31	NW	1712	0.3		P11	NW	243	* 81.8	P31	NW	2566	0.3
		W	1154	0.3		W	1170	0.3			W	2440	* 100.0		W	2088	0.3
		SW	2015	0.3		SW	48	0.3			SW	* 9725	* 100.0		SW	52	0.3
	P12	NW	816	0.3	P32	NW	1064	0.3		P12	NW	1347	* 81.8	P32	NW	1796	0.3
		W	1175	0.3		W	1668	0.3			W	3636	* 100.0		W	2444	0.3
		SW	1378	0.3		SW	39	0.3			SW	* 4815	* 100.0		SW	385	0.3
	P13	NW	837	0.3	P33	NW	142	0.3		P13	NW	1590	* 81.8	P33	NW	157	0.3
		W	1648	0.3		W	1547	0.3			W	2929	* 100.0		W	1347	0.3
		SW	97	0.3		SW	2289	0.3			SW	219	* 100.0		SW	* 7778	0.3
	P14	NW	102	0.3	P34	NW	153	0.3		P14	NW	152	* 81.8	P34	NW	170	0.3
		W	1060	0.3		W	1122	0.3			W	2443	* 100.0		W	3007	0.3
		SW	1667	0.3		SW	81	0.3			SW	2959	* 100.0		SW	* 8230	0.3
	P15	NW	1251	0.3	P35	NW	1787	0.3		P15	NW	1757	* 81.8	P35	NW	1127	0.3
		W	1104	0.3		W	1308	0.3			W	1511	* 100.0		W	1922	0.3
		SW	280	0.3		SW	2244	0.3			SW	658	* 100.0		SW	* 5202	0.3
	P16	NW	847	0.3	P36	NW	462	0.3		P16	NW	1149	0.3	P36	NW	1500	0.3
		W	1403	0.3		W	1574	0.3			W	1894	0.3		W	1800	0.3
		SW	115	0.3		SW	2185	0.3			SW	140	0.3		SW	* 6517	0.3

Table A1. *Cont.*

	Point	Measurement		Simulation	Point	Measurement		Simulation		Point	Measurement		Simulation	Point	Measurement		Simulation
2:00 ~ 2:30 PM	P17	NW	165	0.3	P37	NW	1182	0.3	4:00 ~ 4:30 PM	P17	NW	421	* 81.8	P37	NW	1645	0.3
		W	1330	0.3		W	2046	0.3			W	1917	* 100.0		W	1939	0.3
		SW	63	0.3		SW	53	0.3			SW	234	* 100.0		SW	35	0.3
	P18	NW	191	0.3	P38	NW	74	0.3		P18	NW	1821	* 81.8	P38	NW	78	0.3
		W	1037	0.3		W	1400	0.3			W	1775	* 100.0		W	1726	0.3
		SW	44	0.3		SW	293	0.3			SW	179	* 100.0		SW	3109	0.3
	P19	NW	882	0.3	P39	NW	47	0.3		P19	NW	1235	* 81.8	P39	NW	1295	0.3
		W	1293	0.3		W	1674	0.3			W	3443	* 100.0		W	1938	0.3
		SW	2069	0.3		SW	40	0.3			SW	* 5605	* 100.0		SW	324	0.3
	P20	NW	1053	0.3	P40	NW	1165	0.3		P20	NW	1720	* 81.8	P40	NW	1614	0.3
		W	1109	0.3		W	1804	0.3			W	2501	* 100.0		W	2302	0.3
		SW	1802	0.3		SW	138	0.3			SW	* 8235	* 100.0		SW	351	0.3

*: 4000 cd/m^2 or 45% DGP $\leq$ result.

Appendix B

Table A2. Comparison of the luminance measurements and DGP simulations in Space B. Unit: cd/m_2 (measurement) and DGP (%) (simulation).

	Point	Measurement		Simulation		Measurement		Simulation		Point	Measurement		Simulation	Point	Measurement		Simulation
2:00 ~ 2:30 PM	P1	NW	63	0.3	P15	NW	856	0.3	4:00 ~ 4:30 PM	P1	NW	132	* 82.2	P15	NW	1383	* 82.2
		W	1114	0.3		W	1115	0.3			W	1945	* 100.0		W	2285	* 100.0
		SW	1517	0.3		SW	1333	0.3			SW	* 4677	* 100.0		SW	* 4821	* 100.0
	P2	NW	500	0.3	P16	NW	250	0.3		P2	NW	1176	* 82.2	P16	NW	959	* 82.2
		W	1136	0.3		W	1327	0.3			W	2105	* 100.0		W	2348	* 100.0
		SW	1918	0.3		SW	1981	0.3			SW	* 5899	* 100.0		SW	* 4426	* 100.0
	P3	NW	865	0.3	P17	NW	830	0.3		P3	NW	1408	0.3	P17	NW	1213	* 82.2
		W	1139	0.3		W	1182	0.3			W	2523	0.3		W	2425	* 100.0
		SW	1628	0.3		SW	2058	0.3			SW	3629	0.3		SW	* 5503	* 100.0
	P4	NW	40	0.3	P18	NW	1152	0.3		P4	NW	172	* 82.2	P18	NW	1544	* 79.9
		W	139	0.3		W	1576	* 100.0			W	135	* 100.0		W	2309	* 100.0
		SW	775	0.3		SW	474	* 100.0			SW	* 10,180	* 100.0		SW	340	* 100.0
	P5	NW	939	0.3	P19	NW	732	0.3		P5	NW	1186	* 82.2	P19	NW	1028	0.3
		W	1176	0.3		W	1137	0.3			W	1886	* 100.0		W	1988	0.3
		SW	1886	0.3		SW	1504	0.3			SW	* 6277	* 100.0		SW	* 4041	0.3
	P6	NW	777	0.3	P20	NW	1127	0.3		P6	NW	1345	0.3	P20	NW	1544	0.3
		W	1209	0.3		W	1201	0.3			W	2481	0.3		W	1974	0.3
		SW	2145	0.3		SW	1373	0.3			SW	* 7502	0.3		SW	2912	0.3
	P7	NW	157	0.3	P21	NW	938	0.3		P7	NW	127	* 82.2	P21	NW	1278	0.3
		W	1405	0.3		W	1097	0.3			W	2290	* 100.0		W	2387	0.3
		SW	2079	0.3		SW	1580	0.3			SW	* 4602	* 100.0		SW	* 4581	0.3
	P8	NW	648	0.3	P22	NW	727	0.3		P8	NW	768	* 82.2	P22	NW	1235	0.3
		W	1590	0.3		W	1298	0.3			W	2349	* 100.0		W	2245	0.3
		SW	2889	0.3		SW	1638	0.3			SW	3034	* 100.0		SW	* 4216	0.3

Table A2. *Cont.*

	Point	Measurement		Simulation	Point	Measurement		Simulation		Point	Measurement		Simulation	Point	Measurement		Simulation
2:00 ~ 2:30 PM	P9	NW	976	0.3	P23	NW	1023	0.3	4:00 ~ 4:30 PM	P9	NW	1474	0.3	P23	NW	1271	0.3
		W	1463	* 100.0		W	1288	0.3			W	2267	0.3		W	1986	0.3
		SW	1983	* 100.0		SW	1620	0.3			SW	* 7274	0.3		SW	3350	0.3
	P10	NW	200	0.3	P24	NW	891	0.3		P10	NW	267	* 82.2	P24	NW	1328	0.3
		W	1082	0.3		W	1147	0.3			W	2415	* 100.0		W	2120	0.3
		SW	1582	0.3		SW	1884	0.3			SW	* 5429	* 100.0		SW	* 6176	0.3
	P11	NW	800	0.3	P25	NW	741	0.3		P11	NW	1251	* 82.2	P25	NW	1280	0.3
		W	1140	0.3		W	1304	0.3			W	2062	* 100.0		W	2714	0.3
		SW	1458	0.3		SW	1994	0.3			SW	* 4441	* 100.0		SW	* 6492	0.3
	P12	NW	870	0.3	P26	NW	982	0.3		P12	NW	1451	* 82.2	P26	NW	1344	0.3
		W	1106	0.3		W	1377	0.3			W	2079	* 100.0		W	2237	0.3
		SW	124	0.3		SW	2735	0.3			SW	186	* 100.0		SW	* 5643	0.3
	P13	NW	637	0.3	P27	NW	944	0.3		P13	NW	1126	* 82.2	P27	NW	1570	* 79.9
		W	1251	0.3		W	1644	* 100.0			W	3604	* 100.0		W	1622	* 100.0
		SW	1747	0.3		SW	3140	* 100.0			SW	* 8360	* 100.0		SW	* 12,600	* 100.0
	P14	NW	884	0.3						P14	NW	1060	* 82.2				
		W	1175	0.3							W	2053	* 100.0				
		SW	1723	0.3							SW	* 5195	* 100.0				

*: 4000 cd/m^2 or 45% DGP $\leq$ result.

Appendix C

Table A3. Comparison of the luminance measurements and DGP simulations in Space C. Unit: cd/m^2 (measurement) and DGP (%) (simulation).

	Point	Measurement		Simulation	Point	Measurement		Simulation		Point	Measurement		Simulation	Point	Measurement		Simulation
2:00 ~ 2:30 PM	P1	NW	632	0.3	P21	NW	1366	0.3	4:00 ~ 4:30 PM	P1	NW	412	0.3	P21	NW	2509	* 83.0
		W	1282	* 100		W	* 5502	0.3			W	3130	0.3		W	* 5564	* 100.0
		SW	2999	* 100.0		SW	* 4182	0.3			SW	* 6086	0.3		SW	* 8490	* 100.0
	P2	NW	1439	0.3	P22	NW	3106	0.3		P2	NW	1449	* 83.0	P22	NW	3002	* 83.0
		W	2421	* 100.0		W	* 4252	0.3			W	3034	* 100.0		W	* 5220	* 100.0
		SW	* 4292	* 100.0		SW	* 4218	0.3			SW	* 6018	* 100.0		SW	* 5241	* 100.0
	P3	NW	1865	0.3	P23	NW	2750	0.3		P3	NW	1709	* 83.0	P23	NW	3575	0.3
		W	3495	* 100.0		W	2060	0.3			W	* 5227	* 100.0		W	3741	0.3
		SW	* 4624	* 100.0		SW	3910	0.3			SW	* 6445	* 100.0		SW	1502	0.3
	P4	NW	1242	0.3	P24	NW	2058	0.3		P4	NW	3331	0.3	P24	NW	3141	0.3
		W	3137	* 100.0		W	3047	0.3			W	* 4083	0.3		W	* 5189	0.3
		SW	421	* 100.0		SW	25	0.3			SW	78	0.3		SW	23	0.3
	P5	NW	104	0.3	P25	NW	98	0.3		P5	NW	148	* 83.0	P25	NW	619	0.3
		W	3503	* 100.0		W	2579	0.3			W	* 4130	* 100.0		W	3026	0.3
		SW	2372	* 100.0		SW	82	0.3			SW	* 5163	* 100.0		SW	* 7865	0.3
	P6	NW	2424	0.3	P26	NW	1048	0.3		P6	NW	2807	* 83.0	P26	NW	1958	0.3
		W	3517	* 100.0		W	2075	0.3			W	1880	* 100.0		W	3351	0.3
		SW	2759	* 100.0		SW	3876	0.3			SW	* 7238	* 100.0		SW	* 11,750	0.3
	P7	NW	14	0.3	P27	NW	2464	0.3		P7	NW	27	0.3	P27	NW	3504	0.3
		W	10	* 100.0		W	2790	0.3			W	69	0.3		W	* 6245	0.3
		SW	13	* 100.0		SW	2041	0.3			SW	24	0.3		SW	* 7674	0.3
	P8	NW	2686	0.3	P28	NW	1321	0.3		P8	NW	2413	0.3	P28	NW	2564	0.3
		W	2593	0.3		W	* 4001	0.3			W	* 5005	0.3		W	2939	0.3
		SW	3978	0.3		SW	2111	0.3			SW	* 4470	0.3		SW	* 6688	0.3

Table A3. *Cont.*

	Point	Measurement		Simulation	Point	Measurement		Simulation		Point	Measurement		Simulation	Point	Measurement		Simulation
2:00 ~ 2:30 PM	P9	NW	257	0.3	P29	NW	1500	0.3	4:00 ~ 4:30 PM	P9	NW	508	* 83.0	P29	NW	1450	0.3
		W	2343	0.3		W	* 4606	0.3			W	2577	* 100.0		W	* 5959	0.3
		SW	2271	0.3		SW	2026	0.3			SW	* 8941	* 100.0		SW	* 5650	0.3
	P10	NW	620	0.3	P30	NW	3028	0.3		P10	NW	620	* 83.0	P30	NW	2797	0.3
		W	2701	0.3		W	3189	0.3			W	3474	* 100.0		W	* 4970	0.3
		SW	* 4293	0.3		SW	3540	0.3			SW	* 11,640	* 100.0		SW	* 6339	0.3
	P11	NW	1990	0.3	P31	NW	125.0	0.3		P11	NW	3071	* 83.0	P31	NW	124	0.3
		W	3311	0.3		W	89.0	0.3			W	* 6338	* 100.0		W	90	0.3
		SW	3268	0.3		SW	41.0	0.3			SW	* 6101	* 100.0		SW	79	0.3
	P12	NW	2307	0.3	P32	NW	2528.0	0.3		P12	NW	2093	* 83.0	P32	NW	2041	0.3
		W	3046	0.3		W	3348.0	0.3			W	* 5308	* 100.0		W	* 5389	0.3
		SW	89	0.3		SW	143.0	0.3			SW	125	* 100.0		SW	217	0.3
	P13	NW	2708	0.3	P33	NW	191.0	0.3		P13	NW	2250	* 83.0	P33	NW	313	0.3
		W	* 4914	0.3		W	1820.0	0.3			W	3082	* 100.0		W	* 4086	0.3
		SW	3175	0.3		SW	* 4479.0	0.3			SW	* 8357	* 100.0		SW	* 7851	0.3
	P14	NW	3120	0.3	P34	NW	31.0	0.3		P14	NW	3101	* 83.0	P34	NW	43	0.3
		W	2269	0.3		W	1384.0	0.3			W	* 4161	* 100.0		W	1816	0.3
		SW	2271	0.3		SW	3898.0	0.3			SW	* 6421	* 100.0		SW	* 9640	0.3
	P15	NW	2220	0.3	P35	NW	550.0	0.3		P15	NW	2983	* 83.0	P35	NW	1068	0.3
		W	2651	0.3		W	2909.0	0.3			W	* 4845	* 100.0		W	* 5147	0.3
		SW	2851	0.3		SW	2289.0	0.3			SW	* 6041	* 100.0		SW	3457	0.3
	P16	NW	2103	0.3	P36	NW	1188.0	0.3		P16	NW	1850	0.3	P36	NW	1637	0.3
		W	2826	0.3		W	* 5030.0	0.3			W	3559	0.3		W	* 4860	0.3
		SW	281	0.3		SW	2199.0	0.3			SW	310	0.3		SW	* 4238	0.3

Table A3. *Cont.*

Time	Point	Measurement		Simulation	Point	Measurement		Simulation	Time	Point	Measurement		Simulation	Point	Measurement		Simulation
2:00 ~ 2:30 PM	P17	NW	348	0.3	P37	NW	1491.0	0.3	4:00 ~ 4:30 PM	P17	NW	637	* 83.0	P37	NW	1419	0.3
		W	3754	0.3		W	2809.0	0.3			W	3300	* 100.0		W	* 4050	0.3
		SW	* 5107	0.3		SW	2220.0	0.3			SW	* 10,860	* 100.0		SW	* 4815	0.3
	P18	NW	1274	0.3	P38	NW	1249.0	0.3		P18	NW	1823	* 83.0	P38	NW	2283	0.3
		W	3393	0.3		W	3470.0	0.3			W	* 4144	* 100.0		W	3889	0.3
		SW	* 4695	0.3		SW	* 4282.0	0.3			SW	* 8703	* 100.0		SW	* 4824	0.3
	P19	NW	2950	0.3	P19	NW	3153	* 83.0		P39	NW	2887	0.3				
		W	3289	0.3		W	85.0	0.3			W	* 6217	* 100.0		W	131	0.3
		SW	1904	0.3		SW	93.0	0.3			SW	* 5681	* 100.0		SW	249	0.3
	P20	NW	1399	0.3	P40	NW	33.0	0.3		P20	NW	2565	* 83.0	P40	NW	54	0.3
		W	3358	0.3		W	2294.0	0.3			W	* 4323	* 100.0		W	* 4645	0.3
		SW	* 4803	0.3		SW	38.0	0.3			SW	* 8658	* 100.0		SW	61	0.3

*: 4000 cd/m^2 or 45% DGP $\leq$ result.

References

1. Brainard, G.C.; Hanifin, J.P.; Rollag, M.D.; Greeson, J.; Byrne, B.; Glickman, G.; Sanford, B. Human melatonin regulation is not mediated by the three cone photopic visual system. *J. Clin. Endocrinol. Metab.* **2001**, *86*, 433–436.
2. Cajochen, C.; Zeitzer, J.M.; Czeisler, C.A.; Dijk, D.J. Dose-response relationship for light intensity and ocular and electroencephalographic correlates of human alertness. *Behav. Brain Res.* **2000**, *115*, 75–83. [CrossRef]
3. Lockley, S.W.; Brainard, G.C.; Czeisler, C.A. High sensitivity of the human circadian melatonin rhythm to resetting by short wavelength light. *J. Clin. Endocrinol. Metab.* **2003**, *88*, 4502–4505. [CrossRef]
4. Rea, M.S.; Figueiro, M.G.; Bullough, J.D.; Bierman, A. A model of phototransduction by the human circadian system. *Brain Res. Rev.* **2005**, *50*, 213–228. [CrossRef]
5. Viola, A.U.; James, L.M.; Schlangen, L.J.; Dijk, D.J. Blue-enriched white light in the workplace improves self-reported alertness, performance and sleep quality. *Scand. J. Work Environ. Health* **2008**, *34*, 297–306. [CrossRef]
6. Vetter, C.; Pattison, P.M.; Houser, K.; Herf, M.; Phillips, A.J.; Wright, K.P.; Glickman, G. A review of human physiological responses to light: Implications for the development of integrative lighting solutions. *Leukos* **2022**, *18*, 387–414. [CrossRef]
7. Lu, M.; Hu, S.; Mao, Z.; Liang, P.; Xin, S.; Guan, H. Research on work efficiency and light comfort based on EEG evaluation method. *Build. Environ.* **2020**, *183*, 107122. [CrossRef]
8. Goodman, T.M. Measurement and specification of lighting: A look at the future. *Light. Res. Technol.* **2009**, *41*, 229–243. [CrossRef]
9. Hassan, M.U.; Angelaki, S.; Alfaro, C.V.L.; Major, P.; Styve, A.; Alaliyat, S.A.A.; da Silva Torres, R. Digital Twins for Lighting Analysis: Literature Review, Challenges, and Research Opportunities. In Proceedings of the 36th International ECMS Conference on Modelling and Simulation, Alesund, Norway, 30 May–3 June 2022; ECMS: Caserta, Italy, 2022; Volume 36, pp. 226–235.
10. Ricciardi, P.; Buratti, C. Environmental quality of university classrooms: Subjective and objective evaluation of the thermal, acoustic, and lighting comfort conditions. *Build. Environ.* **2018**, *127*, 23–36. [CrossRef]
11. United Nation General Assembly. Transforming Our World: The 2030 Agenda for Sustainable Development. Available online: https://www.un.org/en/development/desa/population/migration/generalassembly/docs/globalcompact/A_RES_70_1_E.pdf (accessed on 11 October 2023).
12. Solemma. What is ClimateStudio? Available online: https://www.solemma.com/climatestudio (accessed on 5 April 2023).
13. Nielsen, M.V.; Svendsen, S.; Jensen, L.B. Quantifying the potential of automated dynamic solar shading in office buildings through integrated simulations of energy and daylight. *Sol. Energy* **2011**, *85*, 757–768. [CrossRef]
14. Boubekri, M.; Lee, J.; MacNaughton, P.; Woo, M.; Schuyler, L.; Tinianov, B.; Satish, U. The impact of optimized daylight and views on the sleep duration and cognitive performance of office workers. *Int. J. Environ. Res. Public Health* **2020**, *17*, 3219. [CrossRef]
15. Konis, K. A novel circadian daylight metric for building design and evaluation. *Build. Environ.* **2017**, *113*, 22–38. [CrossRef]
16. Bellia, L.; Pedace, A.; Barbato, G. Daylighting offices: A first step toward an analysis of photobiological effects for design practice purposes. *Build. Environ.* **2014**, *74*, 54–64. [CrossRef]
17. Lee, J.; Boubekri, M. Impact of daylight exposure on health, well-being and sleep of office workers based on actigraphy, surveys, and computer simulation. *J. Green Build.* **2020**, *15*, 19–42. [CrossRef]
18. Heerwagen, J.H.; Orians, G.H. Adaptations to windowlessness: A study of the use of visual decor in windowed and windowless offices. *Environ. Behav.* **1986**, *18*, 623–639. [CrossRef]
19. Boubekri, M.; Hull, R.B.; Boyer, L.L. Impact of window size and sunlight penetration on office workers' mood and satisfaction: A novel way of assessing sunlight. *Environ. Behav.* **1991**, *23*, 474–493. [CrossRef]
20. Heschong, L. Daylighting and human performance. *ASHRAE J.* **2002**, *44*, 65–67.
21. Van Den Wymelenberg, K. Patterns of occupant interaction with window blinds: A literature review. *Energy Build.* **2012**, *51*, 165–176. [CrossRef]
22. Wang, N.; Boubekri, M. Design recommendations based on cognitive, mood and preference assessments in a sunlit workspace. *Light. Res. Technol.* **2011**, *43*, 55–72. [CrossRef]
23. Gou, Z.; Lau, S.S.Y.; Qian, F. Comparison of mood and task performance in naturally-lit and artificially-lit environments. *Indoor Built Environ.* **2015**, *24*, 27–36. [CrossRef]
24. Wienold, J.; Christoffersen, J. *Towards a New Daylight Glare Rating*; Lux Europa: Berlin, Germany, 2005; pp. 157–161.
25. Suk, J.Y.; Schiler, M.; Kensek, K. Investigation of existing discomfort glare indices using human subject study data. *Build. Environ.* **2017**, *113*, 121–130. [CrossRef]
26. Veitch, J.A.; Newsham, G.R. Preferred luminous conditions in open-plan offices: Research and practice recommendations. *Int. J. Light. Res. Technol.* **2000**, *32*, 199–212. [CrossRef]
27. Schiler, M. Toward a definition of glare: Can qualitative issues be quantified? In Proceedings of the 2nd EAAE–ARCC Conference on Architectural Research, Paris, France, 4 July 2000.
28. Egan, M.D. *Concepts in Architectural Lighting*; McGraw-Hill: New York, NY, USA, 1983.
29. Wienold, J.; Christoffersen, J. Evaluation methods and development of a new glare prediction model for daylight environments with the use of CCD cameras. *Energy Build.* **2006**, *38*, 743–757. [CrossRef]
30. Solemma. Annual Glare. In ClimateStudio User Guide. Available online: https://climatestudiodocs.com/docs/annualGlare.html (accessed on 5 April 2023).

31. Bechthold, M.; King, N.; Kane, A.O.; Niemasz, J.; Reinhart, C. Integrated environmental design and robotic fabrication workflow for ceramic shading systems. *Autom. Robot. Build. Constr.* **2011**, *9*, 70–75.
32. Motevalian, E. Double Skin Facades Performance: Effects on Daylight and Visual Comfort in Office Spaces. Ph.D. Thesis, University of Southern California, Los Angeles, CA, USA, 2014.
33. Yun, G.; Yoon, K.C.; Kim, K.S. The influence of shading control strategies on the visual comfort and energy demand of office buildings. *Energy Build.* **2014**, *84*, 70–85. [CrossRef]
34. Wagdy, A.; Fathy, F. A parametric approach for achieving optimum daylighting performance through solar screens in desert climates. *J. Build. Eng.* **2015**, *3*, 155–170. [CrossRef]
35. Kong, Z.; Utzinger, M.; Liu, L. Solving glare problems in architecture through integration of HDR image technique and modeling with DIVA. In Proceedings of the 14th Conference of International Building Performance Simulation Association, Hyderabad, India, 7–9 December 2015; pp. 1221–1228.
36. Arango-Díaz, L. Deslumbramiento en ambientes educativos con muro calado en fachada. *Rev. AUS* **2016**, *20*, 62–69. [CrossRef]
37. Shin, C.; Collins, G. Daylight Glare Analysis for an All Glass Cathedral: Integrating Simulation with Common Sense to Improve Visual Comfort. *Proc. SimBuild.* **2016**, *6*, 267–274.
38. Gao, Y.; Dong, J.; Isabella, O.; Zeman, M.; Zhang, G.Q. Daylighting simulation and analysis of buildings with dynamic photovoltaic window shading elements. In Proceedings of the 2017 14th China International Forum on Solid State Lighting: International Forum on Wide Bandgap Semiconductors China (SSLChina: IFWS), Beijing, China, 1–3 November 2017; IEEE: New York, NY, USA, 2017; pp. 52–55.
39. Sherif, A.; Mahmoud, A.; ElSharkawy, M.; Eissa, A. Parametric Configuration of Window Light Shelves for Daylighting of Hospital Patient Rooms under a Desert Clear Sky. *Proceeding PLEA* **2017**, *2*, 3380–3387.
40. Wagdy, A.; Sherif, A.; Sabry, H.; Arafa, R.; Mashaly, I. Daylighting simulation for the configuration of external sun-breakers on south oriented windows of hospital patient rooms under a clear desert sky. *Sol. Energy* **2017**, *149*, 164–175. [CrossRef]
41. Sharma, L.; Kishan Lal, K.; Rakshit, D. Evaluation of impact of passive design measures with energy saving potential through estimation of shading control for visual comfort. *J. Build. Phys.* **2018**, *42*, 220–238. [CrossRef]
42. Bian, Y.; Leng, T.; Ma, Y. A proposed discomfort glare evaluation method based on the concept of 'adaptive zone'. *Build Environ.* **2018**, *143*, 306–317. [CrossRef]
43. Tabadkani, A.; Banihashemi, S.; Hosseini, M.R. Daylighting and visual comfort of oriental sun responsive skins: A parametric analysis. *Build. Simul.* **2018**, *11*, 663–676. [CrossRef]
44. Sharma, L.; Rakshit, D. Visual comfort based algorithmic control for roller shade and assessment of potential energy savings. In *Applications of Solar Energy*, 1st ed.; Tyagi, H., Agarwal, A., Chakraborty, P., Powar, S., Eds.; Springer: Singapore, 2018; pp. 295–316.
45. Ayçam, İ.; Ceylan, Ö. Analysis of Daylight Control in Advanced Façade Systems for Office Module in Ankara. In *Proceedings of the 3rd International Sustainable Buildings Symposium (ISBS 2017)*; Fırat, S., Kinuthia, J., Abu-Tair, A., Eds.; Springer: Cham, Germany, 2018; Volume 13, pp. 359–377.
46. Lee, S.; Lee, K.S. A Study on the improvement of the evaluation scale of discomfort glare in educational facilities. *Energies* **2019**, *12*, 3265. [CrossRef]
47. Pagliolico, S.L.; Verso, V.R.L.; Zublena, M.; Giovannini, L. Preliminary results on a novel photo-bio-screen as a shading system in a kindergarten: Visible transmittance, visual comfort and energy demand for lighting. *Sol. Energy* **2019**, *185*, 41–58. [CrossRef]
48. Mesloub, A.; Ghosh, A.; Touahmia, M.; Albaqawy, G.A.; Noaime, E.; Alsolami, B.M. Performance analysis of photovoltaic integrated shading devices (PVSDs) and semi-transparent photovoltaic (STPV) devices retrofitted to a prototype office building in a hot desert climate. *Sustainability* **2020**, *12*, 10145. [CrossRef]
49. Sun, Y.; Liu, X.; Qu, W.; Cao, G.; Zou, N. Analysis of daylight glare and optimal lighting design for comfortable office lighting. *Optik* **2020**, *206*, 164291. [CrossRef]
50. Srisamranrungruang, T.; Hiyama, K. Balancing of natural ventilation, daylight, thermal effect for a building with double-skin perforated facade (DSPF). *Energy Build.* **2020**, *210*, 109765. [CrossRef]
51. Triantafyllidou, E.F.; Michael, A.G. The impact of installing a concave curved profile blind to a glass window for visual comfort in office buildings. *Procedia. Manuf.* **2020**, *44*, 269–276. [CrossRef]
52. Kiliç, Z.A.; Yener, A.K. Determining Proper Daylighting Design Solution for Visual Comfort and Lighting Energy Efficiency: A Case Study for High-Rise Residential Building. *J. Phys. Conf. Ser.* **2021**, *2069*, 012156. [CrossRef]
53. Özdemir, H.; Çakmak, B.Y. Evaluation of Daylight and Glare Quality of Office Spaces with Flat and Dynamic Shading System Facades in Hot Arid Climate. *J. Daylighting* **2022**, *9*, 197–208. [CrossRef]
54. Mesloub, A.; Ghosh, A.; Touahmia, M.; Albaqawy, G.A.; Alsolami, B.M.; Ahriz, A. Assessment of the overall energy performance of an SPD smart window in a hot desert climate. *Energy* **2022**, *252*, 124073. [CrossRef]
55. Navabi, D.; Amini, Z.; Rahmati, A.; Tahbaz, M.; Butt, T.E.; Sharifi, S.; Mosavi, A. Developing light transmitting concrete for energy saving in buildings. *Case Stud. Constr. Mater.* **2023**, *18*, e01969. [CrossRef]
56. Sui, G.; Liu, J.; Leng, J.; Yu, F. Daylighting performance assessment of traditional skywell dwellings: A case study in Fujian, China. *J. Build. Eng.* **2023**, *68*, 106028. [CrossRef]
57. Sorooshnia, E.; Rashidi, M.; Rahnamayiezekavat, P.; Rezaei, F.; Samali, B. Optimum external shading system for counterbalancing glare probability and daylight illuminance in Sydney's residential buildings. *Eng. Constr. Archit. Manag.* **2023**, *30*, 296–320. [CrossRef]

58. Hassan, F.H.; Ali, K.A.; Ahmed, S.A. Biomimicry as an Approach to Improve Daylighting Performance in Office Buildings in Assiut City, Egypt. *J. Daylighting.* **2023**, *10*, 1–16. [CrossRef]
59. Taleb, H.M.; Moarbes, R. Improving illuminance performance by implementing a kinetic façade system: Case study of office building in Dubai. *J. Asian Archit. Build. Eng.* **2023**, *22*, 2809–2826. [CrossRef]
60. Suk, J.Y.; Schiler, M.; Kensek, K. Development of new daylight glare analysis methodology using absolute glare factor and relative glare factor. *Energy Build.* **2013**, *64*, 113–122. [CrossRef]
61. He, S.; Li, H.; Yan, Y.; Cai, H. Calibrating lighting simulation with panoramic high dynamic range imaging. *J. Build. Perform. Simul.* **2023**, 1–20. [CrossRef]
62. Safranek, S.; Davis, R.G. Sources of error in HDRI for luminance measurement: A review of the literature. *Leukos* **2021**, *17*, 187–208. [CrossRef]
63. Konika Minolta Sensing Americas Inc. CS-150 Luminance and Color Meter. Available online: https://sensing.konicaminolta.us/us/products/cs-150-luminance-and-color-meter/ (accessed on 2 February 2023).
64. Korea Meteorological Administration. Observation and Climate. Available online: https://web.kma.go.kr/w/obs-climate/land/past-obs/obs-by-element.do?stn=108&yy=2023&obs=59 (accessed on 1 March 2023).
65. Konika Minolta Sensing Americas Inc. CL-500A Illuminance Spectrophotometer. Available online: https://sensing.konicaminolta.us/us/products/cl-500a-illuminance-spectrophotometer/ (accessed on 2 February 2023).
66. Suk, J.Y.; Schiler, M.; Kensek, K. Absolute glare factor and relative glare factor based metric: Predicting and quantifying levels of daylight glare in office space. *Energy Build.* **2016**, *130*, 8–19. [CrossRef]
67. Purves, D.; Augustine, G.J.; Fitzpatrick, D. Anatomical distribution of rods and cones. In *Neuroscience*, 2nd ed.; Sinauer Associates: Sunderland, MA, USA, 2001. Available online: https://www.ncbi.nlm.nih.gov/books/NBK10848/ (accessed on 15 January 2023).

sustainability

MDPI

Article

Design Thinking in Lighting Design to Meet User Needs

Kamil Kubiak

Lighting Technology Division, Electrical Power Engineering Institute, Warsaw University of Technology, 75 Koszykowa St., 00-662 Warsaw, Poland; kamil.kubiak@pw.edu.pl

Abstract: In the age of noticing the adverse effects of our activities on the climate, we pay more and more attention to designing sustainable solutions. Assumptions made during such sustainable designs often do not fulfil user needs. Therefore, we frequently encounter discrepancies, such as higher energy consumption and user modifications in the building systems compared to the design and use stages. To mitigate these problems in the lighting design context, the author describes the proposed lighting design method based on the Design Thinking methodology and concerns the commonly applied basic lighting design process. As an example design, an office room interior and lighting design are presented using the Design Thinking method described by the author. The utilised method allowed us to learn about problems that are only sometimes typical for offices and enter them into the design scope. The article contains a detailed analysis and comparison of normative requirements with users' needs, collected from four room users and six guests. The discrepancies between the sustainable design and use stages presented in the article can be considered reduced, and users' interference in the operation of devices will be eliminated due to meeting users' needs and thus achieving sustainable goals.

Keywords: interior lighting; lighting technology; lighting design; design thinking; sustainable design

1. Introduction

In the age of noticing the adverse effects of our economic and social activities on the climate, we pay more and more attention to designing sustainable solutions [1]. The World Business Council for Sustainable Development states that buildings are responsible for at least 40% of energy use in many countries. Lighting accounts for 20–25% of the total electrical use in buildings [2,3]. Worldwide building energy consumption is expected to grow by 45% over the next 20 years [4]. To mitigate this adverse trend, we create various systems that encourage the design of such solutions and try to verify the designs made to meet sustainable design requirements [5]. For example, in 2002, the Energy Performance Building Directive (EPBD) announced new regulatory conditions for all EU countries to decrease the energy needed for heating, cooling, ventilation, and lighting in buildings. Similar LEED-certified (Leadership in Energy and Environmental Design) buildings act in such a way. However, each design requires many decisions and design assumptions. The adopted assumptions seem correct based on the designer's knowledge, applicable standards and regulations, and good practice. However, these standard assumptions sometimes do not fulfil user comfort and satisfaction [6,7]. Because people spend most of their time in buildings and interact with building systems to reach desired comfort, these interactions lead to different energy usage [8–11]. The impact of user actions on various systems, including cooling, heating, etc., has been tested many times [12–14]. For example, energy-efficiency behaviours account for 37% of the variance in electricity consumption between dwellings [5]. Therefore, we observe discrepancies between what we thought we designed and what we designed [12,15–26]. Those discrepancies are more significant for low-energy buildings with more passive design features [27–29]. These discrepancies are the user's emergence and activity; in the literature, it is known as occupant behaviour (OB) [8,11,30–35]. Attempts are being made to reduce this discrepancy and uncertainty [35–38]. Although the user can be specified, it is individual, unusual, and unique [16,39–44]. Many models have been developed to describe the interaction of users with energy-consuming systems, but they are not sufficient [15–17,30,32,35–37,40,41,43–46]. The failure to understand users leads to various consequences, among which the use of the designed device can be mentioned. The developed device's different services cause other effects and results [15]. These consequences may be, for example, electricity consumption, which directly indicates the level of energy efficiency and the assigned certificates, points, etc. Hirst and Goeltz [44] found that less energy was saved than was predicted by an audit.

In every design process, we always have the designer on the one hand and the user on the other. The user waits for a design and has a specific need or problem to be solved. Using his knowledge, standards, and experience, the designer has to optimally solve this need or problem. Standard, precise, repeatable design solutions proposed by designers are becoming less and less sufficient [26,45,47–49]. People and their problems are not as repetitive and predictive as designers believe, and technology might not be used as designers intended [50]. Therefore, there is a field for methods of supporting the design process that will allow reconciling the point of view of the designer and the user. Also, due to the increased awareness of users and the complexity of design undertakings, the need for mutual communication between designer and user, including dialogue, is constantly growing [26,28,45,47–49,51]. Communication is crucial not only in the designer-user relationship but also within the design team. In such a way, "designers are partners with the problem owners" [52]. Users' knowledge helps them to look at the problem differently, generating various designs and concepts for the design team [42]. These led us to move from technology-centred to user-centred methods [53]. Marcolino et al. claim that the early stage of design decision-making significantly impacts the energy consumption of buildings [15]. Thanks to such reconciliation between the parties, the proposed and implemented solutions will be entirely accepted and fully used by the user [54,55]. The user will not experience difficulties in operating and understanding the solution. The user will use all functions of a given solution and, therefore, energy will be consumed as designed. Otherwise, when the user is dissatisfied with the solution and his environment, changes will be made in the system; for example, Reinhart and Voss [56] reported that in 88% of attempts to close the blinds automatically, the users override the control algorithm. Nicol and Humphreys [57] noted, "If a change occurs, such as to produce discomfort, people react in ways to restore their comfort". There is evidence that tailoring the design of a building's elements and systems around occupants may reduce the energy consumption of buildings and increase occupant satisfaction [2,58]. User-centred design (UCD), participatory design (PD), usability engineering (UE), and user experience design (UX) are practical approaches to improving designs based on end-users' needs [42,59]. It is already known that lighting is a significant factor influencing occupants' performance in indoor environments and other aspects [60]. Therefore, lighting will be the scope of the article. So, it is crucial to design lighting in the manner mentioned above.

This article presents the adaptation of the Design Thinking tool for lighting design, based on the example of office lighting, as a tool for designing around users and their needs. The versatility and transparency of this method can ensure its success in lighting design, as is the case in other areas [61–66]. According to the author, this will reduce the discrepancy and uncertainty between the assumptions in the design and actual use, which are linked to energy consumption. Thanks to this approach, we will also experience less user interference in the lighting equipment and its functioning due to dissatisfaction. Thus, we could reduce energy consumption for lighting, which is around 1/6 to 1/5 of worldwide electricity production [67].

2. Methods

2.1. Design Thinking Approach to Design

Design Thinking is a method of creative problem-solving and a tool for creating new products and innovative solutions based on a broad understanding of the problems and needs of users. Tim Brown, CEO of IDEO, presented the most accurate definitions of this method: "Design thinking begins with skills designers have learned over many decades in their quest to match human needs with available technical resources within the practical constraints of business" [62,63]. In Design Thinking, users and stakeholders are involved in the design process from the beginning. Design teams are composed of people with different competencies and roles. Considering people's uniqueness, there is a potential for innovative solutions because they are not standard but individual, unusual, and unrepeatable [65]. This was the reason for the creation of this method in the 1980s and 1990s at Stanford University in California: a need to transfer creative and innovative ideas to the Silicon Valley business world. This trend is still valid worldwide.

Design Thinking is, therefore, the design process, with the main goal of meeting the needs of future users of the designed solution [68–75]. However, for the fulfilment of these needs to be feasible, it must be ensured that the effects obtained using this method are cost-effective, innovative, and technologically feasible [76,77].

Therefore, the main assumptions of the Design Thinking method are:

- focusing on the user;
- creatively looking at the problem from different perspectives;
- prototyping and testing solutions to check the correctness of design.

Due to the universality of Design Thinking, it can be used to solve problems wherever there is no one obvious solution. Such issues happen in engineering, e.g., lighting design. The problems we face in lighting also require a combination of different competencies, i.e., technology, design, business, physiology, or psychology. Considering the above, the potential of the Design Thinking method in lighting design is justified. To sum up, Design Thinking can be used when designing lighting that considers users' needs.

Every design is different, but some general activities are common. The Design Council has developed the 'Double Diamond' model to illustrate this (Figure 1) [78]. The model is divided into four phases: Discover, Define, Develop and Deliver. It maps how the design process passes from points where thinking and possibilities are as broad as possible to situations where they are deliberately narrowed down and focused on distinct objectives (see the vertical extent in Figure 1). We can adopt the 'Double Diamond' model in our lighting design field.

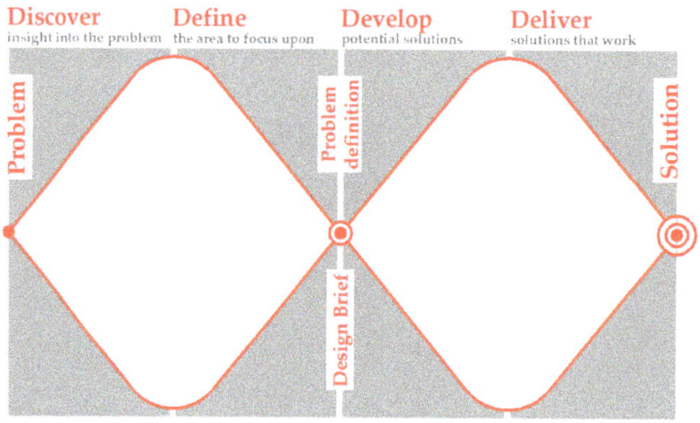

Figure 1. Design Council double diamond model [78].

The double diamond, in more detail, looks as follows (Figure 2a): The Discover phase (the start of a project when the initial problem is known) is a period of discovery, gathering inspiration and insights, identifying user needs, and developing initial ideas for designing a solution to a problem. After that is the Define phase, in which designers try to make a clear brief that frames the fundamental design challenge that constitutes an actual problem. After having a design brief, we move to the Develop period, where we develop solutions and make prototypes and tests. Each is iterated and constitutes a process of trial and error to help designers improve and refine their ideas. The final part of the double diamond model is the Deliver phase, where the resulting product or service (working solution) is finalised and launched. The whole process is iterative, and each stage is also iterative. Each iteration aims to move closer to the final solution [78].

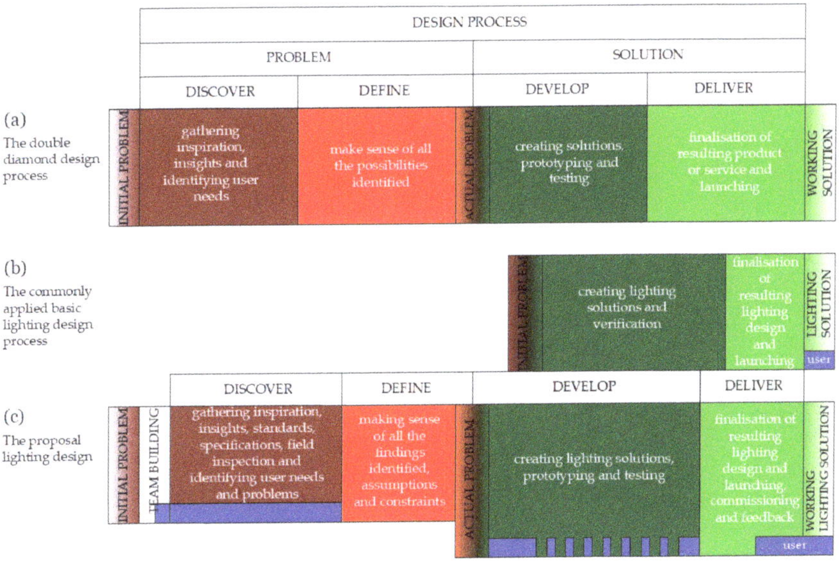

Figure 2. Design process models.

2.2. The Commonly Applied Basic Lighting Design Process

Poland's commonly applied basic lighting design process in a new or modernised facility is similar. It is a part of the entire facility design process related to the electrical installation design. The design office designs independently or by order, and an external company carries out a lighting design based on the technical documentation of the facility and the architectural concept. The electrician, architect, and supervisors then verify that the lighting design proposal presented by the lighting designer meets the requirements. If the conditions are not met, corrections are made. After the final acceptance of the concept, the lighting design, which constitutes the lighting solution, is attached to the facility design and forwarded for implementation. (Figure 2b).

The basis for implementing a lighting solution is the law that imposes the obligation to provide the facility with a lighting installation, e.g., to ensure lighting conditions in the event of a mains power failure, and the specifications of a given project [79,80]. An additional aspect often emphasised is the energy efficiency of the lighting solution [5,81]. The above is an initial problem that the designed lighting intends to solve (Figure 2b).

To find a solution to this problem, the lighting designer takes into account, in addition to the appropriate lighting standard and applicable normative acts, the architect's requirements. However, relevant standards [79] usually give the limit values of lighting parameters without considering whether the values of the lighting parameters fit into a given situation. Moreover, when creating designs, it is often overlooked that lighting is not to be used only as a technical (as for electricians) and decorative (as for architects) element but as one of the elements enabling the proper functioning of the people who use it. As a result, it is forgotten that each user reacts differently to the lighting conditions defined by the designer and has different lighting needs. We can see it in the example of the selection of the colour temperature of light sources. Some people prefer a given room to be lit with a cold colour temperature, and others prefer a warmer one. Such differences in preferences apply to each of the defined lighting parameters. So, when we look at lighting design from the perspective of users' needs, we are faced with a challenge in terms of design. It becomes insufficient to design solutions by a small group of specialists in a given field, who often base design decisions solely on their experience, proven solutions, and habits. Moreover, the constant change in market trends is in opposition to the craftsmanship that the designer achieves. Also, users' attitudes that question solutions that do not suit them based on the knowledge gathered from various sources due to free access to unlimited information deepen the situation. So, an excellent way to deal with the problem is to include its future users in the design process. Therefore, it is vital to consider specifications, legal aspects such as standards, and actual user needs, whether interior lighting or road lighting. We can achieve this effect by basing the design process on the Design Thinking method.

The lack of such an approach currently being used results from the nature of the entire design process, which is characterised by high complexity [82]. Therefore, all optional elements are eliminated, i.e., they do not lead to significant errors or omissions that could cause failures, damages, and legal disputes. Failure to meet the user's needs in a given case has no direct or negative consequences. This fact should be considered regarding user comfort and work efficiency, which are complex to assess and ambiguous during design and at the facility's start-up time. Each person is also subject to the process of adaptation to the prevailing conditions. Nevertheless, this leads to adverse effects in the long run. Moreover, outdoor lighting is not generally treated in such a way that its selection and operation can be influenced. Even simply turning the lighting on/off translates directly into the working conditions and energy consumption.

Many people are involved in the design process, from identifying the initial problem to implementing the final lighting solution. These include lighting designers, architects, industry designers, investors, contractors, and users. The presence of these people in the process, the activities performed, and the moment of their action vary. From the perspective of the subject of this article, what is significant about this commonly applied basic lighting design process is that the user appears only at the end of the entire process (Figure 2b). This is the moment when the lighting solution is already implemented (made, installed, and launched). After entering the facility and starting work, the user needs to change the lighting setting, e.g., due to the approaching dusk. They also become familiar with this solution and learn how to use it. Unfortunately, due to the designer's lack of knowledge of the actual problem, there is a real risk that the lighting, due to its parameters or their selection via the control interface, will not allow for the creation of lighting conditions convenient for a given person, taking into account their preferences and needs. An attempt to "repair" this lighting will lead to various actions, e.g., changing the workstation, purchasing an additional desk lamp, constantly closing the blinds, turning off or covering luminaires that cause glare, having disputes with neighbours about the lighting settings, etc. All this constitutes the problem described in the introduction, which translates into different, often higher energy consumption than assumed in the design, user modifications in the lighting system, and, in general, discrepancies between the design and the usage of the installation, in this case, the lighting installation. This article proposes a solution to this problem, which will be described in the next chapter as a proposal for a lighting design method.

2.3. The Proposal Lighting Design

Considering the lighting design process and its characteristic features described above in the commonly applied basic lighting design process section, a proposal can be made to modify this process. The modifications will aim to eliminate the problems described above. To achieve this, the Design Thinking design tool will be utilised. So, the lighting design process will now be presented as a modification of the currently used process using the Design Thinking methodology (Figure 2c). Industry organisations bringing together lighting designers, such as the Illuminating Engineering Society (IES), the International Association of Lighting Designers, or the Professional Lighting Designers' Association (PLDA), promote the design process in their works, courses for designers, and materials that have features in common with the one proposed by the author. However, a different methodology was used in this work based on Design Thinking, and the resulting differences in approach concern, among others, deep consideration of the end users of the lighting solution.

As before, the reason for implementing a lighting project will also be the applicable law and specifications of a given project and, of course, the energy efficiency of the solution. This will be an initial problem. However, when solving this problem, a lighting designer will consider the appropriate lighting standards, applicable normative acts, and the architect's requirements, specifications, and experience. He will act in a slightly different way and will take action to find out the actual problem.

Initially, a design team of the previously mentioned people is appointed, i.e., a lighting designer, architect, industry designer, investor, and contractor. In addition, the design team will be joined by people such as exemplary users, people responsible for maintaining the lighting installation, specialists in the physiology of vision, and any other people with competencies in the field of lighting design, its impact on humans, and who use, manage, and maintain the lighting installation. Then, a critical stage begins, distinguishing the proposed lighting design process from the currently used one.

First stage—Discover

The first stage that starts the entire process is the Discover stage. This stage is one of the most critical stages of the Design Thinking method. It aims to reach future users of the lighting installation to learn about their needs and actual problems. Various internal, external, individual, and contextual factors influence users' behaviour in a space. Therefore, it is necessary to study them in a multi-disciplinary manner to incorporate many elements such as sociology, psychology, economics, engineering, and design perspectives [9,12,16,83]. To avoid this ambiguity, a good way is to reach future lighting users and examine them in the context of lighting. Such an approach is proposed in this article.

Nowadays, a common practice for many lighting designers is to create solutions based solely on standards, applicable normative acts, specifications, and their own experience, and to define a given problem subjectively [34,40]. Lighting and its performance are strongly associated with users using it individually [47,84,85]. Therefore, when developing solutions with this method, the first stage will be conducting observations, talks, or surveys with users to get to know the actual problem. The result of this stage should be a thorough understanding of the user for whom the final solution is being created. It would help look at the world from the perspective of those that will ultimately use designed lighting. The lighting designer, therefore, will know how the client might be tempted to alter the design and will be more successful in meeting the client's needs [40]. This is a challenging task.

First, it is essential for the lighting designer to observe the users for whom lighting will be designed in their natural environment. Such a method, called in situ observation, offers first-hand evidence of users' behaviour [86]. The lighting designer's observations should examine users' lighting preferences at different times of the day. It is worth paying attention to whether, for example, a given user chooses to work with electric lighting or natural lighting in the morning when starting the day. One should also monitor if the light does not cause glare and the associated feeling of tiredness and sleepiness. The lighting designer should also acquire some subjective impressions and feelings. Thanks to this, he can determine what lighting elements are missing from his perspective. For example, we can sometimes observe that a luminaire is covered with cardboard, so it does not shine on a projector screen. These activities directly prove unrealised lighting needs. Carrying out these activities with the people using the space allows us to obtain information about their life, disposition, age, feelings about the existing lighting installation, and the nature of their work or the time spent in the space during the day. It is critical to determine what level of illuminance, colour, or illumination uniformity users prefer. We should find out their needs regarding the type of lighting—whether they prefer general lighting or maybe localised lighting directly aimed at their workplace. The entire stage allows us to learn about users' needs, features, opinions, and current ways of dealing with lighting. The above activities can be performed most easily in lighting modernisation. However, in the case of new investments, we do not always have access to future users.

Next, we must conduct interview surveys [12,27,66,86–88]. Because of user behaviour diversity, we need an adequate sample to represent the survey population [56,89,90]. During such tasks, both standard and nonstandard users are essential. Nonstandard users allow us to obtain information that we do not even know. Such a user can be a cleaning service person. Although such a person only stays in this area occasionally, he performs essential user activities. In conclusion, we want to find out what the problem is here. This is not only necessary from the perspective of the investor and architect but also, and perhaps mainly, from the perspective of the future user of the lighting installation.

We have many tools that can be used at this stage, for example [76–78]:

- Safari service—an experience of the designed lighting environment by the lighting designer himself. The designer collects information about lighting, uses it, and records his feelings and observations. This activity enables one to understand lighting from the user's point of view, thus raising awareness of strengths and weaknesses.
- User shadowing—the lighting designer observes the user and his environment while using lighting. The observation aims to gather as much knowledge as possible about users' needs and understand their behaviour.
- Interview surveys—personal interview surveys are used to test the user's responses and simultaneously to notice his behaviour.
- The "5 × WHY" technique allows for determining the cause (source) of specific user needs. It examines cause-and-effect relationships leading to the heart of a given problem by asking questions that begin with "Why". This method is also helpful in other stages.

Second stage—Define

In this stage, we need to define the problem that needs to be solved. Following the previous stage, we analyse the responses and observations obtained during the first stage (Discover). In this analysis and synthesis, it is recommended that methods and techniques such as those suggested by Ideo [91] and Stanford [92] be used. It all constitutes essential information. With the help of this information, we can check whether the reason for designing the lighting installation presented at the beginning of the process covers actual problems relating to future lighting users. We should answer the question of what the user exactly needs. How can these requirements be met? It is worth remembering that specific needs are searched for at this stage, not solutions to the problem.

In the second stage, the previously defined problem might appear differently than it is. For example, the initial problem may be changing the layout and luminaires to achieve a transparent effect in a room [93]. After we have asked users in the first stage what they think is causing this lack of transparency, and within different answers, it may turn out that the problem does not lie with lighting and is generated, for example, by the wrong arrangement of office furniture or the lack of access to windows. In this situation, there is a need to redefine the design problem and return to the first stage (Discover), focusing on the need for more access to natural light or furniture arrangement. The multiple redefinitions of the design issue are also the basis of the Design Thinking method. This approach gives the lighting designer the confidence to meet the actual needs of users—not his own. Thanks to such redefinitions, the newly formulated problem fully defines what needs to be done and what the user needs. Therefore, the expressed problem is likely concerning lighting and other aspects, e.g., furniture arrangement, which should be considered in the design (wider solution part of the diagram in Figure 2c). We can then approach lighting design holistically, combining the lighting system's design with other systems. The proper definition of the problem will allow the design of a lighting system that users will ultimately use without any further alterations and with designed energy consumption.

As in the case of the Discover stage, various techniques can be used to define the problem. We may use here [76–78]:

- Empathy map—a way to profile users. Most often, it is made in a diagram divided into six areas. Each area answers different questions: What does the user think and feel? What does the user see? What does the user hear? What does the user say and do? What are the fears and frustrations of the user? What are the needs?
- Creating a persona—creating a non-existent character treated as a representative or archetype of the user group for which the lighting solution is being created through his actions and approach [54].
- Problem mapping—allows us to create an initial outline of its solution by establishing:
  - the owner of the problem, i.e., who the problem concerns;
  - the sources of the problem, i.e., where the problem came from, what are its causes;
  - the symptoms of the problem, i.e., how do you know about the issue;
  - support, or how we can help to solve a given issue.

Third stage—Develop

Having a correctly defined problem, we can move on to creating potential solutions to the problem, which have our user (persona). As is well known, any technical issue can be solved in many ways. The Develop stage is about presenting possible solutions from very different perspectives. This stage occurs in groups and consists of generating as many ideas as possible for solving a defined problem and selecting the best one for implementation according to cost-effectiveness, technological feasibility, sustainability, etc. [54,88]. Ensuring various participants in this stage [31,42,54,94,95] is crucial. Therefore, the Design Thinking method assumes working on a project in interdisciplinary teams. The participation of lighting engineers or architects is essential, but there are also people with little in common with lighting design, e.g., the user or even such people as the cleaning and maintenance service. This allows us to create new and unconventional ideas and, thus, innovative designs aligned with Design Thinking.

The brainstorming method here is recommended and often used in business [78]. Thanks to this technique, each person participating in the project can provide their ideas and develop other people's ideas. However, it should be remembered that, like any technique, brainstorming has specific laws. Compliance with these laws is crucial to achieving the intended goal. The basic principle of brainstorming is knowing that every idea is good. No statement should be criticised—it can be modified to be as creative as possible. In addition, this stage should result in many ideas—a small number will not lead to an optimal solution. Finally, the design team democratically selects the best concepts according to defined criteria, and we can go further on their basis.

However, before implementation, it is essential to prototype the selected solution. The key is to present the solution to the user as soon as possible to verify the design and obtain feedback. When it turns out that we made a mistake in the previous stages, we save time and money on its correction. Regardless of the final result of the work, the prototype must have certain features: be low-budget, simple, and time-saving, and be as close as possible to the target solution [76–78]. It can be a visualisation from a lighting design program such as DIALux or Relux, as often happens when consulting the proposed lighting solutions with the architect and investor. The role of virtual and augmented reality is gradually increasing, which allows us to see the designed space after putting on special glasses [96–100]. Thanks to this, we can present the room model, luminaires, and their arrangement, and most importantly, we can show the visualisation of the lighting effect. However, a prototype can also be used in field trials or demonstration designs, where it is possible to present solutions that can become part of the final design. In lighting design, physical models, such as mock-ups or technical drawings, will not meet their assumptions because it is difficult for users to visualise lighting effects. It is essential that the user can see and imagine what his future lighting will look like.

Immediately after the prototype is made, it is presented to the users, and we ask them for their opinions. This allows for verifying the fulfilment of the needs they define. During tests, it is necessary to explain the operation of the lighting installation. When the user is watching or trying out the effect of the work, we listen to his opinion and ask questions that will help improve the final result. We should refrain from resolving or undermining users' choices during the talks. The created lighting serves the users and cannot be based on the impressions of third parties. We should give them the freedom to get to know the solution and express their opinion. After performing the tests, the comments received from the users should be considered. Testing makes it possible to check whether the entire process has been carried out correctly and whether the users will be satisfied with the final lighting solution. It is crucial to determine whether the solution meets their requirements and what changes should be introduced. The prototype must be improved until the users are satisfied with it. This stage guarantees that users will fully use the selected and implemented lighting concepts as designed. Therefore, there will be limited discrepancies between what we have developed and what we have obtained, and interaction with building systems to reach the user's desired comfort will be limited. Finally, energy consumption will be as determined [12,15–26].

Fourth stage—Deliver

Only when the user's opinion is satisfactory can we proceed to the last stage—design delivery. During this stage, final design documentation is created. Then the documentation is sent for implementation. After the entire lighting installation is completed following the design, it is launched by the contractor. At this point, all lighting location adjustments are made, and parameters are set in the controllers. Then, after the entire interior is finished and the user moves in, there is an essential time during which the user is familiarised with the lighting installation. The operation of the entire installation, control system, and possibilities are presented. In addition, the user is informed of how the lighting installation works, for what purpose, and why. After some time, when the user gains experience using the lighting installation, the feedback phase begins. The user then shares his observations about the operation and service of the lighting installation. This is when corrections can be made, e.g., the orientation of luminaires and control parameters. Only after such a series of actions can we be sure that the lighting installation will work as it should, meet users' needs, and work optimally.

To summarise, the lighting design process presented using Design Thinking stages should be carried out considering that they benefit users (personas). Therefore, the user participates in many project stages (Figure 2c). In each part of the work, one should deviate from the usual patterns and approach the defined problem as creatively as possible. Doing all the stages one by one does not guarantee success, and we may need to go back to one of the previous stages. The stages of creating lighting using the discussed method can be compared to the closed circle shown in Figure 3. The design team moves along this circle, creating an increasingly innovative solution to a given problem.

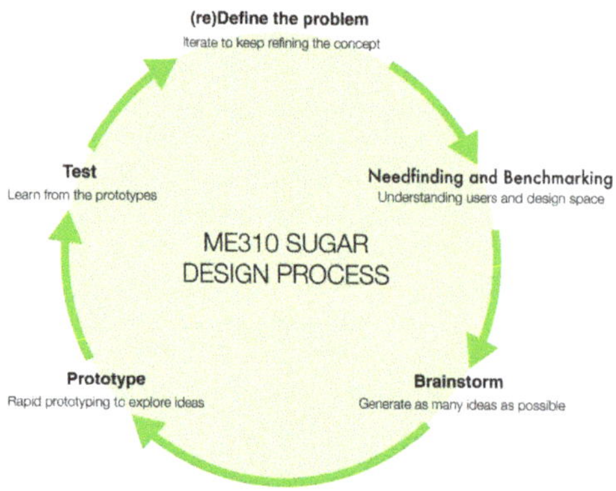

Figure 3. Design Thinking stages and their sequence based on the ME310 Sugar Design Process at Aalto University, Finland.

3. Procedure and Results

The problem that this article is about is the frequently encountered discrepancies between the design and the use of the lighting installation in a building, manifested, in particular, by different, often higher, energy consumption during the use of a building than what was assumed in its design and user modifications to the system during its use.

The author proposes using the Design Thinking method when making lighting designs to solve the above problem in the lighting area. The general form of the Design Thinking method has been described. It describes how lighting designs are carried out and why changes should be introduced. On this basis, the author modified the lighting design process currently in use using the Design Thinking method.

Since the proposed approach to lighting design following the Design Thinking method and its use in lighting design have yet to be described in the scientific literature, this chapter will present an office room whose interior and lighting will be the subject of a project according to the described method. Due to the study nature of the project, not all stages of the process shown in Figure 2c will be used. Nevertheless, it will allow us to draw several valuable conclusions.

The office room for which the lighting design will be developed has the shape of a cuboid 5 m long, 6 m wide, and 3 m high. The room has a suspended ceiling with recessed luminaires and access to four windows with blinds. There are four workstations and five wardrobes in the room. The floor in the room is covered with a grey fabric carpet, and the walls are white. The office room is currently very untidy—it has not been renovated for over 15 years, and the walls and carpet are significantly dirty. Figure 4 shows exemplary photos of the room.

Figure 4. Photos showing the interior of an office room.

The existing lighting installation in the room consists of nine identical raster luminaires with linear fluorescent lamps with electromagnetic ballasts recessed in the suspended ceiling. There is no lighting control system in the room. All luminaires are manually turned on simultaneously. It is also worth noting that the lighting installation has not been changed or renovated since the building was built.

In terms of energy, annual electricity consumption should be determined. However, we need to estimate the expected annual electricity consumption if we do not have actual electricity measurements throughout the year. For this purpose, the installed lighting power in the room should be determined (1). It is based on the following audit:

$$P_i = \text{quantity of lamps in luminaire} \cdot \text{power of one lamp} \cdot \text{quantity of luminaires} + \text{quantity of ballasts} \cdot \text{power of one balast}$$

$$P_i = 4 \cdot 18 \cdot 9 + 18 \cdot 10 = 828 \text{ W} \quad (1)$$

Then, we need to assume the operating time of the installation throughout the year. The lighting can only be turned on completely, making calculations easier. For office rooms, their use is assumed to be 2250 h per year [81]. Hence, the annual electricity consumption will be (2):

$$E = 0.828 \cdot 2250 = 1863 \text{ kWh/y}, \tag{2}$$

The above level of electricity consumption per year is only an estimate. We need to know how users use lighting to get closer to the actual value. The Design Thinking method can also be used for this purpose.

The primary users of the office are office workers working at computer stations—a total of four people. Employees from other rooms and company buildings also occasionally appear in the room. The nature of the work performed in the room can be considered mental work.

Initial problem—the office room has not been renovated for over 15 years, and therefore, the manager plans to modernise (1 to 1) the lighting and refresh the room (paint the walls and ceiling). The manager will do this without the informing or consulting with employees (room users). He will commission an electric company that he is familiar with to perform the work.

In addition to the lighting designer, architect, electrical installation designer, and company owner, the design team also included a representative of the contractor, four regular users, six people visiting the room, a room service technician, and a person from the cleaning service.

The Design Thinking method research tools presented in the described lighting project will only serve as an example. Their purpose will be to enable the designer to learn about the lighting needs and preferences of the users of the designed lighting. As a result, it is expected that the problems mentioned in this article will be reduced.

3.1. *First Stage—Discover*

According to the method description in Section 2.3, the designer will first become acquainted with the space covered by the design and its users. An exemplary research tool called user shadowing and service safari will be used for this purpose.

The user shadowing method was the first used when working with the respondents. The idea of this tool is to observe users while using the product—in this case, the lighting available in the room—without interfering with their behaviour. During the designer's several visits to the room, two research groups were selected among the users. The first group consisted of four employees who stayed in the office room regularly (every day), sitting at their workstations. The second research group consisted of six employees who remained in the room occasionally. Observing their work schedule, it was possible to learn the typical behaviours of people working permanently indoors (the first research group). Of course, each employee has a slightly different action plan based on the tasks assigned, depending on the day. Despite this, some similarities were noticed during the observations.

In terms of room lighting, a fascinating situation was noticed. On a sunny summer day in the office, almost all the blinds were closed, and the electric lighting was on all day. All lighting was turned on, allowing for the completeness of the installation to be checked. Two fluorescent lamps in one luminaire were not working correctly—flashing periodically. This caused repeated irritated glances from users in this direction. Also, during the lunch break, the lighting remained on when users went to the canteen together. A similar situation always occurred among people visiting the room. They were wondering where they could sit and talk to the room's workers. Finding a place to do the work was challenging when there was a need to display materials on a computer screen. In the case of remote presentations, users used a webcam but with a background, often depicting the attractive interior of a modern office. It was noticed that the room's users changed their positions in their armchairs very frequently, especially in the afternoon. Employees adopted various sitting positions—from upright to hunched.

The next tool used at this stage was the Safari service method. In this case, the lighting designer took on the role of the room's user and had the opportunity to better understand all the behaviours, impressions, and interactions with the light in the room that he had previously observed. He was able to spot additional issues that he might not have seen when watching users as a third party. During two full days of work in the room, the designer occupied two characteristic workplaces, one for each day, each located differently around the windows. Together with the regular users of the room, he fulfilled similar duties and tried to behave similarly. Thanks to this, he could feel like a user of the room, which allowed him to examine what, in his opinion, the respondents' needs were and how they could be satisfied. Of course, the observations and feelings were subjective, but they provided another layer of information necessary for the design. The first thing the designer noticed was the pollution of the room, poor technical condition, and mess, e.g., documents in various places. By working in different areas of the room, the designer could learn about the lighting conditions on a sunny day. Opening the blinds at workstations with the monitor facing the window made working impossible. The luminance contrast between the exposed windows and the monitor in their background was too high for a visually comfortable position in places facing opposite directions.

Based on the two previously presented methods, the designer had the opportunity not only to get acquainted with the environment for which he will be carrying out the project but also to define certain typical behaviours and reactions specific to a given interior and its users and certain situations occurring in the room. The next step in understanding users' needs is to learn about their opinions, emotions, and motivations. The following research tool was used to obtain such information: user interviews [86].

Interviews were conducted with everyone employed by the company that stays in the room regularly (4 persons) and stays there occasionally (6 persons). The research was performed individually, i.e., one researcher + one subject, in the room being the subject of the study. The duration of the examination for one person was approximately 20 min. After this time, the test person was changed, with one person tested per day. The tests were carried out in a non-invasive way. At the beginning, the subjects agreed to participate in the study. All people were healthy adults (admitted to work based on a medical certificate). The interviews were divided into two parts—the first one concerned general information and was conducted in the form of a questionnaire, completed independently just before starting the second part—the primary interview. The survey asked about age, eyesight condition, position held, seniority, activities performed, and time spent in the designed room. The second part of the interview was conducted in the form of a live conversation. The person who conducted the interview was the moderator and only had an idea of the talk's direction. The interview was conducted as an example conversation scenario using the questions below. During the interviews, the discussion depended mainly on the respondent, and the questions and their order were adapted to the given person and the answer.

1. Are you satisfied with the work you do?
   Why do you get this impression?
   Are these the only reasons?
2. What is your job?
   What tools do you use while working?
3. Are you satisfied with the room in which you work?
   Why are you satisfied/unsatisfied?
4. What does your dream office look like?
   What would you like to have in it?
   Why does such an office meet your expectations?
   What benefits do you see in such an office?
   What would it affect?
5. What do you feel when you are in the room?
   Why do you feel this way?

If it is variable, what does it depend on?
How do your feelings change after leaving the room (during breaks or at the end of the working day)?

6. Do you think you have enough space in the room?
If yes/no, why do you feel this way?
What do you think? What influences your feelings?
7. Do you think your workplace is well-lit enough?
Why do you think so?
What can you say about the lighting of the entire room?
How would you like your workplace/room lit?
Do the blinds fulfil their functions?
8. Do you think there should be daylight in the room?
Are you satisfied with the amount of daylight in the room?
Why do you think there should be more/less of it?
What impact does adequate daylight have on you?
9. How many hours a day is artificial lighting on?
Does this change throughout the week?
Does it change with particular seasons?

After collecting user information, the designer familiarised himself with the lighting standard for office interiors [79], applicable normative acts, and the investor's specifications. Moreover, he was inspired by photographs of modern office interior design solutions. All this provided an extensive collection of diverse information that describes the design situation from different perspectives. The order used—user first and then standards and specifications—is purposeful and very important here. Thanks to this, the first image of the situation in the designer's head comes from the user, not the standards. It thus remains permanently in memory, giving this information significance. Therefore, knowing the user's needs, we understand why we should use specific technical means. We know their justification, unlike generalising norms.

Completing the first stage of the Design Thinking method allowed us to collect information about the users, the room, and the interactions. Based on them, it was possible to move to the next stage, after which the users' needs would be determined and the resulting actual design problem, referred to as a design challenge in the Design Thinking method, would be defined.

3.2. Second Stage—Define

After all the interviews with the users and observations, it was crucial to create a user persona. A user persona is a character that embodies all user research in an easily identifiable and understandable form. It brings together lots of information about the users to create a single character that represents the group. In this case, since we have two different groups of users—regular and occasional—there was a need for two personas. Such personas cover name, age, occupation, where they live, family, hobbies and interests, likes and dislikes, and most importantly, needs. Among the persona's information, the most important for the project were those related to interior design and lighting (Table 1).

After analysing all user information using the previously mentioned research tools, it was concluded that the office room should be renovated and refreshed in decor. The identified user needs can be divided into two groups: the first group is the needs related to the new interior design, and the second group is the needs directly associated with replacing the existing lighting installation.

Table 1. Personas representing two groups of room users.

Persona	Profile and Mindset	Needs and Aspirations
Peter (regular user)	He is 30 years old; married. A Service Engineer who stays indoors every day. Satisfied with the work performed and with the opportunities for professional development. Irritated by the condition of the room—dirty, messy.	Opportunity to develop. He needs to have a sense of privacy. Comfortable place at the desk, nice interior, visual comfort. He likes the entrance of daylight into the room. He appreciates the ability to adapt the lighting to current needs and situations.
Anna (occasional user)	She is 40 years old; has a husband and a son. A Service Manager who visits the room more than twice a week. She values her high position and the influence she has in the company. She likes freedom of action.	Ease of communication, exchange of views, discussion, and teamwork. She needs a dedicated place to work in the designed room. She likes the entrance of daylight into the room. She needs good lighting in warm colours throughout the entire space.

The most crucial interior design needs are:

- Furniture replacement—more wardrobes and chests of drawers to store items. Rectangular and slightly larger desktops. Adjustable desks and chairs so the employee can decide in which position he wants to work at a given moment. The furniture has bright colours, but its arrangement requires changes to improve communication between employees while maintaining the sense of privacy provided by the current furniture arrangement.
- Supplementing the room's equipment with furniture, enabling the teamwork of the people working in the room with external users. Additional chairs or a sofa and a table where visitors could work.
- Inserting elements that enable the presentation of work results during internal meetings or client conversations. A magnetic board or TV screen allows employees to share their ideas and projects with other team members more quickly.
- Bringing the room to an aesthetic condition (renovation) that will motivate users to perform everyday tasks.
- Room decor that looks good. Plants, paintings, or photo wallpapers can be added to decorate the room's walls.

The most crucial lighting needs are:

- Changing the light sources in the luminaires—the previously used fluorescent lamps negatively affect concentration and irritate the room's users.
- Changing the colour temperature of light sources—the users claim that the light produced in the room is too cold.
- Introducing a lighting control system with the ability of dimming the luminaires—the respondents who use the room want to influence the lighting conditions in and at their workstations and adapt them to the weather outside the building.
- Changing the type of lighting at workstations—apart from general lighting, employees also need luminaires to illuminate their workstations.
- Illumination of the walls and ceiling.
- Highly durable luminaires (light sources)—employees often request to replace burnt-out fluorescent lamps. Unfortunately, based on their statements, it can be concluded that these requests are very rarely fulfilled.

To create a lighting design using the Design Thinking method, in addition to meeting users' needs (personas) examined at the Discover stage, it is also necessary to consider the parameter values listed in the standard [79] and other applicable normative acts that were not in the scope of this work. Therefore, Table 2 compares the normative lighting criteria with the critical opinions of the users (personas) in the interviews. Then, the resulting parameter values for the design are presented, along with their justification. The designed room was given an area type—CAD workstations, conference rooms, and meeting rooms by [79].

Table 2. Comparison of standard values of lighting parameters with user opinions.

Lighting Parameter	Normative Value	User Opinion, Justification	Value Selected
1. Average illuminance	Task area $\geq$ 500 lx Immediate surrounding area $\geq$ 300 lx Background area $\geq$ 100 lx	Users from the first research group agreed they had enough "amount of light" at their workplaces. The measurements showed a higher level than required by the standard, so the requirements increased.	Task area: $\geq$750 lx Immediate surrounding area: $\geq$500 lx Background area: $\geq$200 lx
2. Lighting uniformity	Task area $\geq$ 0.6 Immediate surrounding area $\geq$ 0.4 Background area $\geq$ 0.1	The measured values are much higher than those recommended in the standard. This requirement has also been increased to match the accepted values.	Task area: $\geq$0.7 Immediate surrounding area: $\geq$0.5 Background area: $\geq$0.2
3. Illuminance and lighting uniformity of other surfaces	Walls: $E_m \geq 150$ lx For offices: $U_o \geq 0.1$ Ceiling: $E_m \geq 100$ lx $U_o \geq 0.1$	Users said the walls and ceiling could have been better lit and wanted everything to be a lighter colour throughout the room. The measurements showed lower values than the norm. Therefore, standard lighting values were used, with the possibility of increasing them.	Walls: $E_m \geq 150$ lx; $U_o \geq 0.1$ Ceiling: $E_m \geq 100$ lx; $U_o \geq 0.1$
4. Average cylindrical illuminance and uniformity	For offices: $E_z \geq 150$ lx $U_o \geq 0.1$	The interviews show that good communication and appearance are essential for employees during meetings. Values at the level specified in the standard will be adopted.	$E_z \geq 150$ lx $U_o \geq 0.1$
5. Modelling	0.3–0.6		0.3–0.6
6. Discomfort glare	$R_{UGL} \leq 19$	The research wants to ensure that on sunny days, natural light does not reflect from computer monitor screens and that artificial light does not significantly dazzle them. The value was reduced from the norm because users during interviews claimed that the lighting bothered them.	$R_{UGL} \leq 16$
7. Colour rendering index	$R_a \geq 80$	In remote and on-site meetings, users want to look attractive. They want the light to reflect their appearance better. Light sources other than fluorescent lamps (LED) with a high R_a value will be used.	$R_a \geq 80$

Table 2. *Cont.*

Lighting Parameter	Normative Value	User Opinion, Justification	Value Selected
8. Correlated colour temperature	There are no specific requirements for the correlated colour temperature value. The value depends on the lighting intensity, interior colours and ambient climate.	According to users, the light in the room is currently too cool; they prefer a warmer light colour. Employees want to be able to adjust the light colour to the time of day and circumstances.	T_{cp} = 3000 K controls
9. Luminance distribution	No specific requirements. Too high and too low luminances and contrasts should be avoided. Light colours for the walls and ceiling are advisable.	Users prefer the white colour of the ceiling and walls, without texture. Luminance from natural light disturbs employees at work. According to the respondents, the room has insufficient light when the windows are covered with blinds.	White colours of the room. Blinds transmit light in a diffused manner with limited luminance.
10. Specific requirements	Work with DSE Controllable lighting	When working with artificial light, light from luminaires reflects off computer screens. They want to be able to control the lighting so that it is sometimes brighter and sometimes darker. The respondents want to choose which luminaire should be lit at a given moment.	$L \leq 3000$ cd/m^2, for monitors with $L > 200$ cd/m^2
11. Maintenance factor	There are no specific value requirements.	Users do not like the neglected state of the entire room. They claim they too often have to ask the building maintenance service to replace worn-out fluorescent lamps.	Light sources and luminaires that do not require frequent maintenance, $T \geq 50{,}000$ h
12. Energetic efficiency	There are no specific value requirements.	No user opinions.	Ensure energy efficiency.

Based on Table 2, the following conclusions should be drawn: The values of illuminance and uniformity (items 1 and 2 in Table 2) are currently at a higher level than required by the standard. By carrying out the project without knowing the users' needs, these values would be lowered to match the value in the standard. Therefore, user dissatisfaction with "worse" lighting conditions after modernisation would arise. Their work could become less efficient, and thus, they could take action to increase the lighting intensity by, for example, purchasing additional desk lamps and therefore increasing the electricity demand.

Based on the measurements, the intensity and uniformity of the illumination of the walls and ceiling (item 3 in Table 2) turned out to be lower than recommended by the standard. Therefore, the design should include better lighting on these surfaces. This can be achieved by selecting appropriate luminaires and their installation location and using white wall and ceiling surfaces with high reflectances (item 9 in Table 2). The current luminaires built into the suspended ceiling (a standard solution in offices) and textured walls need more lighting, especially the ceiling. Better illumination of the walls will positively impact reducing luminance contrasts under artificial lighting. It will positively minimise glare and improve the quality of work at monitors (items 6, 9, and 10 in Table 2). Glare occurring in the room comes from, among others, windows and their location. The existing blinds do not fulfil their function because, after they are covered to prevent glare in the room, there is too little light, forcing users to turn on artificial light. This completely contradicts the use of natural light. Additionally, existing electric lighting is also perceived as causing glare. When designing new lighting, you need to consider this and use appropriate luminaires so that the URG determined in the room is low (item 6 in Table 2). Therefore, the UGR value was reduced compared to the norm. New blinds must be selected so that on a sunny day, they eliminate excessive luminance in the room and thus illuminate the room sufficiently so that it is not necessary to turn on artificial lighting.

The colour temperature of light in office rooms is often set at 4000 K (item 8 in Table 2), which is currently the case in this room. However, according to the users, the light in the room is presently too cool and makes them feel bad while working. They often mentioned that the light made them feel overstimulated, which led to irritability. A warm light colour will create a calmer and more pleasant working atmosphere in the room. Therefore, warmer light colours, characteristic of places of rest and not intensive work, should be considered. In addition, employees want to be able to adjust the light colour to the time of day and circumstances. This leads to a lighting control system that allows the colour temperature and intensity to be changed (item 10 in Table 2).

The selection of high-quality luminaires and durable light sources is essential to eliminating the current frequent lighting failures and long waits for their repair, which irritate users (item 11 in Table 2).

Energy efficiency is also essential (item 12 in Table 2). The project must ensure the rational use of electricity. Nevertheless, it must be remembered that the actual energy consumption of the designed lighting will result from the equipment and technical solutions used and how users use them. The Design Thinking method ensures that this use stays consistent with the project's assumptions. As a result, the problems addressed in this article are expected to be reduced. However, based on users' observations of the time the lighting is on during the day and clarification of this during interviews, it can be assumed that the room is used on average 9 h a day. During this time, the lighting is on all the time. Even on sunny days, due to the need to cover the windows and reduce the lighting level due to glare, it is necessary to illuminate the room with artificial lighting. In 2022, there were 251 working days in Poland, and assuming an average of 9 h of lighting per day, we obtain 2259 h. Therefore, this value is consistent with the initial estimates. Nine hours of lighting operation per day throughout the year, regardless of the availability of daylight, gives a justified need to improve the functioning of lighting through energy-saving luminaires, appropriate lighting control, and appropriate distribution of daylight into the interior for room lighting instead of artificial lighting.

The initial problem identified at the beginning of the project, consisting of a simple modernisation of outdated lighting (one-to-one replacement) and refreshing the room, can now be confronted with the results of the Define phase. As the above results show, there are several problems in the room, not only related to lighting, which the designer learned through the Discover phase. In this way, there is an opportunity to implement the project in such a way that the needs of future users are met, and they can fully use it following the project. This gives us a chance to reduce the problems this article is concerned with. Therefore, the actual problem will be broader than initially expected.

The actual problem caused by the failure to meet user needs identified in the Define phase must be appropriately defined as a design challenge. This will allow us to continue working on the project through Design Thinking. This will allow for the proper orientation of subsequent activities performed by the project team. Using the recommendations, the design challenge should first of all be formulated in such a way that:

- It is not too general or limiting;
- It focuses on specific users or groups of users;
- It addresses the needs and problems that were revealed in the Discover stage;
- It is designed in such a way that it allows for the creation of various solutions.

Taking the above into account, the following design challenge was identified as a real problem:

"How can we design the decor and lighting of an office room to provide employees and their guests with an environment conducive to the effective fulfilment of daily business duties?"

As we can see, the design challenge, i.e., the actual problem, differs significantly from the initial problem, which was supposed to concern only replacing the lighting with more economical lighting and refreshing the room.

3.3. *Third Stage—Develop*

This stage aims to create a design that will solve the problem defined at the end of the Define stage. First, ideas for solving the user's (personas) issue will be generated; then, prototypes will be created and tested by users. Ultimately, the final design solution will be made iteratively.

Generating ideas is most often associated with a well-known design tool—brainstorming. In this project, such a tool was used based on the principles of this method. Conducting a creative session with the design team involved presenting potential solutions to the design challenge defined above. From the lighting, it was the choice of luminaires; from the furniture, it was the choice of desks, armchairs, wardrobes, and wall and ceiling finishes. When selecting potential solutions, the need to meet the needs identified in the Define phase was considered. Some solutions were ultimately chosen from the entire pool based on discussion and voting. The selected solutions will then be used to create prototype solutions to the design problem.

Prototyping is another activity in the Design Thinking process. It takes place right after the idea generation stage. Thanks to the created prototypes, the designer can easily and quickly verify whether the solutions developed and selected during idea generation suit a given project.

In this project, three prototypes (from now on referred to as variants) were created. Individual variants are treated as three independent room designs. The variants differ significantly in terms of interior equipment and the selection and arrangement of lighting. For each, details related to the interior design, lighting, and photometric calculations performed to verify the values of individual lighting parameters assumed in Table 2 can be presented. Due to their number, they will not be described in this work. These were the basis for developing variants in terms of lighting and interior design:

- The lighting requirements defined in Table 2, determined by analysing user needs related to lighting and the lighting standard EN 12464-1:2021 [79];
- The defined needs related to room equipment at the Define stage;

- The design solutions are presented at the idea generation stage.

Several variants were chosen so that users could analyse and compare them in detail. This way, we can draw more valuable conclusions. Variant models were made using the lighting design program—Dialux EVO Version 5.12.0.5620. Figure 5 shows one exemplary view of a 3D model for each of the three variants, constituting a prototype. A prototype made in this way allows the user to present the design in the form of realistic photographs or a dynamically changing view using manual control. It is a frequently used lighting design and visualisation tool in the lighting industry. Both designers and users are used to it. Of course, other tools allow us to obtain more realistic models, such as 3ds Max. However, creating models for them is time-consuming and, therefore, expensive, and this is in contradiction with the idea of prototyping, which assumes that the user will be presented with a potential solution to his problem quickly, efficiently, and at low cost. In addition to creating a prototype in lighting design software, field tests were carried out to select the material from which the blinds would be made. The lighting effects that were obtained could not be simulated by a computer, so field tests were conducted.

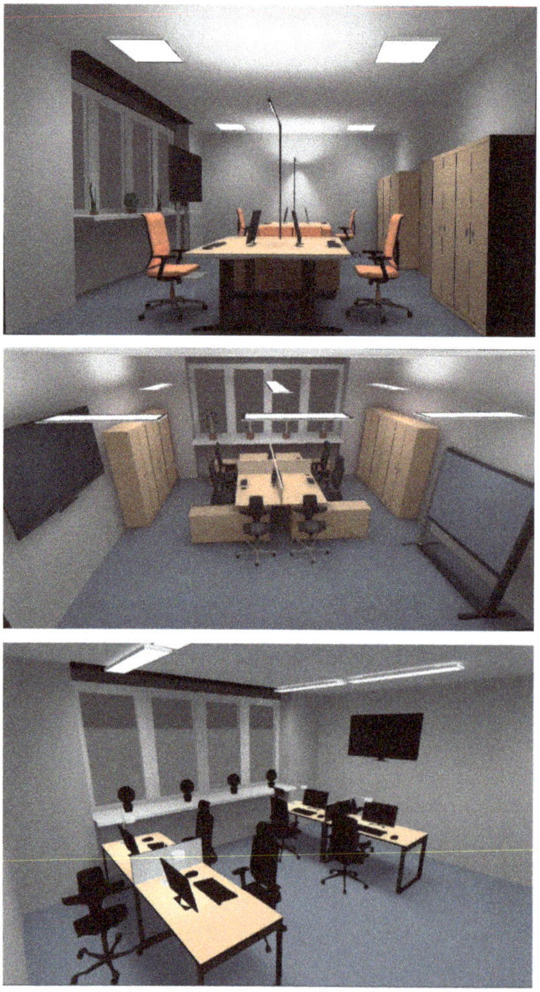

Figure 5. An exemplary view of a 3D model for each of the three prototype variants.

Testing is a crucial design activity according to the Design Thinking method. Prototypes of the designed solutions are presented to and assessed by the users. This allows us to check whether the course of action adopted by the design team is consistent with what users expect. Sometimes, finding an appropriate solution to the problem took time. Testing conclusions regarding prototypes allows for the efficient selection of the final solution to the design problem. Testing is also another opportunity (after the Discovery stage) to understand the user, interact with him, and build relationships. This is due to observations and conversations, sometimes leading to new comments and conclusions.

Taking into account some good advice, as reported by [47], which should be used to ensure that these activities are carried out correctly, it was decided that testing should take into account:

- Showing the prototype and not talking about it—when presenting the prototype, everything was not explained at once, i.e., the testing person was allowed to interpret what he saw, and the designer had to carefully observe and listen to the user's questions about issues related to the presented prototype;
- Creating the impression of an experience to which the user responds—the user had the opportunity to get to know the prototype in all possible ways, thanks to the dynamic change of the presented view;
- More than one prototype allowed the user to compare them and reveal hidden needs.

The project involves creating a lighting solution and introducing changes to the interior design of the office room. Therefore, it was decided that the testing stage would take place as follows:

1. Discussing with users the conclusions regarding the Discover stage and confirming the needs defined at this stage;
2. Presenting variants (3D models) of the designed room using the Dialux Evo software Version 5.12.0.5620 and making the program available to the respondents, which allowed them to independently rotate the views and thus observe the variants from every possible side; presentation of the luminance distribution in the room and a short instruction on how to understand it;
3. Receiving preliminary comments on each of the three presented prototypes;
4. Explaining to users the solutions used in the prototypes and discussing how they will affect the fulfilment of needs;
5. Writing down each prototype's comments received and its advantages and disadvantages.

The tests were conducted on the same users (10 people) interviewed during the Discover stage. Efforts were made to ensure that the tests were carried out in groups, i.e., more than one user participated at one time. Thanks to this, testing was carried out more efficiently, and at the same time, it was possible to conduct discussions between users themselves and between users and the design team. This allowed the design team to receive many constructive comments and observations, which are the most important at this stage.

The results obtained in the testing phase formed the basis for further action. A survey summarising the tests was conducted after a detailed analysis of the conclusions obtained during the testing stage. The survey consisted of arranging the variants in an order consistent with the preferences of a given respondent—the user numbered the prototypes from 1 to 3, marking the design with 1—the most suitable design, 2—less, or 3—the least meeting his needs. The survey results are graphically presented in Figure 6.

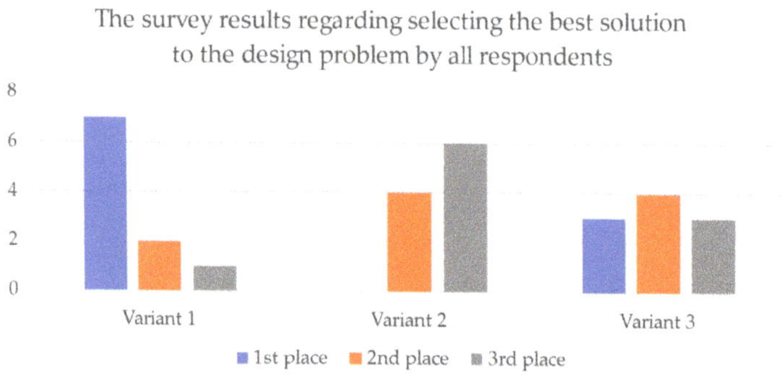

Figure 6. The survey results regarding selecting the best solution to the design problem by all respondents (10 people).

As seen in the chart (Figure 6), variant number I was chosen most often, then variant number III was selected, while the second variant was not chosen as the best. Due to the presence of two characteristic groups of users (regular and occasional), it was decided to check what the results of this test look like for the individual groups. Analysing these results, it turned out that regardless of whether the user is an employee of the room or its guest, variant I is chosen as the best design; however, in the case of type two respondents, variant number III also received the same number of votes. This could be due to the clear separation of the cooperation zone and a more significant amount of space intended for work with guests in variant III, which needed to be added regarding the comments about design I.

Summarising the survey results, variant I was the obvious choice for the best solution to the design problem. However, considering the results obtained in the survey conducted only among guests (the same number of votes for variants I and III), it was worth considering the few advantages of variant III and applying them to the most frequently chosen variant I. Therefore, by adding mobile seats for guests, the design team introduced minor changes in variant I, relating only to the area associated with an additional place for cooperation with visiting room collaborators. Figures 7 and 8 present the final form of the solution to the design problem, i.e., a slightly changed variant I, which considers the respondents' comments obtained during the testing stage.

Figure 7. Visualisation of variant I after corrections motivated by user suggestions—view from the right side of the room.

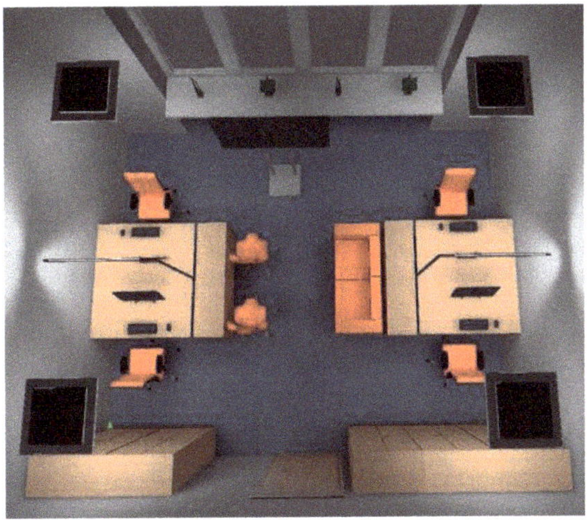

Figure 8. Visualisation of variant I after corrections motivated by user suggestions—top view.

The improved variant I and the unchanged variants II and III were subjected to another survey. As the change was relatively small, the results did not change in the area of selection of regular users of the office room, but this happened in the case of the second group of respondents, i.e., occasional users. Once users accept a solution, they can move on to the next stage.

In terms of energy, the installed lighting power in a room, depending on the number and power of luminaires, is (3):

$$P_{\text{i}} = 4{\cdot}36 + 2{\cdot}69 = 282 \text{ W}, \tag{3}$$

The exact operating time of the lighting installation after the room's modernisation has yet to be discovered. In such a situation, the only savings can be made by reducing the installed lighting power from 828 W to 282 W. This results in an almost 3-fold decrease in power consumption. In addition, there are benefits of automatically turning off the lighting when there are no people in the room, adjusting the lighting level according to current needs, and not turning on the lights on sunny days, thanks to the appropriate introduction of daylight through dedicated blinds. All this has great potential for savings.

3.4. Fourth Stage—Deliver

This stage still needs to be completed as part of the project. The reason was the project's planned completion at the ready documentation stage, awaiting possible implementation. In the case of performing this stage, one would need to follow the steps described in the fourth stage—Deliver in Section 2.3, The Proposal Lighting Design.

To sum up, two critical steps are needed after installing the designed solution. The first one is to adapt the luminaires and the control system to the users' needs, which were determined during work on the project. The second occurs after some time of using the room and assumes making corrections in the installation operation, knowing the users' experience using the facility. Before using the room, there is also a stage of instructing the user on the functioning of the entire installation.

4. Discussion

This article presents the Design Thinking method in lighting design. It aims to minimise the problem of discrepancies between the sustainable design and use of the facility and modifications by users aimed at meeting their needs, which often cause higher energy consumption. Many conclusions can be drawn by analysing the results and experience gathered during the project described in this article. For clarity, the findings will be divided into individual stages of the Design Thinking method.

First stage—Discover

In the lighting project described, based on the Design Thinking method, exemplary research methods were used to collect information about the users and lighting. Other methods and their combinations may replace the techniques used. It is important to emphasise that more than the methods used are needed. Several methods are essential to getting complete results. They should be applied in the appropriate order so that each subsequent method allows for supplementing or clarifying the information obtained from the previous process. This is what happened in this work. Thanks to this, knowledge about the user increases without complicating the situation. In the case of both user observations (user shadowing), user activities performed by the designer (service safari) and interviews, appropriate user representatives (a representative group) should be selected. A representative group, i.e., one for which the addition of another respondent does not introduce a significant change in the available information. In the case of this project, the situation was convenient because the number of people who came to the room regularly or occasionally was small and could be fully taken into account in the research. When the designer finished using the methods described in this work, he felt that he was no longer learning new things, and what he had learned was starting to repeat itself. In another project, choosing the right people will be a challenge. Being "invisible" to users is difficult with the user shadowing method. There is always a change in some user behaviour due to the observer effect (Hawthorne effect). Users initially felt uncomfortable with the designer's presence, but after a few hours, they got used to it and almost did not notice it. This situation can also be expected in other projects. In the service safari method, the value of which is the confrontation of previous observations from the user shadowing method with the designer's feelings, it is difficult to impersonate the user and behave like him. The designer must be trained by one of the employees about the work performed by the employee. However, the novelty effect always occurs and affects the designer's perception. The designer playing the user role is in a new, uncomfortable position. The designer must focus as much as possible on the task. Therefore, the designer is required to feel comfortable in a new environment. To be effective, the interview (questions) should be prepared based on information from the previous stages, such as user shadowing and service safari. The interview is intended to obtain answers to unclear issues observed by the designer.

Regardless of what they are, the research methods should be described as time-consuming, requiring much commitment on both sides (designer and user), cooperation, user self-awareness, designer research skills, and respondents' selection. All this will affect the results obtained, their quality, and the time spent. In this project, the designer already had basic knowledge and experience about the methods used. Additionally, he had no time limit. However, if the designer needs to gain knowledge and experience, the participation of a Design Thinking moderator in the design team is necessary.

The Discover stage allowed us to learn about several problems that are only sometimes typical for offices. They result from the specificity of the room, including equipment, lighting, and users. The critical problem observed during the project was glare, caused by many factors, including inappropriate window covers.

The Discover stage allows us to learn about user needs in a given situation. Thanks to this, it is possible to meet users' needs with appropriate lighting design and, in this process, keep in mind the goal of sustainable development, which is the rational use of electricity for lighting purposes. This approach is supported by the Design Thinking method.

Second stage—Define

User needs identified in the Discover stage verify whether the initially defined problem resulting from the assumed user needs is reflected in reality. In the described case, the scope of work to be carried out had to be changed (deepened) to consider the researched user needs. This is how the actual problem arose. This became the starting point for determining the requirements for the project, i.e., selecting specific parameter values (including lighting). The article focused on the lighting side, so Table 2 contains a detailed analysis and comparison of normative requirements with users' needs. This did not happen in the described case, but it can be expected that it will be challenging to establish requirements when users' needs oppose the normative requirements. It can be expected that the requirements will need to be increased or decreased within the range permitted by standards according to users' needs. In this project, there was a need to increase illuminance and lighting uniformity and decrease the UGR value. Such situations may be challenging to implement due to the demand for additional solutions that increase costs and, sometimes, energy consumption. Of course, reducing parameters and, thus, energy consumption would be desirable. Nevertheless, meeting users' needs is crucial for their proper operation. It guarantees that there is no risk of users introducing modifications that may result in an additional increase in energy consumption, e.g., through non-optimal operation of devices. Introducing a reduction in energy consumption due to the design not meeting users' needs will probably lead to modifications that will eliminate the savings achieved in the design. Therefore, implementing sustainable lighting may be illusory, i.e., real only in the design documentation and not reflected in the use of this installation.

The identified actual design problem had to be skilfully described as a design challenge to enable further practical work. It was not a simple matter. This required the intuition and experience of the design team. No need could be omitted because not all needs would be solved.

Third stage—Develop

Correct execution of the Discover and Define phases allowed for rapid progress through prototyping and testing. The changes needed were cosmetic, decreasing with each iteration and converging to the final solution. However, this was an optimistic case. In practice, a long process should be taken into account. It is often the case that we base decisions on our ideas rather than facts.

Knowing the actual problem and having a specific design challenge, it was necessary to present potential solutions and select those that would be implemented. Generating solutions took a lot of work. Knowledge of tools, considering group dynamics, and how to make choices was needed. Even though the diversity of participants was limited, and they were required to comply with the rules entirely, it still caused problems with mutual communication. Some people use a different language depending on their competencies. Moreover, people specialising in a common area tend to find solutions quickly and do not appreciate comments from people outside their area of competence. Despite these difficulties, solutions were selected for implementation. Due to their non-standard nature (innovation), the resulting solutions will often require field tests to verify their usefulness, durability, etc. This was the case in this project, which involved selecting a new method of covering windows instead of the existing blinds. Choosing the degree of dispersion of the light transmitted through the blinds from the available materials was necessary to minimise glare.

The appropriate method for making prototypes needs to be chosen. In this work, the standard form of the prototype, i.e., visualisations, was chosen. Prototypes such as visualisations are a means of presenting the design and making decisions. However, they still need to be fully addressed to the user. In this work, instead of showing ready-made, static images, the view was presented directly from a computer program so that it could be freely changed. In addition, the presence of a designer was ensured so that the user could receive help and explanations when needed. It was crucial that the prototype be understandable to the user and clearly illustrate the new office room. This is extremely difficult when we present unusual solutions for which the user cannot use their previous knowledge, experience, and associations to interpret them. There was a desire to use all the solutions developed in one design. This forced the selection of solutions and their division into several separate variants. The selection of people for testing is arguable. They may be people who took part in the Discover stage, which was the case in the project, or they may be completely new.

Failure to meet user requirements or the misunderstanding of the prototype results in the need for corrections. It is unknown how many there will be and how long this process will take. Because the design process usually operates within a predetermined time frame, there is little freedom to extend it. Experience, intuition, and good decisions guarantee efficient prototyping and testing. Continuous contact with the user, confirming assumptions, and drawing conclusions is critical to the smooth completion of the entire process. Thanks to this, the testing process for this project took place efficiently and within an acceptable timeframe. However, there is always a risk of incorrect assumptions in the Define phase. When this happens, you must go back into the process and make corrections in the appropriate places. This involves a repetition of actions, this time well-oriented.

Fourth stage—Deliver

Due to the lack of the implementation of this phase of the method in the presented project, documented conclusions cannot be provided. Only recommendations on this issue and areas for review will be given.

The final selection of a given interior design solution and its installation, including lighting, is not the end. Each solution has a specific range of freedom from the installation location and the selection of settings and parameters. These values initially result from the Discover and Define phases. However, the final values are known only later and result directly from a given solution. Only then can a new solution that meets the user's needs be completed and implemented. At this point, implementing the goals of sustainable lighting design and other energy-using systems can be verified. The risk of failing to meet these requirements should be limited because, when using the Design Thinking method, we consider the users' needs for a given space and consider them throughout the design process.

5. Conclusions

To sum up, an approach to designing sustainable solutions that uses standard solutions and models must be revised. People's needs are only partially repeatable. They use the designed devices differently. This project identified the unusual requirements of office employees, which no one talked about or took into account before the planned modernisation. The uniqueness of needs directly results in using devices to ensure comfort, constituting discrepancies between design and use. These, in turn, cause energy consumption to be different than expected.

This article addresses the need for communication between the designer and the user using the Design Thinking method. The designer did not impose solutions on the user, but the results were based on a thorough understanding of his needs. The user was almost always at the centre of the design process. His knowledge allowed designers to look at designing good lighting differently. The user supported the creation of various solutions to meet his needs. The designers' knowledge allowed them to verify these proposals.

Decisions made during design have a crucial impact on future energy consumption, demonstrating the fulfilment of sustainable development goals. The presented project confirmed this. If not for the inclusion of dedicated and user-friendly controls, the lighting would still be left on when employees were absent. If it were not for using appropriate blinds, the lighting would still be turned on on sunny days. This translates into energy consumption and is more than just due to the luminaires and control schedules.

After testing the designed solution, especially after talking to users, there is a guarantee of total acceptance and use of the devices for their intended purpose. It is reasonable that there will be no future problems with their operation, and energy will be consumed as expected in the design. Employee satisfaction, comfort, and efficiency, crucially influenced by lighting, will be high. Therefore, sustainable development will be achieved by meeting people's needs while preserving natural resources.

The presented project successfully used the Design Thinking method, as shown in the literature on other fields.

The problem of discrepancies between the design and use stages presented in the article's introduction can be considered reduced. Users' interference in the operation of devices will be eliminated due to meeting their needs and thus achieving comfortable conditions for them.

The ease of application of Design Thinking in lighting design depends on the case. In the case of lighting modernisation, it is much easier than in the case of new investments; when there are no specific users yet, it is challenging to define them, and we have to use generalisations. The author recommends using the presented method wherever lighting is modernised. There are regular users in such places, and they can take part in the design process. In addition to selecting the desired lighting, the method will indicate other areas related to a given space that should be included in the modernisation process.

Design Thinking and its potential can be used in lighting design and technology. The author sees the possibility of using this method, among other things, for:

- Knowledge about user needs—real, often unrealised, hidden needs, thanks to which we can take them into account when designing lighting and other devices in the field of lighting technology.
- Matching designs—by examining users' needs, a lighting design meets the user's actual needs, which is why it is better used.
- User involvement—we involve users in the design process during needs research and testing. Thanks to this, we can use their knowledge and experience to show our commitment to solving user problems.
- Educational potential—during needs research and testing, we explain to the user how the proposed solution works and why this is and not otherwise. We present what modern technology can offer, thus introducing the user to the basics of lighting, i.e., what is not well known.
- Sustainable resources/environment use—thanks to the knowledge of users' needs, we can choose needed solutions that will be fully used. Thus, we do not introduce options that no one will ever use or luminaires that no one will ever turn on. This allows for better use of resources, thus meeting sustainable development goals.
- Prototyping—having the proposed solutions as a prototype (visualisation, showroom, etc.), we can quickly and inexpensively verify and modify them before implementation.
- Enhance new solutions—new, unusual, non-obvious lighting ideas arise thanks to the inclusion of different people in the design process, often from outside the lighting industry, and getting to know their thoughts and their way of seeing the solution.
- Dealing with the excess of solutions—today, we have many lighting solutions, and thus, it is easy to make the wrong choice. Thanks to the knowledge of the actual needs of the users, choosing the right option is easier and more correct.
- Orientation of science/research—the initiator of the new study is a potential user of their results. The need for research is derived from getting to know user problems and needs.
- The "clash" of lighting standards with users' needs is an actual situation in which the known user requirements will oppose the standards. There is a field for seeking a compromise, and sometimes there may be a need to initiate future changes in the standards.

Funding: The author presents his appreciation for the financial support for publishing the article obtained from the Open Science V program at the Warsaw University of Technology and the Electrical Power Engineering Institute.

Institutional Review Board Statement: The study was conducted in accordance with the Declaration of Helsinki, and approved by the Institutional Review Board (or Ethics Committee) of WARSAW UNIVERSITY OF TECHNOLOGY (Certyfikat_07_24 dated 20 March 2024) for studies involving humans.

Informed Consent Statement: Not applicable.

Data Availability Statement: Dataset available on request from the author.

Acknowledgments: The author would like to thank Karol Szaciłło for his help in preparing the described office project using the Design Thinking method.

Conflicts of Interest: The authors declare no conflicts of interest.

References

1. Janda, K.B. Buildings don't use energy: People do. *Archit. Sci. Rev.* **2011**, *54*, 15–22. [CrossRef]
2. Ander, G.D. *Daylighting Performance and Design*; John Wiley & Sons: Hoboken, NJ, USA, 2003.
3. Phillips, D. *Daylighting: Natural Light in Architecture*; Elsevier: Burlington, MA, USA; Architectural Press: Princeton, NJ, USA, 2004.
4. Gill, Z.M.; Tierney, M.J.; Pegg, I.M.; Allan, N. Low energy dwellings: The contribution of behaviours to actual performance. *Build. Res. Inf.* **2010**, *38*, 491–508. [CrossRef]
5. *Development Climate Energy*; World Business Council for Sustainable Development (WBCSD): Geneva, Switzerland, 2009.
6. Barlow, S.; Fiala, D. Occupant comfort in UK offices—How adaptive comfort theories might influence future low energy office refurbishment strategies. *Energy Build.* **2007**, *39*, 837–846. [CrossRef]
7. Jazizadeh, F.; Kavulya, G.; Kwak, J.-Y.; Becerik-Gerber, B.; Tambe, M.; Wood, W. Human-Building Interaction for Energy Conservation in Office Buildings. In Proceedings of the Construction Research Congress 2012: Construction Challenges in a Flat World, West Lafayette, IN, USA, 21–23 May 2012; pp. 1830–1839.
8. Hong, T.; Chen, Y.; Belafi, Z.; D'Oca, S. Occupant behavior models: A critical review of implementation and representation approaches in building performance simulation programs. *Build. Simul.* **2018**, *11*, 1–14. [CrossRef]
9. Delzendeh, E.; Wu, S.; Lee, A.; Zhou, Y. The impact of occupants' behaviours on building energy analysis: A research review. *Renew. Sustain. Energy Rev.* **2017**, *80*, 1061–1071. [CrossRef]
10. Chen, S.; Zhang, G.; Xia, X.; Chen, Y.; Setunge, S.; Shi, L. The impacts of occupant behavior on building energy consumption: A review. *Sustain. Energy Technol. Assess.* **2021**, *45*, 101212. [CrossRef]
11. Yang, L.; Liu, S.; Liu, J. The Interaction Effect of Occupant Behavior-Related Factors in Office Buildings Based on the DNAS Theory. *Sustainability* **2021**, *13*, 3227. [CrossRef]
12. Fabi, V.; Andersen, R.V.; Corgnati, S.P.; Olesen, B.W. A methodology for modelling energyrelated human behaviour: Application to window opening behaviour in residential buildings. *Build. Simul.* **2013**, *6*, 415–427. [CrossRef]
13. Pegg, I. Assessing the Role of Post-Occupancy Evaluation in the Design Environment—A Case Study Approach. Ph.D. Thesis, Brunel University, Uxbridge, MA, USA, 2007.
14. Firth, S.; Lomas, K.; Wright, A.; Wall, R. Identifying trends in the use of domestic appliances from household electricity consumption measurements. *Energy Build.* **2008**, *40*, 926–936. [CrossRef]
15. Marcolino, L.S.; Kolev, B.; Price, S.; Veetil, S.P.; Gerber, D.J.; Musil, J.; Tambe, M. Aggregating Opinions to Design Energy-Efficient Buildings. In Proceedings of the 8th Multidisciplinary Workshop on Advances in Preference Handling (M-PREF 2014), Quebec City, QC, Canada, 28 July 2014.
16. Yan, D.; O'Brien, W.; Hong, T.; Feng, X.; Gunay, H.B.; Tahmasebi, F.; Mahdavi, A. Occupant behavior modeling for building performance simulation: Current state and future challenges. *Energy Build.* **2015**, *107*, 264–278. [CrossRef]
17. Heydarian, A.; Pantazis, E.; Wang, A.; Gerber, D.; Becerik-Gerber, B. Towards user centered building design: Identifying end-user lighting preferences via immersive virtual environments. *Autom. Constr.* **2017**, *81*, 56–66. [CrossRef]
18. Attia, S.; De Herde, A.; Gratia, E.; Hensen, J.L.M. Achieving informed decision-making for net zero energy buildings design using building performance simulation tools. *Build. Simul.* **2013**, *6*, 3–21. [CrossRef]
19. Hoes, P.; Hensen, J.L.M.; Loomans, M.G.L.C.; de Vries, B.; Bourgeois, D. User behavior in whole building simulation. *Energy Build.* **2009**, *41*, 295–302. [CrossRef]
20. Maier, T.; Krzaczek, M.; Tejchman, J. Comparison of physical performances of the ventilation systems in low-energy residential houses. *Energy Build* **2009**, *41*, 337–353. [CrossRef]
21. Schakib-Ekbatan, K.; Çakici, F.Z.; Schweiker, M.; Wagner, A. Does the occupant behavior match the energy concept of the building?—Analysis of a German naturally ventilated office building. *Build. Environ.* **2015**, *84*, 142–150. [CrossRef]
22. Martinaitis, V.; Zavadskas, E.K.; Motuziene, V.; Vilutiene, T. Importance of occupancy information when simulating energy demand of energy efficient house: A case study. *Energy Build.* **2015**, *101*, 64–75. [CrossRef]
23. Zero Carbon HUB. *Post-Occupancy Evaluation, Rowner Research Project Phase Two*; Zero Carbon HUB: Milton Keynes, UK, 2015.
24. Yang, J.; Santamouris, M.; Lee, S.E. Review of occupancy sensing systems and occupancy modeling methodologies for the application in institutional buildings. *Energy Build.* **2016**, *121*, 344–349. [CrossRef]
25. Calì, D.; Osterhage, T.; Streblow, R.; Müller, D. Energy performance gap in refurbished German dwellings: Lesson learned from a field test. *Energy Build.* **2016**, *127*, 1146–1158. [CrossRef]
26. D'Oca, S.; Fabi, V.; Corgnati, S.P.; Andersen, R.K. Effect of thermostat and window opening occupant behavior models on energy use in homes. *Build. Simul.* **2014**, *7*, 683–694. [CrossRef]

27. Hong, T.; Yan, D.; D'Oca, S.; Chen, C. Ten questions concerning occupant behavior in buildings: The big picture. *Build. Environ.* **2017**, *114*, 518–530. [CrossRef]
28. Bordass, W. *Flying Blind: Everything You Wanted to Know about Energy in Commercial Buildings but Were Afraid to Ask*; Association for the Conservation of Energy: London, UK, 2001.
29. Leaman, A.; Bordass, B. Are users more tolerant of 'green' buildings? *Build. Res. Inf.* **2007**, *35*, 662–673. [CrossRef]
30. Clevenger, C.M.; Haymaker, J. The Impact of the Building Occupant on Energy Modeling Simulations. In Proceedings of the Joint International Conference on Computing and Decision Making in Civil and Building Engineering, Montreal, QC, Canada, 14–16 June 2006.
31. Kleinsmann, M.; Valkenburg, R.; Buijs, J. Why do(n't) actors in collaborative design understand each other? An empirical study towards a better understanding of collaborative design. *CoDesign* **2007**, *3*, 59–73. [CrossRef]
32. Carrara, G.; Fioravanti, A.; Nanni, U. Knowledge-based Collaborative Architectural Design. *Int. J. Des. Sci. Technol.* **2009**, *16*, 1–16.
33. Belessiotis, V.; Mathioulakis, E. Analytical approach of thermosyphon solar domestic hot water system performance. *Sol. Energy* **2002**, *72*, 307–315. [CrossRef]
34. Lee, B.; Trcka, M.; Hensen, J.L.M. Building energy simulation and optimization: A case study of industrial halls with varying process loads and occupancy patterns. *Build. Simul.* **2014**, *7*, 229–236. [CrossRef]
35. He, Y.; Chen, Y.; Chen, Z.; Deng, Z.; Yuan, Y. Impacts of Occupant Behavior on Building Energy Consumption and Energy Savings Analysis of Upgrading ASHRAE 90.1 Energy Efficiency Standards. *Buildings* **2022**, *12*, 1108. [CrossRef]
36. Owens, S.; Driffill, L. How to change attitudes and behaviour in the context of energy. *Energy Policy* **2008**, *36*, 4412–4418. [CrossRef]
37. Steemers, K.; Yun, G.Y. Household energy consumption: A study of the role of occupants. *Build. Res. Inf.* **2009**, *37*, 625–637. [CrossRef]
38. Darby, S. Energy feedback in buildings: Improving the infrastructure for demand reduction. *Build. Res. Inf.* **2008**, *36*, 499–508. [CrossRef]
39. Hensen, J.; Lamberts, R. *Building Performance Simulation for Design and Operation*; Taylor & Francis: Abingdon, UK, 2011; pp. 56–84.
40. Veitch, J.A.; Hine, D.W.; Gifford, R. End users' knowledge, beliefs, and preferences for Lighting. *J. Inter. Des.* **1993**, *19*, 15–26. [CrossRef]
41. Zhang, W.; Tan, S.; Lei, Y.; Wang, S. Life cycle assessment of a single-family residential building in Canada: A case study. *Build. Simul.* **2014**, *7*, 429–438. [CrossRef]
42. Zahedi, M.; Guité, M.; De Paoli, G. Addressing user-centeredness: Communicating meaningfully through design. In Proceedings of the 14th International Conference on Computer Aided Architectural Design Futures, Liege, Belgium, 4–8 July 2011; pp. 513–524.
43. Li, Z.; Jiang, Y.; Wei, Q.T. Survey on Energy Consumption of Air Conditioning in Summer in a Residential Building in Beijing. *J. Heat. Vent. Air Cond.* **2014**, *37*, 46–51.
44. Santin, O.G.; Itard, L.; Visscher, H. The effect of occupancy and building characteristics on energy use for space and water heating in Dutch residential stock. *Energy Build.* **2009**, *41*, 1223–1232. [CrossRef]
45. Ren, Z.; Yang, F.; Bouchlaghem, N.M.; Anumba, C.J. Multi-disciplinary collaborative building design—A comparative study between multi-agent systems and multi-disciplinary optimisation approaches. *Autom. Constr.* **2011**, *20*, 537–549. [CrossRef]
46. Feringa, J. Notes on the Potential of Simulation for Architectural Conception. In Proceedings of the 28th Annual Conference of the Association for Computer Aided Design in Architecture, Minneapolis, MN, USA, 16–19 October 2008; pp. 264–269.
47. Gunay, H.B.; O'Brien, W.; Beausoleil-Morrison, I. A Critical review of state-of-the-art energy and comfort related occupant behavior in office buildings. *Build. Environ.* **2013**, *70*, 31–47. [CrossRef]
48. Zahedi, M.; Guité, M.; de Paoli, G. Dealing with the human-centered approach within HCI projects. In Proceedings of the Design Research Society International Conference, 2010: Design & Complexity, Montreal, QC, Canada, 7–9 July 2010.
49. Dong, B.; Lam, K.P. A real-time model predictive control for building heating and cooling systems based on the occupancy behavior pattern detection and local weather forecasting. *Build. Simul.* **2014**, *7*, 89–106. [CrossRef]
50. Kim, M.J.; Cho, M.E.; Jun, H.J. Developing Design Solutions for Smart Homes Through User-Centered Scenarios. *Front. Psychol.* **2020**, *11*, 516355. [CrossRef] [PubMed]
51. Bogenstätter, U. Prediction and optimization of life-cycle costs in early design. *Build. Res. Inf.* **2000**, *28*, 376–386. [CrossRef]
52. Cross, N. Forty years of design research. *Des. Stud.* **2007**, *28*, 1–4. [CrossRef]
53. Boyarski, D.; Buchanan, R. Brief—Carnegie Mellon University. *Interactions* **2000**, *7*, 20–23. [CrossRef]
54. Zahedi, M.; Poldma, T.; Baha, S.E.; Haats, T. Design Thinking and Aesthetic Meaning-Making: Interlaced Means to Engage in Collaborative Knowledge-Building. In Proceedings of the NordDesign 2012, 9th Conference on Engineering Design and Product Development, Aalburg, Denmark, 22–24 August 2012; pp. 627–637.
55. Dugar, A.M. Slow Design. A new standard for connected Lighting? *Light. India* **2019**, *14*, 11–12.
56. Reinhart, C.F.; Voss, K. Monitoring manual control of electric lighting and blinds. *Light. Res. Technol.* **2003**, *35*, 243–258. [CrossRef]
57. Nicol, J.F.; Humphreys, M.A. Adaptive thermal comfort and sustainable thermal standards for buildings. *Energy Build.* **2002**, *34*, 563–572. [CrossRef]
58. Klein, L.; Kwak, J.-y.; Kavulya, G.; Jazizadeh, F.; Becerik-Gerber, B.; Varakantham, P.; Tambe, M. Coordinating occupant behavior for building energy and comfort management using multi-agent systems. *Autom. Constr.* **2012**, *22*, 525–536. [CrossRef]
59. Bullinger, H.-J.; Bauer, W.; Wenzel, G.; Blach, R. Towards user centred design (UCD) in architecture based on immersive virtual environments. *Comput. Ind.* **2010**, *61*, 372–379. [CrossRef]

60. Romm, J.J. *Lean and Clean Management: How to Boost Profits and Productivity by Reducing Pollution*; Kodansha USA Inc.: New York, NY, USA, 1994.
61. Liedtka, J. Why Design Thinking Works. *Harv. Bus. Rev.* **2018**, *9–10*, 72–79.
62. Brown, T. *Change by Design: How Design Thinking Transforms Organizations and Inspires Innovation*; Harper Business: New York, NY, USA, 2019.
63. Brown, T. Design Thinking. *Harv. Bus. Rev.* **2008**, *6*, 72–79.
64. Dam, R.F.; Siang, T.Y. What is Design Thinking and Why Is It So Popular? Interaction Design Foundation. Available online: https://www.interaction-design.org/literature/article/what-is-design-thinking-and-why-is-it-so-popular (accessed on 10 November 2023).
65. Chasanidou, D.; Gasparini, A.A.; Lee, E. Design Thinking Methods and Tools for Innovation. In *Design, User Experience, and Usability: Design Discourse: 4th International Conference, DUXU 2015, Held as Part of HCI International 2015, Los Angeles, CA, USA, August 2–7, 2015, Proceedings, Part I*; Springer: Cham, Swizterland, 2015; pp. 12–23.
66. Słupik, S.; Kos-Łabędowicz, J.; Trzęsiok, J. Are You a Typical Energy Consumer? Socioeconomic Characteristics of Behavioural Segmentation Representatives of 8 European Countries. *Energies* **2021**, *14*, 6109. [CrossRef]
67. Zissis, G. Energy Consumption and Environmental and Economic Impact of Lighting: The Current Situation. In *Handbook of Advanced Lighting Technology*; Karlicek, R., Sun, C.C., Zissis, G., Ma, R., Eds.; Springer: Cham, Switzerland, 2016; pp. 1–13.
68. Nelson, H.G.; Stolterman, E. *The Design Way*, 2nd ed.; The MIT Press: Cambridge, MA, USA, 2014.
69. Cooper, R.; Junginger, S.; Lockwood, T. Design thinking and design management: A research and practice perspective. *Des. Manag. Rev.* **2009**, *20*, 46–55. [CrossRef]
70. Plattner, H.; Meinel, C.; Leifer, L. *Design Thinking*; Springer: Berlin, Germany, 2011.
71. Brenner, W.; Uebernickel, F.; Abrell, T. Design thinking as mindset, process, and toolbox. In *Design Thinking for Innovation*; Brenner, W., Uebernickel, F., Eds.; Springer: Cham, Switzerland, 2016; pp. 3–24.
72. Micheli, P.; Wilner, S.J.S.; Bhatti, S.H.; Mura, M.; Beverland, M.B. Doing design thinking: Conceptual review, synthesis, and research agenda. *J. Prod. Innov. Manag.* **2019**, *36*, 124–148. [CrossRef]
73. Constain Moreno, G.E.; Collazos, C.A.; Blasco, S.B.; Moreira, F. Software Design for Users with Autism Using Human-Centered Design and Design Thinking Techniques. *Sustainability* **2023**, *15*, 16587. [CrossRef]
74. Magistretti, S.; Bianchi, M.; Calabretta, G.; Candi, M.; Dell'Era, C.; Stigliani, I.; Verganti, R. Framing the multifaceted nature of design thinking in addressing different innovation purposes. *Long Range Plan.* **2022**, *55*, 102163. [CrossRef]
75. Niehaus, M.; Mocan, M. Cultivating Design Thinking for Sustainable Business Transformation in a VUCA World: Insights from a German Case Study. *Sustainability* **2024**, *16*, 2447. [CrossRef]
76. Lockwood, T. *Design Thinking: Integrating Innovation, Customer Experience, and Brand Value*; Allworth Press: New York, NY, USA, 2009.
77. Stickdorn, M.; Schneider, J. *This Is Service Design Thinking*; BIS Publishers: Amsterdam, The Netherlands, 2014.
78. Design Methods for Developing Services. An Introduction to Service Design and a Selection of Service Design Tools. Available online: https://www.designcouncil.org.uk/fileadmin/uploads/dc/Documents/DesignCouncil_Design%2520methods%2520for%2520developing%2520services.pdf (accessed on 10 November 2023).
79. *EN 12464-1:2021*; Light and Lighting—Lighting of Work Places—Part 1: Indoor Work Places. Slovenian Institute for Standardization: Ljubljana, Slovenia, 2021.
80. *EN 1838:2013*; Lighting Applications—Emergency Lighting for Buildings. Slovenian Institute for Standardization: Ljubljana, Slovenia, 2013.
81. *EN 15193-1+A1:2021-09*; Energy Performance of Buildings—Energy Requirements for Lighting—Part 1: Specifications, Module M9. Slovenian Institute for Standardization: Ljubljana, Slovenia, 2021.
82. Flager, F.; Gerber, D.J.; Kallman, B. Measuring the impact of scale and coupling on solution quality for building design problems. *Des. Stud.* **2014**, *35*, 180–199. [CrossRef]
83. O'Brien, W.; Gunay, H.B. The contextual factors contributing to occupants' adaptive comfort behaviors in offices—A review and proposed modeling framework. *Build. Environ.* **2014**, *77*, 77–87. [CrossRef]
84. Cole, R.J.; Brown, Z. Reconciling human and automated intelligence in the provision of occupant comfort. *Intell. Build. Int.* **2009**, *1*, 39–55. [CrossRef]
85. *ANSI/ASHRAE Standard 55*; Thermal Environmental Conditions for Human Occupancy. ASHRAE: Peachtree Corners, Georgia, 2010.
86. Tuniki, H.P.; Jurelioni, A.; Fokaides, P. A review on the approaches in analysing energy-related occupant behaviour research. *J. Build. Eng.* **2021**, *40*, 102630. [CrossRef]
87. Lashinaa, T.; Chraibia, S.; Despenica, M.; Shrubsoleb, P.; Rosemanna, A.; van Loenena, E. Sharing lighting control in an open office: Doing one's best to avoid conflict. *Build. Environ.* **2019**, *148*, 1–10. [CrossRef]
88. Martins, F.; Almeida, M.F.; Calili, R.; Oliveira, A. Design Thinking Applied to Smart Home Projects: A User-Centric and Sustainable Perspective. *Sustainability* **2020**, *12*, 10031. [CrossRef]
89. Haldi, F. A probabilistic model to predict building occupants'diversity towards their interactions with the building envelope. In Proceedings of the BS2013: 13th Conference of International Building Performance Simulation, Chambery, France, 26–29 August 2013.

90. Steg, L. Promoting household energy conservation. *Energy Policy* **2008**, *36*, 4449–4453. [CrossRef]
91. Design Kit: The Human-Centered Design Toolkit. Available online: https://www.ideo.com/post/design-kit (accessed on 10 November 2023).
92. Stanford d.school. *Bootcamp Bootleg*; Hasso Plattner Institute of Design: Stanford, Palo Alto, CA, USA, 2011.
93. Durak, A.; Olguntürk, N.; Yener, C.; Köksal, D.; Gürçınar, Y. Impact of lighting arrangements and illuminances on different impressions of a room. *Build. Environ.* **2007**, *42*, 3476–3482. [CrossRef]
94. Kleinsmann, M.; Valkenburg, R. Barriers and enablers for creating shared understanding in co-design projects. *Des. Stud.* **2008**, *29*, 369–386. [CrossRef]
95. Preece, J.; Rogers, Y.; Sharp, H. *Interaction Design: Beyond Human-Computer Interaction*, 3rd ed.; Wiley: New York, NY, USA, 2011.
96. Bridges, A.; Charitos, D. On architectural design in virtual environments. *Des. Stud.* **1997**, *18*, 143–154. [CrossRef]
97. Zhang, L.; Goßmann, J.; Stevenson, C.; Chi, M.; Cauwenberghs, G.; Gramann, K.; Schulze, J.; Otto, P.; Jung, T.; Peterson, R.; et al. Spatial cognition and architectural design in 4D immersive virtual reality: Testing cognition with a novel audiovisual CAVE-CAD tool. In Proceedings of the Spatial Cognition for Architectural Design Conference, New York, NY, USA, 16–19 November 2011.
98. How Virtual Environments Could Help Architects? Available online: https://www.arch2o.com/virtual-augmented-reality-could-help-architects/ (accessed on 10 November 2023).
99. Krupiński, R. Virtual Reality System and Scientific Visualisation for Smart Designing and Evaluating of Lighting. *Energies* **2020**, *13*, 5518. [CrossRef]
100. Mahmoudzadeh, P.; Afacan, Y.; Adi, M.N. Analyzing occupants' control over lighting systems in office settings using immersive virtual environments. *Build. Environ.* **2021**, *196*, 107823. [CrossRef]

buildings

MDPI

Article

Research on the Spatial Location Design of Guidance Signage Systems to Connect the Space of Transit-Orientated Development Sites Based on Multi-Software Analysis

Xue Ma and Qingjuan Yang *

School of Architecture, Southwest Jiaotong University, Chengdu 611756, China; mxue@my.swjtu.edu.cn
* Correspondence: yqj@home.swjtu.edu.cn

Abstract: In the context of intensive urban development strategies, the Transit-Oriented Development (TOD) model has emerged as a crucial strategy, requiring a high degree of integration of various functional spaces and traffic flows in transport stations. Complex spaces and traffic flows require an effective Guidance Signage System design to ensure efficient guidance and easy transfers. In order to choose the spatial location of the Guidance Signage System in the connecting space of TOD stations and to improve the efficiency of their guidance and visual saliency, three types of connecting space, namely, channel bifurcation nodes, atrium threshold zones, and sunken plaza interface areas were selected. In total, 18 signage system setup areas were derived from the overlapping visual domains of signage and human visual domains using the Depthmap10 and Massmotion11.5 software (six areas in each type of articulation space), which were then combined with eye-tracking technology to analyze the visual saliency of the signage system's layout position on the passengers. The experiments show that the signage location affects passengers' visual attention to a certain extent, and there are differences in the location of high display degrees in different connecting spaces, which require differentiated layouts. This paper summarizes the best spatial location design strategy for the Guidance Signage System according to different connecting spaces, providing more scientific guidance for the spatial location design of the Guidance Signage System for sites connecting spaces. This is essential for enhancing user navigation efficiency, minimizing confusion, and improving overall spatial usability in complex environments.

Keywords: multi-software analysis; TOD; connecting space; guidance signage system; spatial location; eye tracking

1. Introduction

1.1. Background of the Study

In order to achieve low-carbon, ecological, and other construction goals, China promotes compact-oriented urban development. The TOD model, which focuses on the integrated and compact development of rail stations and the surrounding underground space, has received increasing attention as it can effectively solve the problems of urban traffic congestion, environmental pollution, and land waste [1–3]. However, with the promotion of the TOD model for considering a station's connecting space, which links the transport node and the station's commercial complex, the Guidance Signage System has also gradually emerged for use in space location design problems. A complex spatial layout and dense flow of people may easily lead to the signage being obscured, making it enter an individual's line of sight initially, thus making it unable to effectively guide passengers [4].

1.2. Literature Review

In the academic lineage of TOD research, the Guidance Signage System, as an important carrier for optimizing spatial efficiency, has become one of the hot areas of global research in recent years. The review of global TOD research over the past two decades shows that the focus of research in this field has shifted from macroscopic transportation network planning to microscopic spatial behavior intervention and the signage system as a key medium for "space-behavior" coupling, gradually highlighting its design science and practical value. Early research primarily focused on land-use integration and transportation efficiency, but recent studies have increasingly addressed the human-centered design of transitional spaces, particularly the optimization of pedestrian flow and ways of finding systems [5]. A significant surge in TOD-related research occurred post-2015, driven by China's rapid urbanization and the need for sustainable solutions to traffic congestion and land scarcity [6]. Notably, studies have shifted from macro-level urban planning (e.g., station hierarchy and density zoning) to micro-level spatial interactions, such as signage visibility and accessibility [7]. Several studies have highlighted the role of the Guidance Signage System in transportation hubs to improve wayfinding and guidance efficiency [8,9]. However, research on signage positioning within TOD connecting spaces remains fragmented, with the limited integration of human behavioral analytics and dynamic spatial modeling tools. In recent years, some scholars have begun to pay attention to the role of the Guidance Signage System in TOD sites, but their studies are mostly limited to static analysis and lack the consideration of dynamic pedestrian flow lines and human factors [10].

1.2.1. Overview of Connecting Space

Current research on connecting space is mostly in the theoretical direction; Aldo proposed the theory of "intermediate state" and argued that architecture should have a clear middle ground, and intermediary space provides a public place for this contradictory space [11]. Younger describes transitional spaces as follows: transitions between various public spaces should be soft, and transitions from private to public spaces should be clear, easy to access, and give a sense of security [12]. In recent years, this has shifted from theoretical discussion to practical application and has become an important topic in urban planning and transportation engineering. Researchers emphasize the functional composability of articulation space and advocate the integration of transportation, commerce, culture, and other functions into one in order to enhance the efficiency of space utilization and service quality. For example, research on the composite design strategy for the interface space of rail transits and urban complexes has proposed five design strategies for the interface space centered on plazas, building atriums, three-dimensional roofs, underground commercial pedestrian streets, and sky corridors [13].

1.2.2. Research Trends in Wayfinding Systems and Signage Design

Wayfinding systems are an integral part of urban transportation environments, and their design quality has a direct impact on passenger travel efficiency and experience. Research has shown that an efficient wayfinding system should have clarity, consistency, and adaptability [14,15]. Signage system research peaked between 2015 and 2020, focusing on signage visibility, message density, and layout optimization [16,17].

Today, more scholars have conducted a lot of research on the Guidance Signage System. Lam found that the Guidance Signage System has an important influence on wayfinding behavior by studying the layout of airport passenger terminals based on the wayfinding theory [18]. By analyzing the relationship between the location and content of the signage system settings and the wayfinding results, Xiong, Z. pointed out that the optimization of the signage system not only improves the travel experience of the passengers but also effectively alleviates traffic congestion and enhances the overall operational efficiency of the transportation system [9]. K.M. abstracted subway stations into topology maps in order to study the setup method of the Guidance Signage System and pointed out that problems in the design of the signage system can be identified in advance through simulation, and errors and confusion in their actual operation can be reduced [19]. Zhang proposed a guide map for pedestrian transportation networks to determine the best potential locations for signs [20].

However, most of these studies are based on idealized spatial models and do not fully consider the human factors and dynamic changes in the actual environment. For example, during peak hours, signs may be blocked by crowds, leading to the failure of their guidance function [21]. Therefore, how to design a signage system that adapts to complex environments and dynamic needs has become an important direction of current research. In addition, studies have shown that factors such as height, angle, font size, and the color contrast of signs can significantly affect the identification efficiency of passengers [22]. Liu et al. found experimentally that the optimal viewing angle of signs ranges from 15° to 30°, and exceeding this range leads to a significant increase in recognition time [23].

In the layout of Guidance Signage System locations, further consideration of the visual field of pedestrians can more accurately reflect the actual situation and improve the guidance efficiency of signs. Existing studies have shown that it can help to optimize the facility design; for example, Yu investigated the Beijing South Railway Station Comprehensive Transportation Hub based on the visual perception mechanism [24].

Existing research mainly focuses on the definition, composition, design principles, and optimization strategies of connecting space, focusing on theoretical guidance, lacking quantitative analysis, and urgently requiring the introduction of quantitative research methods. Most studies focus on specific environments or single factors, and there are also limitations in analysis methods, lacking the comprehensive consideration of multiple spatial environments, as well as systematic research on the optimization of the Guidance Signage System in complex transportation hubs. In addition, there is a lack of a more comprehensive assessment of the adaptability of the Guidance Signage System in the articulated space of TOD sites, which does not overcome the limitations of a single research method.

1.3. Guidance Signage System Space Location Design Specification and Related Theories

1.3.1. Norms for the Spatial Location of Guidance Signage System

With regard to the layout of the Guidance Signage System, the relevant state departments have issued relevant standards requiring localities to setup the Guidance Signage System at key locations such as starting points, finishing points, turning points, and bifurcation points so as to eliminate pedestrians' doubts about routes. The sign should be controlled at a height within 5° above the horizon, with an offset angle of no more than 15° for static observation and no more than 45° for dynamic observation, i.e., when the pedestrian's head is naturally rotating, and it should be ergonomically designed to remain within a sight height of about 1.5 m.

1.3.2. Pedestrian Visible Area and Visible Area of Guidance Signage System Study

Pedestrians have a semi-curved range of visual fields, including the optimal spatial field of view and the visible spatial field of view. The optimal spatial field of view is the range in which information is most easily accessible when pedestrians are looking straight ahead, while the visible spatial field of view is the range in which pedestrians are able to observe the information under the same conditions [25], which corresponds to the optimal and visible viewing angles of 60° and 120° horizontally and 25° and 50° vertically, respectively, and the clear visibility distance of the Guidance Signage System in this range is 14.10 m and 25.20 m, respectively (Figure 1).

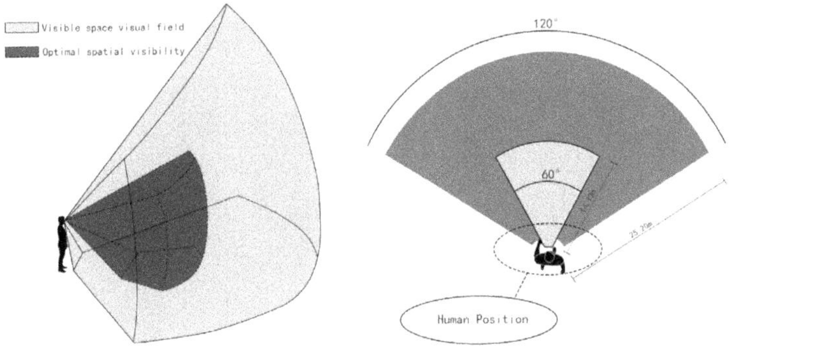

(a) (b)

Figure 1. Pedestrian visibility schematic [25]: (**a**) the best, visible spatial view of the human Guidance Signage System; (**b**) schematic of the visible area of the Guidance Signage System.

According to the visual principle, the visible area of the Guidance Signage System forms a fan-shaped area with the sign as the apex, a vertical offset angle of 5°, and a viewing angle of 140° (Figure 2) [26]; the farthest visible distance for pedestrians is when lifting up the eyes by 5°, and the maximum observable distance of the hanging Guidance Signage System is 17,178 mm [27].

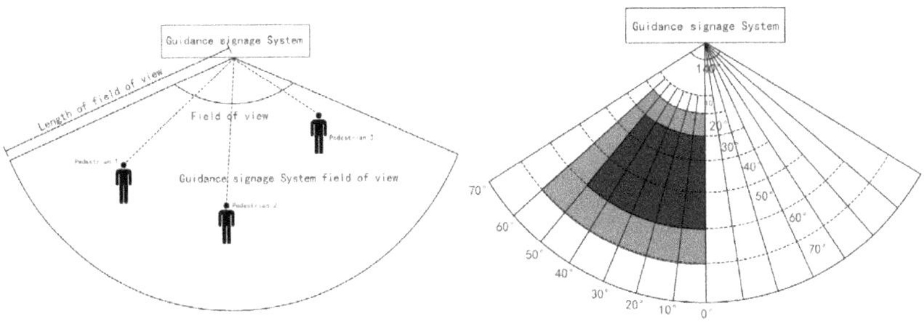

(a) (b)

Figure 2. Schematic diagram of the visual field of a Guidance Signage System [26]: (**a**) display area for Guidance Signage System; (**b**) display area for signage information.

1.3.3. Guidance Signage System

(1) Classification of Guidance Signage System

The Guidance Signage System refers to a specific space and can convey intuitive, clear visual information to achieve effective guidance for passengers through text, graphics, or symbols; it can provide guidance, direction, identification, warning, and other functions, and it is an important tool to provide wayfinding information and help pedestrians to find their way or address their behavior [28].

According to the results of the research summarized, the Guidance Signage System is installed in the following four positions: Hanging, Attachment, Ground-Standing, and Flush-Mounted installations (Table 1). Since the Technical Specification for Guidance Signage System in Public Buildings (GB/T51223-2017) [29] clearly stipulates that the spatial location of the sign should be within a 5° angle upwards of the horizon, when choosing the type of Guidance Signage System, this study focuses on three types of signage: Hanging, Attachment, and Ground-Standing signs and sets them up in a base scene.

Table 1. Classification of Guidance Signage System.

Hanging	Attachment	Ground-Standing	Flush-Mounted

(2) Orientation signage spatial field of view division

In the process of wayfinding, the Guidance Signage System in the connecting space leads to corresponding changes in the visual perception and behavioral habits of individuals due to the spatial variability of the nodes. Therefore, when exploring the visual prominence of the Guidance Signage System, it is necessary to make reasonable adjustments to the location area as well as the type of Guidance Signage System in the experimental scene according to the characteristics of the specific scene space. In the studies of visual scenes by scholars at home and abroad, the visual field plane is usually subdivided into several fixed gaze regions, each representing a specific gaze target. Geoffrey U [30], in his study of visual attention while driving, divided the gaze region into nine regions that do not overlap with each other (Figure 3).

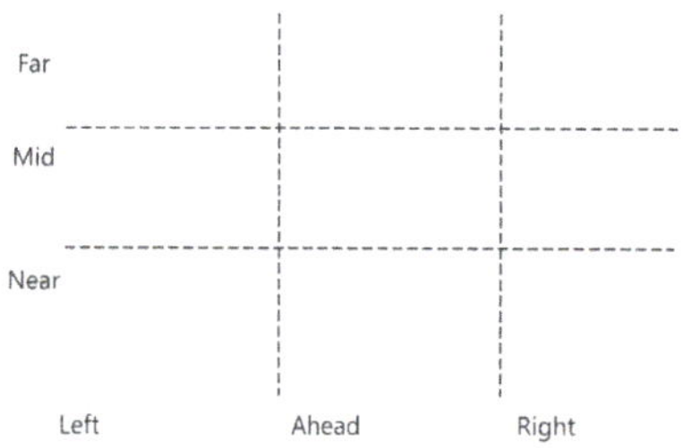

Figure 3. Visual zoning. Geoffrey U's view plane mechanical delineation diagram.

1.4. Research Objective

Given the complexity and subjectivity of spatial cognition, there is a lack of objective knowledge and understanding of the wayfinding process relative to traditional cognitive experimental methods (e.g., questionnaires); ignoring dynamic interfering factors such as crowd density and visual occlusion, the problem of the occlusion of signage during peak hours significantly increases the failure rate of wayfinding, and it is difficult to quantify the impact of such dynamics through questionnaires [31,32]. In this paper, we use the spatial syntax software Depthmap10 and Massmotion11.5 software to analyze the articulation of the spatial visual domain and introduce the method of virtual wayfinding through eye-tracking technology to analyze the influence of the spatial location of the Guidance Signage System on visual saliency. It provides support for the design and visual prominence of the Guidance Signage System in the connecting space of TOD stations, aiming to optimize the layout of the spatial location of the Guidance Signage System from the perspective of passengers' needs and enhance their travel experience and access efficiency.

2. Methods

2.1. Multi-Software Spatial Location Analysis Based on Visual Saliency

2.1.1. Research Hypothesis

Based on the literature and theoretical analyses mentioned above, it is shown that there is a relationship between the setting of the spatial location of the Guidance Signage System and the impact of visual salience. In this regard, this study proposes the following hypothesis: within the same connecting space, there is a difference in the visual perceptual salience of the placement of the Guidance Signage System, i.e., the visual appeal of the Guidance Signage System can be enhanced at certain specific locations. The visual prominence characteristics of the Guidance Signage System vary in different connecting spaces but are also related to spatial location.

2.1.2. Research Step

This study analyzes the effect of the spatial location settings of the guide signage system on the visual salience of pedestrians at TOD sites. The steps of the study are shown in Figure 4.

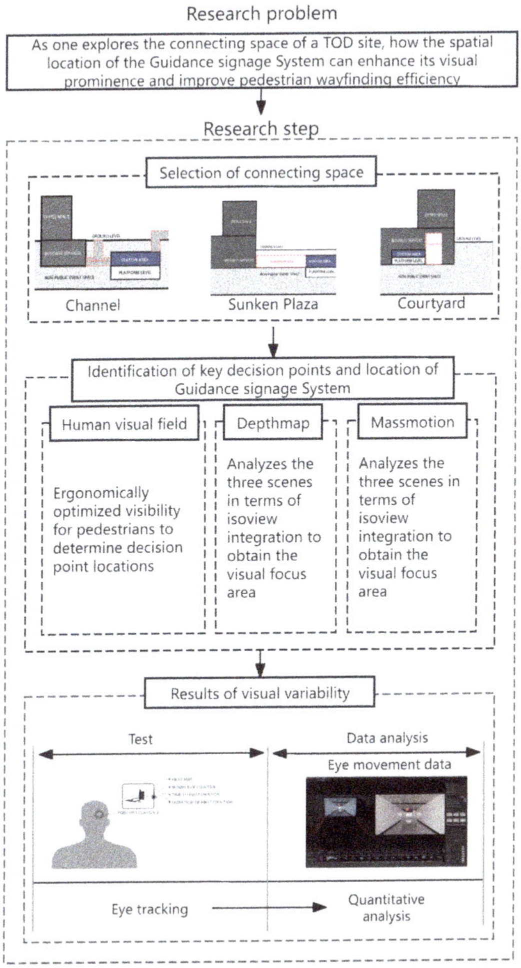

Figure 4. Research steps.

This study is divided into the following three steps: Firstly, three different transition spaces between the rail stations and their neighborhoods are selected as the experimental base scenarios based on existing cases and field research. Secondly, according to the research on the visual field of the Guidance Signage System, passengers' visual perception, and the principle of ergonomics, the visual field simulation of different articulated spatial scenes using Depthmap10 and Massmotion11.5 is used to determine the location of the Guidance Signage System and improve the scene map of different spatial locations of the Guidance Signage System. Thirdly, a simulation experiment is conducted with the help of eye-tracking technology; the variability of the perceived degree of different sign placement locations is analyzed based on the eye-tracking data.

2.2. Scenario Construction

The scenario construction consists of two research steps: the experimental base scenario construction and the multi-software collaborative approach, respectively.

With reference to the Code for Fire Protection Design of Buildings, Code for the Design of Store Buildings, Uniform Standard for Design of Civil Buildings, Code for Fire Prevention in Design of Interior Decoration of Buildings, and other relevant design codes in China for commercial, metro, and public buildings, the Renmin Beilu Station, Chunxi Road Station, and Guanghua Park Station in Chengdu City are used as the basis for constructing the basic model of a connecting space, and this is combined with the results of the research and the summary of the literature to carry out the design. Minjie Li [33] analyzed the congestion points through simulation experiments, counted the stopping positions of passenger flows going to commercial complexes from the underground direction, and finally summarized the key spatial nodes, such as access bifurcations, atrium entrances and exits, and plaza entrances and exits. This was based on the interview results of relevant experts from Chengdu Metro, China Southwest Architecture, China Railway First Survey and Design Institute Group Co., Ltd. (Chengdu, China), and the contents of relevant specifications. Therefore, the size of the channel connecting the space model established in this study was taken as 37.5 m $\times$ 6 m wide (the length of the channel center axis is 37.5 m); the area of 500 m^2 was selected as the base model of the atrium and plaza connecting space, taking into account the fact that the spatial scale affects the flow of passengers. It is concluded by the distribution of passenger flow by Massmotion that the distribution of the density of the people is more uniform when the aspect ratio is 1:1. The business case and site interface space were used as reference models to derive the experimental base scenarios. The final experimental base scenario model was based on 1:1 physical scale modeling to restore the real scenario as much as possible. The specific scenes are shown in Table 2.

Depthmap10, as a spatial syntactic analysis software, performs the graphical analysis of the visibility of the spatial environment, and the results obtained from the drawn axial and convex spatial analysis models reflect the permeability, accessibility, and convenience of space. MassMotion11.5 is a passenger flow simulation software based on social force modeling algorithms developed by Oasys Arup in the UK, which can provide designers with a lot of clear information, such as the way pedestrian access equipment is used, the safety of the pedestrian space, and the phenomenon of crowded spaces. Based on the theory of spatial syntax, Luo combined quantitative data and correlation analysis to explore the impact of spatial accessibility and pedestrian flow on the vitality of underground space by taking the underground space connected to rail transit stations as an example. In this study [34], Ma used Massmotion to simulate the 3D visual time of travelers on the wall, showing the focus of attention of the pedestrian's field of view in his simulated visual time map and proposed a method to simulate the focus of attention of the view [35].

Table 2. Experimental base scenarios.

Experimental Stimulus Material		

Experimental base scenario diagram		
Channel Bifurcation	Atrium Entrance and Exit	Sunken Plaza Entrance and Exit
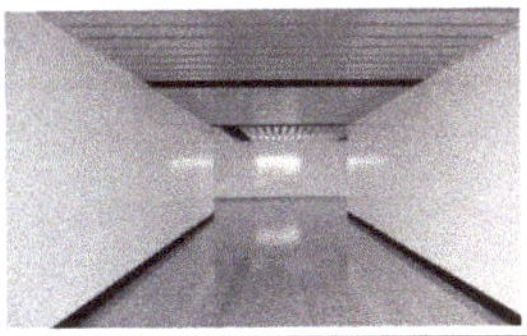	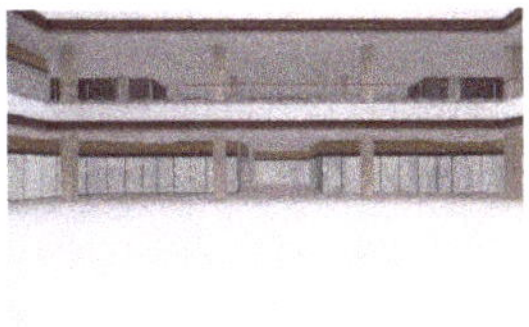	

Based on the above research, this study utilizes Depthmap10 to analyze the three scenarios to obtain the visual focus area and then combines this with the signage, passenger visual field, and Massmotion11.5 passenger flow line for the visual field superposition analysis so as to achieve the design of an appropriate location according to the Guidance Signage System.

2.2.1. Visual Field Overlap Analysis

Combining the above visual area division and the type of Guidance Signage System, a three-dimensional spatial coordinate system was created to divide the visual areas, and according to the nine-grid composition method in photographic composition techniques, a straight line was drawn along each of the X, Y, and Z axes in a single-point perspective scene to form a $3 \times 3 \times 3$ three-dimensional spatial grid, so as to divide the screen into nine independent visual areas (Figure 5), where each of these visual areas corresponds to a different location of the Guidance Signage System in the experimental scene.

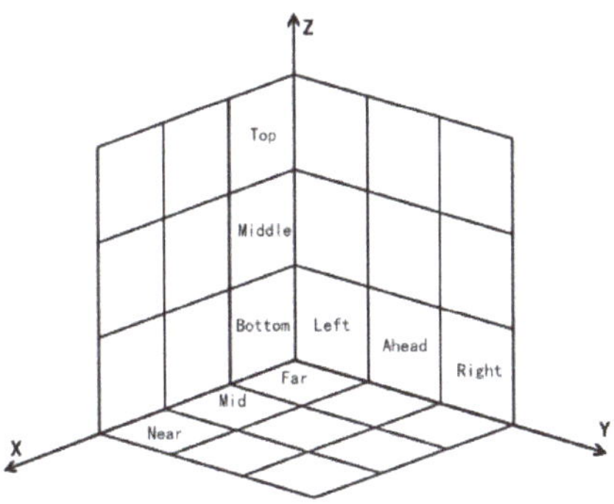

Figure 5. Orientation signage location zoning.

Due to the continuity of the signage setup [36], each path within the connecting space covers several nodes. When laying out the locations of the Guidance Signage System, it is necessary to select suitable points from these nodes as decision points for sign setting. In this study, the locations where pedestrians face difficulties or are most likely to be confused during path selection are defined as decision points.

In this paper, we first employ the spatial syntax software Depthmap10 to analyze the three kinds of connecting spaces in terms of their visual domain before subdividing the spatial plan into a grid, which covers the entire 2D visible area, and when setting the grid size, we balanced the calculation precision and the amount of operations, referring to the research of Zhou Xi [37] and others, selecting a grid size of 1.0 m (the common size is 1–1.5 m). The structure of the visual integration degree calculation is shown in Table 3, where the redder the color, the higher the visual integration degree, which indicates the greater visual visibility of the space. Secondly, the Massmotion11.5 software is used to simulate the passenger flow routes in the three connecting spaces (Figures 6 and 7), and the ground-based Guidance Signage System is laid out in such a way as to avoid the passenger flow routes as much as possible. The results of the visual field analysis of the three kinds of spaces calculated by Depthmap and Massmotion software are shown in Table 3.

Table 3. Analysis of Visibility in Various Connecting Spaces.

A-Channel Isoview Integration	Guidance Signage System Visual Field A1	Pedestrian Visibility A2	Field of View Overlay A3
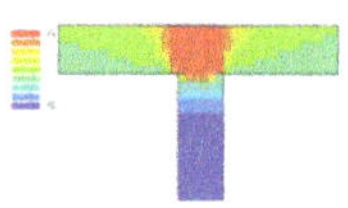	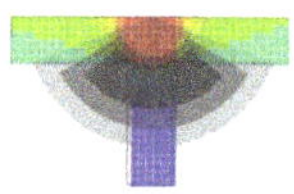	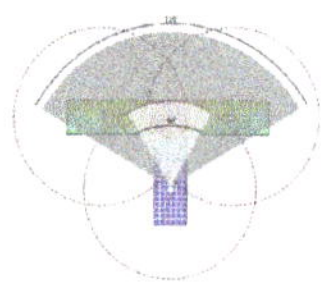	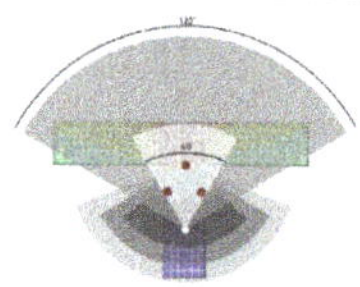
B-Degree of integration of viewsheds such as atriums	Guidance Signage System visual field B1	Pedestrian visibility B2	Field of view overlay B3
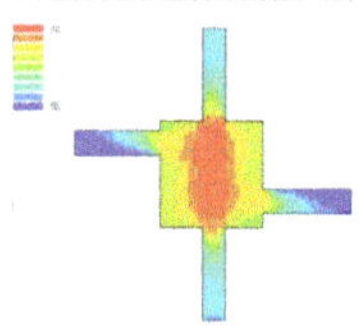	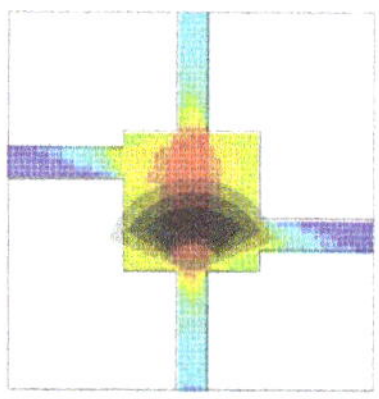	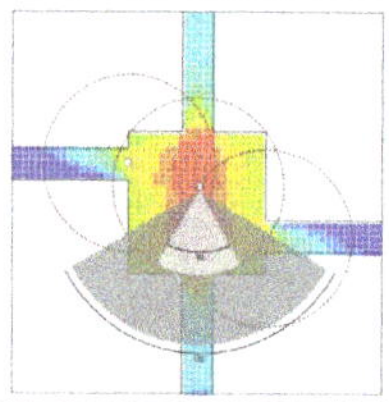	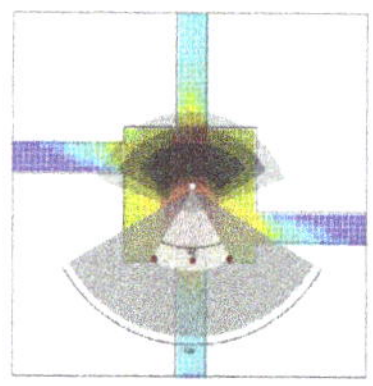
C-Degree of integration of fields of view such as plazas	Guidance Signage System visual field C1	Pedestrian visibility C2	Field of view overlay C3
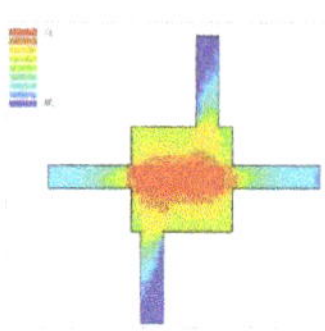	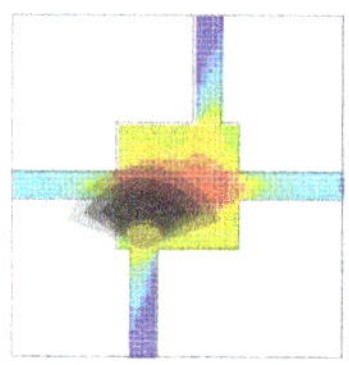	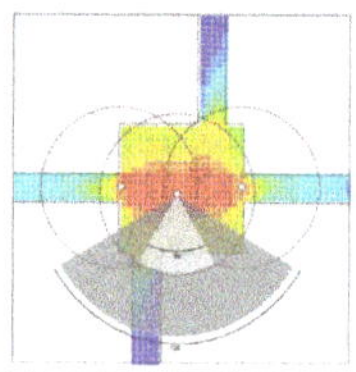	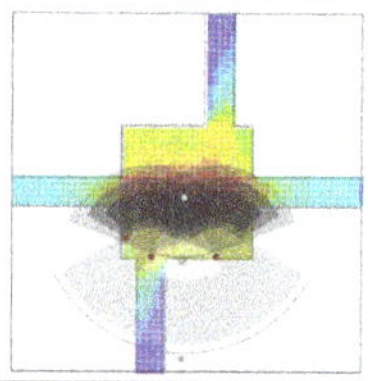

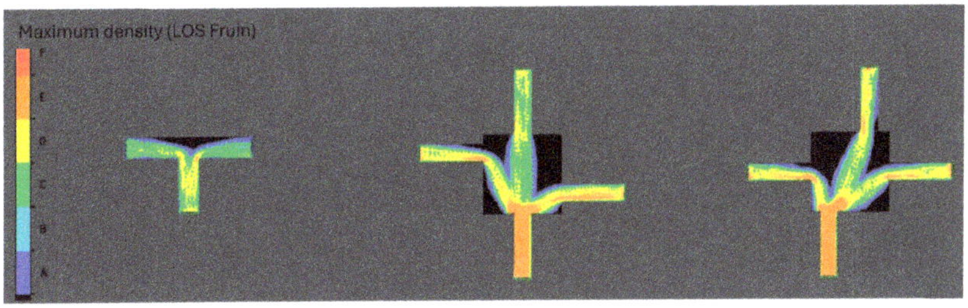

Figure 6. Occupancy Density Distribution Map.

Figure 7. Personnel route flow diagrams(The orange sections show the distribution of people's movement).

2.2.2. Demonstration of Spatial Location of Standard Guidance Signage System

The experiment is based on the above nine different visual areas, the results of the visual field superposition analysis, and the actual application of the spatial layout of the situation, according to the guidance of the hanging type, attached type, and ground type to select the appropriate type of Guidance Signage System arranged in the simulation scene. Based on the literature [38] and research analysis, it can be seen that the signage system used in the process due to the different node spaces makes people produce certain visual and behavioral habit differences; thus, in the study of the signage system of visual salience, due to the space of the different experimental scenes in the location of the setup area to make certain adjustments, in Teng's study of the visual salience of guide signs in underground commercial streets, ergonomics, and human visual characteristics were combined to divide the experimental scene into nine parts that covered the entire visual area, We followed her delineation of visual areas, but unlike Teng's study, the spatial location of the signs mentioned in this study in conjunction with the above was within a 5° angle upward from the horizon, disregarding the three areas below the horizontal line of sight, which were not the focus of our attention. Therefore, six areas were selected as the locations of the Guidance Signage System in the scenes of channel bifurcation (A01–A06), the atrium entrance and exit (B01–B06), and sunken plaza exit (C01–C06), and the corresponding research was carried out using the following specific scenarios.

2.3. Oculomotor Simulation Studies

(1) Experimental equipment

In this experiment, the Tobii Glasses2 ophthalmoscopic eye-tracking device was used as the experimental tool, which mainly consists of three parts: the headgear module, the recording module, and the external data processing, with a sampling rate of 50 Hz, a 1-point calibration, a scene camera with a resolution of 1920 × 1080 pixels, and a camera recording field of view ranging from 82° horizontally to 52° vertically.

(2) Experimental Participants

A total of 25 volunteers were recruited to participate in the simulation trial, including 10 males and 15 females, with an age distribution between 20 and 50 years old, all in good physical condition and without color blindness or color weakness. In the eye-tracking experiment, we strictly followed the ethical requirements of the study to ensure that the rights and interests of all participants were fully protected. All participants were informed of the purpose of the study, the experimental procedure, and the use of the data before data collection. Meanwhile, in order to protect the privacy of the participants, all the data were anonymized or de-identified, and any personally identifiable information will not be disclosed in any research results or public reports.

(3) Experimental stimulus material

The scene diagram in Table 4 was used as the material for the 18-sheet visual experiment. In order to avoid the influence of color on location perception, the experimental scene was set in grey, the logo was a differentiated dark tone, and a blank page was used for the transition between each page. The research experiment is divided into four steps, and the specific flow is shown in Figure 8.

Table 4. Scenarios of different spatial locations for Guidance Signage System.

A-Channel Bifurcation			
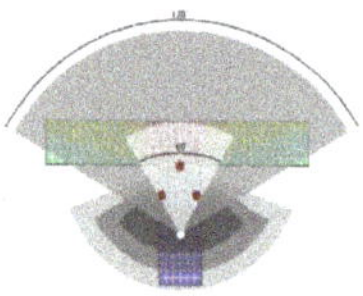	A01	A02	A03
	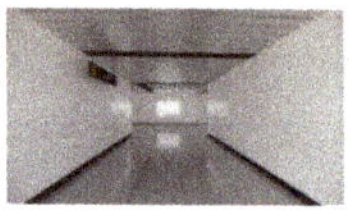	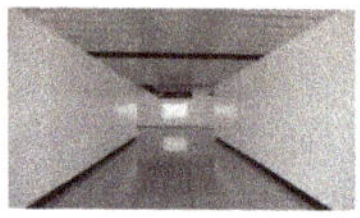	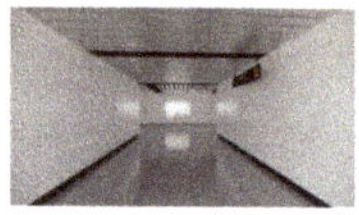
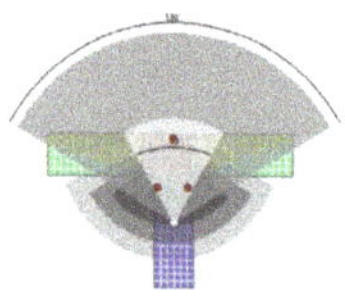	A04	A05	A06
	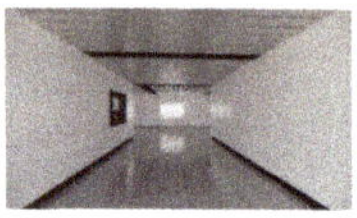	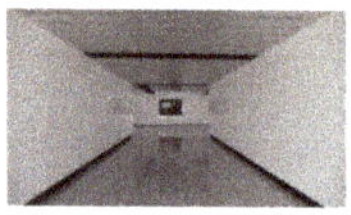	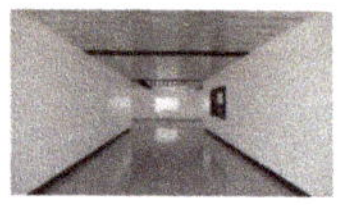
B-Atrium entrance			
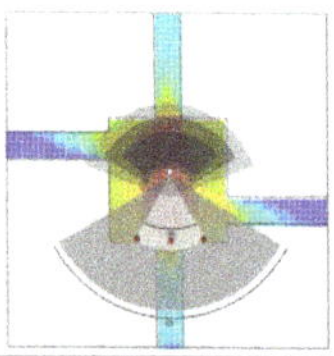	B01	B02	B03
	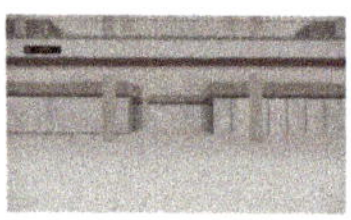		
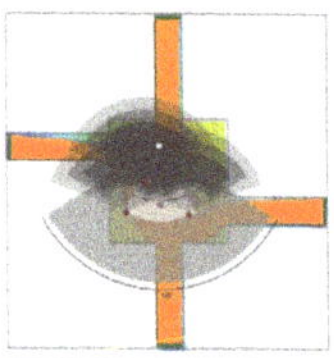	B04	B05	B06
		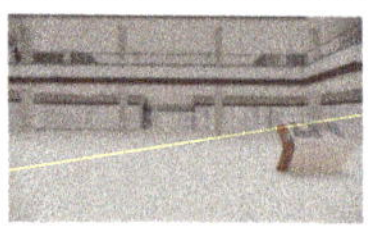	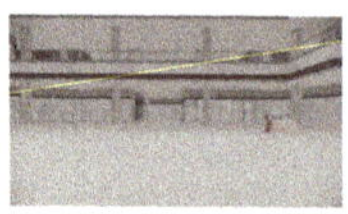

Table 4. *Cont.*

C-Sunken Plaza Entrance and Exit			
	C01	C02	C03
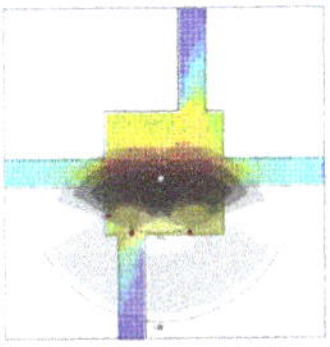			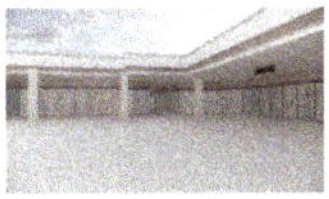
	C04	C05	C06
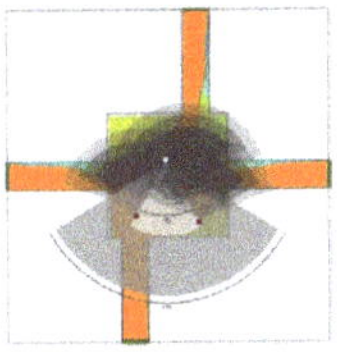			

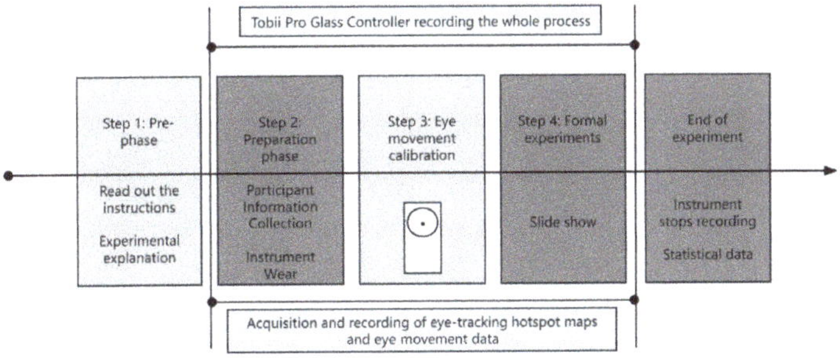

Figure 8. Flow chart of the eye movement experiment.

2.4. Analysis of Results

After the completion of the experiment, the experimental data were collected and sorted out, and two samples that were not qualified due to missing recorded data were excluded; finally, twenty-three qualified sample data were obtained. The qualified data were processed in Tobii Pro Lab software (Version 24.21), the required hotspot maps and eye movement data were output and exported, and the data were summarized and analyzed.

The visualization hotspot map is mainly used to obtain the distribution of eye movement behaviors in space within a certain period of time, which can reflect a large number of regions in the stimulus that attract the subject's attention, has a two-dimensional type of data representation, and has the size of the value measured by the eye-tracking device displayed in different colors, in which the red color indicates the most focused attention and the region of the longest duration of gaze, followed by yellow, and the green color indicates the region of the least focused attention; as can be seen from the resulting visualization hotspot map, darker the red area, the more attention and focus the Guidance Signage System has from the subjects.

As shown in Table 5, which is the hotspot distribution map of different spatial location scenes, with the change in the deployment location, the distribution pattern of the hotspot map presents certain differences, and the comprehensive eye movement view of different spatial location scenes can be found as follows: ① For the same type of Guidance Signage System in different spatial scenes, the closer the layout position is to the pedestrian's line of sight, the more attention it can attract and the more focused it is; the further away it is, the more dispersed it is. ② Combined with the human visual law to analyze the results of the experiment, it can be concluded that when people are observing objects, the line of sight tends to move along the direction from top to bottom and from left to right. The Guidance Signage System on the left side of the visual area is more likely to be noticed than that on the right side.

Table 5. Scene hotspot maps for different spatial locations.

A01	A02	A03
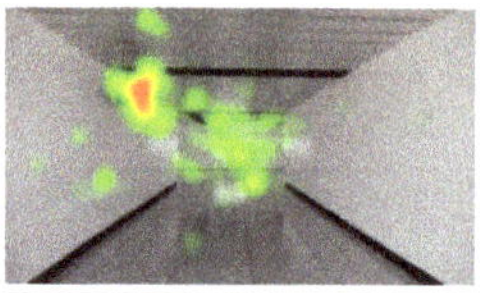	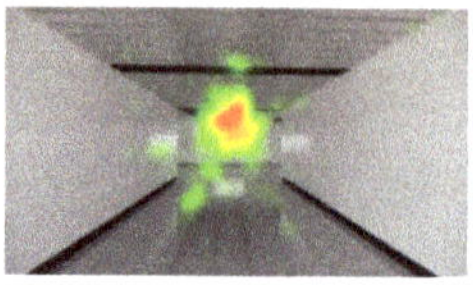	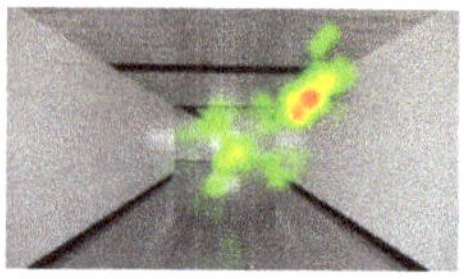
A04	A05	A06
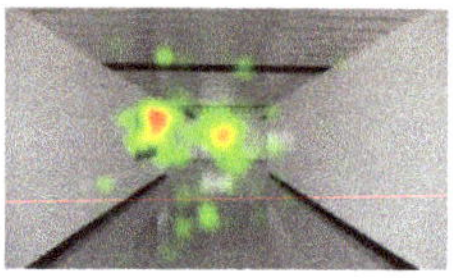	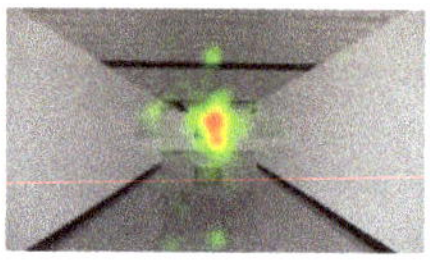	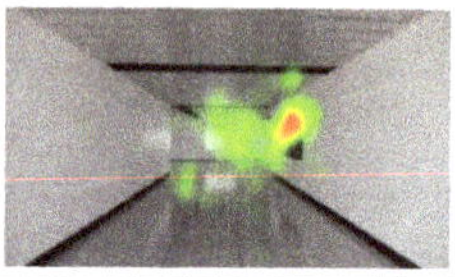
B01	B02	B03
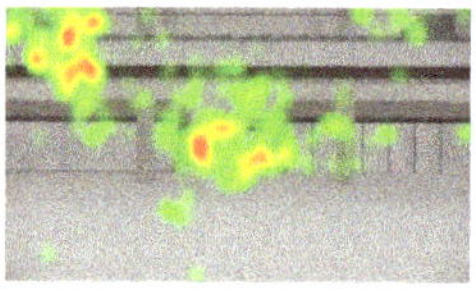	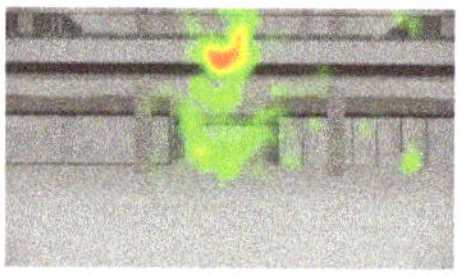	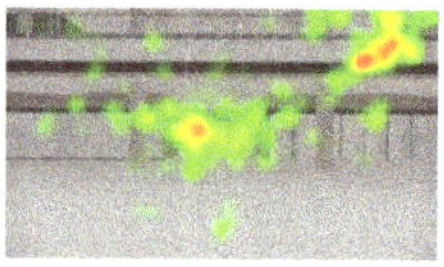
B04	B05	B06
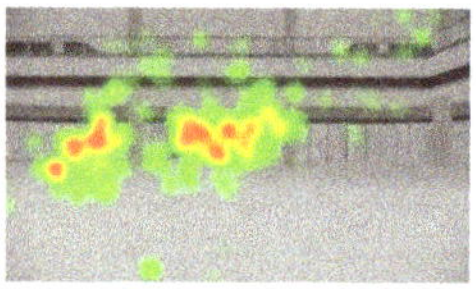	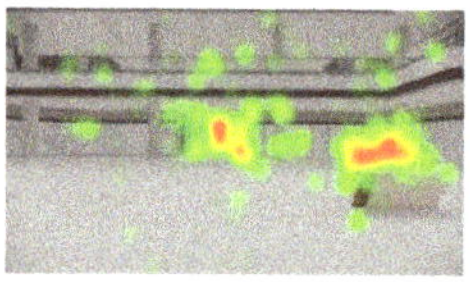	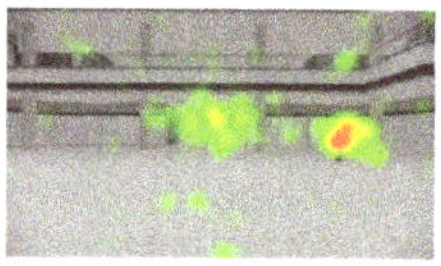
C01	C02	C03
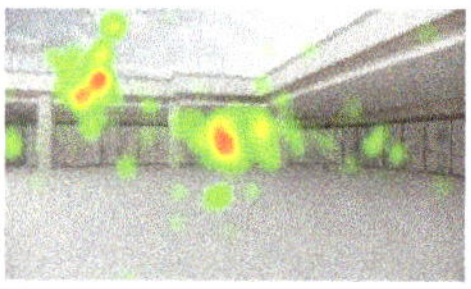		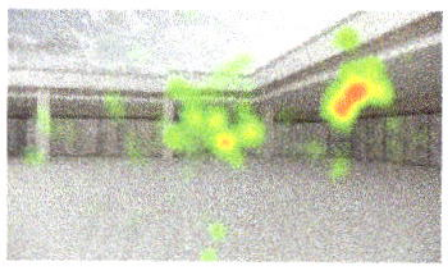
C04	C05	C06
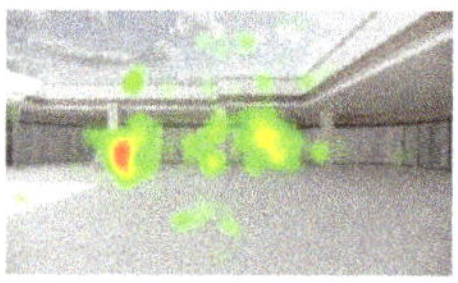	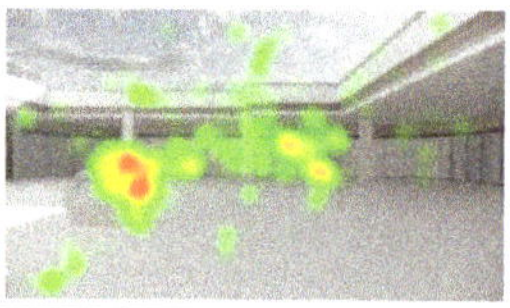	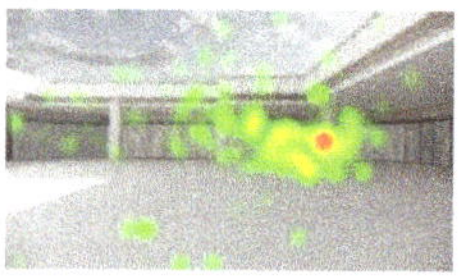

In summary, after analyzing the visual hotspot maps of each scene, it can be found that changes in the layout of the Guidance Signage System directly lead to different distribution patterns of the hotspot maps for each scene, which further confirms the viewpoints of hypothesis one: the location of the Guidance Signage System does have a significant impact on visual saliency. A reference basis is provided for the experimental results.

By analyzing the visualization hotspot map mentioned above, further in-depth investigation with the help of eye-tracking data can be conducted to clarify how the Guidance

Signage System set in different locations affects visual saliency and reveals their roles. Yan Guoli [39] and others, in a review article on eye movement indicators in the field of reading research, emphasized that the selection of appropriate and effective eye movement parameters as the basis of this study is the core of ensuring the quality and depth of eye movement research. Li investigated the effectiveness of prohibited safety sign locations in a coal mine environment by collecting data on the number of gaze points and the time of the first gaze [40]. Li, H.X. selected first gaze time, the total access time, and a number of gaze points as indicators to explore the effectiveness of emergency signage during the evacuation of people in emergency situations [41].This study is an exploratory study [42], so the analysis of eye movement data such as first gaze time, the number of gaze points, duration of first gaze time, and so on (Table 6) during the process of wayfinding indicates the visual perception of the subjects, and the different scenes are classified and sorted in combination with the experimental data (Table 7). Changes in eye movement indices were statistically described and plotted on a line graph (Figures 9 and 10).

Table 6. Mean eye movement data in the region of interest for different spatial location scenarios.

Eye Movement Parameters / Visual Area	Number of Fixation (NF)	Time to First Fixation (TFF)/ms	Duration of First Fixation (DFF)/ms
A01 (upper left)	2.83	130	275
A02 (upper middle class)	3.10	174	475
A03 (upper right)	3.00	200	336
A04 (left and center)	2.70	78	225
A05 (epicenter)	3.80	110	291
A06 (middle right)	3.21	246	429
B01 (upper left)	2.33	187	640
B02 (upper middle class)	2.33	226	344
B03 (upper right)	1.75	96	360
B04 (left and center)	2.19	233	250
B05 (lower left)	4.00	139	274
B06 (middle right)	2.82	80	401
C01 (upper left)	2.25	71	170
C02 (upper middle class)	3.38	186	229
C03 (upper right)	1.89	115	388
C04 (left and center)	2.00	222	641
C05 (lower left)	3.77	148	397
C06 (middle right)	2.14	182	415

Table 7. Concept of eye movement metrics and sorting of eye movement data in areas of interest for different spatial location scenarios.

Eye Movement Index (EMI)	Interpretation of Indicators	Sort
Time to first fixation	Refers to the time it takes for a participant to first gaze at a specific AOI region, with smaller TFF values indicating that the region was noticed by the subject in the shortest possible time.	A04 < A05 < A01 < A02 < A03 < A06
		B06 < B03 < B05 < B01 < B02 < B04
		C01 < C03 < C05 < C06 < C02 < C04
Number of fixations	Refers to the total number of gaze points in a given AOI region, with larger NF values indicating that the region is more noticeable to the sub-fixation ject.	A05 > A06 > A02 > A03 > A01 > A04
		B05 > B06 > B02 = B01 > B04 > B03
		C05 > C02 > C01 > C06 > C04 > C03

Table 7. *Cont.*

Eye Movement Index (EMI)	Interpretation of Indicators	Sort
Duration of first fixation	It refers to the duration of the first gaze in a specific AOI region, and a larger DFF value indicates that this region is more attractive.	A02 > A06 > A03 > A05 > A01 > A04
		B01 > B06 > B03 > B02 > B05 > B04
		C04 > C06 > C05 > C03 > C02 > C01

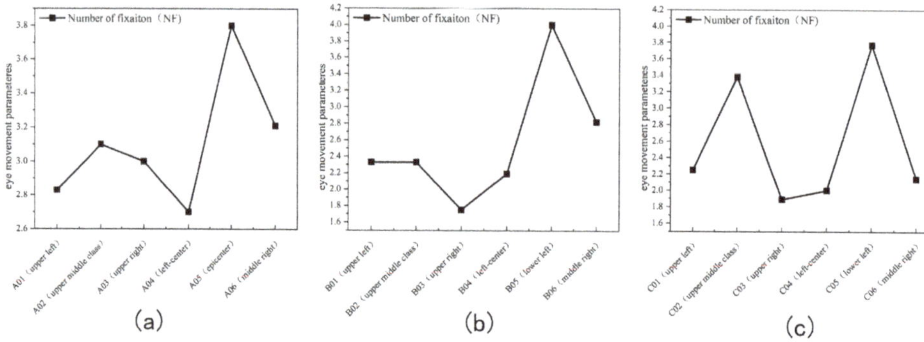

Figure 9. Statistics and comparison of indicators on the number of fixations; (**a**) Number of fixations (NF) for areas AO1 to AO6; (**b**) Number of fixations (NF) for areas BO1 to BO6; (**c**) Number of fixations (NF) for areas CO1 to CO6.

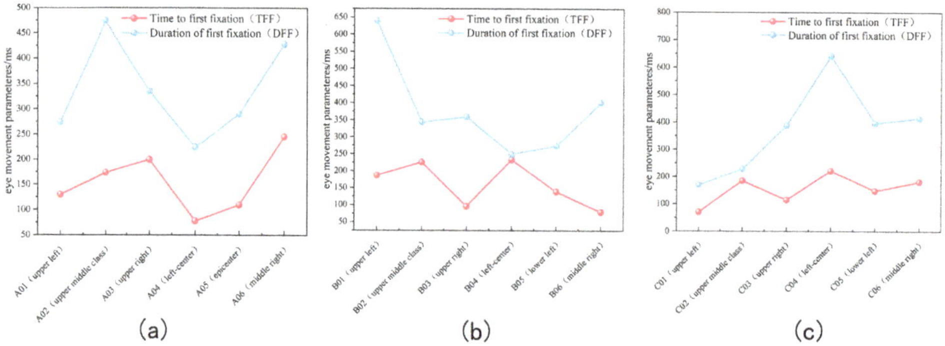

Figure 10. Statistics and comparison of time to first fixation and duration of first fixation indicators; (**a**) Comparison of Time to First Fixation (TFF) and Duration of First Fixation (DFF) for areas AO1 to AO6; (**b**) Comparison of TFF and DFF for areas BO1 to BO6; (**c**) Comparison of TFF and DFF for areas CO1 to CO6.

In the study of the impact of the spatial location of the Guidance Signage System on visual salience, the time to first fixation is a crucial eye movement index. It directly reflects the degree of attention and attraction of the Guidance Signage System's placement and can be used as an important basis for preliminary screening and assessment of the advantages and disadvantages of placement. However, in order to have a more comprehensive understanding of the influence of the location of the Guidance Signage System on visual saliency, it is also necessary to comprehensively consider other eye-movement indicators as well as other factors in the actual scene.

In the scene of the channel bifurcation, the TFF value of area A04 (middle left) is the smallest, which indicates that the Guidance Signage System was quickly noticed by the subjects, while its NF and DFF values were small relative to those of the signs in other locations. For the sign in area A02 (upper middle), although its TFF value is larger than that of area A04 (middle left), its NF and DFF values are the largest. Therefore, on balance, the visual prominence of the Guidance Signage System is greater when it is set in visual area A02 (upper center). In this way, based on the statistical frequency of the three eye movement indicators, it is concluded that in the atrium entrance/exit scenario, it is difficult for pedestrians to quickly notice the Guidance Signage System located above the visual area, which is due to the complex environment of the atrium, and the surrounding arrangement of the commercial area, which, to a certain extent, distracts the visual attention of pedestrians. The significance of the Guidance Signage System located in the visual area of B05 (lower right) is higher, which indicates that the closer the distance of the same type of Guidance Signage System is from the pedestrians' line of sight, the easier it is to attract their attention; in the sunken plaza scenario, the significance of the Guidance Signage System is higher when it is set in the visual area of C05 (lower left). It is also confirmed that in the connecting space of different scenes, the location of the more visually salient Guidance Signage System varies, indicating that there is variability in the distribution of the focus of visual attention between different scenes.

When looking through Figure 10, it can be noticed that the trend of the folds in the two charts (a) and (c) are similar. In these charts, the trends of TFF and DFF are synchronized. For example, in region A03, both TFF and DFF reach higher values, showing the stronger attraction of this region to observers. Regarding other regions such as A01, A06, etc., TFF and DFF show relatively consistent upward or downward trends, which indicates a strong correlation between visual attraction and dwell time in these regions [43]. However, the situation in (b) is different. Although the values of TFF and DFF varied from region to region, there was a clear divergence in their trends. For example, region B04 has relatively low TFF and a high DFF value, implying that although visual attention is rapidly focused, the dwell time is longer. Conversely, region B05 showed the opposite situation, with higher TFF and lower DFF, possibly indicating that although the observer's initial gaze was more rapid, interest in the region did not last long [44].

3. Discussion

From the discussion, it can be concluded that the spatial location design of signage systems for articulated spaces at TOD sites plays a crucial role in enhancing visual salience and guidance efficiency. By combining software tools such as Depthmap10 and Massmotion11.5 with eye-tracking technology, this study identifies the optimal placement strategies for signage in different types of articulated spaces (e.g., passageway bifurcations, atrium entrances and exits, and entrances and exits of sunken plazas). The results of this study emphasize that signage locations should be differentiated according to the spatial characteristics of different articulation spaces in order to maximize visual attention and ensure smooth passenger flow. This study provides a scientific basis for the design of signage systems at TOD sites and practical guidance for the efficient operation and user-friendliness of urban transportation hubs.

3.1. *The Spatial Location of the Guidance Signage System Is Closely Related to Visual Prominence*

3.1.1. Differences in Location Lead to Significant Differences in Visual Saliency

The results show that the Guidance Signage System located in the visual areas A02 (upper middle), A05 (middle), B01 (upper left), B05 (lower right), C02 (upper middle), and C05 (lower left) are more likely to attract the attention of the subjects. In Geoffrey U's study [30], it is clearly stated that the road far ahead and mid-ahead attracted more fixations than any other part of the scene, i.e., the upper middle and center positions; by contrast, the road near the left and road near the right tended to attract fewer fixations than other parts of the scene, i.e., the lower left and lower right positions, which differ from our findings, and the reason for analyzing this may be that Geoffrey U's study may have focused more on the road driving scene, where the spatial layout of the driving scene is relatively simple, and the visual attention of pedestrians is mainly focused on the front of the vehicle. However, this study focuses on the Guidance Signage System within a TOD site; the complexity of the scene and the different functional requirements require pedestrians to make judgments at multiple decision points, leading to differences in the distribution of subjects' visual attention, and the present study focuses more on guiding pedestrians to make quick decisions in complex spaces, while Geoffrey U's study may focus more on the processing of visual information during the driving process. Second, this study used eye-tracking technology, and compared to Geoffrey U's study, the eye-tracking data provided more precise information about visual attention points, allowing for a more detailed and accurate analysis of pedestrian's visual attention. Accordingly, the results show that the hanging and attached Guidance Signage System are setup in the middle of the visual field and above, and the closer the stand-up Guidance Signage System is to the line of sight, the easier it is for them to be gazed at, and they have a higher degree of visual significance, which verifies research hypothesis one. The spatial location occupied by the more visually salient Guidance Signage System varies in different connecting spaces, which validates research hypothesis two. According to human visual patterns and ergonomics, the Guidance Signage System should be placed within the natural line of sight of pedestrians and in accordance with the optimal and visible spatial field of view. It explains why a Guidance Signage System located in the upper center and right in the middle are more likely to be noticed, and these positions are more in line with human visual habits. Secondly, Depthmap's visual field analyses showed that signs are more likely to be seen in areas where there is a high degree of visual field integration. These areas are more visually appealing and attract more eye contact.

3.1.2. Advantageous Locations Should Be Prioritized for Signage System Design

In the process of pedestrian wayfinding, the decision point is the location where passengers need to find information the most; for example, when pedestrians arrive at a decision point, they will be guided by the information guiding signs at that decision point, so the advantageous location is set based on the pedestrian decision point, not for anyone in any location [27]. According to Sijing Teng [38], the upper middle, middle, left middle, and right middle regions are prone to form visual foci, where the upper middle and middle regions are the same as in this study; this suggests that these advantageous locations should be prioritized in the design of our signage system. But the left middle and right middle regions are different from the above study. For example, in some scenes of this study, the Guidance Signage System located in the lower right or lower left position (B05, C05) showed high visual salience, while some Guidance Signage Systems located in the central or upper area (A03, A06) were relatively low. The reason for this may be related to the spatial layout of the specific scene, lighting conditions, and other factors. In addition, the visual preference and behavioral habits of different subjects may also have a certain impact on the experimental results, and the location also needs to be analyzed and selected comprehensively in concert with the expression of color and logo content, whereas the present experiment excluded the design of the logo itself in terms of content and color, and focused on the impact of location on visual perception.

3.2. A Proposal for a Spatial Location Design Method for Guidance Signage Systems Based on Multi-Software Simulation

From the results of the study, the TOD site connecting space Guidance Signage System design should be adapted to local conditions in the design stage for scientific and reasonable design. Priority should be given to areas where pedestrians frequently stop and need to make decisions, such as the upper center, right in the middle, and some specific low areas. These areas are usually high-probability areas of visual focus, and signage systems placed in these locations can maximize visual attention and improve the efficiency of information conveyance. Meanwhile, for different types of articulated spaces, such as passages, stairways, entrances, and exits, the layout of the signage system should be adjusted according to the needs of the spatial function to ensure that the signage system can match the spatial characteristics and meet the cognitive needs of the pedestrians. At present, the existing Guidance Signage System often relies on experience and norms; the lack of a more scientific design and method for the layout cannot meet the needs of passengers to find their way. If the arrangement is inappropriate, it will need to be improved and adjusted at a later stage, which will consume huge manpower and material resources and may also cause unnecessary interference and impact the daily operation of the station. In order to arrange the Guidance Signage System more accurately, this study proposes a multi-software simulation of the proposed method for setting up the spatial location of the Guidance Signage System, which consists of several steps (Figure 11).

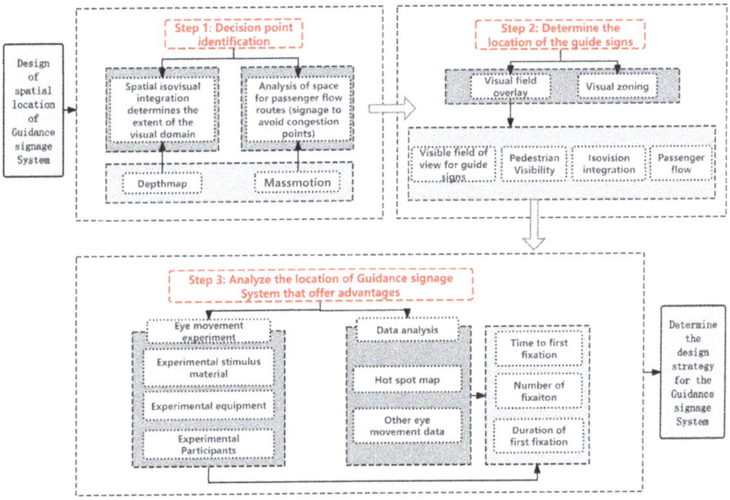

Figure 11. Guidance Signage System space location design process.

Compared with the existing research results [8,9,45], it can be found that although several studies have emphasized the role of the Guidance Signage System in transportation hubs to enhance wayfinding and guidance efficiency, this study is significantly innovative in several aspects. Fei Peng and Zhihua Xiong [8,9] focused on the relationship between the design of the Guidance Signage System and traffic flow in transportation hubs, usually through simulations or questionnaires to evaluate the effectiveness of passenger guidance. Most of these studies have focused on single-function spaces or simple transportation environments and have not explored the signage design in multifunctional and high-density environments in depth. In contrast, our study analyzes the signage layout of multiple articulated spaces in TOD stations by combining software tools such as Depthmap10 and Massmotion11.5 field tests with eye-tracking technology, which provides more accurate and objective data support. Unlike the methodology in Shao's study [45], which relied heavily on traditional design criteria or simulation analysis to explore the effectiveness of the Guidance Signage System, this study examined the effectiveness of the Guidance Signage System through specific spatial scenarios. This study also analyzed the effects of different signage placement locations on the visual saliency of passengers through experiments in specific spatial scenarios. This study combines Depthmap, Massmotion, eye-tracking technology, and other means of analysis, and proposes an all-around analysis method from spatial layout to visual perception. Moreover, in the design of the spatial location of the Guidance Signage System, the visual field of the Guidance Signage System overlaps with the visual perception of the pedestrians, which helps the design of the Guidance Signage System in the connecting space of the TOD station to construct a scientific and reasonable design of the layout at the design stage.

3.3. Limitations and Future Research

There are some limitations to this study. The study did not fully consider the differences in individual characteristics of the subjects, such as the fact that there may be significant differences in visual attention and signage recognition among passengers of different ages, genders, or cultural backgrounds. Future research could explore the responses of different populations to the signage system from the perspective of individual characteristics, which would help provide more personalized guidance for signage system design. In addition, this study ignored the effects of sign color, size, and the surrounding environment when analyzing the effects of the sign system location on the visual salience of pedestrians. The color contrast, size, and environmental conditions of signs can have a significant effect on visual attention; therefore, future research needs to consider the combined effects of these factors in order to optimize signage design. On the other hand, the influence of environmental factors on the effectiveness of signage systems was not fully explored in this study. External conditions such as light and noise may affect sign visibility and passengers' visual responses, and future research should consider how these environmental factors interact with sign design to further improve sign visibility and navigation. By addressing these limitations, future research can not only further improve the design of signage systems but also provide a more comprehensive and flexible solution for signage optimization in transportation hubs, which will promote the efficiency of the transportation system and user experience.

4. Conclusions

This study investigates the visual salience of the Guidance Signage System in different spatial modes of articulation and its impact on passenger wayfinding behavior through the combined use of Depthmap10, Massmotion11.5 software, and eye-tracking technology. The results show that the spatial location of the Guidance Signage System affects its visual attractiveness and wayfinding efficiency to a certain extent. In key connecting spaces, such as passageway bifurcations, atrium entrances, and entrances to sunken plazas, the optimal placement of the Guidance Signage System varies according to the characteristics of the space, presenting a variety of design strategies. These findings not only verify the hypotheses proposed in this paper but also provide an empirical basis for the refinement and scientific design of the spatial Guidance Signage System in the connecting space of TOD stations. This study initially proposes a spatial location design method for the Guidance Signage System in the interface space of TOD stations based on multi-software analysis and clearly points out that the layout of the Guidance Signage System can be optimized to enhance the wayfinding efficiency and traveling experience of passengers, promote the smooth flow of passenger traffic in stations, and improve the operational efficiency and service quality of the whole transport system. Future research can explore the effects of different Guidance Signage System types, colors, sizes, and other factors on visual saliency in order to enhance the practicality and universality of the study.

Author Contributions: Conceptualization, X.M. and Q.Y.; methodology, X.M.; software, X.M.; validation, X.M. and Q.Y.; formal analysis, X.M. and Q.Y.; investigation, X.M. and Q.Y.; resources,

X.M. and Q.Y.; data curation, X.M.; writing—original draft preparation, X.M.; writing—review and editing, Q.Y.; visualization, X.M.; supervision, Q.Y.; project administration, X.M. and Q.Y.; funding acquisition, Q.Y. All authors have read and agreed to the published version of the manuscript.

Funding: This research was funded by China Railway First Survey and Design Institute Group Co., Ltd., Open Research Project "Research on Key Technologies of Railway Station Design Based on TOD Underground Space Integration", [No. 2021KY36ZD(TOD)-01].

Data Availability Statement: The data presented in this study are available upon request from the corresponding author due to restrictions.

Conflicts of Interest: The authors declare no conflicts of interest.

References

1. Calthorpe, P. *The Next American Metropolis: Ecology, Community, and the American Dream*; Princeton Architecture Press: New York, NY, USA, 1993.
2. Gou, M.Z. Experience and Enlightenment of Japan's TOD Station-City Integration Comprehensive Development. *Urban Rail Transit Res.* **2021**, *24*, 15–18.
3. Shiau, T.A.; Huang, M.W.; Lin, W.Y. Developing an Indicator System for Measuring Taiwan's Transport Sustainability. *Int. J. Sustain. Transp.* **2014**, *9*, 81–92. [CrossRef]
4. Sun, L.; Xun, Y.; Teng, S.J. Research into the Visual Saliency of Guide Signs in an Underground Commercial Street Based on an Eye-Movement Experiment. *Sustainability* **2022**, *14*, 16062. [CrossRef]
5. Suzuki, H.; Murakami, J.; Hong, Y.-H.; Tamayose, B. *Transit-Oriented Development in Tokyo*; World Bank Publications: Washington, DC, USA, 2015.
6. Zhang, Y. TOD Planning and Practice in China: A Review. *Urban Plan. Int.* **2018**, *18*, 80–85.
7. Li, X.; Zhang, Y.; Wang, J.; Liu, S.; Chen, Q.; Zhang, T.; Li, M.; Wang, H.; Zhou, L.; Xu, Z. Human-Centered Design in TOD Stations: A Case Study of Shenzhen. *J. Urban Plan. Dev.* **2020**.
8. Peng, F.; Zhang, Z. Detailed Design of Guidance Signage System in Comprehensive Passenger Hub Considering Cognitive Psychology. *China Saf. Sci. Technol.* **2024**, *20*, 165–172.
9. Xiong, Z.H.; Yao, Z.S. Design method of guidance signage in rail transit stations based on wayfinding experiment. *J. Transp. Syst. Eng. Inf. Technol.* **2018**, *18*, 80–85.
10. Xia, Z.; Zhang, Y.; Xu, L. Influencing factors and improvement strategies of TOD performance in rail transit station area. *Planners* **2019**, *35*, 5–12.
11. Eyck, A.V. Place and occasion. *Progress. Archit.* **1962**, *43*, 155.
12. McNeill, D. Fine grain, global city: Jan Gehl, public space and commercial culture in central Sydney. *J. Urban Des.* **2011**, *16*, 161–178. [CrossRef]
13. World Resources Institute. *Creating-Movement and Silence: Creating People-Centered TOD Public Spaces*; World Resources Institute: Beijing, China, 2016.
14. Yang, Q.; Dubey, R.K.; Kalantari, S. PATH-U: A data-driven agent-based wayfinding model incorporating perceived path uncertainty and cognitive strategies in unfamiliar indoor environments. In *Building Simulation*; Tsinghua University Press: Beijing, China, 2024.
15. Devlin, A.S. Wayfinding in healthcare facilities: Contributions from environmental psychology. *Behav. Sci.* **2014**, *4*, 423–436. [CrossRef] [PubMed]
16. Shi, Y.; Duan, Q. A review of the progress in the research on the layout of guidance signage system based on human wayfinding behavior. *Compr. Transp.* **2016**, *38*, 47–56.
17. Zhang, Z.; Jia, L.; Qin, Y. Optimization of guidance signage system layout in subway platforms based on collaborative guidance. *J. Transp. Syst. Eng. Inf.* **2016**, *16*, 146–152. [CrossRef]
18. Lam, W.H.K.; Tam, M.; Wong, S.C.; Wirasinghe, S.C. Wayfinding in the passenger terminal of Hong Kong International Airport. *J. Air Transp. Manag.* **2003**, *9*, 73–81. [CrossRef]
19. Ouyang, K.M.; Liu, S.F. A simulation method for rail transit sign operation. *Int. J. Simul. Model.* **2021**, *4*, 742–753. [CrossRef]
20. Zhang, Z.; Jia, L.; Qin, Y. Location-Allocation model for the design of guidance Guidance signage System for pedestrian wayfinding in public spaces. *IEEE Trans. Intell. Transp. Syst.* **2020**, *22*, 7531–7546. [CrossRef]
21. Mei, X.; Jiang, Y.; Zhang, L. Impact of signage placement on pedestrian wayfinding efficiency in subway stations: A case study of People's Square Station in Shanghai. *World Archit.* **2024**, *10*, 15–22.
22. Shen, Z.F.; Zhang, L.H.; Li, R.; Liang, R. The effects of icon internal characteristics on complex cognition. *Int. J. Ind. Ergon.* **2020**, *79*, 102990. [CrossRef]

23. Liu, Y.; Zhang, X.; Wang, H.; Li, J.; Chen, Z.; Zhang, W.; Liu, S.; Li, Y.; Zhou, F.; Wang, Y. Wayfinding in Complex Environments: The Impact of Signage Design on User Behavior. *J. Environ. Psychol.* **2019**, *65*, 101–112.
24. Yu, D. Simulation Evaluation of Guiding Signage Layout Based on Wayfinding Behaviour in Rail Transit Hub. Master's Thesis, Beijing Jiaotong University, Beijing, China, 2012.
25. Wu, J.; Hu, S.; Chen, Z. Motion space visual perception-based evaluation of navigation signs' distribution. *Tongji Daxue Xuebao/J. Tongji Univ.* **2011**, *39*, 1167–1172+1176.
26. Chen, X. Simulation Modelling and Its Application of Pedestrian Mustering and Evacuation in Rail Transit Hubs. Master's Thesis, Beijing Jiaotong University, Beijing, China, 2018.
27. Ding, L. Research on Visual Field and Layout Design of Hanging Guidance Signs in Subway Stations. Master's Thesis, Beijing Jiaotong University, Beijing, China, 2021.
28. Wen, Q. A Study on the Spatial Environmental Factors That influence Wayfinding in Large Railway Stations. *New Archit.* **2013**, *1*, 103–105.
29. General Administration of Market Regulation of the People's Republic of China. *Technical Code for Signage System of Public Building: GB/T 51223-2017[S]*; China Architecture & Building Press: Beijing, China, 2017.
30. Geoffrey, U.; Peter, C.; Neil, B.; Underwood, J.; Crundall, D. Visual attention while driving: Sequences of eye fixations made by experienced and novice drivers. *Ergonomics* **2003**, *46*, 629–646.
31. Wang, B.; You, T.; Wang, S. The impact of individual sense of direction on wayfinding in passenger hub spaces. *Acta Archit. Sin.* **2017**, *S2*, 103–106.
32. Mi, J.; Xu, L.; Tang, Z. Wayfinding experiments and spatial guidance research in underground public spaces: A case study of People's Square in Shanghai. *Acta Archit. Sin.* **2007**, *12*, 66–70.
33. Li, M.J. Research on the Optimization Design of Large Passenger Flow Line in the Connection Space Between Subway Station and Commercial Complex under TOD Mode. Master's Thesis, Soutnwest Jiaotong University, Sichuan, China, 2023.
34. Lou, K.Q.; Shen, Z.W. Research on Urban Underground Space Vitality based on Space Syntax: A Case Study on the Underground Space Connected by Metro Stations in Chengdu City. *South. Archit.* **2019**, *4*, 116–121.
35. Ma, Z.; Xiao, Y.; Chen, X. Research on Passenger 3D View Focus of Attention at Airport Terminal Based on Massmotion. In *IOP Conference Series: Materials Science and Engineering, Proceedings of the 2019 2nd International Symposium on Traffic Transportation and Civil Architecture (ISTTCA 2019), Chengdu, China, 13–15 December 2019*; IOP Publishing: Bristol, UK, 2020; Volume 780, p. 062047.
36. Shi, Y.; Li, C.; Wang, T. Brief Analysis of Rail Transit Station Wayfinding System in Japan and the Enlightenment. *Urban Rail Transit Res.* **2023**, *26*, 231–234+240.
37. Zhou, X.; Zhang, F.; Zhou, X.; Zhang, F. Sightline analysis and optimisation strategy of underground space based on spatial syntax. *J. Undergr. Space Eng.* **2021**, *17*, 1008–1014.
38. Teng, S. Research on Visual Saliency of Guide Signs in Underground Commercial Street Based on Eye Movement Experiment. Master's Thesis, China University of Mining and Technology, Xuzhou, China, 2021.
39. Yan, G.L.; Xiong, J.P.; Zang, C.L. Review of Eye-movement Measures in Reading Research. *Adv. Psychol. Sci.* **2013**, *21*, 589–605. [CrossRef]
40. Li, L.; Tian, S.C.; Hui, L. On the effectiveness of the coal mine safety signboards for prohibition based on the eye-movement testing apparatus. *J. Saf. Environ.* **2017**, *17*, 1853–1857.
41. Li, H.X.; He, P.P. Application of Eye Tracking Technique in the Study of Effectiveness of Emergency Signs. *Sci. Technol. Eng.* **2018**, *18*, 323–327.
42. Kang, T.H.; Zang, H. The Analysis of Eye-Movement Indexes in Scene Perception: Perspectives from Fixation and Saccade. *Psychol. Sci.* **2020**, *43*, 1312–1318.
43. Li, P.H.; Wang, X.B.; Cong, L. Eye-Tracking Technology Research on the Relationship between Semantic and Visual lmage Based on. *Furnit. Inter. Decor.* **2021**, *6*, 28–33.
44. Li, J.; Chen, L.; Su, Q.; Jiang, Y.; Rong, G. A study on color visual perception of museum exhibition space based on eye movement experiments. *Front. Psychol.* **2024**, *15*, 1431161.
45. Shao, M.; Xie, C.; Sun, L.; Jiang, L. Optimal layout of static guidance information in comprehensive transportation hubs based on pedestrian pathfinding behavior. *Sustainability* **2019**, *11*, 3684. [CrossRef]

buildings

MDPI

Article

Residential Interior Design for People with Special Needs in Thailand Based on Physical Abilities: Age, Gender, and Living Environment Considerations

Pattamon Selanon and Warawoot Chuangchai *

Thammasat University Research Unit in Making of Place and Landscape, Faculty of Architecture and Planning, Thammasat University, Pathum Thani 12121, Thailand
* Correspondence: warawoot.chu@dome.tu.ac.th

Abstract: Residential interior design for people with special needs necessitates more specific research, especially with respect to physical abilities. Previous studies, as opposed to general people, have overlooked the nuanced requirements generated by age, gender, and living environment differences. This study aims to examine physical abilities across different ages, genders, and residencies and identify relationships between in-home mobility and physical abilities among people with special needs. A total of 384 participants (eight different types of disabilities) completed electronic questionnaires. Among young, adult, and older groups, physical abilities and self-support showed statistical differences. Males demonstrated higher physical ability levels but decreased hand–arm movements than females. Greater physical abilities and subjective health were found in urban residents compared to non-urban residents. Additionally, in-home mobility was positively associated with out-of-home mobility, physical abilities, independence, and subjective health (all p values < 0.05). Thus, physical abilities revealed disparities across ages, genders, and resident settings. Also, the greater the mobility inside the house, the more possibilities there are to enhance physical abilities both inside and outside the house, independence, and subjective health. Design recommendations are provided regarding layout, furniture, flooring, and lighting. The consideration of these specific requirements in residential interior design is encouraged.

Keywords: interior architecture; inclusive design; ergonomics in design; indoor mobility; people with disabilities

1. Introduction

People with special needs tend to spend more time at home [1,2], particularly in developing nations where urban facilities and services, such as inaccessible lavatories, a lack of navigation or information signs, and an absence of sloped or uneven pavement, are impeded [3,4]. Currently, the new lifestyle, influenced by the global COVID-19 epidemic, permits substantially more activities at home than was previously conceivable, and this trend continues, such as working or studying, conferencing, purchasing, and entertainment [1,5]. Residential interior design is becoming more important among all populations, including people with special needs, as it affects everyday living and both physical and mental well-being [6,7].

Environmental hazards at home among vulnerable populations, including the elderly and individuals with special needs, have been identified as one of the major flaws in interior design, such as accident-related falls, which can lead to serious injuries and mortality [8,9]. Interior design that restricts mobility within the home (for example, complex layouts, tight corners, narrow doorways, flooring that is too slippery, and a large number of stairs) can be categorized as unsafe design since it includes barriers or obstacles that limit access for users to complete their tasks successfully and safely [6,9,10]. These factors may also hinder some essential daily activities or make self-care activities more difficult, such as toileting, bathing, and cleaning, which reduces their independence and makes them more sedentary, all of which delay or distance healthy living [8,11]. Meanwhile, the global population is aging, and the number of people over the age of 60 is growing [4,12]. Some elderly persons may have some sort of impairment, whether temporary or partially, and it will worsen in the near future as human functions gradually decline with age, resulting in an increase in the number of people with special needs [13–15]. Since housing design is incapable of providing a safe space for people with special needs, this situation will only intensify in the future.

When designing housing spaces for people with special needs, interior design presents additional challenges since it requires a more in-depth understanding and specific requirements based on physical abilities than for the general population [16–18]. However, there is a lack of comprehensive research that examines the relationship between residential interior design and the physical abilities of this population, and even less when multiple types of disabilities are considered. Existing studies have primarily concentrated on general aspects of accessibility and universal design, but they frequently overlook the nuanced needs generated by age, gender, and living environment differences. This knowledge gap necessitates a greater awareness of how physical abilities impact residential interior design and how design interventions should cater to diverse needs. Therefore, the present study aims to examine physical abilities among people with special needs across different ages, genders, and resident settings and investigate relationships between in-home mobility and their abilities. This study will contribute to evidence-based studies and ergonomics in design that promote mobility, safety, and healthy living. By incorporating these considerations into interior design, it is possible to create living spaces that promote physical activity, independence, and overall well-being for people with special needs. Ultimately, the goal is to ensure that their homes provide a safe and healthy environment for sustainable living and an improved quality of life.

2. Materials and Methods

2.1. Participants

A total of 384 participants (with visual, hearing, physical or mobility, intellectual, learning, autism, emotional or behavioral, and multiple disabilities) were recruited for the present study. Males and females who were Thai citizens, aged at least 13 years old, and were officially registered as having one of those disabilities were included. Any individual who could not comprehend the objectives of the research or who could not be reached by the research team during the screening procedure was excluded. Fifty-four participants for each type of disability, except for multiple disabilities, which had six participants, were finalized as the total sample size of the present study. According to the information on their national identification cards, age groups were categorized as 13–14, 15–59, and 60 years old and above for young (n = 126), adult (n = 132), and older (n = 126) participants, respectively. Biological sex served as the basis for categorizing gender groups as males (n = 187) and females (n = 197). Residents of Bangkok, the capital city of Thailand, divided urbanism into those who lived inside and outside, and they were further categorized into urban (n = 235) and non-urban (n = 149) living, respectively (as shown in Figure 1). Prior to agreeing to participate and signing the consent form, all participants and/or their parents or legal guardians were provided with written information regarding the present measurement. They were also given the opportunity to contact the research team with any inquiries they may have had at any point in time. The ethical review board of Thammasat University granted approval for all procedures involving human participants in the present study.

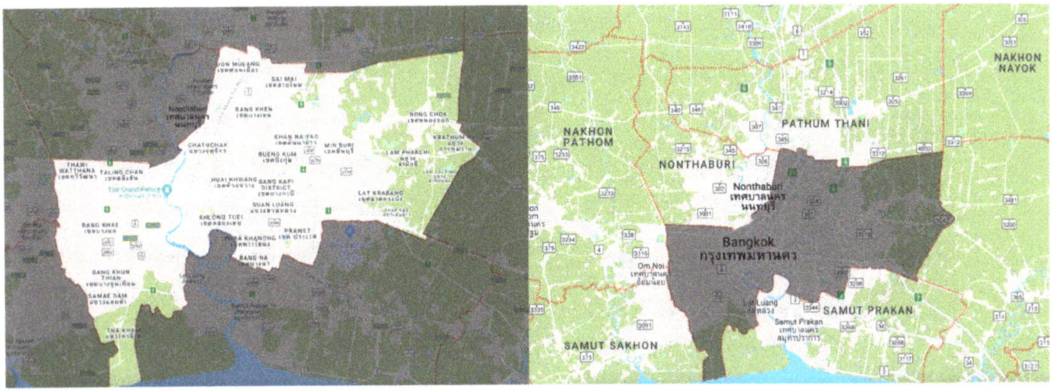

Maps Data: Google ©2023, Snazzy Maps

(a) (b)

Figure 1. Geographical areas of the sample in Thailand. (**a**) Urban residents were referred to as people who lived in Bangkok. The study gathered information from several districts, including Huai Khwang, Sai Mai, Min Buri, Nong Chok, Bang Na, and Thawi Watthana. (**b**) Non-urban residents were referred to as people who lived in suburban and rural areas. Various provinces were observed in the study, including Nonthaburi, Pathum Thani, Samut Prakan, and Nakhon Pathom.

2.2. Measurements

The questionnaire was designed to collect data on health and socio-demographic characteristics, physical performance, and subjective health in the context of everyday living activities in home environments. Health and socio-demographic characteristics were open-ended questions, including age, biological sex, place of living, type of disability (and whether using a wheelchair or not), cause of disability, chronic health conditions, occupation, monthly income in THB, and daily travel use. In addition, physical performances were a closed-ended question and comprised mobility activity levels inside the house, mobility outside the house, postural stability, stepping across different levels, sit-to-stand movement, lifting, hand–arm motion, mobility within the indoor fitness space, physical activity intensity, walking duration, independence, and self-support. Moreover, subjective health was calculated as the average of two rating scale questions: (1) How would you feel about your overall health, apart from the disability? (2) How satisfied would you feel with your health? Scores between 0 and 10 signified the lowest and highest, respectively [3]. The electronic questionnaire was directly sent to all participants via email, with assistance from Thai national organizations and institutes in relation to each type of disability.

2.3. Statistical Analysis

Descriptive statistics are presented as percentage (% of total) in Figure 2. The Kruskal–Wallis test was used to compare three groups of young, adult, and older participants (Figure 3). The present study used the Mann–Whitney U test (for ordinal data), the chi-squared test (for nominal data), or the unpaired t-test (for continuous data) to compare each pair of age (Figures 3 and 4), gender (Figure 5), and residence (Figures 6 and 7) groups. Relationships between mobility within the houses and variables related to their daily living abilities were examined using the Pearson correlation coefficient (Figure 8) for all participants. For comparison results, effect sizes were calculated using eta squared (η^2), phi (φ), and Cohen's d, whereas correlation coefficients (r) were calculated for correlation results [19,20]. All analyses were carried out using IBM SPSS Statistics version 22, and results were considered statistically significant at a p value of less than 0.05 (two-sided).

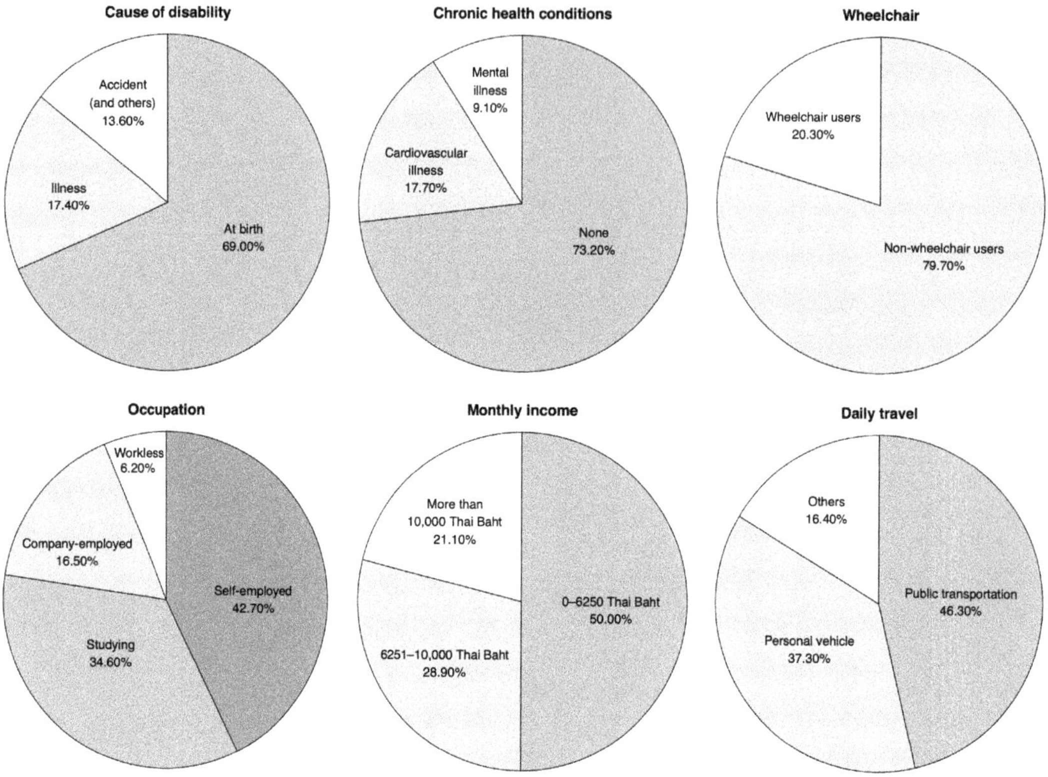

Figure 2. Health and socio-demographic characteristics of all participants.

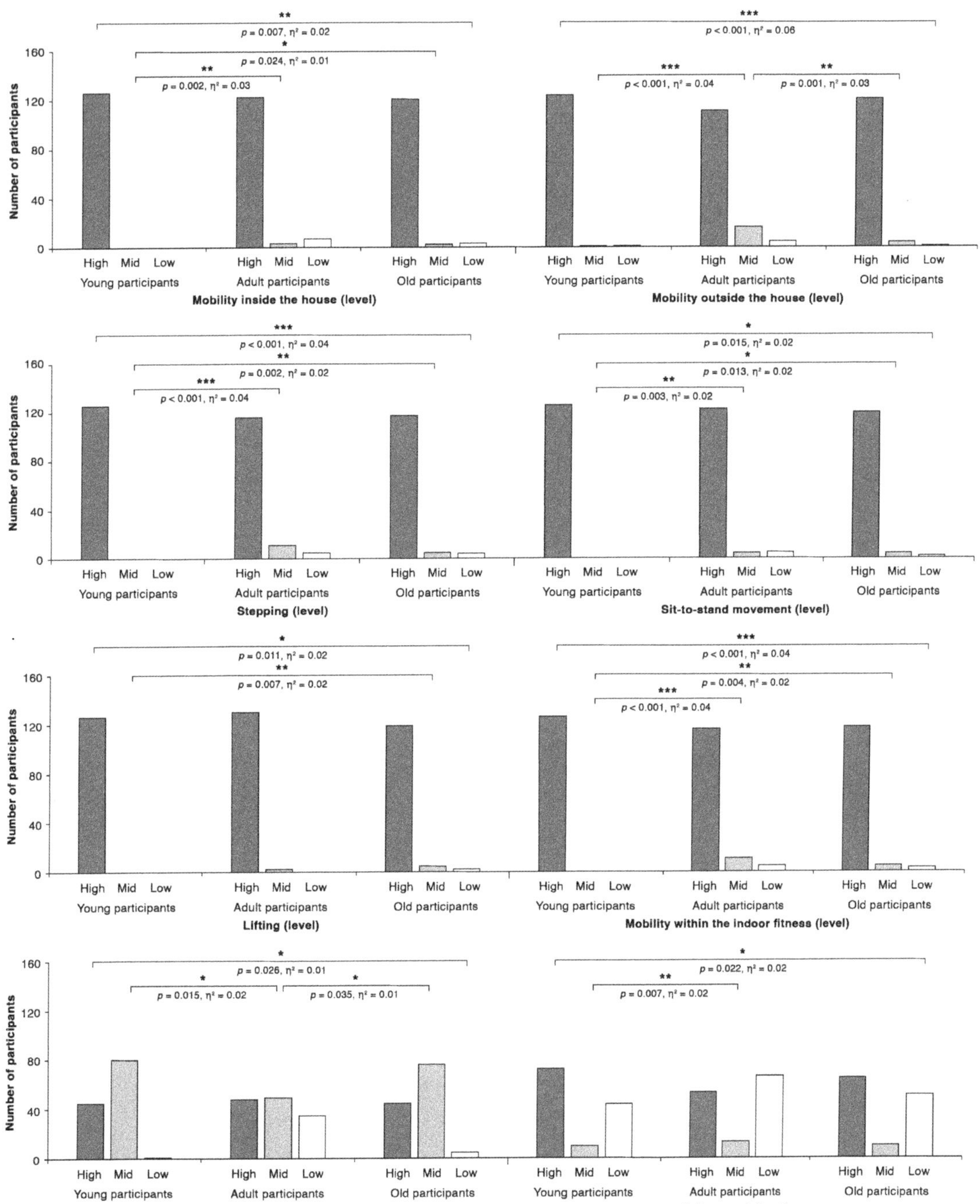

Figure 3. Age considerations: comparison results among young, adult, and older participants. Note: Symbols *, **, and *** denote significance levels at $p < 0.05$, $p < 0.01$, and $p < 0.001$, respectively.

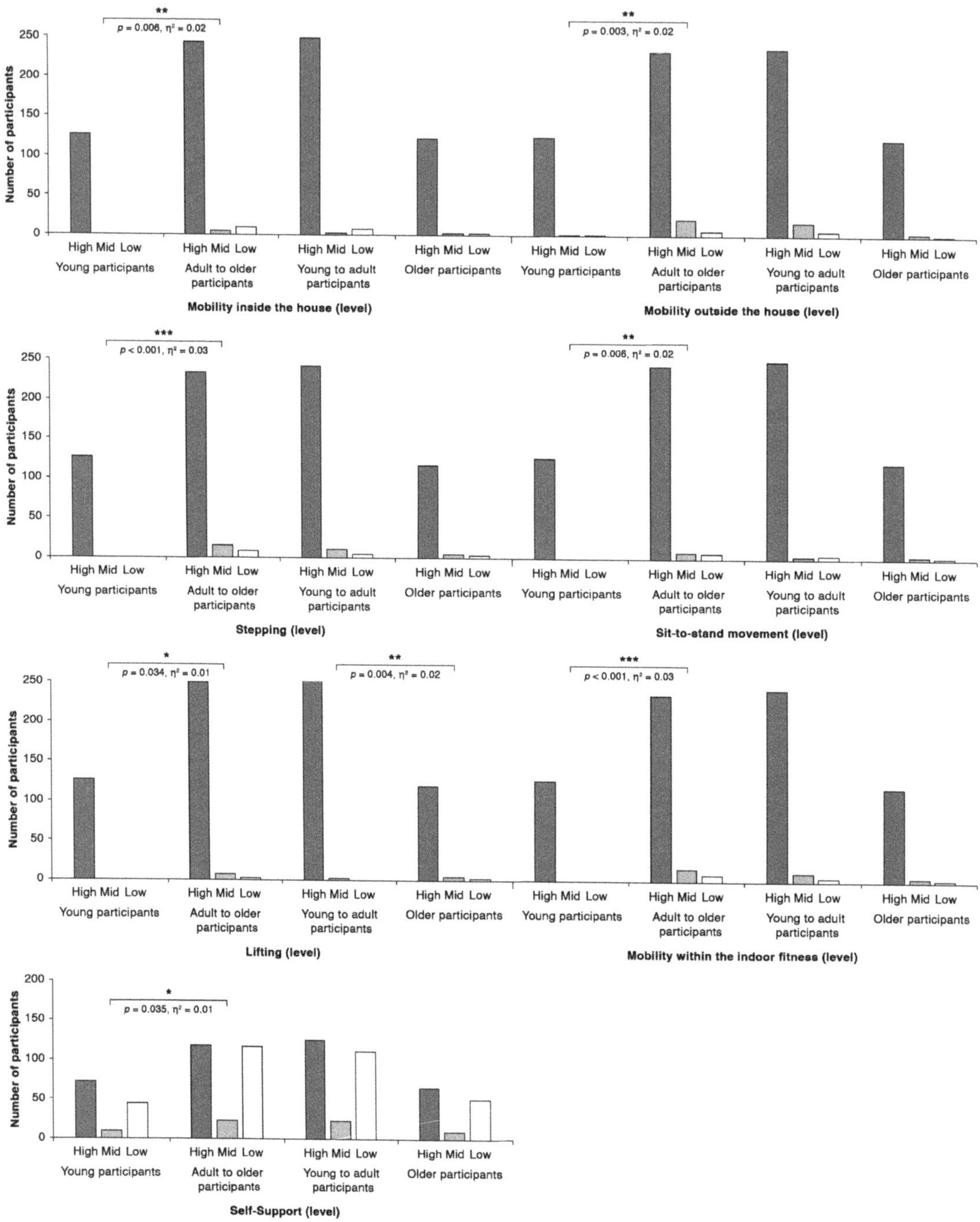

Figure 4. Sub-age groups: comparison results between young and adult to older participants and between young to adult and older participants. Note: Symbols *, **, and *** denote significance levels at $p < 0.05$, $p < 0.01$, and $p < 0.001$, respectively.

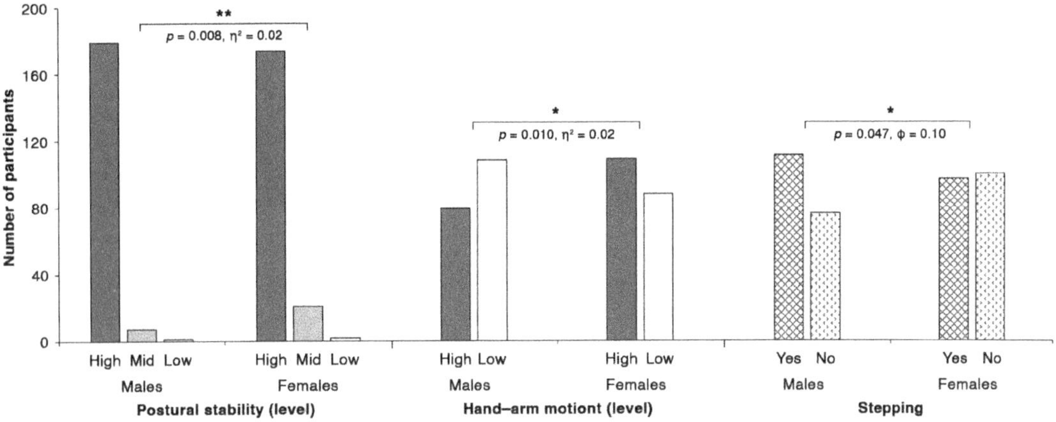

Figure 5. Gender considerations: comparison results between males and females. Note: Symbols * and ** denote significance levels at $p < 0.05$ and $p < 0.01$, respectively.

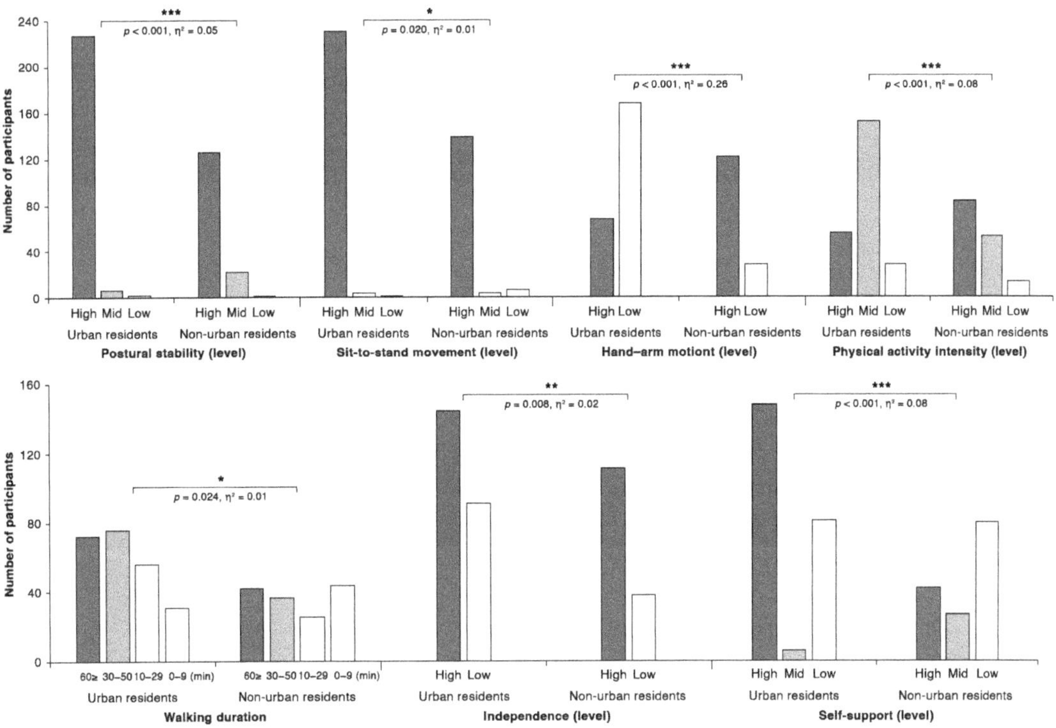

Figure 6. Living environment considerations: comparison results between urban and non-urban living. Note: Symbols *, **, and *** denote significance levels at $p < 0.05$, $p < 0.01$, and $p < 0.001$, respectively.

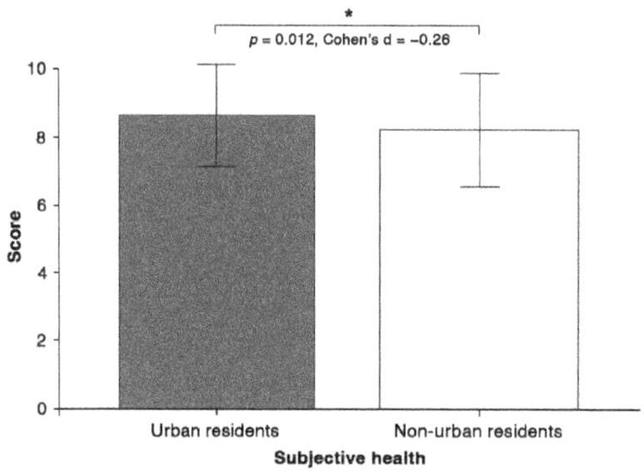

Figure 7. Living environment considerations: comparison results of subjective health between urban and non-urban living. Note: Symbol * denotes significance level at $p < 0.05$.

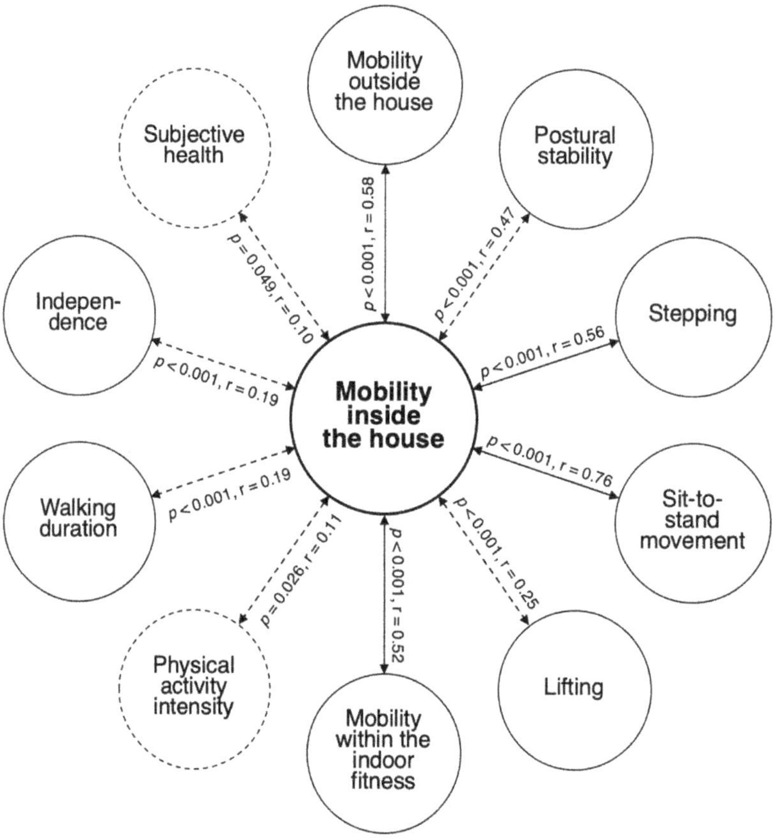

Figure 8. Bivariate correlations between mobility within the house and variables related to daily living abilities. Note: Dashed and continuous circular lines denote significance levels at $p < 0.05$ and $p < 0.001$, respectively. Dashed and continuous lines with arrowheads indicate correlation coefficients (r) of less than and greater than 0.50, respectively.

3. Results

The present study included participants who were mostly born disabled (69.00%); had no serious chronic health conditions (73.20%); did not use wheelchairs (79.70%); worked independently (42.70%); earned approximately THB 6250 (Thai currency) per month (50.00%); and used public transportation (46.30%), such as the sky train, subway, bus, and motorcycle taxi, on a regular basis (as shown in Figure 2).

Figure 3 shows comparisons by age group—young, adult, and older participants—and by pair. Among three age groups, there were statistically significant differences in levels of mobility inside the home ($p = 0.007$ with $\eta^2 = 0.02$), mobility outside the home ($p < 0.001$ with $\eta^2 = 0.06$), step ability ($p < 0.001$ with $\eta^2 = 0.04$), sit-to-stand movement ($p = 0.015$ with $\eta^2 = 0.02$), lifting ability ($p = 0.011$ with $\eta^2 = 0.02$), mobility within indoor fitness space ($p < 0.001$ with $\eta^2 = 0.04$), physical activity intensity ($p = 0.026$ with $\eta^2 = 0.01$), and self-support ($p = 0.022$ with $\eta^2 = 0.02$).

Younger participants were shown to have greater levels of ability than adult participants in terms of mobility within and outside of residences, stepping, sit-to-stand movement, mobility inside of fitness space, physical activity intensity, and self-support (where p values ranged between less than 0.001 and 0.015, and η^2 ranged between 0.02 and 0.04). Younger participants also showed better levels of mobility inside the home, step ability, sit-to-stand movement, lifting ability, and mobility within indoor workout areas (where p values ranged between 0.002 and 0.024, and η^2 ranged between 0.01 and 0.02) than older participants. Additionally, adult participants exhibited differences compared to older participants in mobility outside of residences and physical activity intensity (where p values ranged between 0.001 and 0.035, and η^2 ranged between 0.01 and 0.03), as shown in Figure 3.

Figure 4 demonstrates that the adult to older age group reported greater mobility inside the home, mobility outside the home, step ability, sit-to-stand movement, lifting ability, mobility within indoor fitness space, and self-support (where p values ranged between less than 0.001 and 0.035, and η^2 ranged between 0.01 and 0.03) than the young age group. On the other hand, the younger to adult age group reported greater lifting ability ($p = 0.004$ with $\eta^2 = 0.02$) than the older age group.

Figure 5 compares gender differences in observations. Male participants reported higher levels of postural stability ($p = 0.008$ with $\eta^2 = 0.02$) and were more likely than female participants to have better step ability ($p = 0.047$ with $\varphi = 0.10$). In addition, female participants reported greater hand–arm movement than their male counterparts ($p = 0.010$ with $\eta^2 = 0.02$).

Figures 6 and 7 exhibit comparative results between urban and non-urban living. In comparison to participants who lived in non-urban areas, those who lived in urban areas demonstrated higher levels of postural stability ($p < 0.001$ with $\eta^2 = 0.05$) and sit-to-stand movement ($p = 0.022$ with $\eta^2 = 0.01$), were able to engage in more intense physical activity ($p < 0.001$ with $\eta^2 = 0.08$), could spend more time walking ($p = 0.024$ with $\eta^2 = 0.01$), had more independence in living ($p = 0.008$ with $\eta^2 = 0.02$), were better able to support themselves in daily activity ($p < 0.001$ with $\eta^2 = 0.08$), and gained higher subjective health scores ($p = 0.012$ with Cohen's d = −0.26). On the other hand, non-urban residents moved more fluidly in hand–arm motion than urban residents ($p < 0.001$ with $\eta^2 = 0.26$).

Figure 8 presents mobility within the home, which is correlated to factors pertaining to daily living abilities. All participants reported that their mobilizations within their homes were positively correlated with mobilizations outside of them, with their ability levels for postural control, stepping, sit-to-stand movement, lifting, mobility within indoor fitness spaces, walking duration, and independence (all p values < 0.001 with r varied between 0.19 and 0.76), and they positively correlated with levels of physical activity intensity ($p = 0.026$ with r = 0.11) and subjective health ($p = 0.049$ with r = 0.10).

4. Discussion

Overall, the young participant group exhibited better health in terms of mobility level, physical activity level, and self-help level than those in the adult and older groups. In sub-age considerations, it is more obvious that the young group ranged at a high level in most performances as opposed to the group of adult and older participants, who fell into all levels and varied from low to high levels. Regarding gender, male participants demonstrated greater ability in balancing while being restricted more in the upper extremities, such as the arms or hands, than female participants. Urban residents exhibited higher physical ability, physical fitness, self-reliance, and subjective health (with the exception of hand–arm movement) than non-urban residents. Furthermore, the mobilization inside the home was positively associated with physical performances (including stability, stepping, postural change, and lifting abilities), mobilizations outside and in fitness areas, physical activity, independence, and a positive sense of well-being.

Interestingly, in addition to the older group, the adult group was also found to be inferior to the young group with respect to several categories of physical abilities. Adults with special needs have a significant prevalence of low levels of physical activity [21]. This inactivity corresponds to a downward trend in physical activity and an increase in sedentary behavior with age, both of which contribute to poor physical performance and negative health outcomes [22,23]. Moreover, the adult group had similar characteristics to the older group, wherein it was observed that the older population with intellectual disabilities displayed markedly low levels of physical activity [24]. Unlike the general population, where older individuals were found to be the most vulnerable age due to their restricted movements [6,25,26], the present findings exhibit that adults can be considered the most vulnerable age group, particularly when accompanied by older individuals, among people with special needs. This focuses the attention on interior design for adults when considering people with special needs, who are sometimes overlooked when creating home spaces because adults are more often involved in occupational or workplace design.

According to our findings, there are gender differences among people with special needs that are similar to those found in the general population, particularly among the elderly. The present study discovered that female participants had lower postural stability and stepping ability than male participants, which was consistent with several studies that found that women were more likely to fall due to difficulties in balancing, tripping, or slipping, even within their own homes [8,27,28]. Moreover, the greater movements observed in the arms and hands of the female participants can be attributed to their role as housewives and the everyday housekeeping they perform [29]. The majority of household chores include cleaning, and these tasks usually require the use of the arms and hands [30]. Cleaning occupations, which primarily employ women and/or the elderly [31, 32], emphasize moving the arms and hands more freely and efficiently. These illustrate that home fall prevention strategies and home adjustable fixtures and tools, where range and motion of the arms and hands can be safely attained when people with special needs undertake daily tasks within their houses, should receive more serious consideration in interior design.

Among people with special needs, the present findings found that those who lived in the city had better overall physical abilities than those who lived outside the city. This could be explained by the fact that the urban resident group leads a more active lifestyle, can afford to have more fitness equipment at home, and finds it more convenient to travel to public spaces near their homes where physical activity is more frequently available [33], whereas the rural resident group spends more time on sedentary activities, such as screen time among teenagers [34], and has fewer facilities supported around their neighborhoods. These findings were supported by our health and socio-demographic characteristic results, which revealed that individuals with no chronic health conditions were more likely to be in the urban group than the non-urban group (74.50% and 71.10%, respectively), suggesting the importance of leading an active lifestyle. Urban residents earn a higher income than non-urban residents (median values of THB 9500 and 1000, respectively) and have fewer unemployed people in their group (5.10% and 8.10%, respectively), showing that they have more potential money to spend on health promotion. Both groups were presented as non-wheelchair users, with urban residents reporting 80.00% and non-urban residents reporting 79.20%. However, due to the readiness of the cities, they displayed divergent travel behaviors, with the first group using public transportation (63.50%) and the latter group relying on their own private vehicles (52.20%). Additionally, people with special needs preferred free access to physical activities and fitness equipment, such as at public parks [35], which were more convenient to reach for the urban group because many of them could travel by public transit (e.g., sky train and subway), and these parks provided a greater variety of activities and equipment, whereas the non-urban group was more likely to use neighborhood parks (where public parks were absent), which lacked activities and quality equipment. A closer look at the types of residential living revealed that there were two main types observed in this study: group and private residences. The term group residence refers to residential care homes and centers where residents share living spaces and facilities. Sub-types were also found in age and gender categories, such as only for children, women, and men. In Thailand, the Department of Empowerment of Persons with Disabilities, Ministry of Social Development and Human Security regulates the Home for Protection and Development for Persons with Disabilities, which has locations in Nonthaburi, Pathum Thani, and Samut Prakan, among others. In addition, the term private residence refers to an individual living in a single (or family) home, which can either be owned or rented. As observed within the study, the urban group was more likely to live in a private residence (e.g., a compact unit, apartment, or home), which may be due to their workplaces, whereas the non-urban group was more likely to live in a group residence, which may be due to restricted opportunity and financial constraints. This explanation can help support the subjective health result, which observed that the urban group was more satisfied with their overall health and lives than the non-urban group. Regarding the general population, nevertheless, rural residents were more likely to have higher physical fitness than urban residents [36,37]. Although both general and disabled populations were found to be distinct between home places, there is an inverse relationship and a different rationale when it comes to people with special needs. Thus, interior home design should be approached with prudence, as people with special needs have additional challenges where more confounding factors could impact their capacity to make a good living.

Furthermore, the present findings imply that the greater the degree to which a house provides its users with more opportunities to engage in movements within the home, the more a person's level of physical performance, independence, and subjective health with regard to activities of daily life for people with special needs can be significantly increased. When accessibility and safety are properly occupied, interior housing design can create spaces that encourage people with special needs to move more actively, as physical activity leads to several positive health outcomes. This is supported by previous research, which found that people with disabilities who engaged in high-mobility environments had higher levels of physical activity, more social interactions with their communities (outside of their homes' activities), and better health in both body and mind [38–40]. Conversely, older people who had low mobility (due to environmental restrictions, such as home hazards or inaccessible facilities) were associated with inactive behavior, a fear of falling, low social engagement, and a higher risk of chronic diseases [8,39,41]. Thus, this correlation result emphasizes the importance of interior design, which has a major effect on the well-being of people with special needs. Additionally, it is also expected that successful interior housing design contributes to improving the quality of life of people with special needs and can ensure that their home is a safe, healthy place to live over the long term, which can eventually lead to sustainable living for people with special needs.

4.1. Design Recommendations

4.1.1. Planning and Layout Design

A well-organized planning design should prioritize an uncomplicated space with a non-complex layout. This is advantageous in reducing cognitive loads for older people with special needs, who naturally have a deterioration in cognitive function and make poor decisions [6,42] and for adults with special needs who want to reduce physical effort in mobility and develop self-support. An open-concept layout can provide users with an overview of the entire arrangement, allowing them to recognize their positions within the spaces as proprioception and organize their energy to travel to destination areas and complete tasks [8,43]. This helps people with low levels of mobility (motor function), movement (coordination), and activity intensity in preparing their bodies for performances and supports a sense of autonomy via movement confidence [44].

A clear route and appropriate maneuvering space should be considered, such as planning with minimal junctions or uncrossed circulations, to make daily hygiene routines, housekeeping duties, and work-at-home activities safer and more effective [6,8]. This is carried out to support those who live independently and are responsible for taking care of themselves, as well as those of working age and adults with special needs, in better arranging their repetitive tasks and workloads. Also, it must be ensured that doorways, hallways, and corridors are large enough to accommodate both single (using a wheelchair or walking aid equipment) and multiple (with personal assistance or in the event of paramedic rescue assistance) users and that sharp curves and small corner areas are avoided [45]. Choices for arched openings should be offered between common rooms instead of doors in order to improve accessibility and eliminate the problems caused by difficult openings, damaged hinges, and drooping or heavy doors [6]. This is to ensure that the circulations are accessible and to avoid lifting concerns, as most disabled people, particularly elderly people, have a low lifting ability.

A zonal arrangement should be provided to create proximity and layout continuity by grouping related areas, such as the kitchen, dining, and living spaces, together in order to reduce unnecessary travel and make it easy to understand the functions of the space via the relationships between areas and elements. This helps people with cognitive impairments in staying focused (with selective attention within limited resources) and achieving their tasks [6]. For long distances, a direct approach using furniture configurations for guided circulation channels is recommended. Also, the design should consider including certain resting points in between the traveling path or between rooms so that people with special needs can pause, catch their breath, and relieve weariness while travelling between spaces [6]. This is carried out to support non-urban and non-young groups with low physical intensity so that they can engage in lengthy physical movements, such as mobility. Essential areas, such as bedrooms, bathrooms, and kitchens, should be placed on the same main level (the ground floor is preferred) to avoid the need to navigate stairs [8]. This is carried out to prevent imbalance accidents and fall-related injuries with respect to those that have trouble using stairs, including non-young people and females.

Unusual movements (e.g., bending, reaching, and turning) within spaces, such as in kitchens and bathrooms, must be considered so that these spaces are designed with safety in mind since these motions induce postural inability [46], and these areas also involve slippery surfaces where a high frequency of fall injuries has been reported [8]. Behavioral design to eliminate needless activities or unnecessary tasks can be applied to reduce time spent in those risky areas so that there is less chance for potential harm [6]. Separate tasks that can be completed in other dry spaces can also be an option. These are intended to prevent people with special needs, who lack body balance and have restricted hand–arm motion (females and urban citizens, respectively), from losing their stability and falling.

The design should emphasize the creation of a central location where people with special needs can fully access and conveniently reach these areas for regulating the functions of the house, such as switches for lights, curtains, gate opening, and other frequently used features or wireless control systems [47]. Emergency response devices that are promptly accessible should be installed at multiple locations throughout the home [48]. This assists non-urban residents who have a low level of independence, self-support, subjective health, and walking capacity in having full control of their homes in order to support their daily activities, and this also provides the ability to quickly request assistance if needed.

4.1.2. Built-in and Loose Furniture Design

Built-in countertops, such as those for cooking, working, and gardening, should come with adjustable heights. Built-in storage solutions, such as cabinets and drawers that are easily accessible and at various heights (or vertically) [46], ensure that items are within reach without the requirement for excessive physical exertion, such as kneeling or extending. This adjustable feature helps reduce the overreach of hand–arm movement tasks and supports body balance for males and non-urban residents with special needs, respectively. Lightweight fixtures and non-handle mechanics for opening panels or doors are beneficial for those who suffer from fine motor dysfunction and have low lifting, energy, and hand grip [6]. Open shelves with clear labeling can be considered to make items more apparent, allow individuals to efficiently locate goods, and provide an orderly in-home environment for those with limited vision [43].

Multifunctional furniture that serves several purposes and is flexible enough to accommodate the diverse needs and comfort preferences of users is suggested [49]. This can benefit users with limited budgets and space by saving both money and room for other necessities. When designing or selecting versatile furniture, the biomechanics and aging ergonomics of transfer positions should be considered, such as movements from lying down to sitting or standing up. These transitions can cause centers of mass imbalance in those who have poor sit-to-stand movements, like non-young and non-urban people with special needs. These transitions can potentially induce orthostatic hypotension, which is prevalent among the elderly as well [50].

Ergonomic furniture and management for working at home, such as adjustable desks and chairs, proper lighting, and cable management (to avoid tripping), can make workstations more comfortable and productive [51]. This can help maintain good posture and reduce the incidence of musculoskeletal and office syndrome symptoms. Furniture with rounded edges should be used to reduce unintended bumps and bruising [46], which can assist in safe movements and foster confident mobility throughout the space for those who are unaware of home hazards and may have certain anxieties and fears.

4.1.3. Flooring and Stepping Design

Smooth transitions between different flooring materials and levels should be created, and dangerous surfaces, such as those that are highly frictional and too slippery, should be avoided [43]. This eliminates tripping hazards for people who have abnormal gait patterns (e.g., cerebral palsy, multiple sclerosis, Parkinson's disease, and visual impairments [52]) and makes maneuvering less difficult for those who use wheelchairs and mobility supports. Tactile indications or textures, as well as contrasting hues, should be installed to designate potentially hazardous areas, such as those near staircases, level changes, and floor edges, in order to provide information about the surrounding settings and spaces. This helps guide existing areas and navigate pathways by providing more sensory features for people with visual problems [43,53]. Non-slip flooring materials should be placed in areas where moisture and wet conditions are likely to occur, such as bathrooms, kitchens, and areas close to windows, near exterior doors, and around washing and gardening equipment [43]. This can help reduce the risk of falls associated with a loss of postural balance among disabled people with mobility challenges, women, and those living in remote areas. Flooring that receives all-day sunlight must be specified and built with safe materials because it might generate heat buildup and burn the skin when unintentionally stepped on with bare feet. This should be carried out to protect those with special needs who may be unaware of or are unable to detect the temperature of the floor, such as those who have difficulties directing their stability, have lost foot sensation, or are blind [54].

4.1.4. Lighting Design

Lighting uniformity should be considered to provide clear visibility within the entire space, which, in turn, minimizes accents and shadows. This allows people with limited vision and mobility constraints to comfortably access the space, and those with cognitive disabilities can increase concentration via consistent designs, resulting in more correct responses [6]. Task lighting should be included in areas where precise actions occur, such as kitchen countertops and workstations [55], and dimmable or adjustable lighting elements should be utilized to accommodate functions, sensitivities, and personal preferences. This is carried out to ensure that people with visual problems can accomplish their tasks independently. Motion-activated lighting should be installed in corridors and other high-traffic areas [43]. This eliminates the need for manual switches, making nighttime navigation for people with limited mobility easier and more secure. Indoor lighting design that mimics natural daylight variations should be considered, which can influence circadian rhythm and behavioral activity. This helps improve physiological and mental health by guiding sleep–wake cycles and providing outside alternatives to those who may spend most of their time indoors [56], such as people with physical or mobility impairments or those who are paralyzed.

4.2. Strengths, Limitations, and Further Research

The present study has some strengths. This study represents the first investigation, to the best of our knowledge, in the field of interior design pertaining to people with special needs (eight distinct abilities) in Thailand. There were no reports from any of the participants indicating that the act of completing the questionnaires had any adverse consequences for their mental health, such as feeling stressed or exhausted. Nonetheless, the present study has some limitations. Due to the time restriction, the study had to collect data within the period of 2021–2022, which was the time when COVID-19 occurred, and it was limited in its methods for gathering data among sensitive groups. The figure references of the participants and their homes (e.g., images or drawings) were not disclosed in this study due to privacy concerns regarding the participants' rights and dignity [57,58]. The informed consent statement did not include this request to record and publish any type of image, which could have contributed more to the study's outcomes. The age range window among each group is not equal, with adults having the widest range compared to their younger and older peers. Compared to the other groups, the number of participants with multiple disabilities was relatively small, and those who have multiple disabilities appeared to be more vulnerable than others, which may have an impact on some of the results. The questionnaire completed by individuals other than the intended respondents (such as a relative, guardian, or representative) may yield inaccurate data. Although the impact of these consequences is anticipated to be minimal, it is important that further investigations be carried out with this consideration in mind. For a greater understanding of how design affects people's lives, future studies should consider undertaking visual documentation, in which participants voluntarily contribute photos, images, or video recordings of their activities (e.g., household chores, working, and community socializing) and living spaces (e.g., types of residence, quantity of areas, quality of existing, and its surroundings). Further research should also be balanced in order to be more equal in terms of age range windows and the sample size of each type of disability, and future studies should plan for long-term research to obtain a more comprehensive understanding of the impacts on interior design. Alternatively, each design recommendation provided in this study could potentially be investigated further as distinct topics to specify more precise features and interpretations.

5. Conclusions

Among people with special needs, there were disparities in physical abilities across ages, genders, and places of residence. In the context of the activities of everyday living within the homes, adult participants were found to be the most vulnerable group in addition to older participants compared to young participants. Male and urban participants had better overall physical abilities but were limited in hand–arm movements compared to their female and non-urban counterparts. Moreover, in-home spaces where people with special needs can freely mobilize were associated with greater overall physical abilities, independence, and subjective health. This distinction should be taken into consideration when designing residential interiors since it is interconnected with the degree of mobility within home spaces as well as other physical abilities and an individual's subjective state of health. Based on the findings, design recommendations for both space and element features are proposed. The main suggestions are that the design should emphasize open planning with a non-complex layout, short and uncrossed circulations, resting spots for medium to long travel, built-in furniture with adjustable height, versatile furniture for multipurpose use, even and non-slip flooring, and lighting uniformity and harmony with natural light variations. These recommendations provide additional knowledge for creating residential spaces that require careful consideration. This knowledge can also be broadly applied to pre-aging and healthcare-built environments by using accompanying related criteria (e.g., inclusive design and aging ergonomics). A home's interior design that prioritizes mobility as an aspect of accessibility creates a functional and friendly environment that enhances the physical and mental well-being of its occupants. This promotes independence, boosts self-esteem, and instills confidence, thereby reducing dependence on caregivers or family assistance and mitigating the frustration, stress, and anxiety associated with navigating immobile conditions. For people with special needs, in-home mobility design enables them to age more comfortably and safely without the need for constant support or relocation to assisted living facilities while also maintaining their social connections and the feeling of belonging in their communities, thereby improving their overall quality of life.

Author Contributions: Data collection, funding acquisition, and visualization, P.S.; conceptualization, formal analysis, writing—review and editing, W.C. These authors contributed equally to this work. All authors have read and agreed to the published version of the manuscript.

Funding: P.S. received funding support from the NSRF via the Program Management Unit for Human Resources & Institutional Development, Research and Innovation (Grant number B05F640096). The APC was funded by the NSRF via the Program Management Unit for Human Resources & Institutional Development, Research and Innovation (Grant number B05F640096).

Institutional Review Board Statement: The present study was granted ethical approval by the Human Research Ethics Committee of Thammasat University: Social Sciences (Number 102/2564) and retrospectively registered via the Thai Clinical Trials Registry Committee (TCTR20220806001).

Informed Consent Statement: All participants and/or their parents or legal guardians received all the substantial information along with the questionnaire and gave their written, informed consent to participate in the present study prior to the measurement.

Data Availability Statement: The data reported in the present study are available on reasonable request from the corresponding author. The data are not publicly available due to privacy and ethical restrictions.

Acknowledgments: The authors would like to extend their sincere appreciation to the Thailand Association of the Blind, the National Association of the Deaf in Thailand, the Association of Persons with Physical Disability International, the Association of Persons with Intelligence Disability of Thailand, the Thai Association to Encourage the Potential of Learning Disabled Persons, the Association for the Mentally Ill of Thailand, and the Autistic Thai Foundation for their help and assistance. This study was supported by Thammasat Postdoctoral Fellowship.

Conflicts of Interest: The authors declare no conflict of interest.

References

1. Ralph, K.; Morris, E.A.; Kwon, J. Disability, access to out-of-home activities, and subjective well-being. *Transp. Res. Part A Policy Pract.* **2022**, *163*, 209–227. [CrossRef]

2. Eklund, M.; Leufstadius, C.; Bejerholm, U. Time use among people with psychiatric disabilities: Implications for practice. *Psychiatr. Rehabil. J.* **2009**, *32*, 177–191. [CrossRef] [PubMed]
3. Selanon, P.; Chuangchai, W. Walking activity increases physical abilities and subjective health in people with seven different types of disabilities. *Front. Public Health* **2023**, *11*, 1120926. [CrossRef] [PubMed]
4. Srichuae, S.; Nitivattananon, V.; Perera, R. Aging society in Bangkok and the factors affecting mobility of elderly in urban public spaces and transportation facilities. *IATSS Res.* **2016**, *40*, 26–34. [CrossRef]
5. Coşkun, M.G.; Öztürk, R.İ.; Tak, A.Y.; Sanlier, N. Working from Home during the COVID-19 Pandemic and Its Effects on Diet, Sedentary Lifestyle, and Stress. *Nutrients* **2022**, *14*, 4006. [CrossRef]
6. Chanbenjapipu, P.; Chuangchai, W.; Thepmalee, C.; Wonghempoom, A. A Review Article: Fall Incidents and Interior Architecture—Influence of Executive Function in Normal Ageing. *J. Archit./Plan. Res. Stud.* **2023**, *20*, 139–151. [CrossRef]
7. Alhusban, A.A.; Alhusban, S.A.; Alhusban, M.A. How the COVID 19 pandemic would change the future of architectural design. *J. Eng. Des. Technol.* **2022**, *20*, 339–357. [CrossRef]
8. Chuangchai, W. A Review Article: Environmental Hazards at Home and Ergonomics as Fall Prevention for Elderly Population. *J. Archit./Plan. Res. Stud.* **2017**, *14*, 104438. [CrossRef]
9. Valipoor, S.; Pati, D.; Kazem-Zadeh, M.; Mihandoust, S.; Mohammadigorji, S. Falls in Older Adults: A Systematic Review of Literature on Interior-Scale Elements of the Built Environment. *J. Aging Environ.* **2020**, *34*, 351–374. [CrossRef]
10. Li, J.; Chaulagain, S.; Yost, E. Residents' sense of safety in senior living communities: A conceptual paper. *Int. J. Hosp. Manag.* **2023**, *112*, 103410. [CrossRef]
11. Tripathy, N.K.; Jagnoor, J.; Patro, B.K.; Dhillon, M.S.; Kumar, R. Epidemiology of falls among older adults: A cross sectional study from Chandigarh, India. *Injury* **2015**, *46*, 1801–1805. [CrossRef] [PubMed]
12. Lutz, W.; Sanderson, W.; Scherbov, S. The coming acceleration of global population ageing. *Nature* **2008**, *451*, 716–719. [CrossRef] [PubMed]
13. World Health Organization; World Bank. *World Report on Disability*; World Health Organization: Geneva, Switzerland, 2011; p. 325. ISBN 9789241564182.
14. Verbrugge, L.M.; Yang, L.-S. Aging with Disability and Disability with Aging. *J. Disabil. Policy Stud.* **2002**, *12*, 253–267. [CrossRef]
15. World Health Organization. *WHO Policy on Disability*; WHO: Geneva, Switzerland, 2021; ISBN 978-92-4-002062-7.
16. Struckmeyer, L.; Morgan-Daniel, J.; Ahrentzen, S.; Ellison, C. Home Modification Assessments for Accessibility and Aesthetics: A Rapid Review. *HERD Health Environ. Res. Des. J.* **2021**, *14*, 313–327. [CrossRef]
17. Roos, J.; Koppen, G.; Vollmer, T.C.; Van Schijndel-Speet, M.; Dijkxhoorn, Y. Unlimited Surrounding: A Scoping Review on the Impact of the Built Environment on Health, Behavior, and Quality of Life of Individuals with Intellectual Disabilities in Long-Term Care. *HERD Health Environ. Res. Des. J.* **2022**, *15*, 295–314. [CrossRef] [PubMed]
18. Demirkan, H.; Olguntürk, N. A priority-based 'design for all' approach to guide home designers for independent living. *Archit. Sci. Rev.* **2014**, *57*, 90–104. [CrossRef]
19. Fritz, C.O.; Morris, P.E.; Richler, J.J. Effect size estimates: Current use, calculations, and interpretation. *J. Exp. Psychol. Gen.* **2012**, *141*, 2–18. [CrossRef]
20. Cohen, J. *Statistical Power Analysis for the Behavioral Sciences*, 2nd ed.; Routledge: New York, NY, USA, 1988; p. 567.
21. Dairo, Y.M.; Collett, J.; Dawes, H.; Oskrochi, G.R. Physical activity levels in adults with intellectual disabilities: A systematic review. *Prev. Med. Rep.* **2016**, *4*, 209–219. [CrossRef]
22. Boslaugh, S.E.; Andresen, E.M. Correlates of physical activity for adults with disability. *Prev. Chronic Dis.* **2006**, *3*, A78.
23. Oviedo, G.R.; Guerra-Balic, M.; Baynard, T.; Javierre, C. Effects of aerobic, resistance and balance training in adults with intellectual disabilities. *Res. Dev. Disabil.* **2014**, *35*, 2624–2634. [CrossRef]
24. Hilgenkamp, T.I.M.; Reis, D.; van Wijck, R.; Evenhuis, H.M. Physical activity levels in older adults with intellectual disabilities are extremely low. *Res. Dev. Disabil.* **2012**, *33*, 477–483. [CrossRef] [PubMed]
25. Scott, J.; Abaraogu, U.O.; Ellis, G.; Giné-Garriga, M.; Skelton, D.A. A systematic review of the physical activity levels of acutely ill older adults in Hospital at Home settings: An under-researched field. *Eur. Geriatr. Med.* **2021**, *12*, 227–238. [CrossRef] [PubMed]
26. Rantanen, T.; Eronen, J.; Kauppinen, M.; Kokko, K.; Sanaslahti, S.; Kajan, N.; Portegijs, E. Life-Space Mobility and Active Aging as Factors Underlying Quality of Life among Older People Before and During COVID-19 Lockdown in Finland—A Longitudinal Study. *J. Gerontol. Ser. A* **2020**, *76*, e60–e67. [CrossRef] [PubMed]
27. Bergland, A.; Wyller, T.B. Risk factors for serious fall related injury in elderly women living at home. *Inj. Prev.* **2004**, *10*, 308–313. [CrossRef]
28. Stevens, J.A.; Olson, S. Reducing falls and resulting hip fractures among older women. *Home Care Provid.* **2000**, *5*, 134–141. [CrossRef]
29. Apostoli, P.; Sala, E.; Curti, S.; Cooke, R.M.T.; Violante, F.S.; Mattioli, S. Loads of housework? Biomechanical assessments of the upper limbs in women performing common household tasks. *Int. Arch. Occup. Environ. Health* **2012**, *85*, 421–425. [CrossRef]
30. Öhrling, T.; Kumar, R.; Abrahamsson, L. Assessment of the development and implementation of tools in contract cleaning. *Appl. Ergon.* **2012**, *43*, 687–694. [CrossRef]
31. Kumar, R.; Kumar, S. Musculoskeletal risk factors in cleaning occupation—A literature review. *Int. J. Ind. Ergon.* **2008**, *38*, 158–170. [CrossRef]
32. Wang, M.-H.; Chen, Y.-L.; Chiou, W.-K. Using the OVAKO working posture analysis system in cleaning occupations. *Work* **2019**, *64*, 613–621. [CrossRef]

33. Loucaides, C.A.; Chedzoy, S.M.; Bennett, N. Differences in physical activity levels between urban and rural school children in Cyprus. *Health Educ. Res.* **2004**, *19*, 138–147. [CrossRef]
34. Machado-Rodrigues, A.M.; Coelho-E-Silva, M.J.; Mota, J.; Padez, C.; Martins, R.A.; Cumming, S.P.; Riddoch, C.; Malina, R.M. Urban–rural contrasts in fitness, physical activity, and sedentary behaviour in adolescents. *Health Promot. Int.* **2012**, *29*, 118–129. [CrossRef] [PubMed]
35. Saitta, M.; Devan, H.; Boland, P.; Perry, M.A. Park-based physical activity interventions for persons with disabilities: A mixed-methods systematic review. *Disabil. Health J.* **2019**, *12*, 11–23. [CrossRef] [PubMed]
36. Guthold, R.; Ono, T.; Strong, K.L.; Chatterji, S.; Morabia, A. Worldwide Variability in Physical Inactivity: A 51-Country Survey. *Am. J. Prev. Med.* **2008**, *34*, 486–494. [CrossRef] [PubMed]
37. Chaeroni, A.; Kusmaedi, N.; Ma'mun, A.; Budiana, D. Physical Fitness and Mental Health in Urban and Rural Areas. *Malays. J. Med. Health Sci.* **2021**, *17*, 11–16.
38. Crawford, A.; Hollingsworth, H.H.; Morgan, K.; Gray, D.B. People with mobility impairments: Physical activity and quality of participation. *Disabil. Health J.* **2008**, *1*, 7–13. [CrossRef]
39. Enam, A.; Konduri, K.C.; Eluru, N.; Ravulaparthy, S. Relationship between well-being and daily time use of elderly: Evidence from the disabilities and use of time survey. *Transportation* **2018**, *45*, 1783–1810. [CrossRef]
40. Smith, M.; Calder-Dawe, O.; Carroll, P.; Kayes, N.; Kearns, R.; Lin, E.-Y.; Witten, K. Mobility barriers and enablers and their implications for the wellbeing of disabled children and young people in Aotearoa New Zealand: A cross-sectional qualitative study. *Wellbeing Space Soc.* **2021**, *2*, 100028. [CrossRef]
41. Lord, S.R.; Menz, H.B.; Sherrington, C. Home environment risk factors for falls in older people and the efficacy of home modifications. *Age Ageing* **2006**, *35*, ii55–ii59. [CrossRef]
42. Chuangchai, W.; Raymaekers, S. Design Perspective of Environmental Interventions through the Stroop Effect, Juggling, and Volunteering for Successful Ageing and Cognitive Health: A Review. *J. Archit./Plan. Res. Stud.* **2023**, *20*, 71–84. [CrossRef]
43. Riazi, A.; Ying Boon, M.; Bridge, C.; Dain, S.J. Home modification guidelines as recommended by visually impaired people. *J. Assist. Technol.* **2012**, *6*, 270–284. [CrossRef]
44. Fournier, K.A.; Hass, C.J.; Naik, S.K.; Lodha, N.; Cauraugh, J.H. Motor Coordination in Autism Spectrum Disorders: A Synthesis and Meta-Analysis. *J. Autism Dev. Disord.* **2010**, *40*, 1227–1240. [CrossRef] [PubMed]
45. Koontz, A.M.; Brindle, E.D.; Kankipati, P.; Feathers, D.; Cooper, R.A. Design Features That Affect the Maneuverability of Wheelchairs and Scooters. *Arch. Phys. Med. Rehabil.* **2010**, *91*, 759–764. [CrossRef] [PubMed]
46. Bonenberg, A.; Branowski, B.; Kurczewski, P.; Lewandowska, A.; Sydor, M.; Torzyński, D.; Zabłocki, M. Designing for human use: Examples of kitchen interiors for persons with disability and elderly people. *Hum. Factors Ergon. Manuf. Serv. Ind.* **2019**, *29*, 177–186. [CrossRef]
47. Ramlee, R.A.; Tang, D.H.Z.; Ismail, M.M. Smart home system for Disabled People via Wireless Bluetooth. In Proceedings of the 2012 International Conference on System Engineering and Technology (ICSET), Bandung, Indonesia, 11–12 September 2012; pp. 1–4.
48. Li, K.F. Smart home technology for telemedicine and emergency management. *J. Ambient Intell. Humaniz. Comput.* **2013**, *4*, 535–546. [CrossRef] [PubMed]
49. Kasper, N.V.; Ilvitskaya, S.V.; Petrova, L.V.; Shulginova, O.A. Interior design of habilitation centers for young children. *IOP Conf. Ser. Mater. Sci. Eng.* **2019**, *698*, 033027. [CrossRef]
50. Chuangchai, W.; Pothisiri, W. Postural Changes on Heart Rate Variability among Older Population: A Preliminary Study. *Curr. Gerontol. Geriatr. Res.* **2021**, *2021*, 6611479. [CrossRef]
51. Dirks, S.; Kurth, F. Working from Home in the COVID-19 Pandemic—Which Technological and Social Factors Influence the Working Conditions and Job Satisfaction of People with Disabilities? In Proceedings of the Computers Helping People with Special Needs, Lecco, Italy, 11–15 July 2022; Springer: Cham, Switzerland, 2022; pp. 183–191.
52. Lundh, S.; Nasic, S.; Riad, J. Fatigue, quality of life and walking ability in adults with cerebral palsy. *Gait Posture* **2018**, *61*, 17. [CrossRef]
53. Steinman, B.A.; Nguyen, A.Q.D.; Pynoos, J.; Leland, N.E. Falls-Prevention Interventions for Persons Who Are Blind or Visually Impaired. *Insight* **2011**, *4*, 83–91.
54. Ward, R.E.; Boudreau, R.M.; Caserotti, P.; Harris, T.B.; Zivkovic, S.; Goodpaster, B.H.; Satterfield, S.; Kritchevsky, S.B.; Schwartz, A.V.; Vinik, A.I.; et al. Sensory and Motor Peripheral Nerve Function and Incident Mobility Disability. *J. Am. Geriatr. Soc.* **2014**, *62*, 2273–2279. [CrossRef]
55. Brunnström, G.; Sörensen, S.; Alsterstad, K.; Sjöstrand, J. Quality of light and quality of life—The effect of lighting adaptation among people with low vision. *Ophthalmic Physiol. Opt.* **2004**, *24*, 274–280. [CrossRef]
56. White, M.D.; Ancoli-Israel, S.; Wilson, R.R. Senior Living Environments: Evidence-Based Lighting Design Strategies. *HERD Health Environ. Res. Des. J.* **2013**, *7*, 60–78. [CrossRef] [PubMed]
57. Boxall, K.; Ralph, S. Research ethics and the use of visual images in research with people with intellectual disability. *J. Intellect. Dev. Disabil.* **2009**, *34*, 45–54. [CrossRef] [PubMed]
58. McDonald, K.E.; Kidney, C.A. What Is Right? Ethics in Intellectual Disabilities Research. *J. Policy Pract. Intellect. Disabil.* **2012**, *9*, 27–39. [CrossRef]

buildings

MDPI

Article

Beyond Sleep: Investigating User Needs in Today's Bedrooms

Demet Dincer [1,*], Christian Tietz [2] and Kerem Dalci [3]

1 Interior Architecture, School of Built Environment, University of New South Wales, Sydney 2052, Australia
2 Industrial Design, School of Built Environment, University of New South Wales, Sydney 2052, Australia
3 Orthodontics and Paediatric Dentistry, School of Dentistry, University of Sydney, Camperdown, Sydney 2050, Australia
* Correspondence: d.dincer@unsw.edu.au

Abstract: This study investigates the current use of bedrooms to fill a knowledge gap in the existing sleep studies literature by focusing on user preferences and their needs. Despite substantial research into the optimal physical conditions for sleep, there remains a gap in understanding how individuals use their bedrooms today. As an initial step to bridge this gap, we employed a mixed-method research approach, integrating quantitative and qualitative data results. A survey including multiple-choice and open-ended questionnaire items was undertaken in Australia with 304 participants. Our research findings indicate that 'having a separate bedroom in a house' is the most common arrangement. However, other arrangements, such as living in a bedsit or sharing a bedroom, are indicators of diverse bedroom conditions. In total, 70% of respondents stated that they would like to make changes to their bedrooms for physical, functional, comfort-based, and aesthetic reasons, with comfort ranked as the highest. The majority of respondents stating they would not make any modifications in their bedrooms were found to be owner-occupiers. Mattress quality and privacy emerged as the most significant factors impacting the respondents' bedroom experience, and female participants were reported to place higher importance on cleanliness compared to male participants. Additionally, preferences for sleeping alone or with a partner were found to shift with age. Based on our findings, we suggest that survey-based sleep studies should evaluate the bedroom arrangements before the individual's sleep habits. Further studies are needed to understand how privacy needs influence sleep quality. From a built environment viewpoint, this study emphasizes the need to develop bedroom design solutions tailored to optimize bedroom conditions, particularly for rental tenants who often have limited control over the physical conditions of their sleep environment.

Keywords: bedroom; sleep environment; domestic environment; home space; user needs

1. Introduction

Bedrooms are mostly designated as domestic environments for sleep and the environmental conditions of a bedroom have a direct impact on sleep [1]. However, there is a lack of focus in the sleep literature on the interior characteristics of the sleep environment and users' changing aesthetic and functional preferences. Today, the interplay of various aspects in today's bedrooms has resulted in a highly complicated and multi-layered environment with changing use, ownership, and sharing conditions. The research outcomes reveal a gap in the sleep studies literature concerning bedroom design, with bedrooms noted as environments that lack the controlled conditions found in sleep laboratories [1]. The purpose of this study is to identify and highlight the changing conditions and preferences of bedrooms today, in order to fill a research gap in the current sleep studies literature.

Recent research outcomes provide a thorough understanding of the environmental factors that influence sleep and highlight the optimal physical conditions needed for an ideal sleep environment. Studies indicate that maintaining a temperature between 18 and 28 °C with humidity levels between 40 and 60%, and noise levels below 35 dB are essential. It is also recommended that individuals ensure total darkness and avoid exposure to blue light [2]. Sleep studies also reveal that demographic factors, including gender, marital status, and educational qualifications, have an influence on sleep quality [3–5]. For example, women were reported as having more sleep problems than men, and socio-economic disadvantages (such as low income or low educational qualifications) were found to have an association with reported sleep problems [3]. Compared to married people, unmarried individuals were found to sleep less with a significant difference, both on weekdays and weekends [5], and the presence of children in the household often increases the frequency of insufficient sleep among the adults residing with them [6].

Changes in physical environmental conditions, such as seasonal temperature shifts or alterations in noise types, have different impacts on sleep quality. Research findings indicate that sleep patterns are affected by seasonal and weather changes and the strongest seasonal effects for wake time and sleep duration were found during the spring season [7]. Road, rail, and air traffic noise have different effects on sleep quality, depending on the sound pressure levels and frequencies of the noise [8]. Other types of noise unrelated to traffic, such as the ringing of church bells, were found to increase the frequency of awakenings in individuals [9]. Out-of-laboratory studies also showed that the range of evening light intensity in the domestic environment can shift the circadian phases of individuals [10]. Thermal comfort is significant for sleep quality, and studies suggest having a sleeping-mode control strategy using air conditioners in bedrooms or making arrangements for the optimal bed micro-environments [4]. Another research finding suggests that opening a window or a door leads to better ventilation, and reducing CO_2 levels during sleep implies better sleep efficiency [11].

Studies on changing environmental factors offer guidance for potential modifications to sleep environments to maintain sleep hygiene. For example, overheating in homes during the summertime has become a growing health concern. In their examination of 750 homes, researchers proposed that living rooms or an alternative bedroom space could provide "a safe haven" for sleep [12]. This suggestion might be a solution when there could be an alternative space to function as a sleep environment. Also, inadequate design of domestic spaces should be considered as another contributing factor to overheating in bedrooms [1]. Additionally, there might not be a distinctive difference in the use of bedrooms and living rooms today, aligned with our previous findings, which confirmed that today's bedrooms are multi-functional spaces [13]. A broader range of activities such as working and eating take place in the bedrooms than before [14]. Evidence for the changing use of furniture, such as the bed overtaking the sofa for the first time as the most used piece of furniture in British homes [15], also shows that there is an ongoing change of preferences in the domestic environment.

The new domestic hierarchies and changing user needs are important aspects to consider in the design of today's bedroom spaces [14]. The sleep literature lacks sufficient focus on the spatial characteristics of the sleep environment [1] and users' changing functional preferences. In fact, it is necessary to question how bedrooms function today and to understand the changing user needs beyond the physical environmental parameters for designing a healthy sleep environment.

2. Use of Bedrooms

Social, economic, and political changes have altered the spatial requirements of city dwellers [14]. Rising urban populations caused people to live in smaller spaces and/or crowded buildings [16], which also brought alternative ways to organize the interior spaces, such as having a bed folding away into the wall [17]. The use of bedroom space has also been affected by various factors throughout history, such as the digital economy, pandemics, or work being carried out from home [18,19]. As a result of its expanded use, the bedroom gained growing importance inside the domestic space [20] with new spatial necessities being formed [14,21].

The home has been evaluated as the core of private life, offering the prospect of family interaction and providing privacy for the individual [22]. Within the home, the bedroom is often recognized as the "most private space" [17], as a personal, solitary refuge for the individual. Throughout history, the spatial configuration and functional use of bedrooms have varied within the domestic space. An example of this can be seen in the layouts of buildings constructed to host the working class during the early 20th century in North American cities, such as New York. These layouts relegated kitchens and bathrooms as shared spaces between residents on the same floor, while bedrooms were inhabited as the personal living environment. Similar layouts were observed in the Soviet Union, where a bedroom operated as a private space for residents to dwell individually, and two apartments had one shared bathroom [14]. However, bedrooms were not always the "most private" spaces in a domestic environment. In fact, there were times in history when bedrooms and sleeping were a much more communal affair. Sharing beds with strangers was common for travelers sleeping in inns during the 16th and 17th centuries in Europe and North America. As recently as the 20th century, traveling salesmen used to share beds when bed space was in high demand [17]. The examples of 'publicly sleeping together' mostly occurred outside the home and domestic sphere, yet sharing beds at home with family was and still is not uncommon [23,24]. Hence, having a dedicated bedroom to sleep in was not always the norm, as the idea of associating any specialized function with individual rooms had not yet occurred in the 17th century [25].

Given the historical shifts, understanding the changing needs for today's bedrooms has become more complex, as today's bedrooms are more than spaces hosting sleep. The preferences and social norms change over time; for example, in a US-based study from 2005, middle-class parents had a preference for houses with large master suites located at a distance from other bedrooms in the house [26]. Today, there is an emerging bedroom culture with the bed playing an increasing role [27]. The function of the bed has been discoursed beyond its role to host sleep, highlighting the new user needs of the internet and social media [20,28].

At the 15th Venice Architecture Biennale in 2016, the British Pavilion hosted the Home Economics exhibition, with the theme of 'Life is changing, we must design for it'. The exhibition highlighted the housing crisis as 'a crisis of how we live', addressing the ongoing changes happening in the domestic space. One of the examples provided in their exhibition brief was that it was previously common for a married couple to use twin beds, while today, the double bed is the focus of household requirements for UK bedroom standards. In their proposal entitled "Hours", the authors suggested that modular daybeds would provide a flexible solution in a shared home. Another proposal, "Decades", was a design of a room without any functions, with no predetermined spaces or activities [15]. While we cannot foresee what spatial alterations might be solidified in the future, a need for change is evident. Different lifestyles, such as living alone and nomadic living, can result in dwellings being under-occupied or over-occupied. This often leads to a need for new spatial arrangements such as micro-apartments or co-living spaces [14]. Recent studies have also highlighted a noticeable change in the use of domestic spaces, especially for work-related purposes. This shift has introduced different needs for privacy and sociability [19]. As an example, the sleep environments in care homes are now regarded as living spaces, with research confirming the need for personalization to ensure users' privacy and autonomy [29].

The existing literature on bedrooms mainly focuses on the physical environmental factors [2,30–33], and little attention has been given to user needs. Few studies provide guidance for architects or inform sleep researchers about what characteristics and provisions are needed for an optimal bedroom space today. In fact, the currently defined optimal parameters are also subject to further investigation. For example, checking different configurations of window/door openings showed significant correlations between actigraphy data and questionnaire responses for sleep parameters [11]. It is crucial to approach the use of bedrooms with a comprehensive, open-minded viewpoint, rather than relying solely on previously defined parameters and assumptions on how today's bedrooms function. A number of studies have already highlighted that 'other' design-related aspects might influence the use of bedrooms as sleep environments. For example, one study discussed the significance of "decorations in the bedroom" and the "surrounding environment of the building" as factors affecting bedroom use [34]. These findings are informative in identifying the unconsidered design features that might have an impact on sleep quality. However, these features were constrained by a pre-defined list and disregarded the spatial circumstances of the room. In the study, decoration items were listed as plants, carpet, bookshelf, router, and screen, and the questions about the surrounding environment mainly focused on the view. Additionally, researchers discovered significant disparities in how people utilize their bedrooms, based on their age and gender [35]. Hence, a better understanding is needed of how today's bedrooms function, with whom they are shared, and what factors users believe are important for their bedroom use.

3. Methodology

A survey including multiple-choice and open-ended/text-entry questionnaire items was used to gather data. A mixed-method research approach was applied in this paper, integrating quantitative and qualitative data results. Detailed descriptions of the study design and methodology are provided below.

3.1. Study Design

In preparing our survey, two aspects were defined to have a better understanding of bedroom use in Australia: firstly, functional changes that might be associated with sleep habits, and secondly, user experience and expectations. This forms the second part of the study, following the initial evaluation of its functional use.

To determine the use of bedrooms and user needs, we reviewed the questions in the following sleep studies: the National Sleep Foundation Sleep Survey [36], the Sleep Hygiene Index (SHI) [37], and the Pittsburgh Sleep Quality Index (PSQI) [38]. The National Sleep Foundation Survey included detailed questions about sleep duration and sleep difficulties among key demographic variables including age, gender, education, income, and cultural background. Respondents were queried about their sleep habits, encompassing activities such as watching TV, listening to music, reading, and exercising. There were questions regarding the presence of specific items within the bedroom, such as televisions and computers, which were listed as predefined options. The survey also asked participants about their sleeping arrangements. The PSQI questionnaire mainly focused on sleep duration and sleep quality along with the respondents' health conditions, with one additional question about bed partners or roommates, but made no inquiries regarding the sleep environment. The SHI survey utilized Likert scale questions to assess agreement on bed-related use and habits.

To enhance our understanding beyond the questions from these three surveys, we added a new question about property ownership to better understand potential user needs and possible limitations. We furthermore asked two open-ended questions to assess possible shifts in the current use and perception of bedroom space. This enabled us to explore what adjustments respondents may want, offering insight into current spatial requirements. The survey combined elements from the three surveys above with additional questions, which are structured in two main sections: (1) spatial conditions of bedrooms and (2) user experience and expectations (Appendix A).

At the end of the survey, we asked participants if they would like to provide any photos of their bedroom, along with a disclaimer that no individuals were to be included in these photos. This photo upload option was not mandatory.

3.2. Data Collection and Analysis Method

The data collected for this research were based on the responses from 304 participants about their user experience. The data were non-identifiable as per the negligible risk guidelines and approved by the UNSW Human Ethics Committee. Participants were recruited through an online research panel and the company used weighted randomization based on respondents' demographic data. The online research panel collected data from qualified individuals (18 years and older, Australian residents), offering real-time anomaly detection. The survey invitations provided basic links and information that was non-leading. Respondents received this invitation without specific project details to avoid self-selection bias, and the project details were disclosed later.

A mixed-method research approach was applied in this paper, integrating quantitative and qualitative data results. IBM SPSS Statistics for Windows, Version 28.0. Armonk, NY: IBM Corp was used to analyze data with a statistical significance level of $p < 0.05$. The categorical variables were examined for any significant differences using Fisher's exact test and Pearson's chi-square test. A Kruskal–Wallis test (the nonparametric version of ANOVA) and a one-way ANOVA test were used to evaluate the relationship between the responses and the sociodemographic variables. The qualitative data were examined using a reflexive thematic analysis method and NVivo 12 software was used for evaluation. A reflexive thematic analysis method was used to identify patterns and/or categories [39].

In response to the optional question of sharing a photograph from their bedrooms, 24 survey participants submitted visuals showing their use of the bedroom space. Of these images, 13 of them were relevant to the bedroom research context, and the others either failed to show a bedroom space or were taken from other online resources. Due to the limited number of image submissions, it was not possible to analyze the visual content; however, a small selection of these images is presented in this article.

4. Results

The demographic characteristics including gender, age, ownership, and household income are shown in Table 1, and the sample was representative of the Australian population with regard to age, gender, and geographic location across states. Among the respondents, 49% identified as male and 51% as female. In total, 38 respondents were aged 18–24 (12.5%), 119 were aged 25–44 (39.1%), 101 were aged 45–64 (33.2%) and 46 were aged 65 and above (15.1%). The majority of the respondents (63.5%, N = 193) were owner-occupiers, while 31.3% were identified as rental tenants. A total of 16 respondents selected the 'other' option for their response to this question, reporting that they were 'living with a parent', were 'dependent', or were 'house-sitting'. Of the 304 respondents, 21 people (6.9%) responded that their annual household income was less than AUD 19,999, which is considered below the poverty line in Australia [40]. Furthermore, 75% of the participants who were 45 years of age and above were owner-occupiers.

Table 1. Demographic characteristics of the respondents.

Demographic Characteristics	Variable	Frequency	Percentage
Gender	Male	146	49%
	Female	158	51%
Age	18 to 24 years	38	12.5%
	25 to 44 years	119	39.1%
	45 to 64 years	101	33.2%
	65 and over	46	15.1%
Ownership	Owner-occupier	193	63.5%
	Rental, tenants	95	31.3%
	Other: dependent	16	5.3%
Household Income	Less than AUD 19,999	21	6.9%
	AUD 20,000–49,999	78	25.7%
	AUD 50,000–79,999	71	23.4%
	AUD 80,000–124,999	71	23.4%
	AUD 125,000–199,999	40	13.2%
	AUD 200,000 and over	23	7.6%

The research findings were grouped into three sections for better readability. In the first section (Section 4.1), we presented the data of the survey findings on the spatial characteristics of bedrooms. This provided us with information regarding the bedroom conditions of the household. The second part (Section 4.2) contains data results about the participants' spatial experience and factors affecting their experience. Following that, a third section of the results (Section 4.3) includes the analysis of the open-ended questions. To provide an analysis of the interconnectedness of these sections, a cross-evaluation, and discussion (Sections 5.1 and 5.2) followed the results.

4.1. Bedroom Conditions

The type of domestic environment (e.g., bedsit/studio, apartment, or house), as well as the conditions of the bedroom within the layout (open floor plan where the bedroom is also a living room, or if the bedroom is a separate area allocated for sleep), might give an indication of the different uses of a bedroom. Our findings showed that 4.3% of the participants lived in bedsits or studio apartments, which indicates that there is not a separate bedroom allocated for sleep. The majority, by a large margin (73%), had their own bedroom in their own apartment/house, followed by those who had their own bedroom in a shared apartment/house (20.4%) (Table 2).

Table 2. Results showing the bedroom conditions.

	Variable	Frequency	Percentage
Bedroom Conditions	I live in a bedsit/studio	13	4.3%
	I have my own bedroom in a shared apartment/house	62	20.4%
	I have a shared bedroom (not partner)	5	1.6%
	I have my own bedroom in my own apartment/house	221	72.7%
	Other	3	1%
Bedroom Sharing Conditions (multiple answers possible)	Sleep alone	134	38.1%
	Sleep with a partner	159	45.2%
	Sleep with children	31	8.8%
	Sleep with a pet	28	8.0%

Fisher's Exact Test results showed that there was a significant difference in bedroom conditions between age groups. Accordingly, 25 to 44-year-olds mainly have a bedroom in a flat/house (64.70%), counting more than the respondents stated living in a bedsit/studio (8.40%); 18–24-year-olds primarily sleep in a shared flat/house (50%), higher than having their own bedrooms (39.50%) (Figure 1). Participants having their own bedroom in a shared flat/house was significantly higher among the 25–44-year-olds (45.2%), followed by the 18–24-year-old group (20.6%). Having their own bedroom in their own flat/house was the highest for 45–64-year-olds (38.0%) followed by 25–44-year-olds (34.8%), and significantly higher than the 18–24-year-olds group (6.8%) ($p < 0.001$). Among respondents aged 45 years and older, two individuals reported residing in a bedsit/studio, while one participant indicated sharing their bedroom with a person other than their partner.

In this section of the data results, we extracted the information to find out whether the respondents were the sole occupiers of the bedroom or not. The goal of this question was to determine if there are other users sharing the same bedroom. Five participants (1.6%) mentioned sharing their bedroom with a person other than their partner. Most respondents stated that they sleep with their partners (45%), followed by sleeping alone (38%). In total, 28 participants (8%) mentioned sleeping with a pet, and a similar percentage was observed for those who sleep with children (31 people, with a percentage of 8.80%). Next, 25 of the respondents slept with their partner and children together, 13 with their partner and pets together, and 2 with their partners, children, and pets. The total number of respondents who both responded to "sleep alone" and "sleep with pets" was seven. The responses received from the 'other' category revealed that three people (1%) defined their bedrooms as their cars. One of these three respondents mentioned that they were sleeping in a car with a partner, and two of them with their children.

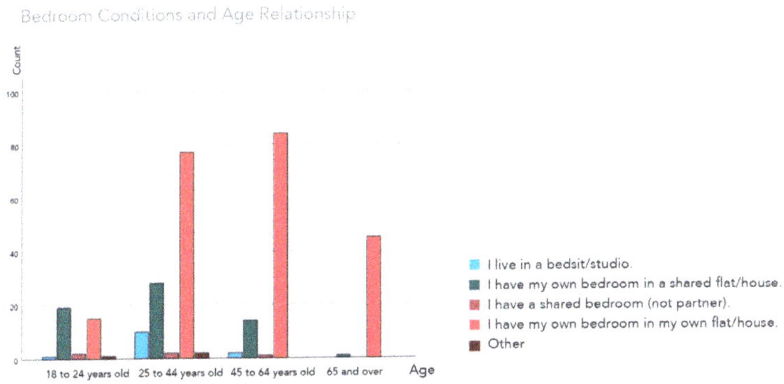

Figure 1. Graph showing the bedroom conditions among different age groups.

Among the 18–24-year-old participants, the majority (68.40%) reported sleeping alone, significantly outnumbering those who preferred sleeping with someone or a pet (31.60%). As age increased to the 25–44-year-old range, the results shifted to 63.00% sleeping with someone or a pet, compared to 37.00% sleeping alone ($p < 0.005$). No significant differences were observed in sleeping preferences beyond the age of 45 (Table 3). Figure 2 illustrates the distribution of bedroom sharing among different age groups.

Table 3. Bedroom sharing conditions and age crosstabulation.

Age–Sleeping Conditions	Sleep with Someone and/or a Pet		Sleeping Alone	
18 to 24 years old	12	31.60%	26	68.40%
25 to 44 years old	75	63.00%	44	37.00%
45 to 64 years old	61	60.40%	40	39.60%
65 and over	22	47.80%	24	52.20%

$p < 0.005$.

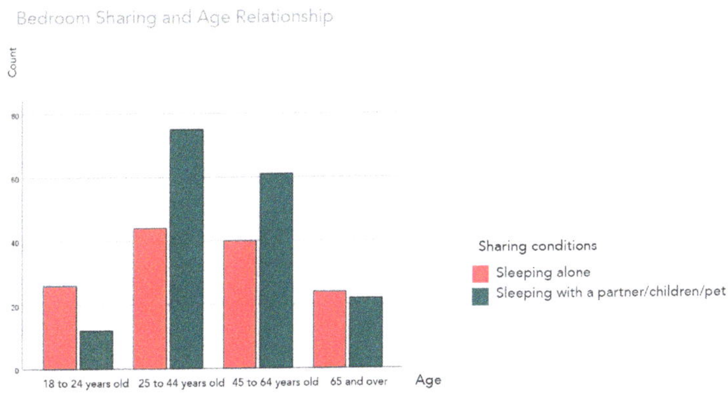

Figure 2. Graph showing the bedroom sharing conditions among different age groups.

There was no significant difference in bedroom sharing conditions between age or gender groups regarding sleeping with children or a pet. According to ownership conditions, owner-occupiers were significantly more likely to report sleeping with another (partner or children), at 60.60%, compared to participants who mentioned sleeping alone (39.40%) ($p < 0.05$). Participants who responded that they sleep with their partners were mainly owner-occupiers (59.60%) when compared to rental tenants ($p < 0.001$).

4.2. Spatial Experience

We asked participants to rate the significance of the following factors to evaluate what respondents prioritize in a bedroom: comfort of mattress and pillows, noise, light, temperature of the room, scents in the room, fresh air and ventilation, cleanliness, and privacy. Table 4 represents the mean values for the eight factors according to their importance, rated on a scale of 1 to 5 points, with 5 indicating the highest level of importance. The findings suggested that mattress comfort was the most significant factor (4.34), followed by privacy (4.2) and noise (4.11); the least important factor was scents in the room (3.03). In fact, scents were identified as "not important at all" by 49 of the 304 respondents. Privacy was rated as "extremely important" by 133 of 304 respondents (43.8%), while it was rated as "not at all important" by 7 (2.3%).

Table 4. The mean results concerning the importance of various factors.

	Mattress Comfort	Noise	Light	Temperature	Scents	Ventilation	Cleanliness	Privacy
Mean	4.34	4.11	3.94	3.99	3.03	3.79	3.96	4.2
Std. Deviation	0.789	0.948	0.971	0.886	1.282	1.006	0.989	0.895

According to the survey results, 60% of the respondents strongly agreed with the statement that "Bedrooms are private spaces", with an overall agreement ratio of 89%. 28% strongly agreed that their bedrooms reflected them and identified with their room, and 42% agreed that they could personalize/change their bedrooms. Personalization ("I can personalize/change my bedroom the way that I want") had a total agreement ratio of 78% from respondents, whereas the bedroom reflecting themselves ("My bedroom reflects me/is a place that identifies me") had a ratio of 62% (Figure 3).

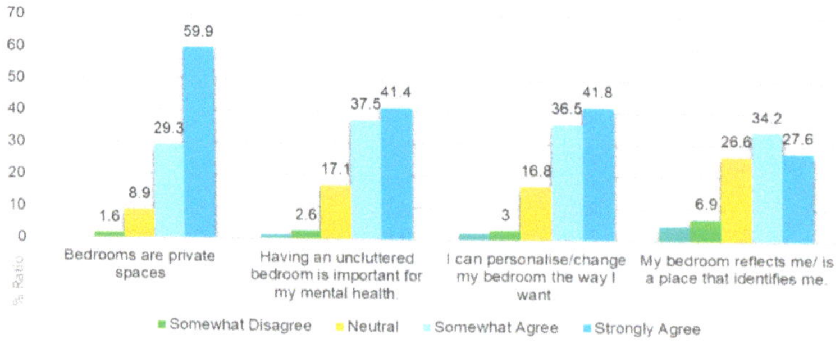

Figure 3. The agreement rankings about bedroom use.

In total, 79% of respondents believed that "Having an uncluttered bedroom is important for (their) mental health", which was not significantly different across age groups. However, a Kruskal–Wallis H Test showed that there was a statistically significant difference in the perception of having an uncluttered bedroom between males and females: $H(1) = 9.957$, $p = 0.002$. While the mean score for males was 3.99 (N = 146), it was 4.30 (N = 158) for females (Figure 4). The importance of cleanliness between genders was also significantly different, as indicated by the Kruskal–Wallis H Test results: $H(1) = 7.736$, $p = 0.005$. Male participants showed much lower agreement with the importance of cleanliness for their sleep environment than female participants (averaging scores of 3.8 and 4.1, respectively, on a 5-point scale assessing the effect of cleanliness on sleep). Regarding the use of a bedroom, female participants (N = 158, mean score: 3.85) referred to their bedroom as a place that reflects/identifies them more than male respondents (N = 146, mean score: 3.6), which was found statistically significant (H(1) = 4.886, p = 0.027).

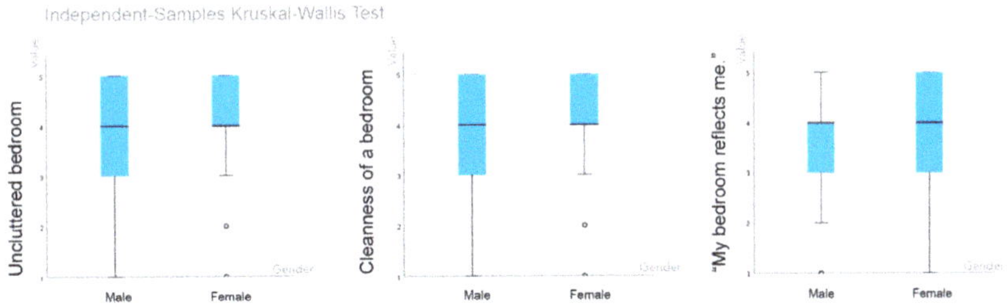

Figure 4. The agreement rankings among different gender groups.

4.3. User Needs and Meaning of the Bedroom

In addition to the multiple-choice questions, we included two open-ended questions to understand different modifications that respondents would like to make in their bedrooms, as well as to examine the current meaning of their bedroom space. The first open-ended question was asking what kind of changes the respondents would like to make in their bedrooms. Firstly, a qualitative analysis software, Nvivo, was employed to examine the word frequencies within the dataset to conduct an initial overview of the data. The words were limited to three letters to exclude articles, synonyms were included, and verbs lacking relevance were excluded (such as make and get). Table 5 represents the results, including the first 20 words.

Table 5. The word frequencies analysis, showing the first 20 words.

Word	Count	Similar Words
bed	43	bed, bedding
bigger	38	bigger, larger
new	28	fresh, modern, new
curtains	21	curtain, curtains, drapes
space	21	space, spaces
better	18	best, better, improved
wardrobe	16	closet, wardrobe, wardrobes
wall	15	wall, walls
room	15	room, suite
bedroom	12	bedroom
light	12	fall, light, lighting, lights
carpet	10	carpet, carpeted, carpets, rugs
blinds	10	blinds, screen, screens
paint	10	paint, painting, pictures
clutter	8	clutter
mattress	1.10	mattress
block	1.00	block, stop, stuff
colour	0.96	colors, colour, colours
different	0.96	different, otherwise
furniture	0.96	furniture

The initial word-frequency analysis results indicate that the participants' responses vary mentioning object-based changes (such as bed, carpet, or mattress) to broader spatial changes (as shown by the descriptive words, such as bigger or better). Therefore, a reflexive thematic analysis method was used to identify patterns and categories, with an aim to achieve a flexible interpretative approach [39]. As a first step, the responses were classified into two categories, which included either "no changes" or a kind of change without any pre-defined classifications. We counted the answers including "none", "nothing", and "no changes" in the categorization of "no changes", as well as statements such as "I have made my bedroom how I want it already", "I am happy with what I have", or "I would not change anything". Among the 304 respondents, 49 mentioned that they were uninterested in making any changes to their bedrooms. This first category resulted in 16% of the respondents indicating that no revisions were needed in their current bedroom spaces (Figure 5).

Figure 5. Photographs by survey participants showing their bedroom environments.

As a second step of the analysis, the responses mentioning a need for change were coded and classified, which resulted in four distinct approaches based on their common characteristics. The first category for the changes was entitled architectural needs, which related to the proportions and physical circumstances of the bedroom space that require a major architectural change. Some examples analyzed in this category were related to the room size, layout, or openings such as windows or doors, or an outside connection. Some of the participant statements classified under this category are as follows: "I wish the room was bigger", "[I prefer having a] door to the bathroom", "more sunlight", "more space, no balcony", "put doors on it", "more space", "have an ensuite", "I'd like to knock out the fireplace", and "I'd remove the inbuilt wardrobe".

The second category focused on "functional needs", encompassing requirements based on objects and other physical improvements in the bedroom space. For example, 15 respondents expressed their need for a "bigger wardrobe" and "more storage space", or a modification to better organize their storage needs as indicated by the following response: "change the clothes storage to be under bed storage to make the room more minimalist". Respondents in this category emphasized activities and practical demands to improve their bedroom space, such as having a bookstand for reading before sleeping or having a dresser for more storage to avoid clutter.

The responses that emphasized "comfort needs", such as "getting a new mattress" or a "humidifier", were classified under a third category. In total, 27 respondents indicated a need for a change in their beds, such as mattress quality, mattress comfort, bed linen, and bed frame as demonstrated by the responses "[I would like to] have the best mattress that provides enough support and comfort". Other answers addressed comfort needs such as having block-out curtains to sleep better, adjustable lighting, or air-conditioning. Noise was also a factor emphasized in responses, with participants reporting needs such as "turn[ing] off exhaust fan noise from the ensuite", "make it more soundproof" and "pad the walls to block all noises", and one respondent mentioned about their bed squeaking. Some of the other comfort-related needs were stated as the room temperature not being at a comfortable level to fall asleep, or lacking security screens to be able to leave the door open in summer. De-cluttering was also mentioned as a need along with better pillows and bed frames.

The final set of responses fell into the category of "aesthetic needs", which pertained not to changes in room layout or comfort requirements, but rather to visual preferences. Most of these responses conveyed the need for color changes in the room, the addition of visual items such as artwork or pictures, or styling preferences for the bedroom space. In total, 16 respondents stated that they wanted to re-paint or recolor their bedroom space and 10 respondents mentioned adding more artwork or pictures. Some of the other examples that addressed aesthetic needs are as follows: "I would change the wall colors and buy furniture to style the way I want if I could", "everything, need a brand new bedroom style and make it my vibe", "make it more cozy", "more modern", and "get rid of a lot of clutter and make it minimalistic".

Our results of this first open-ended question regarding the changes showed that 58 respondents indicated an interest in making physical improvements to their bedrooms (Table 6). The needs regarding comfort (n = 76), such as bed support, block-out curtains, and a need for a humidifier, were reported in greater frequency compared to physical and architectural changes. Responses such as "I'd like to decorate but I can't because I'm renting" demonstrated that being a rental tenant limited their responses not solely to architectural needs, but also to aesthetic appeal. On the other hand, the majority (n = 41 out of 58) of the respondents mentioning that a possible change is needed were identified as owner-occupants. In terms of aesthetic needs (n = 49), the presence of artwork in the bedroom space was seen as a primarily personal room-enhancing factor. The aesthetic needs expressed by the respondents, such as adding artwork and changing the color of the bedroom, are indicative of a further need for personalization, despite the topic not being specifically investigated in the structure of our questionnaire. The percentage of respondents who claimed they would make an aesthetic improvement to their bedroom was evenly spread across ownership groups, with 46% being owner-occupants and 54% being rental tenants and dependents. Respondents also emphasized the importance of storage and having more space in the bedroom with 5% reporting the desire to have better storage options. Room layout was referred to only in a few instances, with responses such as "my partner doesn't like the bed against the window which limits my options" and "I'd like some art for the walls and comfortable seating, so I didn't have to use my bed" (Figure 6).

Table 6. Categories defined by the respondents' answers regarding bedroom modifications.

Category	No Changes	Physical	Functional	Comfort	Aesthetic
Number of respondents	91	58	56	76	49

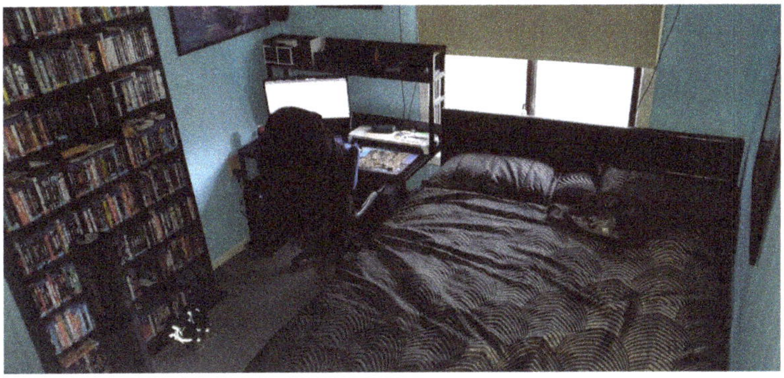

Figure 6. Bedroom photo shared by a survey participant, showing their bedroom, desk, shelves, artwork, and personal items.

The second open-ended question was an inquiry into the meaning of their bedroom space. Most of the respondents defined their bedrooms as a place to sleep (n = 146) and relax (n = 118), with some other related responses such as "a place where I feel comfortable and am able to sleep peacefully", "it's my resting place", and "a place to sleep and rest and refresh for the next day". While sleeping was the main highlighted word, as expected, 39 respondents highlighted the bedroom's privacy aspect. The following are some other examples of how respondents described their bedrooms:

"It's a sanctuary and a safe space to relax"

"[the bedroom is] my identity"

"It's my whole life"

"[the bedroom is] a place to be myself"

"[the bedroom is] my personal space"

"It means a lot. It's my personal quiet space, where I go to relax and feel comfortable".

5. Discussion

This section provides a cross-evaluation of the shared data results and the potential connections between them, followed by a discussion about the implications of this study.

5.1. Cross-Evaluation

Our results showed that sleeping with another or sleeping alone changes with age. As age increases into the range of 25–44, the likelihood of "sleeping with someone or a pet" increases, eventually reaching an equal ratio after 65 years of age. A bedsit/studio arrangement was observed more in the 25–44-year-olds than any other age group, while having a bedroom of their own had the highest value for the 45–64-year-olds. Even though 'having a separate bedroom in a house' remains the most prevalent circumstance (72.7%), other circumstances (27.3%), such as living in a bedsit or sharing a bedroom with others, are important indicators of further possible bedroom-use conditions.

Our findings revealed that mattress comfort was found to be the most significant factor affecting the respondents' bedroom use, followed by privacy and noise. In their responses to open-ended questions, participants mentioned comfort conditions more frequently than practical aspects (such as room size) and aesthetic needs (such as color preferences). Two of the main concerns cited by respondents in relation to comfort conditions were noise and lighting, as indicated by the following respondent comment: "[I would like to] put ceiling fans in, glaze windows to keep noise out, maybe a shutter on the outside noise and light". Based on the findings from other sleep studies about the optimal conditions for a sleep environment, these comfort needs require prompt attention to enhance sleep quality.

In terms of privacy, our study found that respondents regarded privacy in the bedroom as very or extremely important, which is also evidenced by the qualitative results. While there was no significant difference between age groups or gender, respondents sleeping with children reported having less privacy in their bedrooms ($p < 0.005$). Bedrooms were stated as private spaces with a total agreement ratio of 89%, which supported the 83% who conveyed that privacy was a very or extremely important factor. When considered in conjunction with the fact that 62% reported that the bedroom was a place that reflected or identified them, these results further supported that the bedroom was (still) perceived to be a personal and private space.

Most participants in our study were owner-occupiers, sleeping in a separate bedroom space. As anticipated, there was a substantial association between ownership and income ($p < 0.001$), with 73.9% of owner-occupiers reporting an annual income of more than AUD 200,000. There was no significant difference in responses between age groups or ownership statuses regarding the Likert scale agreement question, "My bedroom reflects me". A considerable difference emerged between participants with different ownership statuses regarding the statement "I can personalize/change my bedroom". After these initial results, we conducted a one-way ANOVA Test which showed a statistically significant difference between groups. Owners were more inclined to agree with the statement compared to rental tenants, as reflected in the mean value of 4.24 for owners and 3.91 for rental tenants ($F = 4.467$, $p = 0.016$). A total of 30% of the respondents (n = 91) stated that they did not want to make any changes to their bedrooms. This ratio can be interpreted as respondents being content with their current bedroom conditions. When combined with the answers to the open-ended question on the need for changes in the bedroom space, the results demonstrated that ownership expands opportunities for optimizing the bedrooms to meet physical and/or psychological needs. The majority (n = 70) of those who stated that they would make no modifications were owner-occupiers, accounting for more than rental tenants or dependents.

The results also showed that the "importance of cleanliness", and the "identification of the bedroom space as a reflection of themselves" were more of a consideration for female participants than male participants. Accordingly, having an uncluttered sleep environment is more significant for female participants. On the other hand, there was no difference between age groups regarding how an uncluttered bedroom affects their mental health, with a high agreement across all ages (79%).

5.2. Limitations

This survey was limited to Australian residents. Although Australia is a multicultural society and the participants represent the wider population, we acknowledge that the responses might vary accordingly within different demographics. A larger sample size would have been preferable; however, this sample size was adequate for gaining an initial understanding of current bedroom use in the Australian community. Also, we anticipate that some of our word choices might have generated a different response if the survey had included the wording 'smell' rather than 'scent'. Our initial questions concerning the types of accommodation and the degree of personalization of the bedroom space were limited. In future studies, it would be beneficial to expand the line of questioning to further address and investigate these aspects.

6. Conclusions

The quality and quantity of sleep are influenced by the environment in which sleep takes place, which refers to any room used for sleeping. Given the extensive knowledge that sleep studies have provided to us regarding temperature, humidity, and lighting, we now know more about the ideal conditions for a bedroom. However, bedrooms are not simply rooms with beds, and they are often not controlled environments in which people can make necessary arrangements to aid in better sleep. The way we use and organize our domestic spaces, especially the bedrooms, has changed in response to social, economic, and political shifts. Regardless of these changes, bedrooms are still considered personal and safe places that people would like to use to express their identity.

Most respondents reported a desire to change their bedrooms, and comfort-related changes were reported more frequently than physical or functional adjustments, such as the requirement for block-out curtains or an air humidifier, both of which are known to have a direct impact on sleep. Previous research indicated that the presence of children in the household often increases the frequency of insufficient sleep [6] and our results showed that respondents sleeping with their children reported having less privacy in their bedrooms. Further research is needed to fully understand the influence of privacy on sleep hygiene. In alignment with the results, our suggestions for future studies are as follows:

- Sleep-focused studies: We propose that survey-based sleep research should, as a first step, evaluate the bedroom conditions before the individual's sleep habits. This evaluation should include the accommodation type, and whether there is a dedicated area for sleeping. Having a partner may not always mean sharing a bedroom, and sleeping with children or pets may influence people's sleeping habits in ways that sleep surveys may not have captured. By gathering more data about the sleep environment and

room-sharing conditions, researchers can obtain a more comprehensive understanding of possible environmental problems before focusing on the individual's sleep habits.
- Design studies: There is a need to further investigate the sleep environment from a design and built environment perspective, including the role of bedroom layout and design preferences. With this knowledge, along with what is already known from previous research about the physiological need for sleep, designers can develop healthy surroundings to improve the quality of our sleep. Some of the key outcomes to consider for future studies on sleep environment are the following: (1) Developing layout solutions to minimize clutter to optimize the bedroom space, especially for female users. (2) Investigating the relationship between privacy and sleep and what kind of design strategies can respond to shared bedroom conditions. (3) Developing design strategies on how to improve the sleep environment in terms of sleep hygiene, especially for rental tenants. Our results showed the need to develop more design solutions with a focus on rental accommodations in which individuals may have limited control over their physical environmental conditions.

Author Contributions: Conceptualization, D.D. and C.T.; methodology, D.D. and K.D.; software, D.D. and K.D.; formal analysis, D.D. and K.D.; investigation, D.D. and C.T.; resources, D.D. and C.T.; data curation, D.D. and K.D.; writing—original draft preparation, D.D. and C.T.; visualization, D.D. All authors have read and agreed to the published version of the manuscript.

Funding: This research was funded by the University of New South Wales, School of Built Environment.

Data Availability Statement: The original contributions presented in the study are included in the article, further inquiries can be directed to the corresponding author.

Conflicts of Interest: The authors declare no conflict of interest.

Appendix A

Table A1. Detailed explanations for survey questions.

Demographic Characteristics	Question	Question Type	Answer Choice
Age	Which group are you in?	Multiple choice, allow one answer	• 18–24 years old • 25–44 years old • 45–64 years old • 65 and over
Gender	How do you define your gender?	Multiple choice, allow one answer	• Male • Female • Prefer not to say
Ownership	This question is about housing. Are you a/an	Multiple choice, allow one answer	• Owner-occupier • Rental tenant • Other: please specify
Household Income	What is your total household income considering last year and before tax and other deductions are taken out?	Multiple choice, allow one answer	• Less than AUD 19,999 • AUD 20,000–49,999 • AUD 50,000–79,999 • AUD 80,000–124,000 • AUD 125,000–199,999 • AUD 200,000 and over
Spatial Characteristics and Conditions of Use	**Question**	**Question Type**	**Answer Choice**
Bedroom Conditions	Which of the following categories best describes where your bedroom is?	Multiple choice, allow one answer	• I live in a bedsit/studio • I have my own bedroom in a shared flat/house • I have a shared bedroom (not partner) • I have my own bedroom in my own flat/house • Other (please specify)
Bedroom Sharing	Which of the following apply to you: Most of the nights, I sleep (/with)...	Multiple choice, allow multiple answers	• I sleep alone • I sleep with my partner • I sleep with children • I sleep with a pet

Table A1. *Cont.*

User Experience	Question	Question Type	Answer Choice
Factors Affecting the Experience	The following questions are about your experience in your bedroom. Please indicate how important the following items are in general.	Matrix question, Likert scale	• Comfort of mattress and pillows • Noise • Light • Temperature of the room • Scents in the room • Fresh air and ventilation • Cleanliness of the room • Privacy
User Experience	Please rate the following statements regarding your use of your bedroom in general.	Matrix question, Likert scale	• Bedrooms are private spaces. • Having an uncluttered bedroom is important for my mental health. • I can personalize/change my bedroom the way I want. • My bedroom reflects me/is a place that identifies me.
User Preferences (changes)	If you could change your bedroom, what types of changes would you like to do?	Open-ended question, text entry	
Meaning	What does a bedroom mean to you?	Open-ended question, text entry	

References

1. Emmitt, S. Exploring the nexus between bedroom design and sleep quality in a warming climate. *Urban Clim.* **2023**, *51*, 101635. [CrossRef]
2. Caddick, Z.A.; Gregory, K.; Arsitescu, L.; Flynn-Evans, E. A review of the environmental parameters necessary for an optimal sleep environment. *Build. Environ.* **2018**, *132*, 11–20. [CrossRef]
3. Arber, S.; Bote, M.; Meadows, R. Gender and socioeconomic patterning of self-reported sleep problems in Britain. *Soc. Sci. Med.* **2009**, *68*, 281–289. [CrossRef]
4. Lan, L.; Tsuzuki, K.; Lian, Z.W. Thermal environment and sleep quality: A review. *Energy Build.* **2017**, *149*, 101–113. [CrossRef]
5. Hale, L. Who has time to sleep? *J. Public Health* **2005**, *27*, 205–211. [CrossRef]
6. Adults, A.; Chapman, D. Household demographics and perceived insufficient sleep. *J. Community Health* **2012**, *37*, 344–349. [CrossRef]
7. Mattingly, S.M.; Grover, T.; Martinez, G.J.; Aledavood, T.; Robles-Granda, P.; Nies, K.; Striegel, A.; Mark, G. The effects of seasons and weather on sleep patterns measured through longitudinal multimodal sensing. *Digit. Med.* **2021**, *4*, 76. [CrossRef]
8. Basner, M.; Muller, U.; Elmenhorst, E. Single and combined effects of air, road, and rail traffic noise on sleep and recuperation. *Sleep* **2011**, *34*, 11–23. [CrossRef]
9. Brink, M.; Omlin, S. An event-related analysis of awakening reactions due to nocturnal church bell noise. *Sci. Total Environ.* **2011**, *409*, 5210–5220. [CrossRef]
10. Burgess, H.; Molina, T. Home lighting before usual bedrime impacts circadian timing: A field study. *Photochem. Photobiol.* **2014**, *90*, 723–726. [CrossRef]
11. Mishra, A.; van Ruitenbeek, A.; Kort, H. Window/door opening-mediated bedroom ventilation and its impact on sleep quality of healthy, young adults. *Indoor Air* **2018**, *28*, 339–351. [CrossRef]
12. Drury, P.; Watson, S.; Lomas, K. Summertime overheating in UK homes: Is there a safe haven? *Build. Cities* **2021**, *2*, 970–990. [CrossRef]
13. Dincer, D.; Tietz, C.; Dalci, K. An investigation into sleep environment as a multi-functional space. *Buildings* **2023**, *13*, 406. [CrossRef]
14. Martella, F.; Amann Alcocer, A. An emergent housing approach: The bedroom as the contemporary minimum living cell. *Home Cult.* **2022**, *18*, 229–254. [CrossRef]
15. Self, J.; Williams, F.; Bose, S. *Home Economics. Five New Models for Domestic Life. Venice Architecture Biennale 2016. British Pavillion Catalogue*, 1st ed.; The Spaces: London, UK, 2016; ISBN 978-0-9573914-8-2. Available online: https://creativedundee.com/wp-content/uploads/2016/10/Home-Economics-Pamphlet-Proof-EN_v4.pdf (accessed on 7 April 2024).
16. Nasreen, Z. Tracking the Rise of Room Sharing and Overcrowding, and What It Means for Housing in Australia. The Conversation, The Conversation Media Group. 2019. Available online: https://theconversation.com/tracking-the-rise-of-room-sharing-and-overcrowding-and-what-it-means-for-housing-in-australia-107265 (accessed on 23 May 2023).
17. Fagan, B.; Durrani, N. *What We Did in Bed*; Yale University Press: London, UK, 2019.
18. Wright, L. *Warm and Snug: The History of the Bed*; Routledge & Kegan Paul: London, UK, 1962.
19. Garber, M. Finding Privacy during the Pandemic. The Atlantic. 2020. Available online: https://www.theatlantic.com/culture/archive/2020/03/finding-privacy-during-pandemic/608944/ (accessed on 23 May 2023).
20. Colomina, B. The 24/7 bed. In *Work Body Leisure*; Verzier, M.O., Axel, N., Eds.; Hatie Cantz: Rotterdam, The Netherlands, 2018; pp. 189–204.
21. Lincoln, S. Bedroom Culture: A Review of Research. *Space Place Environ.* **2016**, *3*, 421–439.

22. De Macedo, P.F.; Ornstein, S.; Elali, G.A. Privacy and housing: Research perspectives based on a systematic literature review. *J. Hous. Built Environ.* **2022**, *37*, 653–683. [CrossRef]
23. Sommer, R. *Personal Space: The Behavioural Basis of Design*; Prentice-Hall: Englewood Cliffs, NJ, USA, 1969.
24. Riis, J.A. *How the Other Half Lives: Studies among the Tenements of New York*; Bedford Books of St. Martin's Press: Boston, MA, USA, 1996.
25. Rybczynski, W. *Home: A Short History of an Idea*; Pocket Books: London, UK, 2001.
26. Vollmer, J.; Schulze, P.; Chebra, J. The American master bedroom: Its changing location and significance of the family. *Inter. Des. Educ. Counc. J. Inter. Des.* **2005**, *31*, 1. [CrossRef]
27. Martella, F.; Enia, M. Towards an urban domesticity. Contemporary Architecture and the blurring boundaries between the house and the city. *Hous. Theory Soc.* **2021**, *38*, 402–418. [CrossRef]
28. Colomina, B. The Century of the Bed. In *The Century of the Bed*; ARGE: Wien, Austria, 2014; pp. 10–23.
29. Nettleton, S.; Buse, C.; Martin, D. 'Essentially it's just a lot of bedrooms': Architectural design, prescribed personalisation and the construction of care homes for later life. *Sociol. Health Illn.* **2018**, *40*, 1156–1171. [CrossRef]
30. Kawada, T.; Suzuki, S. Instantaneous change in sleep stage with noise of a passing truck. *Percept. Mot. Ski.* **1995**, *80*, 1031–1040. [CrossRef]
31. Kim, M.; Chun, C.; Han, J. A Study on Bedroom Environment and Sleep Quality in Korea. *Indoor Built Environ.* **2010**, *19*, 123–128. [CrossRef]
32. Lockley, S.W.; Evans, E.E.; Scheer, F.A.; Brainard, G.C.; Czeisler, C.A.; Aeschbach, D. Short-wavelength sensitivity for the direct effects of light on alertness, vigilance, and the waking electroencephalogram in humans. *Sleep* **2006**, *29*, 161–168.
33. Wassing, R.; Schalkwijk, F.; Lakbila-Kamal, O.; Ramautar, J.R.; Stoffers, D.; Mutsaerts, H.J.M.M.; Talamini, L.M.; Someren, E.J.W. Haunted by the past: Old emotions remain salient in insomnia disorder. *Brain* **2019**, *142*, 1783–1796. [CrossRef]
34. Xu, X.; Lan, L.; Shen, J.; Sun, Y.; Lian, Z. Five hypotheses concerned with bedroom environment and sleep quality: A questionnaire survey in Shanghai city, China. *Build. Environ.* **2021**, *205*, 108252. [CrossRef]
35. Bjorvatn, B.; Waage, S.; Pallesen, S. The association between insomnia and bedroom habits and bedroom characteristics: An exploratory cross-sectional study of a representative sample of adults. *Sleep Health* **2018**, *4*, 188–193. [CrossRef] [PubMed]
36. Adams, R.; Appleton, S.; Taylor, A.; McEvoy, D.; Antic, N. *Report to the Sleep Health Foundation 2016 Sleep Health Survey of Australian Adults*; The Adelaide Institute for Sleep Health & The University of Adelaide: Adelaide, Australia, 2016.
37. Sleep Hygiene Index (SHI). Available online: https://novopsych.com.au/assessments/health/sleep-hygiene-index-shi/ (accessed on 1 December 2022).
38. Pittsburgh Sleep Quality Index. Available online: https://www.sleep.pitt.edu/instruments (accessed on 2 November 2022).
39. Braun, V.; Clarke, V. *Successful Qualitative Research: A Practical Guide for Beginners*; Sage Publications: London, UK, 2013.
40. ACOSS (The Australian Council of Social Service) and UNSW Sydney. Poverty in Australia. Available online: https://povertyandinequality.acoss.org.au/poverty/#:~:text=Our%202022%20Poverty%20in%20Australia,a%20couple%20with%202%20children (accessed on 15 January 2024).

buildings

MDPI

Article

Evaluating Patients' Preferences for Dental Clinic Waiting Area Design and the Impact on Perceived Stress

Milad Emami [1], Mahdieh Pazhouhanfar [1,*] and Jonathan Stoltz [2]

[1] Department of Architecture, Faculty of Engineering, Golestan University, Gorgan 4918888369, Iran; mld6521@gmail.com

[2] Department of People and Society, Swedish University of Agricultural Sciences, Alnarp, 234 22 Lomma, Sweden; jonathan.stoltz@slu.se

* Correspondence: m.pazhouhanfar@gmail.com or m.pazhouhanfar@gu.ac.ir

Abstract: The waiting area in dental clinics is a known source of stress for patients. However, positive distractions and comfortable design elements might reduce stress levels and provide mental stimulation while awaiting treatment. While ambient elements might play an important role in patient well-being, few studies have directly explored the impact of various design attributes of dental clinic waiting areas on patients' perceived stress. This study intended to bridge that gap and employed a discrete choice experiment method where 250 participants randomly selected a block and evaluated six pairs of computer-generated images of dental clinic waiting areas created based on eight interior design attributes with various levels: ambient lighting, false ceiling, window size, window view, indoor plants, wall shape, wall material, and seating options. Each visitor chose their preferred option in relation to its presumed effect on stress mitigation. The results suggest that the presence of ambient lighting, a false ceiling with a nature design, large window dimensions, a nature window view, green indoor plants, a curved wall shape, natural wall materials, and mixed seating options all can contribute to mitigating patients' perceived stress. Additionally, our findings indicate that age, gender, and education might influence choices across some levels of these variables. Overall, these results might assist architects and designers in shaping clinic environments mitigating patient stress during visits. Further studies would be needed to validate our findings and should also consider additional design attributes, more immersive stimuli presentation technologies, as well as potential differences across cultural contexts.

Keywords: stress; dental clinic; health; healthcare; discrete choice model

1. Introduction

Stress is a significant issue in dental settings, especially for individuals who visit oral surgery clinics [1]. Dental fear, stemming from previous negative encounters, apprehension about pain, and concerns about losing control during treatment, frequently causes patients to refrain from seeking dental care [2,3]. This avoidance can lead to unfavorable oral health results [4]. Additionally, dental anxiety can hinder treatment, as patients may involuntarily move or flinch during procedures, which can complicate the dentist's efforts [5]. Furthermore, this anxiety can considerably affect quality of life. Stress is a reaction that affects an individual psychologically, physiologically, and often behaviorally as they respond to situations that can either enhance or impair well-being [6].

In the past few decades, there has been a change in the focus of healthcare environment design research. Instead of just reducing the negative aspects of the physical environment, the emphasis is now on creating and improving positive experiences. Ulrich's supportive design theory [7] demonstrates this shift by emphasizing three key elements that should be considered in the physical environment of any healthcare facility: empowering a sense of control, fostering social support, and providing access to positive environmental stimuli while minimizing exposure to negative factors. Understanding the impact of the physical environment on patient well-being allows healthcare facilities to establish healing effects for patients and a pleasant working atmosphere for medical staff. This approach has the potential to lower staff stress levels and enhance patient safety and overall wellness [7].

The waiting area in dental clinics, where patients typically spend time before treatment, is a well-known source of stress [8]. These spaces are important in healthcare environments as they can either help alleviate or worsen the stress experienced by users [9], impacting patient anxiety levels, comfort, treatment outcomes, and their overall perception of care [9]. When patients perceive the waiting room as comfortable and containing positive distractions, it can lead to not only reduced stress levels but also increased mental stimulation [10]. Apart from the typical architectural features found in healthcare settings, various ambient elements also contribute to the well-being of patients in the waiting room. However, there have been few studies focusing on how the interior design of waiting areas in dental clinics influences patients' perceived stress. By examining influential environmental factors and significant components of interior architecture, this research aims to understand the extent to which waiting area attributes might impact perceived stress.

1.1. *Interior Design Attributes Related to Perceived Stress*

1.1.1. Ambient Lighting

Nowadays, a common method used for treating individuals with depression and mood disorders is light therapy. It can be said that light therapy is one of the best treatments for improving sleep disorders, depression, and reducing stress and anxiety levels [11]. Past studies have focused on the influence of interior lighting on the perceived stress from the environment and indicate that light perception affects patients and improves their physical and mental issues [12]. Other researchers [13] in their study aiming to investigate the impact of daylight and interior lighting on circadian rhythm, length of stay, and pain levels of hospitalized patients concluded that daylight is positively associated with the reduction of length of stay and pain levels in hospital patients. Interior lighting should be accompanied by daylight, and when the hospital building is properly combined with daylight and lighting, it will be beneficial for the patients' health and the hospital's well-being [14]. Additionally, eastern and southern patient rooms are better for patients and contribute to their improvement [14]. Therefore, interior lighting is one of the therapeutic environmental factors and has a tremendous impact on individuals' mental state and stress [15]. Bright and cool light during the day supports the suppression of melatonin, while dim and warm light in the evening can help minimize melatonin suppression before sleep [16].

1.1.2. False Ceiling

Designing the ceiling as well as designing walls and using boards in healthcare spaces can be equally effective. This is because the ceiling, like walls, can play a crucial role in reducing stress and anxiety and creating calmness, especially during treatment when a patient is lying down and facing upwards, such as in a dental unit. Ulrich showed in a study that one of the factors that can have a positive impact on patient recovery through distraction is the use of natural landscape images on boards, wall panels, and ceilings in healthcare spaces [17]. Images of water, plants, animals, and other elements of nature can influence individual preferences [18]. Dutro et al. [15] believe that natural images on operating room ceilings can alleviate patients' pain and stress, and false ceilings with flower and leaf patterns and tree designs with accent lighting are suggested on embossed patterns. They state that adding small lights on the ceiling that gently turn on and off during long night hours simulates stars and creates a positive distraction. Additionally, exposure to water and sky images, similar to green landscapes, might be beneficial for the well-being of patients [19].

1.1.3. Window Size

Window dimensions not only impact lighting conditions but also influence users' perception of space [20,21]. A study by [20] found that medium and large windows demonstrate how small changes in their size can affect spatial perception. Their research, which examined three different window sizes, showed that both the window size and the type of space significantly affect participants' spatial perception and their satisfaction with the outside view, with larger windows contributing to better spaces. The study also revealed a significant interaction between window size and space type regarding satisfaction with the view, suggesting that the window size preference depends on the type of space where the windows are situated. Additionally, when it comes to assessing visibility levels, windows in smaller spaces received higher ratings than those in larger spaces. The researchers concluded that window size influences individuals' perception of a space, and the type of space influences size preferences. Specifically, medium to large window sizes are preferred for creating more enjoyable, interesting, stimulating, well-lit, spacious, and satisfying spaces. Moreover, individuals generally prefer brighter spaces with larger windows or minimal furniture [22].

1.1.4. Window View

In a now classical study, Ulrich reported how a view with visible trees significantly improved patients' well-being and recovery after surgery compared with a view facing buildings [23]. Since then, the significance of the quality of the window view has attracted considerable attention in fields such as architecture, urban health, and property valuation. Since spaces lacking a view from the window are generally less preferred by people [24,25], a well-designed architectural plan typically considers both interior design and outdoor views holistically [22,23]. Natural landscapes and scenic views might serve as favorable environmental cues that contribute to positive sensory experiences and notably reduce stress levels [17]. Such landscapes might inspire positive thoughts and mitigate negative emotions like fear and anger, ultimately alleviating individuals' stress [26]. Hartig and his team [27] further assert that when individuals are under high stress, exposure to nature elicits positive feelings and diminishes negative thoughts and emotions, while also reducing psychophysiological activity, helping in coping with stress and agitation. Multiple studies indicate a positive link between views of natural landscapes from windows and mental health enhancements and overall well-being [18], with researchers believing that nature has the potential to positively impact a wide array of mental health outcomes [28]. Conversely, a study on individuals' blood pressure revealed that natural landscapes are less stressful compared to urban landscapes [26]. Consequently, images of natural landscapes, as opposed to urban landscapes, promote greater tranquility [29] and improved mental health [18], suggesting that urban nature can be used as a "nature-based solution" to enhance public health [30].

1.1.5. Indoor Plants

It is well established that exposure to natural scenes might reduce physiological stress and increase positive emotions. Especially, trees and greenery seem beneficial in this regard [31]. It is therefore interesting to consider enriching indoor spaces with greenery to induce stress-reducing responses in visitors. Additionally, colorful flowers used for room decoration and artificial flower pots might also have a positive effect on individuals' mood states [32], and combinations of colorful flowers are often preferred [33]. However, it has been suggested that people who are stressed often have a lower tolerance for high-diversity environments, rich in such things as, e.g., colorful flowers [34]. Depending on patient stress levels, different preferences might thus be expressed regarding indoor plants and whether diversity and color or a more cohesive and uniform green indoor vegetation is preferred.

1.1.6. Wall Shape

The link between interior design—particularly wall shapes—and perceived stress in healthcare settings has been extensively researched. Studies indicate that curvilinear environments are more effective for stress recovery than linear ones. For instance, Li et al. [35] found that participants in curvilinear spaces experienced better stress recovery, as evidenced by physiological measures such as salivary cortisol and heart rate variability, along with subjective stress assessments. Likewise, Shemesh et al. [36] showed that minimal wall curvature correlates with heightened physiological stress responses, implying that excessive linearity may increase stress levels. Aesthetic preferences also influence perceptions, with research suggesting that curved environments are seen as more calming and less stressful, in line with the biophilia hypothesis [37]. Overall, incorporating curvilinear designs in healthcare environments can improve patient comfort and alleviate stress, underscoring the significance of thoughtful architectural design for enhanced mental health outcomes.

1.1.7. Wall Materials

Researchers suggested that the least an architect can do to support people's well-being is to use natural elements and materials in the design. Natural elements and materials might create a sense of comfort and relaxation, as evidenced by decreased blood pressure and increased pulse rate, improve mental health, and reduce stress levels [38]. In addition, studies have shown that wood has antimicrobial properties against some pathogens and wooden materials, due to their natural appearance and biophilic effects on humans, present a nature-based topic for construction [39]. Additionally, maintaining cleanliness in indoor spaces and design elements from accumulated bacteria and microbes using antibacterial technology in synthetic materials and incorporating nanotechnology and materials in hospital buildings enhance their efficiency in terms of economics, environmental, health, and beauty. The positive effects include moderating indoor air humidity fluctuations, inducing positive emotions in individuals, inhibiting bacteria, and having positive psychological effects on human health and well-being [40]. Overall, it has been suggested that wood can contribute to healthy buildings and be beneficial for both the environment and people [41]. In addition, natural features are associated with stress-reducing effects, while the view of a brick wall has been associated with worse recovery after surgery compared to a view with visible greenery.

1.1.8. Seating Options

Finally, another potentially influential factor in reducing stress is the furniture and its arrangement. Essentially, the comfort and tranquility of an individual are significantly impacted by the layout of the environment [42]. Studies indicate that individuals with relatively strong social interactions tend to experience better mental and physical health and a greater sense of peace compared to those who are isolated and withdrawn. Social interactions can play a critical role in reducing stress and aiding in the recovery process of patients [26]. Therefore, designing and organizing furniture in waiting areas and other environments where people spend extended periods in a manner that encourages group interactions can be highly effective in achieving this objective. A study by Devlin [43], aimed at exploring individuals' visual preferences for waiting spaces in hospital environments by assessing the priority of five different seating arrangements, suggests that there is no compelling evidence for the primacy of a specific arrangement of chairs in a doctor's office waiting room. However, the seating choices reveal a preference for end seats over middle seats. While a social seating arrangement in a doctor's office may offer a sense of comforting familiarity to individuals, it is important to note that not all individuals visit the doctor with someone else. Therefore, a layout designed to encourage social interaction may not always be suitable [44], especially for psychiatric patients, and may not yield desirable results. This consideration may not apply universally to patients who typically seek primary care from doctors, as social interaction may not always be expected or desired in such cases. It has been suggested that highly social environments are the least preferred by patients who experience high levels of stress and fatigue [43,44].

In summary, the literature discussed above emphasizes the influence of various interior design factors on people's perception of healthcare environments, particularly in relation to perceived stress. The identified factors include ambient lighting, false ceilings, window size, window view, indoor plants, wall shape, wall materials, and seating options. Figure 1 summarizes the theoretical model that formed the basis for our study.

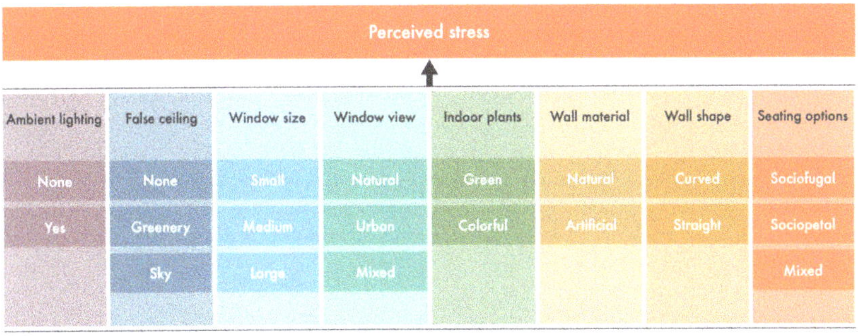

Figure 1. A summary of the theoretical model behind the study, with eight environmental design attributes at different levels hypothetically affecting patients' perceived stress while waiting for dental treatment.

1.2. Research Aims

There is a lack of research regarding the combined influence of the described factors on visitors' perceived stress. This study aimed to address this gap by combining these different factors in specific scenarios and asking participants to indicate their preferences among discrete choices, which, as far as we are aware, presents a novel approach. The main objective was to enhance understanding of how the interior design of dental clinics, involving these eight specific attributes at varying levels, might impacts patients' perceived stress. Primary research questions could thus be summarized as:

1. Which interior design attributes are the most important to consider in mitigating patients' perceived stress in dental clinic waiting areas?
2. Which levels of these attributes are generally most efficient in mitigating patients' perceived stress?
3. Which combinations of attributes/levels are most efficient in mitigating patients' perceived stress?

2. Materials and Methods

The primary aim of this research was to examine and assess how the interior design features of a dental clinic impact patients' perceived stress. To accomplish this, the discrete choice method was utilized which analyzes individual preferences based on their selections. Figure 2 summarizes the different steps of method, typical for the design of a discrete choice experiment.

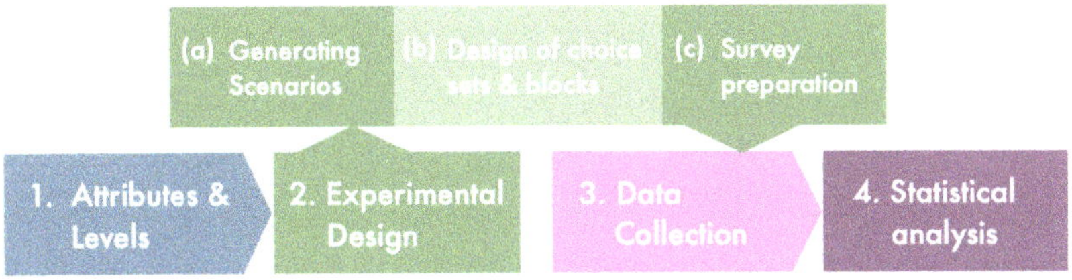

Figure 2. A diagram describing the main steps of the method, typical for the design of a discrete choice experiment.

2.1. Discrete Choice Method

The discrete choice method is used in various fields such as architecture, economics, marketing, and transportation due to its ability to analyze consumer behavior and decision-making [44–50]. This method, a subset of selection modeling, is a modern technique for gauging people's preferences. Originating from the works of Louviere and Hensher [51], and others, it provides an effective way to understand how individuals or communities make choices within certain contexts through surveys. The key objective is to uncover the structure of preferences and emphasize the relative importance of different attributes [52]. By allowing researchers to simulate various options in a controlled environment, the method helps identify influential attributes in a given scenario and facilitates the ranking of these attributes in comparison to each other. Its success is attributed to its strong theoretical foundation, drawing from Lancaster's consumer theory and McFadden's theory of random utility models [53]. Based on probabilistic choices, random utility theory posits that individuals choose the option that maximizes their utility within a set of alternatives, and the utility of a specific option depends on the utility associated with each of its attributes. In practice, a questionnaire presents participants with a series of options characterized by specific attributes organized into choice sets.

2.2. Study Area

The data for this study were collected inside three dental clinics in the city of Shiraz (Figure 3) during the afternoon shift (second shift) in an area of approximately 80 square meters, which included the waiting areas, reception, restroom, water cooler, staff rest area, dentist's room, treatment area, and instrument sterilization area. The selection of Shiraz as the study area was based on several key factors. Firstly, Shiraz is a major urban center with a diverse population, providing a representative sample for the study. Secondly, the city has a well-established healthcare infrastructure, including numerous dental clinics, which facilitates the collection of reliable data. The choice of the three specific dental clinics was made to ensure a comprehensive representation of different types of dental practices within the city. These clinics were selected based on their patient volume, range of services offered, and accessibility, ensuring that the study results would be both relevant and generalizable to similar urban settings.

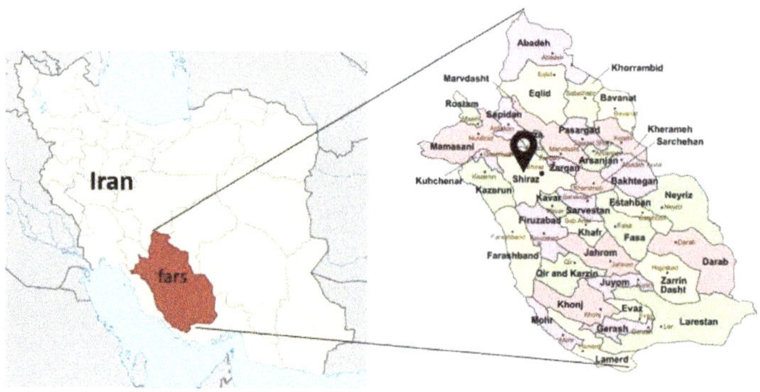

Figure 3. Study location, in Fars province, Iran.

2.3. Participants

The study involved a diverse group of participants in relation to gender, age, and educational attainment (Table 1). Of the total participants, 126 were male (50.4%), and 124 were female (49.6%). The age distribution was as follows: 15–35 years (40%), 35–55 years (42%), and 55–75 years (17.8%). The majority of participants were in the 35–55 age group. In terms of educational level, the breakdown was as follows: Undergraduate and diploma (57.6%), Bachelor of Science (28%), Master of Science (8%), and Doctorate (6.4%).

Table 1. Individual characteristics of the study participants.

	Classification	Frequency (%)
Gender	Male	50.4
	Female	49.6
Age	15–35 years old	40
	36–55 years old	42.2
	56–75 years old	17.8
Education	Undergraduate and Diploma	57.6
	Bachelor of Science	28
	Master of Science	8
	Doctorate	6.4

2.4. Questionnaire

To assess the perceived stress in patients, initial conversations were held with individuals present in the clinic's waiting area, presenting the research objectives. Individuals who were willing to participate completed a satisfaction survey form and then went on participating in the study. Here, the survey was divided into two sections: the initial section centered on a visual discrete choice experiment (DCE), while the subsequent section was comprised of close-ended inquiries. The purpose of the latter was to collect data regarding the varied backgrounds and traits of the participants, encompassing demographic details like gender, age, and education. Within the DCE segment of the survey, participants were asked to randomly select a block containing six pairs of computer-generated images of dental clinic scenarios. From the selected series of scenario pairs, they were then tasked with selecting the option that they believed would induce the least perceived stress while waiting for dental treatment.

2.5. Experimental Design

2.5.1. Attributes and Their Levels

The initial phase of the experimental design involved identifying the relevant attributes and their respective levels. Attributes here refer to the interior design variables used to define the options presented to the participants in the experiment and were selected based on their proposed relevance in mitigating perceived stress in individuals. Based on the presented literature review, eight attributes were chosen as the independent variables defining a dental clinic waiting area. Table 2 summarizes these eight attributes and their corresponding levels, together with references to the literature supporting each attribute's potential relevance in relation to individuals' perceived stress.

Table 2. Eight interior design attributes in different levels included in the study.

Attributes	Levels	References
Ambient lighting	Yes None	[13–16,42]
False ceiling	Virtual sky Virtual greenery None	[15,22]
Window size	Small Medium Large	[17,18]
Window view	Natural Urban Mixed	[19–22,36]
Indoor plants	Green Colorful	[32–34,43]
Wall shape	Curved Straight	[33,39]

Table 2. *Cont.*

Attributes	Levels	References
Wall material	Natural (bamboo) Artificial (bricks)	[38]
Seating options	Sociofugal Sociopetal Mixed	[34,35,44]

The Ambient lighting of the space in study had two levels: the presence of ambient lighting indicated in the ceiling and its absence. The False ceiling attribute had three levels: a false ceiling with a virtual sky pattern, a false ceiling with a virtual greenery pattern, and the absence of a false ceiling. The Window size was considered at three levels: windows of large, medium, and small size. The Window view of the waiting area also had three levels: views of natural scenery, views of an urban space, and views of a mixed nature and urban space. Indoor plants had two levels: green or colorful, represented by pots with green or colorful plants, respectively, placed by the window in the waiting area. The Wall shape of the window wall of the waiting area had two levels: a flat wall with a traditional straight-edged/sharp corner and a rounded wall with a curved corner transition. Similarly, the Wall material of the waiting area feature wall also had two levels: a natural material (bamboo) and an artificial material (brick). Finally, the Seating options had three levels: a social option facilitating interaction among individuals (sociopetal), a non-social arrangement facilitating privacy (sociofugal), and a combination of the two (mixed). Examples of the representation of these attributes and levels in the experiment are shown in Figure 4.

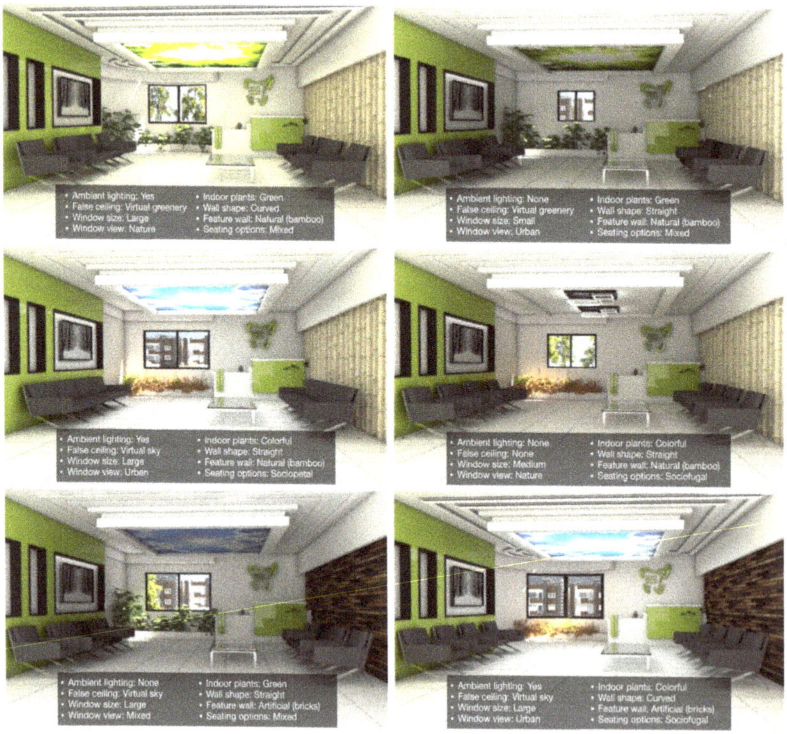

Figure 4. Examples of representations of the included attributes at various levels used in the study (see Table 2 for a detailed list of these attributes and levels).

2.5.2. Scenarios and Choice Sets

The subsequent step in experimental design involved creating various scenarios and sets of choices for these scenarios. Using the eight attributes and their potential levels, the total number of possible combinations was determined using the complete factorial experiment method and Equation (1): full factorial (number of combinations) = Π Zk Equation (1), where Z denotes the number of levels of an attribute, and k represents the number of attributes with the same levels of Z. Consequently, a complete factorial experimental design yields 1458 probabilistic scenarios. This method allows for the examination of individual characteristics' effects as well as the bilateral and multilateral interactions among different attribute levels. While increasing the number of scenarios enhances the comprehensiveness of user responses, it also escalates the time and cost involved. Thus, adopting a fractional factorial design, with a reduced number of combinations, can still yield reliable results for primary and select interaction effects. Employing SAS software version 9.2, the number of scenarios was optimized, resulting in 72 scenarios.

These 72 scenarios were divided into six blocks, each containing six choice sets comprising three alternatives (scenario 1, scenario 2, and none). These blocks were randomly assigned to participants. Participants were instructed to select the scenario that induced the least perceived stress for them or to indicate no preference for either option. Including the "none" option aimed to alleviate pressure on participants to make a choice and minimize the likelihood of exaggerated estimates of individual preferences. Visual representations of the options for each choice task were created using SketchUp(2019) and Photoshop software (2020). This visual approach is effective for diminishing external influences during the decision-making process in Discrete Choice Experiments (DCEs). The images were designed to provide viewers with a realistic perspective, allowing them to evaluate the simulation as if they were physically present. Figure 5 illustrates two images forming a choice set. The left image (Figure 5, Option 1) displays a setting without ambient lighting, a false ceiling with a virtual sky pattern, a small window with an urban view, colorful indoor plants, a straight-edged window wall, a feature wall with an artificial material (bricks), and a sociofugal seating pattern. In contrast, the right image (Figure 5, Option 2) presents a space with ambient lighting present, a similar virtual sky false ceiling, a large window with an urban view, green indoor plants, a curved window wall shape, a feature wall with natural material (bamboo), and a sociofugal seating layout.

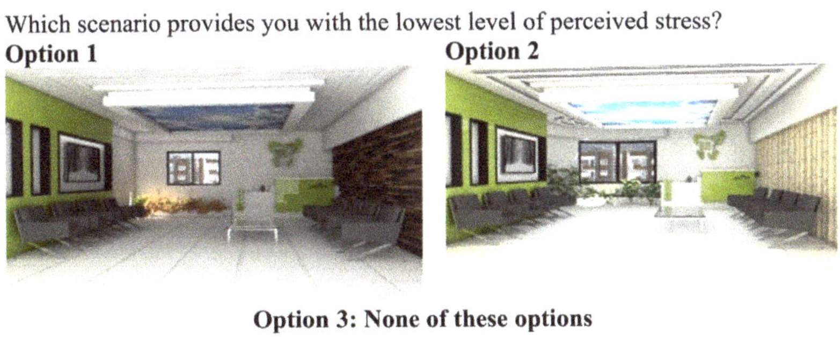

Figure 5. An example of a task chosen for the choice experiment with three available options.

The survey took place between February and March 2022, during which discrete choice surveys were distributed to individuals aged 18 and above at three dental clinics (see Section 2.2.). Upon introduction to the survey's objectives and tasks, participants used the interviewer's computer for data collection while following health protocols. With consent, participants randomly selected a block, each containing six sets with three options (Option 1, Option 2, and None; see Figure 5). They were then instructed to choose the scenario from each set that induced the lowest perceived stress, resulting in each participant evaluating

12 different clinic designs. In total, 250 questionnaires were collected, and the data were subsequently compiled into an Excel file.

2.6. Data Analysis

Random utility theory was utilized to examine the data collected from the choice experiment. A mixed logit (ML) model was used to account for variations in individuals' decision-making processes when selecting options. Attribute levels were grouped and represented using dummy codes to calculate their impacts.

2.6.1. Modeling Choice Data

DCE is commonly utilized in research to evaluate individuals' preferences by presenting them with several options in a choice set and requesting them to indicate their preferred choice. Lancaster's characteristics theory of value [51] suggests that each commodity's value is the sum of its constituent attributes, leading to the development of multiple discrete choice models based on this theory and McFadden's random utility [54]. These models aim to assess the influence of causal factors on individuals' decision-making [55]. It is hypothesized that individuals select the option with the highest utility, influenced by the attractiveness of that option's features [56]. The utility functions assigned to person n to choose alternative i from the available options are as follows:

$$U_{ni} = Vni + \varepsilon ni \quad (1)$$

The random utility theory, which is used to study people's decision-making, comprises two components. The initial component pertains to the observable aspects, denoted as (deterministic) Vni.

$$Vni = \beta i\ Xnmi \quad (2)$$

Within this formula, βi symbolizes the parameter coefficient vector for alternative i = (1,. . ., I), while $Xnmi$ denotes the observable attribute levels of i for individual n in choice situation m. Nevertheless, there exist unobservable factors, εni, which impact individuals' decisions. Therefore, as the second component, εni is a random error that, in a multinomial logit (MNL) model, is uniformly and independently distributed according to a Gumbell distribution [57].

The chance of individual n selecting option i is represented as $Lni(\beta)$, where Lni is a function determined by parameter β, and it is defined in the following manner:

$$L_{ni}\ (\beta) = \frac{exp(V_{ni}\ (\beta))}{\sum_{j=1}^{j} exp\ (V_{ni}\ (\beta))} = \frac{exp(X_{nmi}\ \beta)}{\sum_{j=1}^{j} exp(X_{nmi}\ \beta)} \quad (3)$$

In this context, $Vni(\beta)$ represents the deterministic (systematic) segment of the utility function, contingent on the fixed parameter β. Despite its benefits, the MNL model has various restrictions, such as the potential for individuals to assign differing importance to distinct attributes. The mixed logit model tackles these challenges by accommodating diversity; it serves as an expanded adaptation of the MNL, enabling parameters for all, or a subset of, variables to vary among respondents, instead of remaining consistent across all respondents.

By acknowledging potential disparities, such as variations in the sociodemographic characteristics of respondents, unrestricted substitution patterns, and correlations among unobserved elements, it has become one of the most widely utilized models for estimating random utility models in diverse fields [58]. The approach for determining the probability of selecting an alternative using the mixed logit model is as follows:

$$Pni = \int Lni(\beta) f(\beta)\ d\beta \quad (4)$$

Pni represents the likelihood of individual *n* selecting alternative *i*, while *Lni*(β) denotes the likelihood of individual *n* choosing alternative i in the MNL model, which is determined by parameter b. The function $f(\beta)$ is a density function defined by constant parameters b such that β = b, 1, $\beta \neq$ b, and 0.

2.6.2. The Mixed Logit Model

The PandaBiogene software (3.1.14) was used to estimate both the Multinomial Logit (MNL) and Mixed Logit (ML) models. The sample consisted of 250 respondents, each making six choices, resulting in a total of 1500 observations. After an initial estimation, variables with insignificant variability were identified and treated as fixed parameters. Out of the seven variables, two were designated as fixed, while the remaining five were considered random. The randomly estimated parameters (βn) for each variable *n* represent the mean effect, with the standard deviation indicating the level of variability in the sample for that specific variable [59]. In the final ML model, random parameters were estimated based on a normal distribution, allowing for equally positive and negative variations in the variables depending on the respondent. The estimation process involved the use of 200 Halton draws. Various criteria, including a pseudo-R2 or R2 (McFadden) value of 0.132, were used in logistic regression to assess the model fit, indicating satisfactory model performance.

3. Results

3.1. Parameter Values

The results presented in Table 3 illustrate the significance, or weights, attributed to each attribute as parameter values. Among these, it was found that the presence of ambient lighting (β19 = 2.43, p = 0.00) and mixed seating options (β8 = 2.35, p = 0.00) had the most substantial positive impact on individuals' selections. Following these, attributes such as a large window (β18 = 2.24, p = 0.00), green indoor plants (β2 = 1.54, p = 0.00), a virtual greenery false ceiling (β2 = 1.54, p = 0.00), and a virtual sky false ceiling (β1 = 1.36, p = 0.00) were highlighted for their potential importance. Attributes like a curved wall (β14 = 1.35, p = 0.00), a natural feature wall material (bamboo; β4 = 1.14, p = 0.00), a nature window view (β12 = 1.06, p = 0.00), and a medium-sized window (β17 = 0.8, p = 0.00) were noted to have a moderate impact on utility. Lastly, an artificial feature wall material (bricks; β5 = 0.37, p = 0.00) and a straight-edged window wall (β15 = 0.26, p = 0.04) were found to have the least utility value in study.

In this model, colorful indoor plants (β10 = −1.92, p = 0.00), no false ceiling (β3 = −1.54, p = 0.00), a small window (β16 = −1.53, p = 0.00), a mixed (urban and natural) window view (β13 = −0.57, p = 0.00), sociopetal seating (β6 = −0.61, p = 0.00), and sociofugal seating (β7 = −0.24, p = 0.02) were all identified as negative parameters, indicating general dissatisfaction regarding potential stress mitigation among participants, although these results were statistically significant.

Table 3. Parameter estimates of the ML model.

Attributes/Level	Param.	Value	p	Random Param.	p	Age	Education	Gender
Constant (no choice)	B0	−1.31	0.00	-	-	-	-	-
False ceiling						-	-	-
Sky	B1	1.36	0.00	0.57	0.02	-	-	-
Greenery	B2	1.54	0.00	0.47	0.02	-	-	-
Feature wall						-	-	-
Natural (bamboo)	B4	1.14	0.00	0.58	0.00	0.00	-	-
Seating options							-	-
Sociopetal	B6	−0.61	0.00	0.4	0.01	0.00	-	-
Mixed	B8	2.35	0.00	0.69	0.02	0.00	-	0.00

Table 3. *Cont.*

Attributes/Level	Param.	Value	p	Random Param.	p	Age	Education	Gender
Indoor plants							-	
Green	B9	3.33	0.00	1.81	0.00	0.00	-	0.04
Colorful	B10	−1.92	0.00	−1.11	0.00	0.00	-	-
Window view						-	-	-
Urban	B11	0.97	0.00	0.65	0.02	0.00	-	-
Wall shape						-	-	-
Curved	B14	1.35	0.00	0.58	0.00	-	-	-
Window size						-	-	-
Small	B16	−1.53	0.00	−0.8	0.00	-	-	-
Large	B18	2.24	0.00	2.18	0.00	-	-	-
Fixed parameters	Param.	Value	p	Random st. Dev	p	Age	Education	Gender
False ceiling				-	-			
None	B3	−1.54	0.00	-	-	0.01	0.01	0.01
Ambient lighting				-	-		-	-
Yes	B19	2.43	0.00	-	-		-	-
Feature wall				-	-		-	-
Artificial (bricks)	B5	0.37	0.00	-	-	0.00	-	-
Seating options				-	-		-	-
Sociofugal	B7	−0.24	0.02	-	-	0.00	-	-
Window view				-	-		-	-
Natural	B12	1.06	0.00	-	-	0.00	-	-
Mixed	B13	−0.57	0.01	-	-	0.00	-	-
Wall shape				-	-	-	-	-
Straight	B15	0.26	0.04	-	-	-	-	-
Window size				-	-	-	-	-
Medium	B17	0.80	-	-	-	-	-	-

Note: The results were statistically significant in $p < 0.05$.

3.2. Random Parameters

The analysis indicated that all parameters demonstrated a mean positive effect with standard deviations at a significance level of $p < 0.01$, indicating considerable heterogeneity among the respondents. Therefore, these parameters should be considered fixed rather than random. It is important to note that colorful indoor plants (−1.11, $p = 0.00$) and small window size (−0.80, $p = 0.00$) both had noteworthy negative effects on the utility function. Regarding the random parameters, large window (2.18, $p = 0.00$), green indoor plants (1.81, $p = 0.00$), mixed (sociopetal and sociofugal) seating options (0.69, $p = 0.02$), urban window view (0.65, $p = 0.02$), curved window wall (0.582, $p = 0.00$), natural feature wall material (bamboo; 0.58, $p = 0.00$), virtual sky false ceiling (0.57, $p = 0.02$), and virtual greenery false ceiling (0.47, $p = 0.02$), were all statistically significant at $p < 0.05$, indicating diversity in the preferences of the respondents across these levels and attributes. Regarding sociopetal seating options, the negative coefficient (−0.61) indicates that, on average, the variable "sociopetal" has a negative effect on the dependent variable. This means that as the value of "sociopetal" increases, the likelihood of the outcome decreases. However, the positive standard deviation (0.40, $p = 0.02$) suggests that there is variability in the effect of sociopetal seating across individuals. This means that while the average effect is negative, for some individuals, the effect could be less negative or even positive.

Furthermore, individual characteristics were used as control variables in the mixed logit analysis [51], and sociodemographic factors were taken into consideration to address variations in preferences within the study. The model incorporating age, gender, and education significantly influenced heterogeneity among some variables.

4. Discussion

This study involved a discrete choice experiment with computer-generated images of dental clinic interiors. Individuals were asked to choose their preferred option among pairs of images based on their estimated stress-reducing influence. First, we discuss in more detail around each of the attributes presenting an important influence over participants' choices in study. This is followed by a discussion around the observed heterogeneity among certain influential parameters, i.e., attributes and levels for which there was a particular variation in how they influenced the choices of our participants. We conclude with a discussion around some important limitations of our study and finally present our conclusions and some potential implications of our results.

4.1. *Attribute Influence on Choices*

The findings highlight existing ambient lighting and mixed seating options (i.e., a mix of sociofugal and sociopetal options) as the most positively influential factors in our participant's decision-making regarding potential environmental stress mitigation. Furthermore, attributes such as a false ceiling with virtual greenery, a large window, a natural window view, green indoor plants, a curved window wall, a natural feature wall material, were all deemed highly desirable in relation to stress mitigation. On the contrary, the absence of a false ceiling, a small window, a mixed window view, colorful indoor plants, and a sociofugal seating pattern were all found to have a relatively negative impact on choices in our study. In the following, our results in relation to each attribute are discussed in more detail.

4.1.1. Ambient Lighting

Ambient lighting may have a profound impact on the perceived stress levels of dental clinic visitors according to our findings. This is in line with earlier research suggesting that appropriate lighting can lead to shorter hospital stays and greater patient satisfaction [14]. It also aligns with the existing literature emphasizing the role of natural light in promoting positive emotional responses and reducing anxiety in healthcare settings [17]. Our findings here highlight the need for careful consideration of lighting design in dental clinics, balancing the benefits of natural illumination with the calming effects of softer ambient lighting. Future research should further explore the nuanced effects of various lighting types and designs, aiming to optimize patient comfort and well-being in clinical environments.

4.1.2. False Ceiling

False ceilings, particularly those featuring natural motifs such as sky or vegetation patterns, emerge in our study as a potentially significant factor influencing patient comfort and stress levels. This study found a strong preference for false ceilings with sky or greenery motifs, which aligns with the findings of Dutro et al. [15], who reported that ceilings adorned with floral and leaf patterns, as well as tree designs, effectively can alleviate pain and reduce stress among patients. The psychological impact of visual elements containing natural features in healthcare settings is well documented. Li and Samuelson [19] noted that exposure to images of the sky can enhance patients' health and well-being, suggesting that such designs can create a more uplifting and calming atmosphere in line with our findings here. This might be particularly important to consider in dental clinics, where patients often experience anxiety related to their treatment. Arguably, incorporating natural designs into false ceilings not only beautifies the space but might also foster a sense of connection to nature, which has been shown to have therapeutic benefits [56]. In conclusion, the expressed preference in this study for false ceilings with natural motifs underscores the importance of thoughtful visual design in healthcare environments, preferably using calming natural features as an aesthetic foundation, in line with what has been suggested by other studies and biophilic theories [31].

4.1.3. Window Size and View

In our study, large windows were preferred over small or medium-sized windows. This is consistent with the findings of Mascoso et al. [20], which indicates that larger windows contribute to creating environments that are perceived as more pleasant, interesting, and spacious. Larger windows allow for increased natural light, which not only brightens the space but also might enhance the overall ambiance. Such brightness is associated with improved mood and reduced feelings of stress and anxiety among patients. The presence of natural light has been shown to positively influence individuals' spatial perception, making them feel more connected to the outside world and less confined within the clinical environment [22]. According to previous research, the parameter of window display greatly affects user preferences. This suggests that users favor spaces with larger, unobstructed windows that offer better visibility [57]. Furthermore, users consistently choose rooms with the most natural views [58]. These findings are also consistent with Ozdemir's [59] study, which showed that office occupants with larger windows and natural views are more satisfied with their work environment. A preference for a natural window view in the present study thus aligns with the findings of other researchers showing that natural landscapes compared to urban landscapes promote greater relaxation [29] and improve mental health [18], and that the use of nature and natural views in healthcare environments increases patient satisfaction [60,61]. While urban landscapes with buildings commonly are seen as stress-inducing [26], our results here thus add to the evidence regarding the importance of taking the presence of outdoor natural features into consideration also from the interior design perspective and incorporate such features, where possible, into the architectural design.

4.1.4. Indoor Plants

In our study, participants expressed a clear preference for environments enriched with green indoor plants. This preference is supported by existing research that demonstrates the stress-reducing effects of plants in healthcare settings [44,53]. Studies have shown that exposure to green plants can lead to decreased anxiety and pain levels in patients, enhancing their overall well-being [61,62]. However, it is noteworthy that some studies, such as those by Helen [33], suggest that colorful flower combinations may be preferred over greenery. This indicates that while green plants have proven benefits, individual preferences can vary, and some patients may find vibrant flowers more appealing. It has been suggested, however, that particularly stressed individuals might find diverse environments, with a lot of variation regarding, e.g., species and colors, more stressful while preferring more cohesive and uniform settings [34]. This is supported by our results here, where green indoor plants were clearly preferred over options with colored plants. Future research could further explore such preferences and potential differences among individuals to optimize the design of clinical environments.

4.1.5. Wall Shape and Wall Material

Our findings indicate a preference for curved walls over straight-edged walls in relation to potential stress mitigation. This result is particularly relevant given the contrasting views in the existing literature regarding wall shapes in healthcare settings. Our findings here are consistent with the reports of Li et al. [35] and Shemesh et al. [36] in demonstrating a clear preference for curved walls over straight-edged walls in mitigating perceived stress. Participants reporting lower perceived stress in curvilinear environments might align with the biophilia hypothesis, suggesting that natural forms, such as curves, are inherently more calming and less stressful [37]. A preference for curvilinear designs again underscores the importance of thoughtful architectural design in healthcare settings to enhance patient comfort and reduce stress. Our findings suggest that incorporating curvilinear and natural elements in healthcare settings might improve mental health outcomes, perhaps by fostering a sense of connection with nature. To optimize healthcare environments for patient well-being, future research should continue to explore relationships between architectural shapes in relation to perceived stress. In addition, there was a clear preference in our study for a feature wall with a natural material (bamboo) over an artificial material (bricks). This might again be seen in relation to ideas such as the biophilia hypothesis and the stress reduction theory, suggesting a calming effect from environmental features reminding of evolutionary supportive conditions [31].

4.1.6. Seating Options

Finally, Seating options, categorized into sociofugal, sociopetal, and mixed, were examined for potential influence on patients' perceived stress. Our findings reveal that exclusively sociofugal options did not significantly influence stress, while a mixed arrangement showed a notably positive effect. Sociofugal seating, which discourages interaction by creating physical separation, may offer some patients a sense of privacy. However, this layout might also lead to feelings of isolation, which could be detrimental for some individuals' perception of stress. In this regard, DevLin [43] suggests that the layout of a doctor's office seating can provide a supportive atmosphere socially. On the other hand, sociopetal seating promotes social interaction, potentially providing comfort for some patients but increasing anxiety for those who prefer solitude in a clinical environment. The tendency for exclusively sociopetal seating options to be the least preferred in relation to stress mitigation in our study is in line with other studies suggesting that stressed individuals might be bothered by highly social settings and often prefer solitude [45,46]. Mixed seating options, allowing patients to select their level of social engagement, appeared in our study to be the most preferred option in relation to stress mitigation. This offered flexibility might cater to individuals' varying preferences and foster a sense of control, essential in a healthcare setting often associated with anxiety. These results align with the principles of supportive design, which emphasize the importance of patient choice and comfort in reducing stress [17].

In summary, based on our findings regarding the studied attributes, Figure 6A illustrates an example of a generally more restorative, i.e., stress-reducing, dental clinic waiting area, while Figure 6B show a relatively more stress-inducing environment. In the first case (Figure 6A), ambient lighting is present, there is a false ceiling displaying a natural pattern (greenery), there is a large window with a nature view, green indoor plants, a curved window wall, a natural material feature wall (bamboo), and mixed seating options (supporting both sociopetal and sociofugal preferences). In the second case (Figure 6B), ambient lighting is missing, there is no false ceiling, there is a small window with an urban view, colorful indoor plants, a straight-edged window wall, a feature wall with an artificial material (bricks), and only sociopetal seating options. Although these examples illustrate general preferences according to our findings, many attributes also showed considerable variation among individuals in our study.

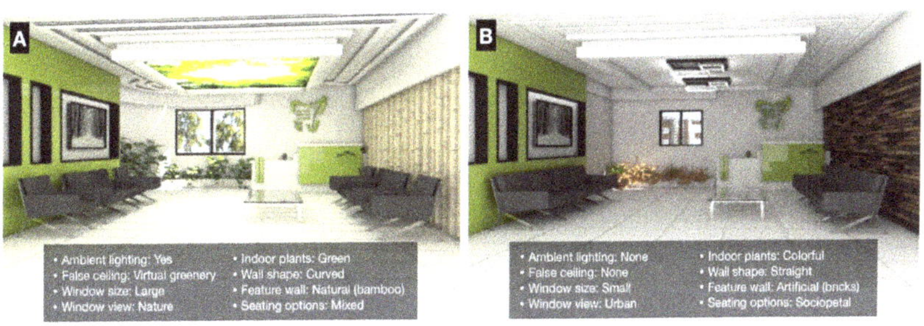

Figure 6. Examples of attributes in a generally more stress-reducing (**A**) and a relatively more stress-inducing (**B**) dental clinic environment according to our study findings.

4.2. Heterogeneity among Attributes

In the mixed logit model, random parameters were examined to identify variations in preferences and choices among participants. Virtual sky and virtual greenery false ceilings, natural feature wall material (bamboo), the sociopetal and mixed seating options, green and colorful indoor plants, urban window view, curved window wall, and large window size all displayed substantial diversity in how they were perceived in relation to potential stress mitigation. Furthermore, the analysis of preferences based on age, gender, and ethnicity using random parameters in the mixed logit model indicated significant differences in how these demographic factors influenced patient choices. First, age significantly influenced preferences for certain design attributes. Older participants tended to prefer larger windows and natural views, which may be linked to a desire for comfort and connection to nature. In contrast, younger participants showed a stronger preference for more modern design elements, such as vibrant colors and innovative seating arrangements. Gender differences were also notable. Female participants expressed a higher preference for green indoor plants and natural materials, suggesting a greater inclination towards elements that promote a calming environment. Males, on the other hand, were more favorable toward bold design choices, such as sociopetal seating arrangements that encourage interaction.

Although our study primarily focused on age and gender, the results also suggest that education might also play a role in preferences, particularly regarding the choice of wall materials and colors. Participants from different educational backgrounds may have varying levels of associations with certain design elements, impacting their stress perception in dental clinics. It is thus recommended for future similar studies to also consider similar demographic factors as background variables, as they might potentially influence people's preferences.

4.3. Study Limitations

In research that examines people's preferences, particularly when spatial elements are involved, researchers commonly employ visual representations to depict the scenarios under scrutiny and gauge their impact on the utility function [52]. In this study, computer-generated images were utilized to showcase a combination of eight variables. These images were assessed based on their influence on the perceived stress of individuals in a dental clinic environment. However, this study faced some limitations. Primarily, the discrete choice experiment method, while useful, imposes constraints on the number of attributes that can be included. While an increase in the number of scenarios generally improves the quality of participant responses, it is advisable to limit the alternatives in the selection set to ensure obtaining the most reliable results due to time and cost limitations.

In our study, respondents were presented with two scenarios in the questionnaire and asked to choose the option that induced the least sense of perceived stress or indicate if neither option was preferred. Although there was a fourth option where both scenarios were perceived as equally preferred, the focus in our study remained on those who selected one of the images. It is believed that including the fourth alternative would not have significantly altered the outcomes in our study, but it could be considered in future research. An important omission from this study is that it did not assess participants' initial stress levels in any formal way. This made it impossible to investigate whether individual stress levels might affect choices regarding environmental preferences. Future similar studies should consider assessing the initial perceived stress of the participants with a suitable tool to assess the impact of perceived stress on attribute preferences.

Ultimately, more comprehensive, and diverse studies are necessary to ensure dependable results. Such studies could also include other potentially important attributes that were not considered in our study here, such as, e.g., differences between different wall colors (not only materials), the presence of other people, etc. It would also be interesting for such further investigations to use potentially more immersive options for stimuli presentation, such as, e.g., virtual reality (VR) solutions. The potentially increased realism offered by such technologies might improve the ecological validity of similar studies.

5. Conclusions

To design better dental care environments, it is crucial to understand the impact of specific interior design attributes in dental clinic waiting areas on patients' perceived stress. This study aimed to identify key interior design attributes that could mitigate waiting patients' perceived stress. A discrete choice experiment was carried out, and our findings suggest the potentially critical role of ambient lighting in mitigating stress of waiting patients. Our study suggests that well-designed ambient lighting, preferably in combination with a false ceiling displaying a natural motif, might be useful to reduce stress and enhance patient comfort. Furthermore, a large window with a natural view appears as a preferred option with regards to general stress mitigation, as opposed to small windows displaying urban elements. Designers might also consider curved wall shapes over straight edges, feature walls with natural materials, as well as enrichment of the waiting area with green indoor plants. Finally, by providing mixed seating options, supporting social interactions as well as privacy, the design might cater to the varied needs among individuals and offer a sense of control important to reduce the stress often associated with dental clinic visits.

Author Contributions: Conceptualization, M.P. and M.E.; methodology, M.P. and M.E.; software, M.P.; validation, M.P.; formal analysis, M.P. and M.E.; investigation, M.P.; resources, M.P.; data curation, M.E.; writing—original draft preparation, M.E.; writing—review and editing, M.P. and J.S.; visualizations, M.E. and J.S.; supervision, M.P.; project administration, M.P. All authors have read and agreed to the published version of the manuscript.

Funding: This research received no external funding.

Institutional Review Board Statement: The study was conducted in accordance with the Declaration of Helsinki.

Informed Consent Statement: Informed consent was obtained from all subjects involved in the study.

Data Availability Statement: The data presented in this study are available on request from the corresponding author. The data are not publicly available due to institutional policies that restrict data sharing.

Acknowledgments: This article was inspired by the first author's master's thesis in architecture, titled "Design of dental clinic approach to reduce stress based on the discrete choice model", completed at the Department of Engineering, Faculty of Architecture, Golestan University, Iran.

Conflicts of Interest: The authors declare no conflicts of interest.

References

1. Shindova, M.P.; Belcheva, A.B. Dental Fear and Anxiety in Children: A Review of the Environmental Factors. *Folia Med.* **2021**, *63*, 177–182. [CrossRef] [PubMed]
2. Alahmari, A.A.; Khan, A.A.; Alamri, F.A.; Almuzaini, Y.S.; Alradini, F.A.; Almohamadi, E.; Alsaeedi, S.; Asiri, S.; Motair, W.; Almadah, A.; et al. Hajj 2021: Role of Mitigation Measures for Health Security. *J. Infect. Public Health* **2022**, *15*, 1350–1354. [CrossRef] [PubMed]
3. Hatipoğlu, Ö. Factors Affecting the Decision-Making of Direct Pulp Capping Procedures amongst Turkish Dental Practitioners. *Eur. Endod. J.* **2018**, *3*, 167–173. [CrossRef] [PubMed]
4. Armfield, J.M. Development and Psychometric Evaluation of the Index of Dental Anxiety and Fear (IDAF-4C+). *Psychol. Assess.* **2010**, *22*, 279–287. [CrossRef]
5. Murad, O.; Al-Dassean, K.A.; Al Neweiri, A.M.; Murad, H.O.; Murad, B.O. The Arabic Version of the Fear of COVID-19 Scale: Psychometric Properties and Relationship to Future Anxiety in Jordanians. *Cogent Psychol.* **2022**, *9*, 2064730. [CrossRef]
6. Baum, A.; Fleming, R.; Singer, J.E. Understanding Environmental Stress: Strategies for Conceptual and Methodological Integration. In *Advances in Environmental Psychology (Volume 5)*; Routledge: London, UK, 2020; pp. 185–205.
7. van Hoof, J.; Rutten, P.G.S.; Struck, C.; Huisman, E.R.C.M.; Kort, H.S.M. The Integrated and Evidence-Based Design of Healthcare Environments. *Archit. Eng. Des. Manag.* **2015**, *11*, 243–263. [CrossRef]
8. Beukeboom, C.J.; Langeveld, D.; Tanja-Dijkstra, K. Stress-Reducing Effects of Real and Artificial Nature in a Hospital Waiting Room. *J. Altern. Complement. Med.* **2012**, *18*, 329–333. [CrossRef]
9. Liddicoat, S. The Therapeutic Waiting Room: Therapist and Service User Perspectives on the Psychologically Supportive Dimensions of Architectural Space. *Health Environ. Res. Des. J.* **2020**, *13*, 103–118. [CrossRef]

10. Kearns, R.A.; Neuwelt, P.M.; Eggleton, K. Permeable Boundaries? Patient Perspectives on Space and Time in General Practice Waiting Rooms. *Health Place* **2020**, *63*, 102347. [CrossRef]
11. Demirkol, M.E.; Namlı, Z.; Tamam, L. Efficacy of Light Therapy on Non-Seasonal Depression and Inflammatory Markers. *Eur. J. Psychiatry* **2019**, *33*, 104–111. [CrossRef]
12. Simon, M.; Maben, J.; Murrells, T.; Griffiths, P. Is Single Room Hospital Accommodation Associated with Differences in Healthcareassociated Infection, Falls, Pressure Ulcers or Medication Errors? A Natural Experiment with Non-Equivalent Controls. *J. Health Serv. Res. Policy* **2016**, *21*, 147–155. [CrossRef] [PubMed]
13. Ergan, S.; Shi, Z.; Yu, X. Towards Quantifying Human Experience in the Built Environment: A Crowdsourcing Based Experiment to Identify Influential Architectural Design Features. *J. Build. Eng.* **2018**, *20*, 51–59. [CrossRef]
14. Alboadam, H.S.M.; Mohamed, Y.M. Role of Daylight in Local Hospitals Design. In Proceedings of the IOP Conference Series: Earth and Environmental Science, Baghdad, Iraq & Istanbul, Turkey, 1–2 June 2021; Volume 779.
15. Dutro, A.R. *An Art-Light Mosaic Light Distraction for the Pediatric Healthcare Environment*; Virginia Polytechnic Institute and State University: Blacksburg, VA, USA, 2016.
16. Houser, K.W.; Boyce, P.R.; Zeitzer, J.M.; Herf, M. Human-Centric Lighting: Myth, Magic or Metaphor? *Light. Res. Technol.* **2021**, *53*, 97–118. [CrossRef]
17. Ulrich, R.S. Effects of Interior Design on Wellness: Theory and Recent Scientific Research. *J. Health Care Inter. Des.* **1991**, *3*, 97–109.
18. Chang, C.; Oh, R.R.Y.; Le Nghiem, T.P.; Zhang, Y.; Tan, C.L.Y.; Lin, B.B.; Gaston, K.J.; Fuller, R.A.; Carrasco, L.R. Life Satisfaction Linked to the Diversity of Nature Experiences and Nature Views from the Window. *Landsc. Urban Plan.* **2020**, *202*, 103874. [CrossRef]
19. Li, W.; Samuelson, H. A New Method for Visualizing and Evaluating Views in Architectural Design. *Dev. Built Environ.* **2020**, *1*, 100005. [CrossRef]
20. Moscoso, C.; Chamilothori, K.; Wienold, J.; Andersen, M.; Matusiak, B. Window Size Effects on Subjective Impressions of Daylit Spaces: Indoor Studies at High Latitudes Using Virtual Reality. *LEUKOS J. Illum. Eng. Soc. N. Am.* **2021**, *17*, 242–264. [CrossRef]
21. Abd-Alhamid, F.; Kent, M.; Wu, Y. Assessment of Window Size and Layout Impact on a View Quality Perception in a Virtual Reality Environment. *Leukos* **2024**, *20*, 239–260. [CrossRef]
22. Bokharaei, S.; Nasar, J.L. Perceived Spaciousness and Preference in Sequential Experience. *Hum. Factors* **2016**, *58*, 1069–1081. [CrossRef]
23. Ulrich, R.S. View through a Window May Influence Recovery from Surgery. *Science* **1984**, *224*, 420–421. [CrossRef]
24. Jiang, S.; Powers, M.; Allison, D.; Vincent, E. Informing Healthcare Waiting Area Design Using Transparency Attributes: A Comparative Preference Study. *HERD Health Environ. Res. Des. J.* **2016**, *10*, 49–63. [CrossRef] [PubMed]
25. Ko, W.H.; Kent, M.G.; Schiavon, S.; Levitt, B.; Betti, G. A Window View Quality Assessment Framework. *Leukos* **2022**, *18*, 268–293. [CrossRef]
26. Ulrich, R.S.; Simons, R.F.; Losito, B.D.; Fiorito, E.; Miles, M.A.; Zelson, M. Stress Recovery during Exposure to Natural and Urban Environments. *J. Environ. Psychol.* **1991**, *11*, 201–230. [CrossRef]
27. Hartig, T.; Mitchell, R.; De Vries, S.; Frumkin, H. Nature and Health. Annual Review of Public Health. *Annu. Rev.* **2014**, *35*, 207–228.
28. Soga, M.; Akasaka, M. Multiple Landscape-Management and Social-Policy Approaches Are Essential to Mitigate the Extinction of Experience. *Landsc. Urban Plan.* **2019**, *191*, 103634. [CrossRef]
29. Grassini, S.; Revonsuo, A.; Castellotti, S.; Petrizzo, I.; Benedetti, V. Processing of Natural Scenery Is Associated with Lower Attentional and Cognitive Load Compared with Urban Ones. *J. Environ. Psychol.* **2019**, *62*, 1–11. [CrossRef]
30. van den Bosch, M. Ode Sang Urban Natural Environments as Nature-Based Solutions for Improved Public Health—A Systematic Review of Reviews. *Environ. Res.* **2017**, *158*, 373–384. [CrossRef]
31. Ulrich, R.S. Aesthetic and Affective Response to Natural Environment. In *Behavior and the Natural Environment*; Springer: Berlin/Heidelberg, Germany, 1983; pp. 85–125.
32. Sumarsono, S.; Mesra, M.; Zulkifli, Z. Artificial Flower as a Space Decoration Media Characteristics of North Sumatra Local Culture. In Proceedings of the 4th International Conference on Innovation in Education, Science and Culture, ICIESC 2022, Medan, Indonesia, 11 October 2022.
33. Hoyle, H.; Norton, B.; Dunnett, N.; Richards, J.P.; Russell, J.M.; Warren, P. Plant Species or Flower Colour Diversity? Identifying the Drivers of Public and Invertebrate Response to Designed Annual Meadows. *Landsc. Urban Plan.* **2018**, *180*, 103–113. [CrossRef]
34. Stoltz, J. Layered Habitats: An Evolutionary Model for Present-Day Recreational Needs. *Front. Psychol.* **2022**, *13*, 914294. [CrossRef]
35. Li, Z.; Huang, X.; White, M. Effects of the Visual Character of Transitional Spaces on Human Stress Recovery in a Virtual Reality Environment. *Int. J. Environ. Res. Public Health* **2022**, *19*, 13143. [CrossRef]
36. Shemesh, A.; Talmon, R.; Karp, O.; Amir, I.; Bar, M.; Grobman, Y.J. Affective Response to Architecture–Investigating Human Reaction to Spaces with Different Geometry. *Archit. Sci. Rev.* **2017**, *60*, 116–125. [CrossRef]
37. Tawil, N.; Ascone, L.; Kühn, S. The Contour Effect: Differences in the Aesthetic Preference and Stress Response to Photo-Realistic Living Environments. *Front. Psychol.* **2022**, *13*, 933344. [CrossRef] [PubMed]
38. Tsunetsugu, Y.; Miyazaki, Y.; Sato, H. Physiological Effects in Humans Induced by the Visual Stimulation of Room Interiors with Different Wood Quantities. *J. Wood Sci.* **2007**, *53*, 11–16. [CrossRef]

39. Zhong, W.; Schröder, T.; Bekkering, J. Biophilic Design in Architecture and Its Contributions to Health, Well-Being, and Sustainability: A Critical Review. *Front. Archit. Res.* **2022**, *11*, 114–141. [CrossRef]
40. Häyrinen, L.; Toppinen, A.; Toivonen, R. Finnish Young Adults' Perceptions of the Health, Well-Being and Sustainability of Wooden Interior Materials. *Scand. J. For. Res.* **2020**, *35*, 394–402. [CrossRef]
41. Mahmoud, M.T.; Ahmed, H.A.; Al-Din, M.S.K. The Impact of Nanotechnology on Exterior and Interior Finishing Materials in Hospitals. *Int. J. Mech. Eng* **2022**, *7*, 12.
42. Moran, T. Hospital Hotel Crain's Detroit Business. *Detroit* **1993**, *9*, 11.
43. Devlin, A.S. Seating in Doctors' Waiting Rooms: Has COVID-19 Changed Our Choices? *Health Environ. Res. Des. J.* **2022**, *15*, 41–62. [CrossRef]
44. Arneill, A.B.; Devlin, A.S. Perceived Quality of Care: The Influence of the Waiting Room Environment. *J. Environ. Psychol.* **2002**, *22*, 345–360. [CrossRef]
45. Grahn, P.; Tenngart Ivarsson, C.; Stigsdotter, U.K.; Bengtsson, I.L. *Using Affordances as a Health Promoting Tool in a Therapeutic Garden*; Routledge: London, UK, 2010; ISBN 0203853253. Available online: https://www.taylorfrancis.com/books/mono/10.4324/9780203853252/innovative-approaches-researching-landscape-health-catharine-ward-thompson-simon-bell-peter-aspinall (accessed on 19 September 2024).
46. Pálsdóttir, A.M.; Stigsdotter, U.K.; Persson, D.; Thorpert, P.; Grahn, P. The Qualities of Natural Environments That Support the Rehabilitation Process of Individuals with Stress-Related Mental Disorder in Nature-Based Rehabilitation. *Urban For. Urban Green.* **2018**, *29*, 312–321. [CrossRef]
47. Hashemi Kashani, S.M.; Pazhouhanfar, M. Role of Physical Attributes of Preferred Building Facades on Perceived Visual Complexity: A Discrete Choice Experiment. *Environ. Dev. Sustain.* **2023**, *26*, 13515–13534. [CrossRef]
48. Chang, K.G.; Chien, H. The Influences of Landscape Features on Visitation of Hospital Green Spaces—A Choice Experiment Approach. *Int. J. Environ. Res. Public Health* **2017**, *14*, 724. [CrossRef] [PubMed]
49. Sezavar, N.; Pazhouhanfar, M.; Van Dongen, R.P.; Grahn, P. The Importance of Designing the Spatial Distribution and Density of Vegetation in Urban Parks for Increased Experience of Safety. *J. Clean. Prod.* **2023**, *403*, 136768. [CrossRef]
50. Shayestefar, M.; Pazhouhanfar, M.; van Oel, C.; Grahn, P. Exploring the Influence of the Visual Attributes of Kaplan's Preference Matrix in the Assessment of Urban Parks: A Discrete Choice Analysis. *Sustainability* **2022**, *14*, 7357. [CrossRef]
51. Louviere, J.J.; Hensher, D. Design and Analysis of Simulated Choice or Allocation Experiments in Travel Choice Modeling. *Transp. Res. Rec.* **1982**, *890*, 11–17.
52. Brownstone, D. Applied Choice Analysis: A Primer. *J. Am. Stat. Assoc.* **2007**, *102*, 390. [CrossRef]
53. Lancaster, K.J. A New Approach to Consumer Theory. *J. Polit. Econ.* **1966**, *74*, 132–157. [CrossRef]
54. McLaughlan, R. Psychosocially Supportive Design: The Case for Greater Attention to Social Space Within the Pediatric Hospital. *HERD Health Environ. Res. Des. J.* **2018**, *11*, 151–162. [CrossRef]
55. Todorova, A.; Asakawa, S.; Aikoh, T. Preferences for and Attitudes towards Street Flowers and Trees in Sapporo, Japan. *Landsc. Urban Plan.* **2004**, *69*, 403–416. [CrossRef]
56. Bonham-Corcoran, M.; Armstrong, A.; O'Briain, A.; Cassidy, A.; Turner, N. The Benefits of Nature-Based Therapy for the Individual and the Environment: An Integrative Review. *Ir. J. Occup. Ther.* **2022**, *50*, 16–27. [CrossRef]
57. Woo, M.; Jafarifiroozabadi, R.; MacNaughton, P.; Mihandoust, S.; Kennedy, S.; Joseph, A. Using Discrete Choice Methodology to Explore the Impact of Patient Room Window Design on Hospital Choice. *J. Patient Exp.* **2022**, *9*, 23743735221107240. [CrossRef] [PubMed]
58. van Oel, C.J.; Mlihi, M.; Freeke, A. Design Models for Single Patient Rooms Tested for Patient Preferences. *Health Environ. Res. Des. J.* **2021**, *14*, 31–46. [CrossRef] [PubMed]
59. Ozdemir, A. The Effect of Window Views' Openness and Naturalness on the Perception of Rooms' Spaciousness and Brightness: A Visual Preference Study. *Sci. Res. Essays* **2010**, *5*, 2275–2287.
60. Golbabaei Pasandi, H.; Mahdavi, S.; Solmaz Talebi, S.; Jahanfar, S.; Shayestefar, M.; Hossein Ebrahimi, M. Investigating the Prevalence of Hearing Loss and Its Related Factors in Professional Drivers in Shahroud City, Iran. *Int. J. Occup. Saf. Ergon.* **2022**, *28*, 1994–1999. [CrossRef] [PubMed]
61. Elsadek, M.; Xie, J.; Jie, G.; Xiaofeng, S.; Clinic, D. The Impact of Indoor Plants on Patient Recovery: Physiological and Psychological Effects in Dental Clinics. *HortScience* **2023**, *58*, 1376–1382. [CrossRef]
62. Gu, J.; Liu, H.; Lu, H. Can Even a Small Amount of Greenery Be Helpful in Reducing Stress? A Systematic Review. *Int. J. Environ. Res. Public Health* **2022**, *19*, 9778. [CrossRef]

buildings

MDPI

Article

Modulating Perception in Interior Architecture Through Décor: An Eye-Tracking Study of a Living Room Scene

Weronika Wlazły and Agata Bonenberg *

Faculty of Architecture, Poznan University of Technology, 61-131 Poznan, Poland; weronika.wlazly@put.poznan.pl
* Correspondence: agata.bonenberg@put.poznan.pl; Tel.: +48-605408200

Abstract: The visual perception of interior architecture plays a crucial role in real estate marketing, influencing the decisions of buyers, interior architects, and real estate agents. These professionals rely on personal assessments of space, often drawing from their experience of using décor to influence how interiors are perceived. While intuition may validate some approaches, this study explores an under-examined aspect of interior design using a mobile eye-tracking device. It investigates how decorative elements affect spatial perception and offers insights into how individuals visually engage with interior environments. By integrating décor into the analysis of interior architecture, this study broadens the traditional scope of the field, demonstrating how décor composition can modulate spatial perception using eye-tracking technology. Results show that effective styling can redirect attention from key architectural elements, sometimes causing them to be overlooked during the critical first moments of observation commonly known as the "first impression". These findings have important implications for interior design practice and architectural education.

Keywords: interior architecture; interior design; home staging; décor; eye-tracking; eye movement; modulation; control; real estate marketing

1. Introduction

Visual perception of interior architecture plays a fundamental role in real estate marketing [1,2]. Buyers, interior architects, designers, home stagers, and real estate agents are guided through business decisions by subjective and personal assessments of the space. Professionals usually have broad individual experience regarding how to use décor to improve interior perception and direct attention toward or away from specific elements within a space [3,4]. Home staging practitioners operate based on subjective beliefs and experiences, assuming that décor influences perceptions of the interior [5]. Although intuition supports the validity of some of these ideas, the concept has not yet been explored scientifically. Therefore, the scope of this research is to objectify interior perception and demonstrate how human attention can be modulated through intentional design choices, whether the goal is to highlight certain features or distract from less-desirable aspects. To obtain data that can facilitate the exploration of this issue, an eye-tracking study was conducted to analyze sight paths in a sample living room scene. The scene was studied in three different scene variants, and the changes in sight paths were observed. The knowledge gained from this study contributes to understanding the efficiency of home stylists' actions in the real estate market. This can confirm or disconfirm the validity of a home stager's usefulness in renting or selling properties. Such an approach can be applied to both interior design and home staging in professional settings to optimize design outcomes.

1.1. Research Background

The eye-tracking method is a well-established tool in visual marketing research, as demonstrated by Wedel et al. [6]. The eye-tracking method was selected due to its ability to offer objective, real-time insights into visual attention and user engagement. By directly measuring where users focus, the duration of their attention, and what captures their interest, eye-tracking eliminates the biases inherent in self-reported data from surveys or interviews. This method quantifies visual behavior, making it more reliable than subjective techniques such as observational studies, while remaining more human-centered than computational models [7]. Furthermore, in the real estate context, it provides immediate feedback, whereas other methods depend on delayed user recall, which may not accurately capture real-time reactions. The main limitation of this method is the lack of insight into the emotions that accompany the visual engagement of the subjects, which is easy to obtain through surveys and questionnaires. In architecture literature, the application of eye-tracking methods in architecture, interior architecture, and interior styling have been analyzed, confirming a growing number of studies utilizing the eye-tracking method. These studies either focused on examining architectural structures themselves or on placing these structures in the broader context of other elements, both architectural and non-architectural. For example, a study by Lisińska-Kuśnierz and Krupa [8] explored the use of eye-tracking to analyze urban spaces, demonstrating its effectiveness in understanding how individuals visually interact with their surroundings. Another study by Rusnak [9] further confirmed the applicability of eye-tracking in architectural design, particularly when dealing with cultural heritage monuments. This study analyzed how the addition of logos and the use of specific colors affected the perception of existing monuments. Before conducting the eye-tracking experiment, a detailed survey was distributed to experts on architecture and cultural heritage sites. The findings not only validated the opinions of these experts, but also provided new insights into how visual stimuli, such as logos and colors, influence the perception of historical architecture. This underlines the high value of eye-tracking in architectural research, as it not only supports expert opinions but also offers additional data-driven perspectives on how architectural elements are experienced visually. An eye-tracking device provides valuable insights into interior elements that capture the most visual attention. Both stationary and mobile eye-tracking devices are commonly used in such studies. For instance, Tuszyńska-Bogucka et al. [10] employed a stationary eye-tracker to demonstrate that different spatial arrangements evoke varied reactions. Contrastingly, Rusnak and Ramus [11] conducted an on-site eye-tracking study of the Warsaw Uprising Museum's interior, using a mobile eye-tracker to identify the most visually engaging elements. Their research aimed to identify the elements within a museum's interior that captured the most visual attention from visitors. The applicability of the eye-tracking method to interior design is also confirmed through research conducted by Göktaş et al. [12]. In their study, a kitchen environment was analyzed to gather valuable data to improve the design of kitchen furniture. Using a mobile eye-tracker, the researchers identified kitchen areas that attracted the most visual attention. This information proved instrumental in guiding the decision-making process related to furniture design, ensuring that both functionality and aesthetic appeal were considered based on where users naturally focused their gazes. The key findings showed that eye-level components and those made from various materials consistently attracted the most visual interest. Additionally, lighting and the strategic placement of accessories play critical roles in drawing attention, helping designers optimize both the aesthetic appeal and functionality of their products. Regarding studies exploring interior architecture and different design approaches, the closest related research was conducted by Bo Hyeon Park and Kyung Hoon Hyun [13]. They examined the combinations of colors and materials used in interior design, recognizing the gap in the field due to the lack of well-established principles or rules governing the application of interior design styles. As noted by the authors, "no specific principles or rules have been established" [13]. Their research emphasized the style of interior architecture rather than focusing solely on the visual impact or presence of individual styling components.

A study has been conducted on the perception of greenery in interior spaces, focusing on various arrangements of greenery, as presented in the paper, "Occulographic Study on Greenery in Interiors of Public Libraries" [14]. This research explored the potential for consciously designing greenery placements to capture and direct attention more effectively. An analysis of the preceding studies reveals that the research presented in this study introduces a new perspective within the field of interior architecture, as no research has been conducted on the modulation of perception in interior architecture through décor. This study expands the traditional scope of interior architecture by integrating decorative elements into the analysis of spatial perception, thereby deepening scientific research in this area.

To obtain a broader perspective, the VOSViewer application was used to visualize a bibliometric network of terms. A keyword map (Figure 1) was created based on 79 search results for "eye tracking interior" in the Web of Science search engine. A total of 457 keywords were analyzed, with a minimum occurrence threshold of one. The graph shows the largest set of connected items, consisting of 388 keywords. Figure 2 zooms in on the connections for the keyword "interior design", revealing 43 links. This suggests that the established connection for the terms "eye tracking" and "interior design" is not very strong, further highlighting the novel character of the research area.

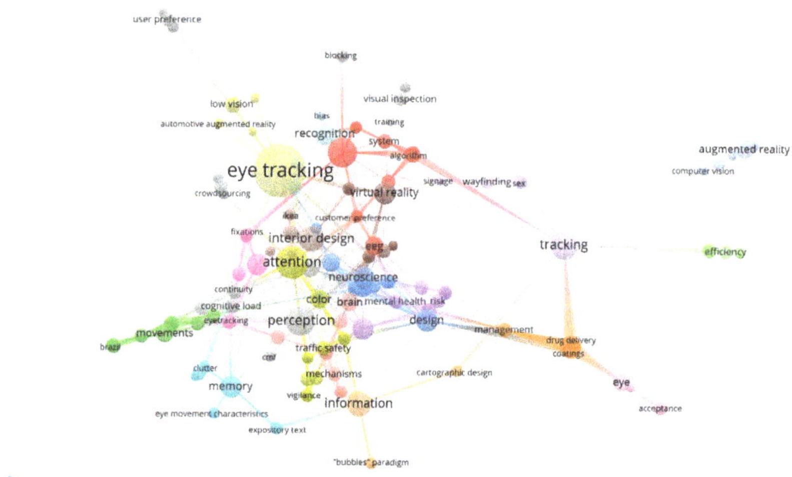

Figure 1. Keywords map, created in VOSViewer. The set of 392 connected items. Source: W. Wlazly.

The keyword map shows the current trends in eye-tracking research on interior design. Regarding graphics, this study explores interior design in the context of perception, design evaluation, on-site experience, color, and the impact of décor. However, this study does not explore the emotions and judgements of the study participants.

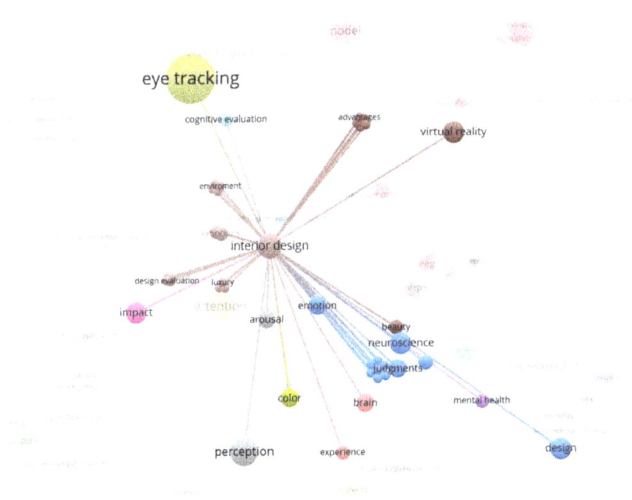

Figure 2. Keywords map, zoomed to keyword "interior design", presenting 43 links, created in VOSViewer. Source: W. Wlazly.

1.2. Aims of Study

The objective of the current study is to investigate whether it is possible to modulate perception in interior architecture through décor, and, in doing so, to provide scientific insights into the impact of décor styling in design processes.

Décor refers to the arrangement and selection of elements within a space that contribute to its aesthetic, functional, or symbolic attributes [15]. These elements may include furniture, textiles, vases, plants, tableware, paintings, artwork and similar elements. According to the *Collins Dictionary*, décor denotes the style or mode of decoration, construction, and ornamentation. Interestingly, the term is also used in the context of theater, where it refers to scenic decoration and scenography [15]. This dual association is particularly relevant in the context of home staging, where the purpose is to create an illusion, the right mood, and atmosphere for the clients.

The working hypothesis proposes that gaze points and fixation distribution, measured with an eye-tracking device during the first 8 s of visual exposure, are influenced by the composition of the décor. This study highlights the importance of décor placement in real estate marketing, leading to more informed business and design decisions. This could also initiate a discussion on the potential for modulating perception in interior spaces using not only a designer's taste and experience, but also scientific methods.

2. Materials and Methods

An eye-tracking method was employed to demonstrate how décor modulates perception in interior architecture [16]. As a research approach, eye-tracking offers valuable insights into how the audience perceives and interacts with a space, providing data-driven evidence to guide design decisions [17]. This study was conducted at the Faculty of Architecture of the Poznan University of Technology. A mobile eye-tracking device, Pupil Invisible Glasses (200 Hz), was used. The results were developed using iMotions 9.3 software. The authors chose a mobile eye-tracking device to record true on-site experiences, maintaining a real sense of distance from the objects being viewed and viewing them on a real scale.

The study consisted of three phases, in all of which each participant was expected to participate. As a result, there were supposed to be three recordings for each participant.

Initially, there were 20 participants, all without significant visual impairment or who were wearing lenses. However, the results of the study were aggregated for 18 participants with all three recordings. Recordings of two participants were excluded, due to the following reasons:

One participant took part only in the first phase of the study; therefore, there was only one recording for this participant.

One recording was missing due to an unknown technical issue, so the other two recordings of this participant were also excluded from the study.

The object of this study was a living room. In each phase of the study, the stimuli were the same static scene, arranged differently (Table 1).

Table 1. Study scenes' characteristics (by Authors).

Scene	1. Basic living room scene (control trial)	2. Styled living room scene	3. Styled living room scene

Independent variable (interior design)	Interior architecture elements: -floor, -wall, -table, -four chairs		
Independent variable characteristics	Elements of the interior in minimalist modern style. Gray floor made of ceramic tiles, and dark Gray panel wall. Modern light wooden furniture.		
Dependent variable (décor/styling)	Dependent variable absent	Décor arrangement on the table consisting of the following Dominant elements: textiles (also on two of the chairs), flower composition in a vase. Complementary elements: candles, two glasses, walnuts.	Décor arrangement on the table consisting of the following Dominant elements: textiles, candles, flower composition in a vase. Complementary elements: walnuts, apples
Dependent variable characteristics	Dependent variable absent	Composition: Harmonious dispersal Colors: Subdued colors, smooth color transitions.	Composition concentrated taller elements Colors Vivid, expressive colors, contrast.

An exposure time of 8 s per stimulus was chosen based on earlier research by Rusnak and Szewczyk [11,18], who revealed that, during their study, participants moved their heads after the first 8.6 s. Based on this research, a time lapse of 8 s was used to record the first impression.

The research scenario was prepared in advance and implemented during the study. The study began with an introduction to interior architecture and a description of home staging. The participants were informed that the scenes were elements of a residential space. They were told that they were going to observe the arrangement of a sample living room scene—part of a potential house—in different setups. However, the participants were unaware of the study's aims. The task given to participants at the beginning of each phase was as follows: "Evaluate the aesthetics of the scene." The task was intended to enable the participants to explore the scenes freely without pointing out any elements.

There was a marked place within 2 m of the scene, which was the same for each phase (Figure 3) The distance was determined to be similar to the distance from which the scene would be observed under the conditions of apartment presentation. The lighting conditions were identical for all participants during all phases.

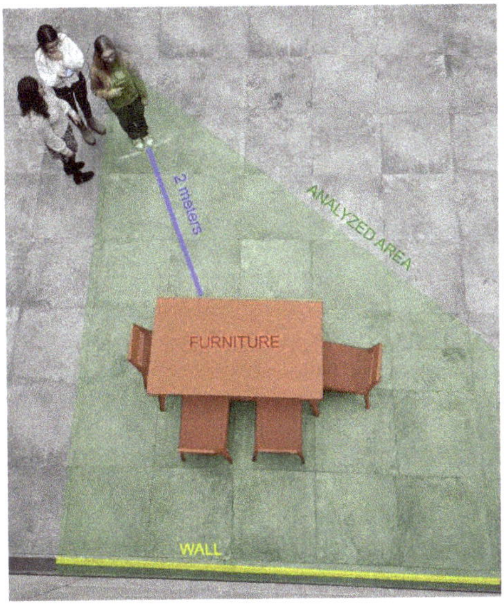

Figure 3. Study conditions. (by Authors).

The study was conducted in the afternoon and lasted for 2 h. No external stimuli were allowed to affect the participants' stress or tiredness levels. Two scholars carried out the study; one led the study, and the other supervised the equipment. Although the viewing angle was wider, the study group had to concentrate on a section of the space presenting the scene, which was considered in the subsequent analysis of the recording material.

3. Results

The results of the eye-tracking study are presented graphically in the form of heat maps (Figures 4–6) and Areas of Interest (AOI) (Figures 7–9), together with the parametric data for gaze- and fixation-based metrics in the second table in Section 3.2.

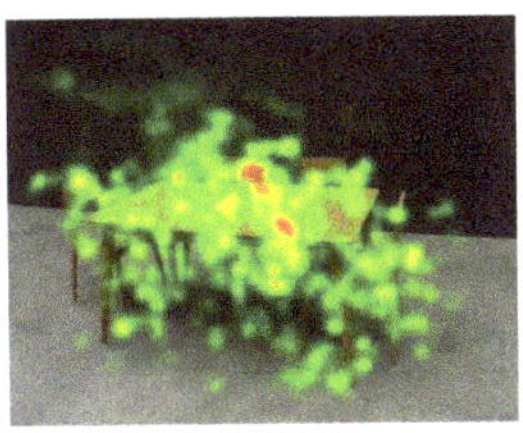

Figure 4. Scene 1. Basic living room scene (control trial). Aggregated Heat Map for 18 participants. Source: Authors.

Figure 5. Scene 2. Styled living room scene. Aggregated Heat Map for 18 participants. Source: Authors.

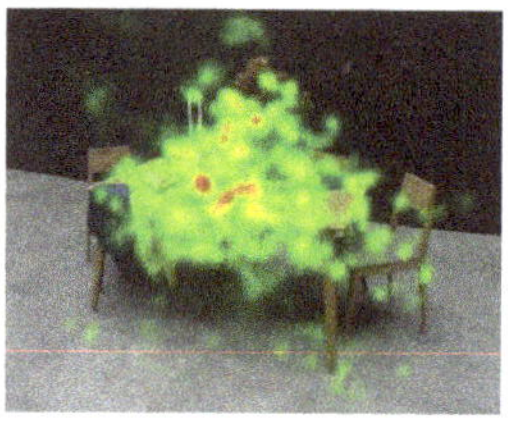

Figure 6. Scene 3. Styled living room scene. Aggregated Heat Map for 18 participants. Source: Authors.

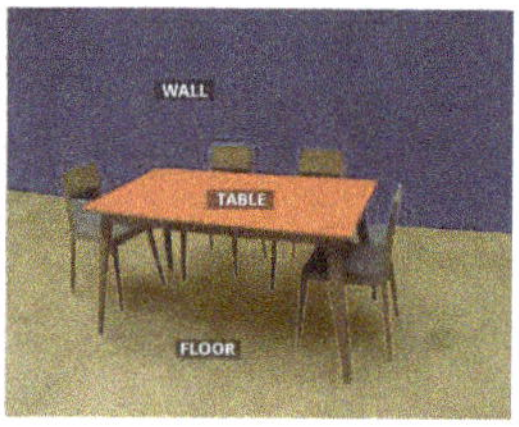

Figure 7. Scene 1—Areas of Interest: furniture with no décor. Control trial. Source: Authors.

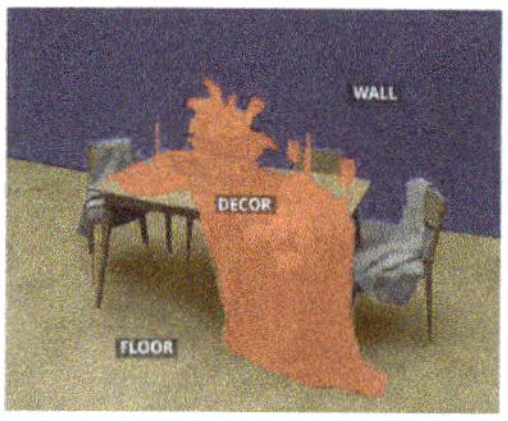

Figure 8. Scene 2—Areas of Interest: furniture with décor, version 1, subdued colors. Source: Authors.

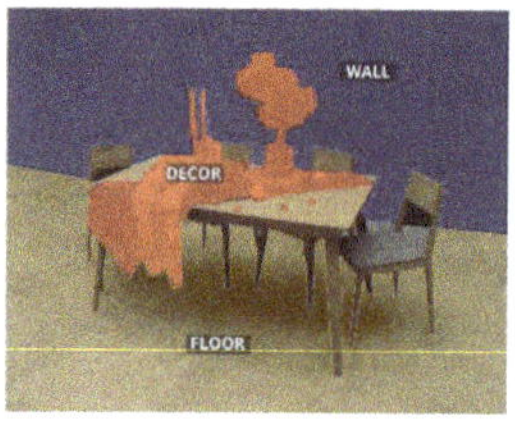

Figure 9. Scene 3—Areas of Interest: furniture with décor, version 2, expressive colors. Source: Authors.

3.1. Heat Maps

A heat map is a graphical display of the visual concentration areas. In this case, the heat maps represent the parts of the scene that were most often looked at, as they are heat maps of the gaze data. They are readable not only for scientists, but also for people who may be interested in the results of the study, such as interior designers, home stagers, realtors, and real estate agents who are not familiar with scientific terms. The presented heat maps of gaze data were aggregated for 18 participants using iMotions 9.3 software. Green indicates low attention, while yellow to red shows where high attention is present (Figures 4–6).

Attention in scene 1 (Figure 4) was more widespread in comparison to the other two scenes (Figures 5 and 6), where attention tended to concentrate mostly on the composition placed on the table. There were two areas with the highest attention in Scene 1 (Figure 4). They were the edges of the table—one on the edge closest to the observer, and the other on the parallel edge. Little attention was paid to walls; however, it is clear that they spread horizontally. Both flower compositions attracted our attention (Figures 5 and 6). Nevertheless, more attention was paid to the composition composed of contrasting white and brown elements (Figure 5) despite them being slightly shorter and more in harmony with the rest of the arrangement. The appearance of the décor arrangements on the table resulted in an almost complete distraction from the floor (Figures 5 and 6). Therefore, attention to the floor was visible in a significant way only in Scenario 1 (Figure 4). The heat maps show that placing the décor on the table resulted in visual attention focused mainly on the décor. All high-attention areas were placed on the décor arrangement, and little attention was paid to the other elements of the scenes. Placing the décor changed the way in which the architectural elements of the scene were perceived.

3.2. Areas of Interest

To further analyze the collected data, we defined AOI that enabled the analysis of gaze and fixation data for a particular region within the stimuli. Three AOIs were defined for each scene (Figures 7–9) based on the variables listed in Table 1. Two of them were independent variables and were present in all scenes: the wall and floor. The third AOI was located in the table area and was defined differently in the first scene compared to the rest. In the first scene, it was defined as a table (Figure 7), whereas in the other scenes it was defined as a décor arrangement (Figures 8 and 9). AOIs defined as décor arrangements are larger in size than those in the AOI Table (Table 2).

Table 2. AOI Size (px2) in each scene. (by Authors).

	1. Basic Living Room Scene (Control Trial)			**2. Styled Living Room Scene**			**3. Styled Living Room Scene**		
AOI	TABLE	WALL	FLOOR	DÉCOR	WALL	FLOOR	DÉCOR	WALL	FLOOR
Size (px2)	27,578	196,090.75	181,664.01	82,585.14	170,765.19	152,849.22	42,472.55	171,831.32	185,148.77

The stimulus duration was the same stimulus duration seen in the study phase, namely 8 s. Because the scenes were static, the AOI duration in all cases was the same as the stimulus duration. There was a high score for valid data for all scenes, between 98 and 99. Aggregated data from 18 respondents were used for the AOI analysis.

Gaze- and fixation-based metrics for all scenes are presented in Table 3. In particular, similarities were sought in the parameters of Scenes 2 and 3 and in the changes in Scenes 2 and 3 in relation to the first scene. A subsequent result of the research was an increase in the number of fixations on the table area after adding décor. Additionally, expected changes would be, among others, in the dwell count, dwell time, AOI hit time, and time to first fixation (TTFF) on the AOI. These changes would confirm this study's assumption that it is possible to control perception by introducing eye-catching elements.

Table 3. Areas of interest gaze and fixation-based metrics. (by Authors).

	Scena 1			**Scena 2**			**Scena 3**		
	TABLE	WALL	FLOOR	DÉCOR	WALL	FLOOR	DÉCOR	WALL	FLOOR
Size (px2)	27,578	196,090.75	181,664.01	82,585.14	170,765.19	152,849.22	42,472.55	171,831.32	185,148.77
Gaze based metrics									
Respondent count	18	17	15	18	14	13	18	18	15
Respondent ratio (%)	100	94.44	83.33	100	77.78	72.22	100	100	83.33
Dwell count	4.22	3	4.07	5.06	4.43	2.08	5.94	3.17	3.53
Revisit count	3.22	2	3.07	4.06	3.43	1.08	4.94	2.17	2.53
AOI hit time (ms)	1147.56	1422.47	1027.87	287.78	1005.43	3791.38	811.72	1661.67	2291.8
Dwell time (ms)	2021.81	2039.6	2670.65	5239.31	1598.97	620.77	3119.13	1858.5	1390.2
Dwell time (%)	24.18	24.23	31.23	61.95	18.9	7.3	36.21	21.62	16.17
First dwell duration (ms)	434.33	704.88	1046.93	1165.33	558	300.23	516.72	599.33	329.93
Last dwell duration (ms)	619.61	949.18	992.73	1487.72	494	516	592.06	697.67	760.87
Skip count	1.39	0.53	1.27	1.44	1.29	1.31	1.83	0.72	1.93
Fixation based metrics									
Respondent count	16	17	13	18	14	6	16	15	10
Respondent ratio (%)	88.89	94.44	72.22	100	77.78	33.33	88.89	83.33	55.56
Dwells with fixations	2.62	2	2.54	2.5	2.86	1.5	4	2.33	2.1
Revisit count	1.62	1	1.54	1.5	1.86	0.5	3	1.33	1.1
Fixation count	7.31	6.24	9.46	15.44	6.64	2.33	12.56	6.6	3.9
TTFF AOI (ms)	1761.97	2251.65	1653.81	1568.47	2462.04	3469.33	1151.41	1725.43	3948.65
TTFF max. (ms)	2532.86	2612.11	3555.69	1568.47	3830.47	6794.28	1971.69	2864.36	6014.36
Dwell time (ms)	1137.03	957.91	1377.27	2041.53	913.64	309	1981.16	1071.93	493.95
Dwell time (%)	13.52	11.47	16.09	24.21	10.8	3.64	22.97	12.44	5.73
Fixation duration (ms)	155.43	137.2	143.44	122.26	168.57	169.17	167.64	168.45	126.01
First fixation duration (ms)	136.03	144.56	156.15	94.11	145	185.33	179.28	137.7	105.1
Last fixation duration (ms)	190.41	124.47	182.38	116.81	217.5	168.67	161.16	241.13	132.9
Dispersion (deg)	0.47	0.51	0.45	0.46	0.48	0.44	0.51	0.53	0.46

The first dwelling duration for the table area, AOI Décor in Scenes 2 and 3 compared to the AOI Table in Scene 1, was longer. The first dwelling duration in Scenes 2 and 3 was shortened for the wall and floor AOI. Respondents in Scenes 2 and 3 paid attention to individual AOIs in the following order: décor, wall, and floor. Contrastingly, for Scene 1, the control trial indicated an order of table, floor, and wall. Although the table was empty, it was the first area to be observed; however, the order of the architectural elements of the space was altered. When décor was added, the floor lost its importance. The number of fixations increased significantly in the table area after the décor arrangement was added. The number of fixations was 2.1 times more for AOI Décor in Scene 2 than for AOI Tables in Scenes 1, and 1.7 times more for AOI Décor in Scene 3 than for AOI Table in Scene 1. This shows that placing a décor arrangement, thereby styling the scene, results in a more focused eyesight on the area that was styled, and thus there is a conscious change. However, this is not related to fixation duration, which, despite the differences in each scene, does not exhibit a common tendency to change the arrangement compared to the first scene. For Scenes 2 and 3, the TTFF for the AOI Décor was shorter than the TTFF for the AOI Table in Scene 1. The TTFF reduction confirms the possibility of controlling the elements that first attract attention. The dwell time was similarly extended for the AOI Décor in Scenes 2 and 3, being 1.75 times longer compared to the dwell time for the AOI Table in Scene 1. After adding the décor arrangement, the number of people whose fixations were on the walls and floor decreased. The decrease in the number of fixations on the wall and floor AOIs was greater for Scene 2 than for Scene 3. The number of people with fixations on the décor was the same or increased in relation to Scenario 1. It follows from this that architectural elements of the space that were independent variables lost their importance in Scenes 2 and 3 compared to Scene 1.

3.3. Results Summary

The hypothesis that gaze points and fixation distribution during the first impression (within an 8 s time frame) are influenced by the composition of the décor has been confirmed. The arrangement of décor in an architectural interior has been proven to be a key factor in guiding the distribution of sight paths, as measured using an eye-tracking device. The data show that, in Scene 3, adding prominent vertical elements including candleholders with candles and a vase with flowers resulted in a predominantly vertical orientation of the gaze points. However, this was somewhat predictable. In contrast, Scene 2 yielded an unexpected result: despite the tablecloth creating a diagonal line in the décor arrangement, the gaze points did not follow this direction. Instead, they remained focused on a table, drawn from the various other elements present. Such a pattern would have been difficult to detect without eye-tracking studies. This confirms the usefulness of the chosen research method, as it reveals previously unnoticed patterns of visual attention. A comparison between scenes demonstrates how décor modulates perception, with different sight paths and gaze points emerging when observing the same space. Additionally, the décor in Scenes 2 and 3 diverted attention from the architectural elements, such as the walls and floor, which are visually important in Scene 1.

4. Discussion

4.1. Overview of Findings

Based on the publications referenced in Section 1.1 "Research Background", the use of eye-tracking technology for studying the perception of interior architecture is fully validated [19–21]. Given the lack of similar research on modulating perceptions by adding decorative interior styling elements, this study offers new insights into styling as a key component of the design process. The findings show that effective décor styling can significantly divert attention from key architectural elements to the point where these elements may go unnoticed during the critical first few seconds of observation—often referred to as the "first impression". This study confirmed that décor plays a crucial role in shaping how people perceive interior spaces. By introducing decorative elements, perception can be deliberately guided toward specific areas, shifting attention away from fixed architectural features. Therefore, décor should be recognized as a powerful tool for managing and enhancing perceptions of spaces. Although this concept has long been intuitively applied in practice, it has now been scientifically validated. This research also supports the effectiveness of real estate marketing strategies such as those employed by home stagers who use décor to enhance property presentations.

4.2. Results Contribution and Application

This study supports key concepts in real estate marketing and the home staging industry, providing evidence that can be used to demonstrate the value of home staging services to clients [22]. The results, illustrated through visual tools such as heat maps, clearly show how strategically placed décor can guide attention and improve the first impressions of a space. These findings have the potential to significantly enhance interior design education [23,24]. The insights gained from this study serve as a valuable resource for interior architecture curricula, helping with understanding the importance of visual perception and learning how to consciously influence it in a purposeful way.

4.3. A Comparison of the Findings with the Literature

The aforementioned works in the Research Background demonstrate the applicability of eye-tracking methodology in both architectural studies [8,9] and interior design research [11,12]. This study further substantiates the utility of mobile eye-tracking devices in exploring the perception of interior spaces. Additionally, it supports the thesis of Gotkas et al. [12], which aims to facilitate decision-making in design processes, with the expectation that the insights gained from this study will contribute to informed decision-making in home staging practices. Furthermore, this research confirms the potential for purposeful arrangement to guide conscious visual pathways [14]. Finally, it expands the existing body of knowledge regarding home staging, a field that has previously been underexplored [3]. This article, unlike the existing literature, places particular emphasis on the presence of decorative elements. It introduces a new direction in interior architecture research, focusing on real estate marketing.

4.4. Recommendation for Future Research

This study examined the impact of a décor arrangement by comparing two different setups of the same living room scene and one scene without décor. Further research on the perception of interior architecture and the role of décor should delve more deeply into the individual elements of the composition, exploring their presence and visual appearance. Additionally, such studies could be expanded to include emotional responses elicited by décor, providing a more comprehensive understanding of whether attracting attention in specific instances positively or negatively affects the observer [23,24].

In future research, a comparison between expert expectations and the actual results of an eye-tracking study [25–27], as mentioned in Section 1.1, "Research Background", could be valuable. This would further validate our findings and offer deeper insight into the effectiveness of décor in influencing perceptions.

Author Contributions: Conceptualization, A.B. and W.W.; methodology, W.W.; validation, A.B.; formal analysis, W.W.; investigation, W.W.; data curation, W.W.; writing—original draft preparation, A.B. and W.W.; writing—review and editing, A.B.; visualization, W.W.; supervision, A.B.; funding acquisition, A.B. All authors have read and agreed to the published version of the manuscript.

Funding: This work was supported by the Poznan University of Technology, research funding number: 0113/SBAD/2402.

Institutional Review Board Statement: This study was approved by the Ethics Committee for Research Involving Humans at Poznan University of Technology, ZR-023/2024, 8 November 2024.

Informed Consent Statement: Informed consent was obtained from all subjects involved in the study.

Data Availability Statement: Data will be made available on request.

Conflicts of Interest: The authors declare no conflicts of interest.

References

1. Navickas, V.; Skripkiūnas, T.; Tanas, J.; Trojanek, M. The influence of architecture on real estate market value: A methodological framework. *J. Int. Stud.* **2020**, *4*, 38–53. [CrossRef]
2. Cho, J.Y.; Lee, E.-J. Impact of interior colors in retail store atmosphere on consumers' perceived store luxury, emotions, and preference. *Cloth. Text. Res. J.* **2017**, *35*, 33–48. [CrossRef]
3. Bonenberg, A.; Wlazły, W. Home staging, ie preparing real estate for sale. Discussion of the definition. *Space Form* **2022**, *52*, 43–60. [CrossRef]

4. Suh, J.; Cho, J.Y. Analyzing individual differences in creative performance: A case study on the combinational ideation method in the interior design process. *J. Inter. Des.* **2018**, *43*, 9–23. [CrossRef]
5. De-Juan-Ripoll, C.; Nolé, M.L.; Montañana, A.; Llinares, C. Relationship between visual Attention patterns and subjective evaluations in housing sales information: A study using eye-tracking technology. *Buildings* **2024**, *14*, 2106. [CrossRef]
6. Wedel, M.; Pieters, R. A review of eye-tracking research in marketing. *Rev. Market. Res.* **2017**, *4*, 123–147. [CrossRef]
7. Indraprastha, A.; Shinozaki, M. Computational models for measuring spatial quality of interior design in virtual environment. *Build. Environ.* **2012**, *49*, 67–85. [CrossRef]
8. Lisińska-Kuśnierz, M.; Krupa, M. Suitability of eye tracking in assessing the visual perception of architecture—A case study concerning selected projects located in Cologne. *Buildings* **2020**, *10*, 20. [CrossRef]
9. Rusnak, M. Applicability of eye trackers in marketing activities related to historical monuments. Comparison of experts' predictions and visual reactions of non-professionals. *J. Cult. Herit.* **2021**, *49*, 152–163. [CrossRef]
10. Tuszyńska-Bogucka, W.; Kwiatkowski, B.; Chmielewska, M.; Dzieńkowski, M.; Kocki, W.; Pełka, J.; Przesmycka, N.; Bogucki, J.; Galkowski, D. The effects of interior design on wellness—Eye tracking analysis in determining emotional experience of architectural space. A survey on a group of volunteers from the Lublin Region, Eastern Poland. *Ann. Agric. Environ. Med.* **2020**, *27*, 113–122. [CrossRef]
11. Rusnak, M.; Ramus, E. With an eye tracker at the Warsaw Rising Museum: Valorization of adaptation of historical interiors. *J. Herit. Conserv.* **2019**, *58*, 78–90.
12. Göktaş, O.; Ergin, E.; Çetin, G.; Özkoç, H.H.; Fırat, A.; Gazel, G.G. Investigation of user-product interaction by determining the focal points of visual interest in different types of kitchen furniture: An eye-tracking study. *Displays* **2024**, *83*, 102745. [CrossRef]
13. Park, B.H.; Hyun, K.H. Analysis of pairings of colors and materials of furnishings in interior design with a data-driven framework. *J. Comput. Des. Eng.* **2022**, *9*, 2419–2438. [CrossRef]
14. Bonenberg, A.; Wlazły, W. Occulographic study on greenery in interiors of public libraries. *Space Form* **2023**, *56*, 67–84. [CrossRef]
15. Collins English Dictionary. Decor. Available online: https://www.collinsdictionary.com/dictionary/english/decor (accessed on 6 December 2024).
16. Duchowski, A.T. *Eye Tracking Methodology: Theory and Practice*; Springer: London, UK, 2017; ISBN 978-1-84628-608-7.
17. Rusnak, M.; Chmielewski, P.; Szewczyk, J. Zmiany w postrzeganiu prezbiterium przy różnej długości nawy. Kościół tunelowy w badaniach okulograficznych. *Architectus* **2019**, *2*, 73–83. [CrossRef]
18. Rusnak, M.; Szewczyk, J. Eye tracker as innovative conservation tool. Ideas for expanding range of research related to architectural and urban heritage. *J. Herit. Conserv.* **2018**, *54*, 25–35.
19. Zou, Z.; Ergan, S. Where do we look? An eye-tracking study of architectural features in building design. In *Advances in Informatics and Computing in Civil and Construction Engineering*; Mutis, I., Hartmann, T., Eds.; Springer: Cham, Switzerland, 2019. [CrossRef]
20. Lavdas, A.A. Eye-tracking applications in architecture and design. *Encyclopedia* **2024**, *4*, 1312–1323. [CrossRef]
21. Li, N.; Zhang, S.; Xia, L.; Wu, Y. Investigating the visual behavior characteristics of architectural heritage using eye-tracking. *Buildings* **2022**, *12*, 1058. [CrossRef]
22. Guo, F.; Ding, Y.; Liu, W.; Liu, C.; Zhang, X. Can eye-tracking data be measured to assess product design?: Visual attention mechanism should be considered. *Int. J. Ind. Ergon.* **2016**, *53*, 229–235. [CrossRef]
23. Rusnak, M.A.; Rabiega, M. The potential of using an eye tracker in architectural education: Three perspectives for ordinary users, students and lecturers. *Buildings* **2021**, *11*, 245. [CrossRef]
24. Aalto, P.; Steinert, M. Emergence of eye-tracking in architectural research: A review of studies 1976–2021. *Archit. Sci. Rev.* **2024**, *4*, 1–11. [CrossRef]
25. Lim, J.Z.; Mountstephens, J.; Teo, J. Emotion recognition using eye-tracking: Taxonomy, review and current challenges. *Sensors* **2020**, *20*, 2384. [CrossRef] [PubMed]
26. Kim, J.; Kim, N. Quantifying emotions in architectural environments using biometrics. *Appl. Sci.* **2022**, *12*, 9998. [CrossRef]
27. Pretorius, M.C.; van Biljon, J.; de Kock, E. Added value of eye tracking in usability studies: Expert and non-expert participants. In *Human-Computer Interaction, Proceedings of the Second IFIP TC 13 Symposium, HCIS 2010, Held as Part of WCC 2010, Brisbane, Australia, 20–23 September 2010*; Forbrig, P., Paternó, F., Mark Pejtersen, A., Eds.; IFIP Advances in Information and Communication Technology; Springer: Berlin/Heidelberg, Germany, 2010; Volume 332. [CrossRef]

buildings

MDPI

Article

From Space to Satisfaction: Investigating Architectural Interior Determinants of Quality Work Environments

Vesna Krizmanić [1] and Aleksandra Milovanović [2,*]

[1] Archinova Architecture, 11050 Belgrade, Serbia; vesna.krizmanic@archinova.rs
[2] Faculty of Architecture, University of Belgrade, 11120 Belgrade, Serbia
* Correspondence: alekmil@arh.bg.ac.rs

Abstract: This study investigates the relationship between architectural attributes of workspaces and employee satisfaction, situating its inquiry within the broader context of post-pandemic hybrid work models and the evolving understanding of territoriality in organizational environments. Drawing on the social production of space and the multidimensional framework of Quality of Work Life (QWL), this research employs a quantitative, questionnaire-based methodology across three diverse corporate settings in Belgrade, Serbia. A total of 124 participants took part in the survey, representing the logistics (Fercam), IT (UBConnect), and healthcare (Medigroup) sectors. The survey integrates validated instruments—the Workspace Characteristics Scale (WCS) and the Workspace Satisfaction Scale (WSS)—to assess the impact of 12 variables and spatial features, reflecting functional, sensory, and sociological dimensions. Multiple regression analyses reveal that aesthetics, flexibility, and a sense of belonging consistently emerge as significant predictors of workspace satisfaction, with sector-specific variations observed across the cases. Notably, the findings underscore the importance of aligning spatial design with organizational values and user expectations while also highlighting the nuanced roles of sociological and psychological dimensions. The results offer practical relevance for architects and interior designers, providing insights into how design strategies tailored to user needs and sector-specific cultures can foster more effective workplace environments. This study concludes that evidence-based, multidimensional design strategies are essential for fostering well-being and productivity in contemporary work environments, demonstrating the value of aligning design with user expectations to enhance organizational outcomes. It recommends further research into the interplay between flexibility and belonging as determinants of workplace satisfaction.

Keywords: workplace design; quality of work life; spatial satisfaction; employee experience; user-centered design

1. Introduction

1.1. Background and Relevance

Time spent in contemporary workspaces—particularly in the aftermath of COVID-19 and the integration of hybrid work models, alongside companies' ongoing efforts to manage the challenges of hybridization and reintegration into the physical workplace—has brought renewed attention to the influence of architectural and spatial elements of both territory and interior environments on employee satisfaction and productivity. Concurrently, the context of the information age and the distinction between the space of places and the space of flows, as theorized by Manuel Castells [1], foregrounds the notion of territoriality and the emergent dynamics shaped by informational flows. This raises critical questions about the definition of territory in contemporary work environments and how such territories can be meaningfully connected to employees. In this study, territory is not understood solely in terms of physical or architectural features. Instead, space is conceptualized through the lens of Henri Lefebvre's theory of the social production of space [2], which frames territoriality as an identity-bearing dimension. In his seminal work *The Production of Space* (originally published in 1974 and translated into English in 1991), Lefebvre proposes that space must be understood as a social construct rather than as a purely geometric, mathematical, material, or physical product. He shifts the analytical focus from a semiotic conception of space as a coded message to its active construction and sociopolitical embodiment. Accordingly, this research integrates the sociological dimension into the analysis of the physical interior of the workspace. This study forms part of a broader conceptual model intended to inform and support the design of work environments that promote both the mental and physical well-being of users. It also incorporates contemporary examples of workspace interiors to ground the findings in current architectural practice.

The design of the workplace and its influence on employee satisfaction has been explored through a range of variables, including performance [3], motivation and satisfaction [4], and employee work performance and well-being [5]. These dimensions are often examined in conjunction with aspects of the physical environment, such as spatial configuration [6], workspace size [7], aesthetic impression [8], furniture quality [9], the adaptability of the space to individual needs [10], distance between workstations [11], and layout organization [12]. Other relevant factors include flexibility privacy [13,14]. Knight and Haslam [15] demonstrated through experimental research that employee participation in the design of the workspace significantly contributes to productivity, well-being, and a sense of dignity and the research connection of psychological comfort and autonomy and organizational identification. Complementing this, Nikolaeva and Russo [16] (p. 206) argue that "there is an individualized interpretation of physical space that shapes perceptions of value, autonomy, comfort, identity, and dignity." Given that employees with higher levels of professional capacity tend to seek greater autonomy in their work [17], it becomes increasingly important to understand and address the diverse structures and needs of different user profiles. In line with the above, workplace satisfaction metrics play a critical analytical role in shaping the framework for both the design and post-occupancy evaluation of contemporary work environments, serving not only as indicators of user experience but also as essential tools for informing evidence-based spatial and organizational strategies.

1.2. Conceptual Framing and Theoretical Context

Quality of Work Life (QWL) research has evolved significantly over the past five decades, transitioning from basic workplace condition assessments to comprehensive multidimensional frameworks that capture the complex interplay between work environment factors and employee well-being, becoming increasingly important within the scope of architectural design. QWL refers to the dimension of life quality that is influenced by work, extending beyond mere job satisfaction to encompass the broader context of the work environment and its developmental potential [18]. The concept of QWL is considered a subcategory of the overall quality of life (QoL) and has been studied independently since the late 1970s [19].

The evolution of QWL research can be traced through distinct phases, each characterized by different conceptual approaches and measurement strategies. During the 1980s and 1990s, researchers began to move beyond simple job satisfaction measures to develop more comprehensive frameworks for understanding work–life quality [20]. Efraty and Sirgy [21] conducted seminal work during this period, focusing on both work-related and personal factors that could improve or reduce QWL, establishing important theoretical foundations for subsequent research. Their approach was particularly notable for incorporating life satisfaction as a broader outcome measure, recognizing that workplace experiences extend their influence beyond organizational boundaries. This approach is supported by a series of studies aimed at strengthening the conceptual framework of Quality of Work Life (QWL) and its theoretical consolidation, including the following advancements: (1) a systematic investigation on how specific job characteristics could be measured and modified to enhance overall work–life quality [22], (2) an understanding of QWL antecedents by investigating positive work experiences and their relationship to both personal and organizational factors, providing empirical evidence for the multidimensional nature of work–life quality [23], and (3) the development of metrics that could assess various work-related factors systematically [24]. These studies established the foundation for contemporary measurement approaches by demonstrating that QWL could be reliably assessed through structured instruments that captured multiple dimensions of workplace experience. In this context, Simon A. Easton and Darren L. Van Laar from the University of Portsmouth developed the Work-Related Quality of Life (WRQoL) Scale as a targeted instrument for measuring work–life quality. The World Health Organization [25] (p. 1) further defines quality of life as "an individual's perception of their position in life in the context of the culture and value systems in which they live and in relation to their goals, expectations, standards, and concerns".

The development of psychometrically sound QWL measurement instruments represents a critical advancement in the field, with several studies providing a comprehensive validation of multidimensional scales, including (1) a comprehensive measurement index that identifies four primary QWL dimensions—work stress, work occupy, job and career satisfaction, and working conditions [26]—and (2) a case study examination of the relationship between QWL and job satisfaction, providing empirical evidence for the practical application of QWL measurement in organizational settings [27].

In this emerging theoretical framework, QWL metrics become a vital analytical tool in architectural design, offering a structured means of assessing how spatial environments contribute to employee well-being, satisfaction, and performance. Recent research has emphasized the importance of measurable spatial features in optimizing workplace outcomes from following lenses: (1) the development of a biophilic design index demonstrating significant improvements in concentration, stress recovery, and reduced presenteeism through natural materials and circadian lighting systems [28], (2) a longitudinal analysis identifying spatial flexibility, ergonomic adaptability, and informal collaboration zones as predictors of job satisfaction, physical comfort, and innovation [29], and (3) the formulation of a Functional Comfort Index (FCI) integrating metrics such as acoustic zoning, thermal autonomy, and daylight sufficiency, showing strong correlations with overall QWL scores in sustainable office settings [30]. Collectively, these studies establish architectural design as a multidimensional contributor to QWL and highlight the potential for data-driven spatial strategies to support holistic workplace well-being. Following this line of reasoning, this manuscript addresses the impact of workspace characteristics on quality of work life, understood as an integral component of the broader concept of quality of life.

1.3. Research Design and Paper Outline: Bridging the Gap Towards Integrated Framework of QWL Measurment

An insight into recent research on QWL measurement in architectural and interior design reveals an increasingly diverse yet methodologically dispersed understanding of spatial determinants influencing workplace well-being. Many studies demonstrate stronger focus on building physics (e.g., environmental comfort) over aesthetics [31], including air quality, lighting conditions, acoustic quality, thermal comfort [32], and workstation ergonomics [33]. Simultaneously, others highlight design-related dimensions [34], such as spatial layout, color harmony, material selection, and aesthetic coherence [35].

Beyond physical parameters, a scope of the literature examines psychosocial and emotional variables, including emotional health outcomes [36], perceived harmony and enchantment [37], the meaningfulness of work [38], and interpersonal relationships [39]. Still, important spatial constructs such as personal space [33], collaborative zones [40], quiet zones for concentration, and inspirational environments for creativity [41] are addressed only intermittently.

The literature sporadically acknowledges workplace flexibility, notably in terms of spatial adaptability [42]. However, flexibility is rarely examined in conjunction with aesthetic preference or a sense of belonging, despite their critical role in shaping employee satisfaction and spatial identity [37]. Similarly, dignity, understood through lenses of equality and control in workspace environments, is only implicitly addressed through constructs like organizational culture and management support [39,43].

This evidences a significant research gap, as no prior study has developed an integrated framework capable of capturing the combined effects of flexibility, belonging, and aesthetics across diverse workspace typologies. Moreover, although certain studies applied multidimensional assessments—such as TIHPS-based system models [44] or wearable multimodal integrations [45]—none link these data explicitly to spatial properties of architectural environments.

In response, this study proposes a QWL model that operationalizes spatial constructs through discrete and measurable categories, aiming to (1) bridge the gap between functional, psychological, and experiential dimensions of the workspace and (2) provide a comprehensive tool for the evidence-based evaluation of spatial quality in contemporary workplaces. Central to this investigation are two guiding questions: RQ1—which spatial characteristics of the workspace most strongly predict employee satisfaction across diverse organizational settings? RQ 2—how can these findings inform practical design strategies that enhance employee satisfaction, well-being, and broader organizational outcomes?

To address these research questions, this study employs a quantitative, questionnaire-based approach implemented across three distinct corporate settings in Belgrade, Serbia. These cases, selected for their diversity in organizational structure and spatial typology, allow for the exploration of sector-specific variations and broader patterns in workspace experience. Situated within this urban and organizational context, this research also responds to a broader regional gap: the number of studies addressing the measurement of QWL in the field of architecture and interior design remains limited across Serbia and Eastern Europe. Existing research in Serbia has primarily focused on aspects such as employees' work–life balance [46], while spatial qualities of the work environment within architectural and interior design domains have been largely overlooked. Although certain Eastern European countries, such as the Czech Republic, have developed QWL measurement instruments—most notably the Subjective Quality of Working Life index (SQWLi) [47]—their integration into architectural and interior design research remains limited. In this context, the present study, through a case study in Belgrade, aims to address this existing gap, contributing to the originality of this research and enhancing its contextual value within the regional body of knowledge on QWL in architectural and design disciplines.

The structure of the paper is as follows. Section 2 outlines the methodological framework, including the design of the questionnaire, description of the observed workspaces, characteristics of the sample, and analytical procedures. Section 3 presents the results of the multiple regression models used to assess the predictive value of each spatial attribute, along with general insights from value-based questions. Section 4 offers a critical discussion of the findings, including a comparative interpretation of the three cases and broader reflections on workspace design implications. Finally, Section 5 concludes this paper with key takeaways, methodological considerations, and recommendations for future research.

2. Materials and Methods

This study employed a quantitative, questionnaire-based research design to explore the relationship between specific spatial characteristics and user satisfaction across three distinct corporate work environments. The methodological framework is organized into three key components. First, the design of the questionnaire is presented (Section 2.1), detailing two core instruments—the Workspace Characteristics Scale (WCS) and the Workspace Satisfaction Scale (WSS)—followed by demographic and general value-based items. Second, the implementation process is outlined (Section 2.2), including a description of the selected workspaces, sample composition, and procedural steps for data collection. The spatial characteristics measured by the WCS were conceptually grounded in the authors' prior design experience and theoretical considerations, rather than collected through a separate observational phase. Third, the analytical approach is described (Section 2.3), focusing on a series of multiple regression models used to evaluate the predictive potential of spatial variables on workspace satisfaction. Together, these elements offer a comprehensive foundation for understanding the empirical basis of the study.

2.1. *Questionnaire Design*

The questionnaire was structured into four distinct sections, each addressing a specific aspect of the research framework. The first section focused on collecting general background information about the respondents, including company name and location, organizational role (ranging from executive management to entry-level positions), departmental affiliation, gender, educational background, professional field, total years of work experience, tenure within the company, basic psychological self-assessment, and the type of workspace they occupy (individual office, shared office for 2–8 people, assigned desk in an open-plan office, or flexible/unassigned seating in a shared open-plan setting). The second and third sections formed the analytical core of the questionnaire and included two validated instruments: the Workspace Characteristics Scale (WCS) and the Workspace Satisfaction Scale (WSS). These tools were designed to assess key spatial attributes and the overall satisfaction of respondents with their work environment. Finally, the fourth section comprised a set of general value-based questions intended to capture broader attitudes toward the workspace and its perceived impact. This final segment served as a complementary layer of insight, enriching the interpretation of the primary scale-based responses. An overview of both the initial and revised versions of the scales, along with the list of general and demographic questions, is provided in Appendix A.

2.1.1. Instrument 1: Workspace Characteristics Scale (WCS)

The WCS was theoretically grounded in Lefebvre's triadic concept of space [2]—spatial practice, representations of space, and representational spaces—providing a socio-spatial lens for assessing workspace quality. The scale consisted of 12 subscales, each designed to measure a distinct spatial characteristic of the work environment. Aspects of spatial practice, referring to the physical and functional routines within space, were assessed through the following: (1) personal space, (2) distance from other employees, (3) areas for social interaction, (4) variability in organization (the possibility of choosing a place to work within the workspace), (5) spaces for quiet and focused work, and (6) spaces for creative work and inspiration. Representations of space, as the formal, planned spatial order, were captured through the following: (7) aesthetic preference (the importance and consistency of aesthetic appearance), (8) dignity (equality and control), and (9) work flexibility (the possibility of choosing the time and place of work outside the workspace). Finally, representational spaces, reflecting users' lived, emotional, and symbolic experience, were measured via (10) the sociological dimension of space, (11) a sense of belonging, and (12) views and open outlook. The full scale comprised 38 items rated on a 10-point scale (1 = least descriptive of the workspace; 10 = most descriptive), with 10 items reverse-coded. Each subscale included between two and four items.

For the purpose of conducting the survey in Workspace 3, the legal department from that office required specific adaptations of the WCS. As a result, the third subscale—the sociological dimension of space, consisting of four items related to interpersonal relationships and social dynamics within the company—was removed due to internal corporate constraints. However, the key spatial indicator of sociological relevance—spaces for informal social interaction—remained included and assessed, thereby preserving the spatial validity of the construct. The excluded items were not directly related to spatial characteristics but rather to perceived interpersonal relations, which extend beyond the architectural scope of this study and align more closely with psychological or organizational research. Additionally, the decision to exclude interpersonal relationship questions was shaped by company-specific sensitivities, which may reflect a broader contextual limitation when applying comprehensive tools in real organizational environments. These adaptations, while methodologically justified, are acknowledged as potential limitations affecting the scope of measurement, and they highlight the complexity of implementing spatially grounded instruments in applied workplace settings.

In addition to this modification, 7 other items were removed from different subscales due to internal policy and procedural constraints, resulting in a total reduction of 11 items. Despite these removals, each retained subscale included at least two to three items, ensuring conceptual representation and preserving internal coherence for comparative analysis across cases. This selective adaptation introduces a potential instrumentation bias, as one site was evaluated using a slightly modified tool. Nevertheless, the modifications are transparently documented and interpreted as a methodological limitation arising from context-specific constraints.

An additional four-item scale aimed at evaluating alignment with company values was also introduced to explore potential connections with workspace satisfaction. However, due to corporate policy in Workspace 3, responses to these items were not obtained. As a result, these data were excluded from analysis, and the final research focus was methodologically limited to spatial aspects of the workspace environment.

2.1.2. Instrument 2: Workspace Satisfaction Scale (WSS)

The WSS consisted of four items/questions designed to assess respondents' satisfaction with their company's workspace. The scale focused on the spatial influence on quality of work life (QWL), modeled after established scales such as Working Conditions (WCS) as part of QWL—Satisfaction with Quality of Work Life. Respondents rated each item on a scale from 1 to 10: S1. The workspace fully meets the needs of my job description for successfully completing work tasks. S2. I feel good and comfortable in my workspace. S3. My work performance in this space is excellent. S4. I am satisfied with the aesthetic elements of my company's workspace.

2.1.3. General Value-Based Questions

The fourth section of the questionnaire was designed to explore respondents' broader attitudes and personal values related to the workspace. Using a 10-point Likert scale (1—strongly disagree, 10—strongly agree), participants rated their level of agreement with a series of statements concerning the symbolic, functional, and psychosocial dimensions of the work environment. Topics included the perceived impact of workspace quality on job satisfaction and performance, alignment between spatial design and organizational values, social dynamics in the office, and the role of aesthetics and natural elements. Additionally, this section incorporated multiple-choice questions addressing preferred office typologies and the estimated proportion of time respondents require for focused work versus team collaboration. These insights served to complement the core analytical scales and informed further reflections on spatial preferences and behavioral needs in different organizational contexts.

2.2. *Questionnaire Implementation*

2.2.1. Observatory Framework—Workspaces

This research was conducted across the workspaces of three companies based in Belgrade, Serbia, each operating in distinct sectors. A primary criterion for case selection was the direct involvement of one of the study's authors in the architectural design of all three work environments. This ensured privileged access to the design rationale, spatial documentation, and post-occupancy conditions, thereby supporting a robust and accurate analysis. Following this, the cases were selected purposefully to reflect diversity in both organizational structure and spatial typology. The selected companies include the following:

- Case 1: Fercam—A logistics and transportation firm with a traditional organizational structure and defined fixed office zones each for team for 75 employees. The organization of work is team-based (with five or six employees in a team), so the organization of the space has prioritized teams and groups of workstations consisting of 5 (6) desks in an open space. The accompanying functions include three kitchens with seating areas and visual dividers and greenery, separated from the workspace, which also represent informal zones for socialization and interaction, harmonized materialization and chlorite, and the space also contains three formal meeting rooms (two medium-sized and one large room).
- Case 2: UB Connect—A multinational engineering and telecommunications company characterized by hybrid work practices and closed office spatial organization; there is no designated flexible work zone, instead the concept of traditional workspaces has been applied where each employee has their own workspace, and the area is organized with a larger number of individual offices, without a large open space. The number of employees using the space is 80 (84 workstations were designed), but the work concept changed after 2020, so the space is regularly used by 60 employees (the others work from home or from other locations). The space also includes three closed meeting rooms of various sizes, an entrance hall with seating, a dining area with a kitchen, without additional facilities. Individual offices are defined by the project task for management (four separate offices for one, and one office for two people from the company's management), three offices are designated for engineering teams with eight employees each, and two for ten each. The structure is clear, engineering-oriented, and team-oriented and organized.
- Case 3: Medigroup—A private healthcare provider operating within a highly structured and function-specific office environment. The flexible work zone is organized in most of the space with 40 workstations; while 45 positions are planned with fixed workstations, only the lockers for laptops and personal belongings are personalized. The number of employees using the space is 130. Work desks in the flexible part of the space are intended to be used on a 'first come, first served' basis, while only the accounting, IT service, and management departments have their personalized work

desks due to the nature of their work. The space also includes four formal closed meeting rooms of various sizes, one open meeting area, two phone booths designated for phone calls, designed lounge seating areas, a dining area with a kitchen, and a space for entertainment and relaxation (billiards, table tennis, video game console), along with a large terrace for outdoor leisure. Individual offices are defined for management (four separate offices for one person each, and one office for two persons from the company's management).

The focus on Belgrade was a deliberate strategy to ensure a shared cultural, legislative, and economic context across cases, which allowed for a more coherent comparison of spatial and organizational influences on employee experience. Additionally, practical feasibility was a determining factor in finalizing case selection. All three workspaces are located within the same commercial office complex, providing a consistent external setting (Figure 1). Notably, Case 1 and Case 3 share a matching structural layout, which enables a particularly meaningful spatial comparison. While additional companies that met the primary inclusion criterion (i.e., having been designed by the same author) were approached, not all granted permission for survey implementation. Therefore, the final selection represents a balance between conceptual relevance, contextual control, and empirical feasibility.

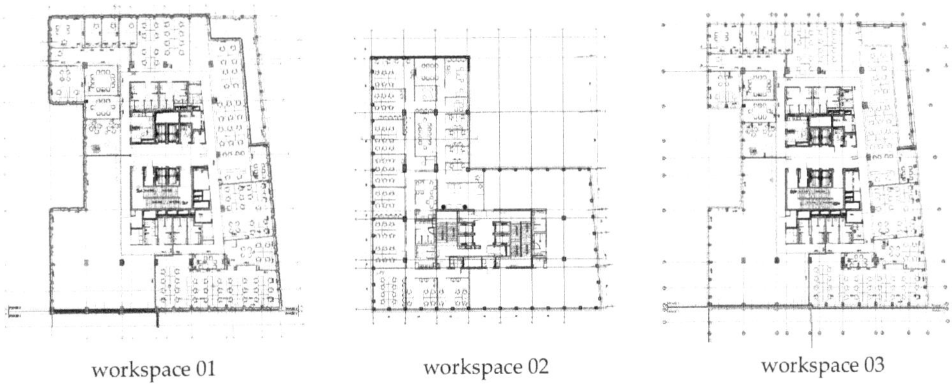

Figure 1. Organizational structure and spatial typology of analyzed workspaces. Source: Authors.

2.2.2. Research Sample

The research sample consisted of a total of 124 respondents, among whom gender data were collected for 86 participants (56% identified as female). Respondents were surveyed from three different workspaces, with the following structure: 49 employees from the first workspace (Fercam, out of a total of 70 employees, i.e., 70%), 37 from the second (UB Connect, out of 80 employees, i.e., 46%), and 38 from the third (Medigroup, out of 130 employees, i.e., 29%). Different modes of workspace usage (permanent desk, occasional visits, and hybrid work) are further described in the case study analysis.

Prior to deployment, the questionnaire was informally pre-tested with colleagues experienced in workplace research to ensure clarity, contextual appropriateness, and technical functionality. Minor wording adjustments were made without altering the original constructs.

2.2.3. Data Collection Procedure

The survey was conducted online using the platform soscisurvey.de. Respondents first answered demographic questions, followed by the items from the Workspace Characteristics Scale (WCS) and the Workspace Satisfaction Scale (WSS), and finally, general questions. The WCS and WSS items were mixed and presented in random order. The questionnaire took up to 15 min to complete, and no formal compensation was provided for participation. No personal or identifiable data were collected from respondents, ensuring complete anonymity. The survey was conducted between 4 March and 15 March 2024.

The questionnaire was distributed via email to all employees in the three participating companies by either the HR manager or the company director, whose role was limited strictly to logistical facilitation. The invitation included clear information about the purpose of the research, ensured voluntary participation, and contained an informed consent statement. Neither the HR managers nor the company directors had access to the responses or any aspect of the implementation or results.

While this approach helped protect participant anonymity and minimize response or social desirability bias, it is acknowledged that the need to obtain company-level approval and the presence of organizational hierarchies may represent a potential source of sampling-related limitations.

2.3. Questionnaire Analysis

The data from the WCS and WSS were processed in four iterations, using IBM SPSS v26 software for each round. Each iteration involved conducting a multiple regression analysis with the WCS subscale scores as predictor variables and the overall WSS score as the criterion variable. An "enter all together" method was applied in all regression models.

The first iteration included the complete dataset, combining responses from all three workspaces. The remaining three iterations involved separate analyses for each individual workspace: iteration two presented results for the first workspace, iteration three for the second, and iteration four for the third workspace.

Due to the shortened version of the questionnaire used in Workspace 3—MediGroup, the predictor variables in iterations one and four differ to those in iterations two and three. Specifically, iterations one and four were conducted using the modified scales, while iterations two and three used the original version (see Instrument Sections).

It is important to note that the results from iterations two, three, and four are not statistically comparable, as the statistical significance of differences between the models was not tested. Although this limits statistical conclusions regarding differences between the workspaces, it does not prevent interpreting the models through a case study methodology.

To ensure the reliability of the regression models, multicollinearity was assessed by calculating Variance Inflation Factor (VIF) scores for all predictor variables. All VIF values were below the commonly accepted threshold of 5, indicating acceptable collinearity levels. Additionally, standardized beta coefficients (β) are reported in the result tables to indicate the relative strength of each predictor.

To evaluate the internal consistency of the measurement instruments, Cronbach's alpha was calculated for the full dataset. The WCS showed strong internal reliability ($\alpha = 0.857$), while the WSS achieved a similarly robust reliability score ($\alpha = 0.871$). These results confirm that both scales meet the accepted threshold of 0.70, supporting the reliability of the instruments applied in the study.

Demographic and general question data were processed using basic frequency and percentage calculations.

3. Results

3.1. Models for Predicting Workspace Satisfaction

3.1.1. Iteration 1: Synthesis Sample

Iteration 1 included data from all respondents (a total of 124 cases), and the modified WCS (28 items) was used. A multiple regression model was created with 11 predictor variables and 1 criterion variable. The set of 11 predictors explained 82.8% of the variance in the criterion variable (R^2adj = 0.828).

The significant predictors identified were aesthetics (β = 0.474), a sense of belonging (β = 0.194), and flexibility (β = 0.141), all showing a positive relationship with the criterion variable. This indicates that respondents who gave higher ratings on these subscales reported greater satisfaction with their workspace.

Standardized beta coefficients, as well as the original correlation coefficients between each predictor and the criterion variable, are presented in Table 1.

Table 1. First iteration model parameters.

Spatial Characteristics of the Workspace (Features or Properties)	β Coefficient	t	*p*	Original Correlation
distance	0.052	1.103	0.272	0.480
views and open outlook	0.080	1.932	0.056	0.377
sociological dimension	0.021	0.362	0.718	0.668
spaces for social interaction	0.026	0.561	0.576	0.433
personalization	0.194	2.931	0.004	0.806
sense of belonging	−0.004	−0.098	0.922	0.126
variability	0.141	2.695	0.008	0.672
flexibility	0.474	6.304	0.000	0.872
aesthetics	0.022	0.542	0.589	0.151
dignity	0.068	1.212	0.228	0.579
spaces for focused work	0.083	1.523	0.131	0.676
spaces for creative work	0.052	1.103	0.272	0.480

3.1.2. Iteration 2: Fercam

Iteration 2 included data from respondents in Workspace 1—Fercam (49 cases in total), using the original version of the WCS. A multiple regression model was created, including 12 predictor variables and 1 criterion variable. This set of 12 predictors explained 68.1% of the variance in the criterion variable (R^2adj = 0.681).

The only statistically significant predictor was aesthetics (β = 0.558), which showed a positive relationship with the criterion variable. This indicates that respondents who rated the aesthetics of their workspace also more highly reported greater satisfaction on the WSS.

Standardized beta coefficients, as well as the original correlation coefficients between each predictor and the criterion variable, are presented in Table 2.

Table 2. Second iteration model parameters.

Spatial Characteristics of the Workspace (Features or Properties)	β Coefficient	t	*p*	Original Correlation
distance	0.034	0.274	0.785	0.538
views and open outlook	0.037	0.367	0.715	0.297
sociological dimension	0.065	0.508	0.613	0.592
spaces for social interaction	−0.013	−0.107	0.915	0.608
personalization	−0.040	−0.376	0.708	0.301

Table 2. *Cont.*

Spatial Characteristics of the Workspace (Features or Properties)	β Coefficient	t	*p*	Original Correlation
sense of belonging	0.000	0.001	0.998	0.741
variability	0.073	0.710	0.481	0.294
flexibility	0.033	0.280	0.780	0.434
aesthetics	0.558	3.483	0.001	0.837
dignity	0.011	0.117	0.906	0.322
spaces for focused work	0.060	0.468	0.642	0.506
spaces for creative work	0.246	1.743	0.089	0.705

3.1.3. Iteration 3: UB Connect

Iteration 3 included data from respondents in Workspace 2—UB CONNECT (a total of 37 cases), using the original WCS. A multiple regression model was created, consisting of the same 12 predictor variables and 1 criterion variable as in Iteration 2. This set of 12 predictors explained 81.4% of the variance in the criterion variable (R^2adj = 0.814).

A large number of variables emerged as significant predictors: distance (β = 0.296), open view (β = 0.236), a sense of belonging (β = 0.785), flexibility (β = 0.314), aesthetics (β = 0.334), dignity (β = −0.282), spaces for focused work (β = 0.425), and spaces for creative work (β = −0.444).

The most influential predictor was a sense of belonging, while the weakest was open view. The directions of the standardized beta coefficients suggest that respondents who perceived their workspace as offering greater distance, more open views, a stronger sense of belonging, greater flexibility, stronger aesthetic influence, and more spaces for focused work reported higher levels of workspace satisfaction. In contrast, a stronger protection of dignity and a higher number of spaces for creative work showed a negative association with workspace satisfaction in Workspace 2.

Standardized beta coefficients, as well as the original correlation coefficients between each predictor and the criterion variable, are presented in Table 3.

Table 3. Third iteration model parameters.

Spatial Characteristics of the Workspace (Features or Properties)	β Coefficient	t	*p*	Original Correlation
distance	0.296	2.877	0.008	0.414
views and open outlook	0.236	2.889	0.008	0.499
sociological dimension	−0.082	−0.625	0.537	0.676
spaces for social interaction	−0.113	−0.683	0.500	0.707
personalization	−0.283	−1.968	0.060	0.613
sense of belonging	0.785	3.761	0.000	0.752
variability	0.018	0.228	0.821	0.374
flexibility	0.314	2.377	0.025	0.526
aesthetics	0.334	2.234	0.034	0.824
dignity	−0.282	−2.518	0.018	0.237
spaces for focused work	0.425	2.836	0.009	0.579
spaces for creative work	−0.444	−2.500	0.019	0.618

3.1.4. Iteration 4: Medigroup

Iteration 4 included data from respondents in Workspace 3—MEDIGROUP (a total of 38 cases) and used the modified WCS, as in Iteration 1, which included all respondents. A multiple regression model was created with 11 predictor variables and 1 criterion variable, consistent with the structure of Iteration 1.

This set of 11 predictors explained 93.7% of the variance in the criterion variable (R^2adj = 0.937). The variables that emerged as significant predictors were a sense of belonging (β = 0.382), flexibility (β = 0.248), spaces for creative work (β = 0.234), and dignity (β = 0.181).

All of the significant predictor variables were positively associated with the criterion score, suggesting that respondents who expressed stronger agreement with the items in these subscales tended to report higher satisfaction with their workspace.

Standardized beta coefficients, as well as the original correlation coefficients between each predictor and the criterion variable, are presented in Table 4.

Table 4. Fourth iteration model parameters.

Spatial Characteristics of the Workspace (Features or Properties)	β Coefficient	t	*p*	Original Correlation
distance	0.007	0.11	0.913	0.454
views and open outlook	0.042	0.843	0.407	0.346
sociological dimension	0.04	0.541	0.593	0.558
spaces for social interaction	−0.089	−1.374	0.181	0.493
personalization	0.382	3.812	0.001	0.925
sense of belonging	−0.078	−1.369	0.183	0.069
variability	0.248	3.73	0.001	0.763
flexibility	0.138	1.244	0.225	0.888
aesthetics	0.181	3.073	0.005	0.224
dignity	0.118	1.41	0.17	0.728
spaces for focused work	0.234	3.564	0.001	0.736
spaces for creative work	0.007	0.11	0.913	0.454

3.2. *General Insights: Attitudes Toward Workspace and Work Performance*

In addition to the WCS and WSS, respondents answered seven general questions. The responses were analyzed across all three workspaces. The first four questions were formulated as rating-scale items, where respondents were asked to indicate their level of agreement on a scale from 1 to 10—with 1 indicating complete disagreement and 10 indicating complete agreement. The remaining three questions were multiple-choice items.

Responses to the first four items are presented in Table 5.

The first item, "A high-quality workspace significantly affects my job satisfaction and work efficiency", was strongly endorsed: over 30% of respondents selected the highest rating (10), while 86.3% responded on the positive end of the scale (6 or higher).

The second item also showed a strong positive trend, with 76.6% of respondents choosing a score of 6 or above. Notably, the proportion of extremely positive responses was slightly lower, while moderate positive ratings increased.

Table 5. Distribution of respondents' rating on four items of general questions.

A High-Quality Workspace Significantly Influences My Job Satisfaction and Work Efficiency.										
rating	1	2	3	4	5	6	7	8	9	10
% of respondents	0.8	1.6	1.6	1.6	8.1	8.9	8.9	19.4	17.7	31.5
The chaos of a workspace leads to chaotic work outcomes.										
rating	1	2	3	4	5	6	7	8	9	10
% of respondents	2.4	3.2	4.0	2.4	11.3	8.1	13.7	9.7	16.1	29.0
The appearance of the workspace influences my decision to work for a given company.										
rating	1	2	3	4	5	6	7	8	9	10
% of respondents	12.1	7.3	9.7	10.5	12.9	10.5	11.3	8.1	6.5	11.3
A stronger connection to nature in the workspace, more greenery, and natural materials would help me achieve better work results.										
rating	1	2	3	4	5	6	7	8	9	10
% of respondents	1.6	1.6	5.6	2.4	11.3	7.3	11.3	14.5	10.5	33.9

A shift occurred with the third item, "The appearance of the workspace influences my decision to work in a given company", where only 47.6% of respondents positioned themselves on the positive side of the scale. The remainder leaned toward the negative, and there was a near-equal distribution between minimum and maximum agreement—suggesting ambivalence or divided opinion.

The fourth item, "Connection to nature, more greenery, and natural materials would help improve my work performance", received the strongest endorsement: 77% of respondents selected scores on the positive end of the scale, with 33.9% giving the maximum score (10), indicating a high valuation of biophilic design elements.

Responses to the fifth general question (multiple choice) are presented in Table 6, while results for the sixth and seventh general questions—also multiple choice—were analyzed separately for each workspace and are summarized in Table 7.

Table 6. Responses to the fifth general question.

If I Could Choose, It Seems That I Would Achieve the Best Work Results and Personal Satisfaction in:	% Responses
an individual office	17.7
an office shared with up to 6 employees	54.0
an open-plan office with my own desk	12.9
the type of office does not affect my work performance or satisfaction	15.3

Table 7. Responses to the sixth and seventh general questions.

It Is Essential for Me to Have the Possibility of Quiet, Undisturbed, and Focused Work Within the Workspace.	For Teamwork with Colleagues in a Shared Physical Space, the Following is Essential to Me:	
% Responses	Alternative	% Responses
	Workspace 1	
22.4	0–20% of total working time	22.4
44.9	20–50% of total working time	30.6
26.5	50–80% of total working time	36.7
6.1	80–100% of total working time	6.2

Table 7. *Cont.*

It Is Essential for Me to Have the Possibility of Quiet, Undisturbed, and Focused Work Within the Workspace.		For Teamwork with Colleagues in a Shared Physical Space, the Following is Essential to Me:
% Responses	**Alternative**	**% Responses**
	Workspace 2	
18.9	0–20% of total working time	35.1
40.5	20–50% of total working time	37.8
32.4	50–80% of total working time	27
8.1	80–100% of total working time	0
	Workspace 3	
21.1	0–20% of total working time	34.2
34.2	20–50% of total working time	36.8
34.2	50–80% of total working time	21.1
10.5	80–100% of total working time	7.9

When comparing average scores on the shortened WSS, Workspace 1 showed the highest overall satisfaction, while Workspace 3 had the lowest. To statistically examine these differences, a univariate analysis of variance (ANOVA) was conducted, with workspace as the independent variable (three levels: Workspaces 1, 2, and 3), and the WSS score as the dependent variable. The results of the ANOVA revealed a significant difference among the three workspaces, $F(2, 121) = 7.492$, $p < 0.001$, accounting for approximately 11% of the variance in workspace satisfaction.

Additional questions (K1–K4), posed only to respondents in Workspaces 1 and 2, are synthesized in Table 8.

Table 8. Responses to the additional questions.

How Satisfied Are You with Your Job Overall, Excluding the Workspace?										
rating	1	2	3	4	5	6	7	8	9	10
% of respondents	2.3	0	3.5	2.3	10.5	1.2	12.8	27.9	23.3	16.3
The values promoted by my company are also my personal values.										
rating	1	2	3	4	5	6	7	8	9	10
% of respondents	1.2	1.2	3.5	4.7	12.8	7.0	16.3	22.1	15.1	16.3
How satisfied are you with the social relationships within the company?										
rating	1	2	3	4	5	6	7	8	9	10
% of respondents	0.0	1.2	2.3	3.5	7.0	10.5	14.0	24.4	18.6	18.6
How satisfied are you with the respect and recognition you receive within the organization?										
rating	1	2	3	4	5	6	7	8	9	10
% of respondents	1.2	1.2	2.3	2.3	10.5	3.5	15.1	16.3	17.4	30.2

4. Discussion

4.1. Comparative Insights: Original Correlations Between Spatial Characteristics of the Workspace

The diagram (Figure 2) presents a comparative insight into the original correlations (i.e., the correlations between each independent variable and the dependent variable without controlling for the influence of other predictors—low correlations: $r < 0.3$, moderate: $0.31 < r < 0.7$, and high: $r > 0.71$—demonstrate a remarkably strong relationship across nearly all predictor variables—spatial elements) between spatial characteristics of the workspace—understood as experiential features or properties—and the perceived quality of work life across the three distinct workplace environments (W1, W2, and W3) and a synthetic sample (S).

Figure 2. Comparative insights into original correlation of spatial characteristics of the workspace (features or properties) for Workspace 1 (W1), Workspace 2 (W2), Workspace 3 (W3), and the synthesis sample (S).

Several key observations emerge from the cross-insight. Attributes such as distance, views and open outlook, flexibility, and aesthetics demonstrate consistent relevance across all four datasets (W1, W2, W3, and S), highlighting their foundational role in shaping employee satisfaction. These variables are not only frequently represented but also exhibit robust correlation strength, indicating their stability as predictors across differing organizational typologies of workspace and spatial settings. Among these, aesthetic preference shows the strongest influence ($\beta = 0.474$).

Each workspace exhibits distinct emphases: (1) Workspace 1 is more strongly associated with distance, views, and aesthetics, suggesting a spatial layout conducive to individual focus and perceptual comfort, (2) Workspace 2 (W2) shows the most diverse profile, with relevance extending to spaces for creative work, personalization, dignity, and the sociological dimension—reflecting the complex demands of hybrid, team-based environments that combine autonomy with social dynamics—and (3) Workspace 3 (W3) places emphasis on a sense of belonging, spaces for focused work, and flexibility, indicating the importance of concentration and psychological safety in service-oriented settings.

Additionally, it is worth noting that W1 and W3 share a matching structural layout as both are situated within the same office complex and follow a comparable architectural configuration. This spatial congruence may partially explain the similarity in the prominence of certain predictors such as distance, flexibility, and focused work zones observed across their datasets. In contrast, W2, which features a distinct enclosed-office typology, demonstrates a more diversified set of significant predictors, including creative zones and dignity, likely reflecting its differentiated spatial logic and hybrid organizational practices. These structural differences underscore the nuanced ways in which workspace typology interacts with organizational routines and user needs, reinforcing the importance of context-specific interpretations when evaluating spatial determinants of satisfaction.

The inclusion of the sociological dimension reveals moderate correlations in W2 and in the synthetic sample, indicating its relevance in organizations with stronger relational cultures. Although not among the most dominant predictors, it adds important nuance by capturing the symbolic and interpersonal dynamics of space. Conversely, attributes such as variability, spaces for social interaction, personalization, and dignity show lower correlation frequencies across the datasets. While these may enhance satisfaction in particular cultural or operational contexts, they do not emerge as universal determinants of QWL. Their weaker representation suggests that they may function as contextual amplifiers rather than foundational drivers of satisfaction.

Although some variables, such as personalization, variability, dignity, and spaces for social interaction, appeared with lower correlation frequencies across the datasets, this does not negate their theoretical relevance. Their weaker statistical representation may stem from contextual and organizational constraints that limit their expression or perception. For example, personalization may be constrained in flexible or standardized desk-sharing environments, reducing its salience for respondents. Similarly, the sociological dimension—which showed moderate correlations in W2 and in the synthetic sample but was excluded from the analysis in W3 due to company-imposed limitations—was not consistently significant in predictive models. Importantly, while the sociological dimension was removed, the subscale related to spaces for social interaction was retained and considered particularly relevant, as it captures spatially anchored opportunities for informal encounters and collective use, thereby maintaining a link to the social affordances of space.

The synthesis sample consolidates the cross-contextual relevance of aesthetics, flexibility, distance, and a sense of belonging, reaffirming their dominant role in shaping employee satisfaction. It also reflects the moderate recognition of the sociological dimension, further validating the idea that spatial quality is both functional and relational. This confirms that while organizational cultures and spatial policies vary, certain spatial features consistently underpin user satisfaction and perceived performance across settings.

4.2. Synthesis of Key Insights: General Reflections on Workspace Satisfaction and Design Implications

The responses to the general value-based questions confirm that over 50% of employees across all three workspaces rated their agreement with the following statement a score of 8 to 10: "A high-quality workspace significantly impacts my job satisfaction and work efficiency". However, workspace quality was not considered a decisive factor in choosing to work for a specific company.

An exceptionally high percentage (33.9% rating it 10) agreed that "Connection to nature within the workspace, more greenery, and natural materials would help me achieve better work outcomes". This result provides a clear design directive for future spatial planning.

Questions 6 and 7, related to office type and the percentage of time required for team vs. focused work, did not yield directly analyzable insights in relation to workspace satisfaction. However, they proved useful as a pilot tool for future surveys. Interestingly, a large portion of employees in all three workspaces indicated the need for 20–50% of the workday to be dedicated to focused work (44.9%, 40.5%, and 34.2%, respectively). Meanwhile, teamwork needs exceeding 50% of work time were reported by 42.9% of respondents at Fercam and 27–29% in the other two spaces. This aligns with the previously identified connection between higher satisfaction and a greater need for teamwork.

Focused work may be supported not only through isolated physical spaces but also via communication rules within defined zones. However, for social interaction, a designated and supportive environment is essential. While this study does not explore Questions 6 and 7 in detail per workspace, the results suggest that sector-based analysis could support the further reconfiguration or redesign of working environments.

Workspace influence on performance and satisfaction is clearly substantial, even if not a primary criterion in job selection. The findings suggest that it is crucial to foster a sense of belonging (feeling at home, safe, and part of the organization), flexibility (in time, location, and communication), and aesthetics (orderly, harmonious, and pleasant spaces with coordinated colors, materials, and furniture). Meanwhile, variability (i.e., the ability to choose different work zones depending on task type) showed the least importance across the sample.

The strong demand for team collaboration indicates that more team-oriented spaces are needed or that seating arrangements (e.g., clusters, team-based layouts) should be aligned with actual team dynamics. The activity-based working model—where workspace changes depending on task—has not yet been adopted in the corporate systems surveyed. It is recommended that such models be introduced first through corporate procedures, and only then reflected spatially.

The survey results show that dignity was neither perceived as threatened nor as a significant factor influencing satisfaction. This may reflect the relatively egalitarian and non-intrusive nature of contemporary workspaces, which tend to ensure equitable access and reduce surveillance.

Although anecdotal conversations with employees suggested that personalization would be important, the survey did not confirm this. In none of the three workspaces did personalization emerge as a significant factor influencing satisfaction, even though it appeared somewhat relevant in Workspace 2 based on the raw correlation.

Only 5.6% of respondents gave a score below 5 to the statement that "A high-quality workspace significantly impacts my satisfaction and work efficiency," supporting the theoretical premise that workspace design contributes positively to user satisfaction and perceived performance.

4.3. Limitations and Future Directions

This study offers a multifaceted contribution to the growing literature on user satisfaction in workspaces, both conceptually and methodologically. Conceptually, it advances existing research landscape by operationalizing spatial constructs through a multidimensional QWL framework. While previous studies (presented in the introductory section) have typically focused on isolated dimensions—such as physical–spatial arrangements (spatial practice) [7,10,11], managerial planning and control (representations of space) [9], or user perceptions and emotional responses (representational spaces) [5,8]—this research integrates all three into a unified model grounded in Lefebvre's triadic conception of space. The application of the WCS across three diverse organizational settings provides new insights into how specific spatial features—such as distance, flexibility, aesthetics, and a sense of belonging—consistently correlate with perceived satisfaction.

Importantly, this study also identifies spatial variables—such as personalization, dignity, and variability—that, despite strong theoretical relevance, did not emerge as statistically significant predictors in all contexts. These findings nuance prevailing assumptions in the literature and suggest that such attributes function more as contextual amplifiers rather than universal determinants of satisfaction, particularly in flexible or standardized work environments.

Methodologically, this study contributes by introducing a validated and theoretically grounded scale (WCS), which was tested for internal consistency (Cronbach's $\alpha = 0.857$) and paired with the WSS (Cronbach's $\alpha = 0.871$). This study also openly acknowledges and discusses key methodological limitations, including differences in questionnaire versions and the non-inclusion of certain subscales in Workspace 3 due to organizational constraints. Rather than compromising the study's integrity, this transparent treatment of limitations provides a precise methodological framework for future researchers and supports reproducibility and critical engagement with the model in other empirical contexts.

While this study was conducted within a uniform cultural and legislative context (Serbia), which facilitated internal coherence across cases, this also imposes certain limitations on its external validity. Organizational and spatial experiences are shaped by broader sociocultural norms, which may vary across regions and sectors. Although the spatial variables used are derived from interdisciplinary frameworks and not specific to the Serbian context, caution should be exercised when generalizing the findings. Future studies are encouraged to apply the proposed model in different cultural and organizational environments to assess its broader applicability and identify potential context-specific adaptations.

5. Conclusions

The objective of this survey was to examine the relationship between specific spatial elements and satisfaction with work life and perceived effectiveness. The results underscore a strong connection not only to spatial attributes, but also to organizational identity. Workspaces that embody the values of the organization (through aligned messages, codes, and materials) and that are designed through an integrated approach—one that synthesizes cultural, sociological, and psychological dimensions into an aesthetically and materially coherent environment—have a higher potential to foster satisfaction, effectiveness, and productivity. Such spaces succeed in realizing their full potential by forming a two-way relationship with their users.

A key research direction for future studies could be examining the balance between flexibility and belonging. For instance, one could investigate how options for spatial and temporal flexibility relate to psychologically measured feelings of organizational belonging.

Finally, a significant challenge of survey-based research lies in the interdisciplinary complexity of the topic. Influences from various management models, organizational cultures and structures, team dynamics, and employees' current psychological and work status all interact with the perception of satisfaction and effectiveness at the moment of response, potentially introducing variability that cannot be easily controlled within a single survey.

Looking ahead, this research affirms the need to place users at the center of workspace design—moving beyond efficiency metrics to embrace spatial strategies that reflect organizational values and support diverse modes of work. As hybrid and flexible work models continue to reshape organizational cultures and employee expectations, spatial design must evolve accordingly—bridging physical form and lived experience. Recognizing space as a socially constructed and symbolically mediated entity is essential for designing work environments aligned with organizational structures and contemporary work practices.

Author Contributions: Conceptualization, V.K.; methodology, V.K.; validation, A.M.; formal analysis, V.K.; investigation, V.K. and A.M.; resources, V.K.; writing—original draft preparation, A.M.; writing—review and editing, V.K.; visualization, A.M. All authors have read and agreed to the published version of the manuscript.

Funding: This research received no external funding.

Institutional Review Board Statement: This study did not require ethical approval. Questionnaires were conducted with clear clarification on the study rationale, full respect of the participants' anonymity, and in accordance with the Code of Conduct for Scientific Research (21 February 2018, National Council for Scientific and Technological Development, Serbia) and the Code of Professional Ethics at the University of Belgrade (10 July 2016, Gazette of the University of Belgrade, No. 193).

Informed Consent Statement: Informed consent was obtained from all subjects involved in the study.

Data Availability Statement: The original contributions presented in the study are included in the article, and further inquiries can be directed to the corresponding author.

Acknowledgments: The authors would like to express their sincere gratitude to the management teams of the three participating companies (Fercam, UB Connect, and Medigroup) for their valuable support and collaboration throughout the research process. Our special thanks also go to all employees who generously contributed their time and insights by participating in the questionnaire. Their engagement was essential to the success of this study.

Conflicts of Interest: Author Vesna Krizmanić was employed by the Archinova Architecture Company. The remaining authors declare that the research was conducted in the absence of any commercial or financial relationships that could be construed as a potential conflict of interest.

Abbreviations

The following abbreviations are used in this manuscript:

QWL	Quality of Work Life
WCS	Workspace Characteristics Scale
WSS	Workspace Satisfaction Scale

Appendix A

The following questionnaire was administered to participants in this study. It comprises four sections: general respondent information (Appendix A.1), the Workspace Satisfaction Scale (WSS) (Appendix A.2), and the Workspace Characteristics Scale (WCS) (Appendix A.3), and general attitudes toward workspace values and needs (Appendix A.4). Participants were asked to respond using a Likert scale (1–10) unless otherwise noted. Note: The modified version of the scale (used in the third workspace) does not include the following items: K1–K4 (from Section 2), 3.1–3.4, 6.3, 7.3, 8.1, 8.2, 10.1, and 10.2 (from Section 3), and Q.1, Q.3, Q.4, and Q.9 (from Section 4).

Appendix A.1. General Respondent Information

To understand respondents' background and organizational context.

- Company name and location
- Position in company hierarchy (Top management; Middle management; Employee; Independent expert; Entry-level)
- Department or division within the company
- Gender
- Professional field
- Level of education
- Total years of work experience
- Years of experience in the current company
- Psychological profile: I consider myself an introvert, extrovert, or neither
- Current type of workspace:
    a. Individual office for one person
    b. Shared office for 2–8 employees
    c. Open-plan office with assigned desk
    d. Open-plan office with unassigned desks (hot-desking)

Appendix A.2. Workspace Satisfaction Scale (WSS)

To evaluate how satisfied participants are with their workspace and its influence on work quality. Please rate the following statements on a scale from 1 (strongly disagree) to 10 (strongly agree):

- R.1. The workspace fully supports the requirements of my job
- R.2. I feel comfortable and at ease in my workspace

- R.3. I perform well in my current workspace
- R.4. I am satisfied with the aesthetic elements of the workspace

Additional questions on general job satisfaction (beyond spatial context):

- K.1. Overall, how satisfied are you with your job (excluding spatial aspects)?
- K.2. The company's values align with my own
- K.3. I am satisfied with social relationships within the company
- K.4. I feel respected and valued in the organization

Appendix A.3. Workspace Characteristics Scale (WCS)

To examine the influence of physical and perceptual features of the workspace. Please rate the following statements on a scale from 1 (strongly disagree) to 10 (strongly agree):

- 1. Distance from colleagues
  (1—Too close/10—At a comfortable distance)
  1.1. The physical distance between me and my coworkers at my workstation is comfortable.
  1.2. Sometimes I feel that the proximity of colleagues bothers me during work.
  1.3. Conversations or noise from coworkers disrupt my work performance.
- 2. View—Visual openness
  (1—No view or visually closed/10—Visually open view)
  2.1. I have an open view toward other parts of the workspace or the outside.
  2.2. My direct line of sight is blocked by a wall or partition.
- 3. Sociological dimension of the space
  (1—I do not enjoy in-person interaction/10—I enjoy in-person interaction)
  3.1. I feel good when I have regular face-to-face contact with colleagues.
  3.2. Virtual meetings and online collaboration can fully replace physical presence.
  3.3. I would still come to the office primarily for social interaction, even if my tasks could be done remotely.
  3.4. My colleagues and I create the atmosphere of the workplace and bring life to the space.
- 4. Informal social spaces
  (1—Not needed/10—Very much needed)
  4.1. The workspace should offer more spaces for informal interaction with coworkers and clients.
  4.2. I enjoy spending time with colleagues even outside of work.
  4.3. I am comfortable holding meetings in informal areas, not just in formal meeting rooms.
- 5. Personal space and personalization
  (1—I do not need to personalize/10—Personalization is essential)
  5.1. I need personal space in the workplace to feel more comfortable.
  5.2. I enjoy personalizing my workspace with decorations or objects.
  5.3. When I can design my environment the way I like, it positively affects my performance. (scale 1–5)
- 6. Sense of belonging
  (1—I do not feel I belong/10—I strongly feel I belong)
  6.1. My workspace allows me to feel like part of the organization.
  6.2. I feel "at home" in my current workspace.
  6.3. I feel safe and secure in my workplace.
- 7. Spatial variability (within the office)
  (1—I dislike switching spaces/10—I enjoy variability)
  7.1. I appreciate being able to choose among different types of work areas depending

on the task.
7.2. I feel more productive when I have a fixed desk.
7.3. When I change my workspace, it takes time to adapt and resume work.

- 8. Flexibility of work (outside the office)
(1—No flexibility/10—Full flexibility)
8.1. I have the option to work from various locations beyond the office (e.g., home or third spaces).
8.2. I have flexible working hours.
8.3. The workspace allows me to quickly reach colleagues needed for communication.
- 9. Aesthetic preference
(1—Poor aesthetics/10—Excellent aesthetics)
9.1. My workspace gives me a sense of clarity, order, and coherence.
9.2. I find the color palette, furniture, and materials aesthetically pleasing.
9.3. My workspace feels cheerful and pleasant.
9.4. My workspace is tidy, without clutter or excess papers and materials.
- 10. Equality and control (dignity)
(1—The space undermines dignity/10—The space supports dignity)
10.1. I feel constantly monitored in my workspace.
10.2. I feel my workspace is worse than the spaces of my colleagues.
10.3. I have no control over who enters my workspace.
10.4. My workspace is arranged so that colleagues can see what I am doing on my computer.
- 11. Spaces for quiet and focused work
(1—I cannot concentrate/10—I can fully concentrate)
11.1. I can focus and work quietly in my current workspace.
11.2. I have access to another space where I can isolate myself to work quietly if needed.
- 12. Spaces for creative work and inspiration
(1—The space is not stimulating/10—Very stimulating)
12.1. I get my best ideas in this workspace.
12.2. I need spaces that support creativity and inspiration at work.
12.3. My workspace is a source of inspiration.

Appendix A.4. General Attitudinal Questions

Evaluation of value-based positions and individual needs regarding workspace design.

Please rate each of the following statements on a scale from 1 to 10 (1—strongly disagree, 10—strongly agree):

- Q.1. The workspace and office building reflect the company's image and should embody its values.
- Q.2. A high-quality workspace significantly impacts my satisfaction and work efficiency.
- Q.3. Being surrounded by positive, hardworking colleagues in my work environment helps me achieve better results.
- Q.4. Employees who are higher in the company hierarchy should have better workspaces and working conditions.
- Q.5. A chaotic workspace leads to chaotic work results.
- Q.6. The appearance of the workspace influences my decision to work for a particular company.
- Q.8. Connection to nature in the workspace—more greenery and natural materials—would help me achieve better work outcomes.

- Q.9. Workplace flexibility is important to me—I want to choose when and where I work.
- Q.10. If I could choose, I believe I would achieve the best work results and personal satisfaction in:
  A. A private office
  B. An office shared with up to 6 employees
  C. An open-plan office with my own designated desk
  D. An open-plan office with shared desks and full flexibility (no assigned desk)
  E. The type of office does not affect my work performance or satisfaction
- Q.11. Quiet, undisturbed time for focused work is essential to me in the workplace. Please indicate what percentage of your total work time you need for focused work:
  0–20% of total work time
  20–50% of total work time
  50–80% of total work time
  80–100% of total work time
- Q.12. For teamwork with colleagues in a shared physical workspace, I personally need:
  0–20% of total work time
  20–50% of total work time
  50–80% of total work time
  80–100% of total work time

References

1. Castells, M. *The Rise of the Network Society*, 2nd ed.; John Wiley & Sons: Chichester, UK, 2011.
2. Lefebvre, H. *The Production of Space*; Blackwell: Oxford, UK, 1991.
3. Parker, S.K.; Turner, N. Work Design and Individual Work Performance: Research Findings and an Agenda for Future Inquiry. In *Psychological Management of Individual Performance*; John Wiley & Sons: Chichester, UK, 2002; pp. 69–93.
4. Hackman, J.R.; Oldham, G.R. *Work Redesign*; Addison-Wesley: Reading, MA, USA, 1980.
5. Warr, P.; Nielsen, K. Wellbeing and Work Performance. In *Handbook of Well-Being*; DEF Publishers: Salt Lake City, UT, USA, 2018.
6. Windlinger, L. Complex Effects of Work and Office Design. In Proceedings of the EFMC, Zürich, Switzerland, 26–27 June 2007; pp. 91–98.
7. Veitch, J.A.; Charles, K.E.; Farley, K.M.; Newsham, G.R. A Model of Satisfaction with Open-Plan Office Conditions: COPE Field Findings. *J. Environ. Psychol.* **2007**, *27*, 177–189. [CrossRef]
8. Xie, J.; Sawyer, A.O.; Ge, S.; Li, T. Subjective Impression of an Office with Biophilic Design and Blue Lighting: A Pilot Study. *Buildings* **2023**, *13*, 42. [CrossRef]
9. Binyaseen, A.M. Office Design Features and Future Organizational Change toward Supporting Sustainability. *Buildings* **2024**, *14*, 260. [CrossRef]
10. Schaumann, D. Human Behavior Adaptability in Responsive Buildings: An Exploratory Study in Workplace Settings. *Buildings* **2024**, *14*, 1830. [CrossRef]
11. Fisher, A.; Ucci, M.; Smith, L.; Sawyer, A.; Spinney, R.; Konstantatou, M.; Marmot, A. Associations between the Objectively Measured Office Environment and Workplace Step Count and Sitting Time: Cross-Sectional Analyses from the Active Buildings Study. *Int. J. Environ. Res. Public Health* **2018**, *15*, 1135. [CrossRef] [PubMed]
12. Zhuang, D.; Zhao, X.; Gan, V.J.; Yang, Y.; Shi, X. Comparative Investigation of Office Layout Influences on Occupant Satisfaction from Priori and Posteriori Perspectives. *Buildings* **2022**, *12*, 1327. [CrossRef]
13. Sundstrom, E. *Work Places. The Psychology of the Physical Environment in Offices and Factories*; Cambridge University Press: Cambridge, UK, 1986.
14. Kupritz, V.W. Privacy in the Workplace: The Impact of Building Design. *J. Environ. Psychol.* **1998**, *18*, 341–356. [CrossRef]
15. Knight, C.; Haslam, A.H. Your Place or Mine? Organizational Identification and Comfort as Mediators of Relationships between the Managerial Control of Workspace and Employees' Satisfaction and Well-being. *Br. J. Manag.* **2010**, *21*, 717–735. [CrossRef]
16. Nikolaeva, R.; Russo, S.D. Office Design and Dignity at Work in the Knowledge Economy. In *Dignity and the Organization*; Palgrave Macmillan: London, UK, 2016; pp. 197–220.
17. Hodson, R. *Dignity at Work*; Cambridge University Press: Cambridge, UK, 2001.

18. Varghese, S.; Jayan, C. Quality of Work Life: A Dynamic Multidimensional Construct at Work Place–Part II. *Guru J. Behav. Soc. Sci.* **2013**, *1*, 91–104.
19. Hackman, J.R.; Oldham, G.R. Motivation through the Design of Work: Test of a Theory. *Organ. Behav. Hum. Perform.* **1976**, *16*, 250–279. [CrossRef]
20. Abdullah, N.A.C.; Zakaria, N.; Zahoor, N. Developments in Quality of Work-Life Research and Directions for Future Research. *SAGE Open* **2021**, *11*. [CrossRef]
21. Efraty, D.; Sirgy, M.J. The Effects of Quality of Working Life (QWL) on Employee Behavioral Responses. *Soc. Indic. Res.* **1990**, *22*, 31–47. [CrossRef]
22. Baba, V.V.; Jamal, M. Routinization of Job Context and Job Content as Related to Employees' Quality of Working Life: A Study of Canadian Nurses. *J. Organ. Behav.* **1991**, *12*, 379–386. [CrossRef]
23. Hart, P.M. Predicting Employee Life Satisfaction: A Coherent Model of Personality, Work, and Nonwork Experiences, and Domain Satisfactions. *J. Appl. Psychol.* **1999**, *84*, 564. [CrossRef]
24. Igbaria, M.; Parasuraman, S.; Badawy, M.K. Work Experiences, Job Involvement, and Quality of Work Life among Information Systems Personnel. *Manag. Inf. Syst. Q.* **1994**, *18*, 175–201. [CrossRef]
25. World Health Organization. *WHOQOL: Measuring Quality of Life*; WHO: Geneva, Switzerland, 1994.
26. Almarshad, S.O. A Measurement Scale for Evaluating Quality of Work Life: Conceptualization and Empirical Validation. *Trends Appl. Sci. Res.* **2015**, *10*, 143. [CrossRef]
27. Othman, A.; Mok, C.L. Relationship between Quality of Work Life (QWL) and Job Satisfaction: A Case Study of Enterprise "XYZ" in Malacca. In Proceedings of the International Conference on Human Capital Development (ICONHCD 2009), Kuantan, Pahang, Malaysia, 25–27 May 2009.
28. Haristianti, V.; Raja, T.M. Implementation of a Biophilic Design Approach in Office Space Design as an Attempt to Increase Employee Well-Being. *KnE Soc. Sci.* **2025**, *10*, 32–39. [CrossRef]
29. Shirvanisaadatabadi, E.; Gharaei, A. The Relationship between Architects' Quality of Work Life and Their Productivity: A Case Study. *J. Sustain. Dev.* **2025**, *8*, 60–70. [CrossRef]
30. Vischer, J.C.; Wifi, M. The Effect of Workplace Design on Quality of Life at Work. In *Handbook of Environmental Psychology and Quality of Life Research*; Springer International Publishing: Cham, Switzerland, 2016; pp. 387–400.
31. Dávalos Quevedo, M.V.; Luna-Navarro, A.; Pottgiesser, U.; Blum, U. Methods and Approaches for Evaluating Occupant Satisfaction with Office Space Design: A Systematic Review. *Ergonomics* **2025**, 1–21. [CrossRef]
32. Baharetha, S.; Hassanain, M.A.; Alshibani, A.; Ouis, D.; Gomaa, M.M.; Ezz, M.Z. A Post-Occupancy Evaluation Framework for Enhancing Resident Satisfaction and Building Performance in Multi-Story Residential Developments in Saudi Arabia. *Architecture* **2025**, *5*, 8. [CrossRef]
33. Borsos, Á.; Zoltán, E.S.; Pozsgai, É.; Cakó, B.; Medvegy, G.; Girán, J. The Comfort Map—A Possible Tool for Increasing Personal Comfort in Office Workplaces. *Buildings* **2021**, *11*, 233. [CrossRef]
34. Song, G.; Zhu, S. Interior Design Assessment System Based on Computer Vision and Multimedia. *Comput.-Aided Des. Appl.* **2024**, *21*, 264–278. [CrossRef]
35. Ezz, M.S.; Mahdy, M.A.F.; Baharetha, S.; Hassanain, M.A.; Gomaa, M.M. Post Occupancy Evaluation of Architectural Design Studio Facilities. *Front. Built Environ.* **2025**, *21*, 1549313. [CrossRef]
36. Bodin Danielsson, C.; Theorell, T. Office Design's Impact on Psychosocial Work Environment and Emotional Health. *Int. J. Environ. Res. Public Health* **2024**, *21*, 438. [CrossRef] [PubMed]
37. Jacobi, S.L.; Bjørner, T. Architects' Perception of Quality of Life—Impact, Practice, and Barriers. *Architecture* **2024**, *4*, 267–280. [CrossRef]
38. Batuchina, A.; Medžiūnienė, I.; Lecaj, R. Multidimensional Scale of Meaningful Work: Construction and Validation. *Front. Psychol.* **2025**, *16*, 1578825. [CrossRef]
39. Soundarya Priya, M.G.; Anandh, K.S. Exploring the Determinants of Quality of Work Life in the Construction Industry: A Quantitative Approach. *Buildings* **2024**, *14*, 1607. [CrossRef]
40. Tsakonas, K.G. Measuring the Value of Workspace Architectural Design: Construction of the Workspace Communication Suitability Index (WOCSIT). Ph.D. Thesis, Massachusetts Institute of Technology, Cambridge, MA, USA, 2008.
41. Wijesooriya, N.; Brambilla, A.; Markauskaite, L. Biophilic Quality Matrix: A Tool to Evaluate the Biophilic Quality of a Building during Early Design Stage. *Clean. Prod. Lett.* **2025**, *8*, 100094. [CrossRef]
42. Vande Moere, A.; Arko, S.; Drasilova, A.S.; Ondráček, T.; Pigliautile, I.; Pioppi, B.; Pisello, A.L.; Prochazka, A.J.; Roncancio, P.A.; Schaumann, D.; et al. The Adaptive Workplace: Orchestrating Architectural Services around the Wellbeing of Individual Occupants. *arXiv* **2024**, arXiv:2403.16595.
43. dos Santos Ribeiro, E.; da Silva, E.K.V.; de Albuquerque Jatoba, L.; Andrade, W.N.; Miranda, L.N. Quality of Life at Work of Nurses in Public Hospital Institutions. *Enfermería Glob.* **2021**, *20*, 489–501.

44. Walke, F.; Aden, A.; Winkler, T.J. Work System Quality Assessment: A TIHPS Grounded Measurement Model. In Proceedings of the 57th Hawaii International Conference on System Sciences, Honolulu, HI, USA, 3–6 January 2024.
45. Voigt, J.; Hohn, J.; Mut, E.; Schreiber, C.; Geisler, S.M.; Pinta, P.S.; Hamm, A.; Stutzer, F.; Kosakyan, H.; Burkhardt, J.-M.; et al. Conceptional Framework for the Objective Work-Related Quality of Life Measurement through Multimodal Data Integration from Wearables and Digital Interaction. In Proceedings of the 19th Conference on Computer Science and Intelligence Systems (FedCSIS), Belgrade, Serbia, 8–11 September 2024; pp. 61–68.
46. Berber, N.; Slavić, A.; Gašić, D. Employees' Work–Life Balance in the Contemporary Business Environment: Evidence from Serbia. In Proceedings of the XIX International May Conference on Strategic Management—IMCSM24 Proceedings, Bor, Serbia, 31 May 2024; pp. 24–37.
47. Štěpánek, M. Regional Differencies in Quality of Working Life in the Czech Republic. *Časopis výzkumu a aplikací v profesionální bezpečnosti* **2022**, *15*. Available online: https://www.josra.cz/vydani/clanek/regional-differencies-in-quality-of-working-life-in-the-czech-republic (accessed on 1 May 2025).

sustainability

MDPI

Article

An Exploration of Biophilic Design Features within Preschool Interiors

Ibtisam Mohammed [1,*], Zeynep Onur [2] and Çiğdem Çağnan [3]

1 Department of Architecture, Faculty of Engineering, University of Duhok, Duhok 42001, Iraq
2 Department of Architecture, Faculty of Architecture and Fine Arts, International Final University, Girne 99320, Cyprus; zeynep.onur@final.edu.tr
3 Department of Architecture, Faculty of Architecture, Near East University, Nicosia 99138, Cyprus; cigdem.cagnan@neu.edu.tr
* Correspondence: ibtsam_11@yahoo.com

Abstract: This study examined the incorporation of biophilic elements in Duhok preschool interiors and integrated them into the Biophilic Interior Design Matrix (BID-M). This approach holds great promise for the development of restorative environments. The Biophilic Design Matrix (BDM) is based on Kellert's list of biophilic design attributes, adjusted to suit preschool interiors. Photos and plan drawings were employed as methods to analyze 59 interior spaces within six preschools, both private and public. The biophilic matrix revealed the presence of 30% of biophilic attributes in the analyzed spaces, with an average score of 16.45 out of 53 total points. Individual scores ranged from 8 to 22 for each space, highlighting variations in biophilic design within the six preschools. Notably, the study identified a lack of biophilic features that foster human–nature relationships in the majority of preschools, albeit minimally observed in some spaces. Conversely, the environmental features scored higher than the average. This research highlights the importance of enhancing biophilic design elements, especially those that strengthen human–nature relationships. The results enhance biophilic design in educational settings, promoting nature-centered, restorative environments for future interventions. Furthermore, we propose an innovative approach for assessing and incorporating biophilia in interiors, recognizing its positive impact on human health and wellbeing.

Keywords: biophilia; preschool interiors; biophilic interior design matrix; biophilic design

1. Introduction

Modern living has made nature more and more difficult for individuals to access, but the yearning to connect with nature still exists. Children are particularly affected by this, as they have to stay indoors while separated from their parents, which is especially true for them. This lack of connection with nature is concerning, as studies have shown that there is an inherent need for humans to be connected with nature, known as biophilia. Biophilia is defined as an inherent need to be connected with nature [1]. Biophilia is a natural need to connect with nature, which has become essential for children's cognitive functioning and wellbeing [1–3]. Biophilia is not just a desire but a natural need that has been found to be essential for children's cognitive functioning and wellbeing. Including nature in urban environments created for children is one possible solution to address this issue. Biophilic design strategies require consideration of the people using the place, its location, and its function. Plants are one way to incorporate nature into space, but there are other ways to incorporate nature [4,5]. To aid interior designers, Kellert [6] proposed six biophilic elements that could be implemented in a given space. To detect and evaluate biophilic elements, a reliable technique is required. The six elements include environmental features, natural shapes and forms, natural patterns and processes, light and space, place-based relationships, and human–nature relationships [7].

A few studies have been conducted to create a trustworthy coding method for determining biophilic contents in various interior environments. Based on Kellert [6], the matrix has 6 biophilic design elements and 52 biophilic features. Twenty of the original seventy-two features were left out of the matrix. This was due to their inability to be visually analyzed and their lack of connection to the interior space [7]. The BID-M was created by McGee and Marshall-Baker [8] in order to evaluate the impact of biophilic features on playrooms in child healthcare facilities on wellbeing [7]. Because evidence-based design has been more popular as an interdisciplinary method of gathering and sharing knowledge since the beginning of the 1980s, designers have had more and more reasons to incorporate biophilic design into interiors [9]. As more research is conducted on the individual attributes and wider applications of the matrix scoring are made to show greater trends and research opportunities, this initial quantification of biophilia will continue to expand. Additionally, there is a need to find more evidence-based arguments in favor of integrating biophilia into interior spaces like play areas. Future research is necessary to establish the best applications for each feature. The development and testing of numerous attribute-friendly features of design could potentially provide insight into the benefits of biophilia and how to best use it in interiors.

In light of the above, this study aims to address the following research questions:

1. What is the average presence of biophilic attributes in the analyzed spaces, and how does this align with the overall score and percentage of biophilic attributes' presence?
2. How can the Biophilic Interior Design Matrix (BID-M) be used to identify biophilic design features in preschool interiors?
3. What is the extent of biophilic element incorporation within the interiors of Duhok preschools, based on the six biophilic design elements proposed by Kellert?

These research questions will guide the exploration of biophilic design features within the interiors of Duhok preschools, utilizing the Biophilic Interior Design Matrix (BID-M) and Kellert's proposed biophilic design elements as foundational frameworks. By addressing these research questions, this study aims to contribute to the understanding and application of biophilic design principles in preschool interiors, paving the way for creating healthier and more nurturing environments for young children.

The study used a photographic methodology to document biophilic attributes for preschool interiors in Duhok, using the Biophilic Interior Design Matrix to address the study's objective. The BID-M assisted in quantifying the numerous biophilic design attributes present in preschool settings. The goal of the current research was to explore the existence of biophilic features in the interiors of Duhok preschools and include them in the BID-M based on the six elements proposed by Kellert [6].

Biophilic design, despite its potential benefits, remains an area of interior design expertise that has received relatively little attention. By addressing the problem of integrating biophilic design features in interior design, this research contributes to bridging this gap and sheds light on the potential of utilizing the BID-M instrument for the identification and implementation of biophilic design elements [9].

By examining and incorporating biophilic features into the BID-M, this study aims to provide valuable insights into the biophilic attributes present in preschool interiors, highlighting their significance in promoting a healthier and more nurturing environment for young children. The findings of this research will contribute to the understanding and application of biophilic design principles in the context of preschool interiors, ultimately aiming to enhance the wellbeing and cognitive development of children.

Moving forward, this research serves as a foundation for further exploration and integration of biophilic design elements in interior design practice. By recognizing the potential of the BID-M as a tool for biophilic design assessment, designers and practitioners can more effectively incorporate nature-inspired elements into interior spaces, creating restorative and enriching environments for users.

2. Literature Review

2.1. The Concept of Biophilic Design and Its Benefits

Biophilic design is an architectural and interior design concept that aims to bring humans closer to nature by integrating natural components into the built environment. The notion behind the concept is that humans have an inbuilt connection to nature and that exposure to natural elements may improve health and wellbeing [10]. This is to say that there is a connection between nature and human wellbeing.

Biophilic design has both health and financial benefits, and neglecting nature can lead to a loss of profit [11,12]. According to an argument made by Browning et al. [11] (p. 3), "incorporating nature into the built environment is not just a luxury but a sound economic investment in health and productivity, based on well-researched neurological and physiological evidence". Biophilic design can lead to increased healing, learning, property values, and productivity while decreasing absenteeism, anxiety, and incarceration expenses. It is important to value and safeguard the environment [6]. Restorative environmental design (RED) is a strategy that aims to minimize and mitigate adverse impacts on the natural environment and foster beneficial contact between people and nature in contemporary buildings and landscapes. Kellert [6] defined it as a low-environmental-impact approach. Interior environment designers play a significant role in the development of interior/exterior links and features that can improve user exposure to biophilia, which can have positive effects on health [13,14] and wellbeing [15,16]. Restorative environmental design connects sustainability and biophilia, generating globally conscious individuals who value nature and desire to return it to its ideal state. The addition of biophilic design and imitating natural settings is clearly the next step in the sustainable design movement [17]. However, there is insufficient support for best practices for developing the interior environments that people naturally choose [18]. Biophilic design is an approach to creating healthy, sustainable, and adaptable environments. It also has benefits such as improved air quality, reduced stress levels, and increased productivity, as highlighted by Cacique, Maria, and Sheng-Jung Ou [19]. The term "biophilia" refers to the emotional bonds that people develop with other living species. The organic or naturalistic component, which combines direct and symbolic aspects of nature, and the place-based or vernacular dimension, which stresses the integration of architecture and landscapes with the environment and culture of an area, are the two dimensions of biophilic design. This architectural method generates a sense of place and identity, which can increase people's loyalty and responsibility for their built environment and heritage.

2.2. Sustainability and Biophilia

The concept of sustainability centers on the idea of addressing present demands without risking future generations' capacity to meet their own needs. The close relationship between humans and the natural environment is an important part of sustainability. Biophilia is a technique that provides a sustainable design strategy that includes reconnecting humans with nature [20]. Sustainable design balances human needs with natural and cultural environments, whereas biophilic design examines how the environment affects our physiology and psychology. Sustainable architecture is created in harmony with the environment and natural resources. Biophilia and sustainability can be achieved by preserving natural and cultural environments and using resources in site and building design to encourage exploration. This knowledge can be used to modify space to enhance human experiences. Biophilic design addresses sustainability through tactile, emotional, and experiential approaches, creating buildings that are resource-saving, energy-efficient, and improve the socialization, productivity, health, and wellbeing of their occupants [21].

2.3. Biophilic Interior Design and Tools

Biophilic design is a novel approach to interior design that tries to incorporate natural aspects into the built environment. Designers may create environments that are not only visually beautiful but also give multiple advantages to the occupants by including biophilic features such as plants, natural sunlight, and organic materials. These biophilic characteristics not only improve the aesthetic attractiveness of the interior space but also improve the physical and emotional wellbeing of those who live in these places. Plants are one biophilic component that may be introduced into interior spaces. Indoor plants have been proven to provide several advantages to people. Plants have been shown in studies to enhance air quality by lowering levels of carbon dioxide and other pollutants.

The fundamental dimensions of biophilic design can be categorized into six key elements: environmental features, natural shapes and forms, natural patterns and processes, light and space, place-based relationships, and evolved human–nature relationships. These six elements encompass over 70 specific design attributes that contribute to creating a stronger connection between humans and nature. A Yale professor has extensively categorized biophilia into these elements and identified 72 specific features within them. For instance, within the environmental features element, air is recognized as one of its attributes. This comprehensive list of biophilic design attributes in landscape and architecture was developed based on the extensive exposure of the professor to a diverse range of authors and researchers in the field [6,18]. The Terrapin Bright Green List of 14 Patterns of Biophilic Design is an alternative tool that emphasizes recognizable patterns, nature–health linkages, and nature–design relationships [22]. Kellert's original list of 72 attributes is limited by a shorter list of 14 attributes. Leadership in Energy and Environmental Design (LEED) has raised awareness of sustainable design but has not provided holistic design strategies for biophilic design. Innovators in biophilic design, such as WELL Building and the Living Building Challenge, reference Kellert's original list of design attributes. Kellert's vocabulary for interior designers was extended further, allowing them to assist in WELL Building and the Living Building Challenge [18].

2.4. Identifying Biophilic Features in Interior Spaces

Color, light, and materiality are all connected design elements that occur together in an interior environment. Nevertheless, it is unknown how these three features are being applied in relation to biophilic features and how existing research supports biophilic features [23]. McGee and Marshall-Baker created the BDM to identify biophilic features in 24 children's healthcare interiors. A study in South Atlantic state Child Life play areas found consistent biophilic attributes in some settings while lacking in others [8]. The BDM serves as a valuable instrument for evaluating the presence and effectiveness of biophilic design in healthcare environments for children.

Marte et al. tested the BID-M in urban playrooms by coding photographs from 45 children's playrooms in Manhattan residential buildings. They evaluated overall and component inter-rater reliability, finding high overall reliability but low reliability in some components [7]. Designers' perceptions of biophilic design were explored by developing an updated Biophilic Interior Design Matrix (BDM) as a design tool. The results showed that practitioners had a better understanding of biophilic design, enabling designers to incorporate nature-based features in indoor settings [18].

McGee and Park examined biophilic interior design utilizing the BID-M vocabulary and its essential elements. They found that practitioners incorporated a range of biophilic attributes into their work, with color preferences being the top attribute. Abstraction of nature, composition, natural light, and natural materials were the top attributes that practitioners and the literature review had in common. Additional research is needed to fully understand how biophilic design can be variedly included for the best natural interior settings [23]. McGee et al. evaluated how Chinese design professionals perceived biophilia and the BID-M, as well as their reliability and validity. A pre- and post-questionnaire was used with 101 interior designers and architects, and the Chinese designer's translation of the BID-M was seen as reliable, valid, and helpful in the design process [24].

2.5. The Importance of Biophilic Design in Preschool Interiors

The use of natural features such as plants, natural materials, and lighting is an essential part of biophilic design in preschool settings. Awad et al. investigated the significance of incorporating biophilic design principles into elementary school design regulations, such as lighting, daylight, materials, and ventilation, to enhance the wellbeing and health of children [25]. Meanwhile, Ondul et al. aimed to design a "Refuge Pod" piece of furniture using a biophilic design approach in a preschool setting while taking the demands of the space into consideration for both furniture and space. The research revealed that children are imaginative and like to play in tiny areas [26], underscoring the significance of establishing personal and engaging places.

In a study conducted by Yanez et al., young children from urban and rural areas were compared in terms of their biophilia and attitudes towards nature. The results revealed no significant variation based on geographical region. The study identified common themes, including the definition of nature, awareness of environmental effects, and adherence to natural rules. The researchers suggested that cognitive development and personal preferences might be more accurate indicators in this context [27]. In another study by Ünal and Özen, the benefits of exposing preschool children to nature through biophilic design in built environments were investigated. The researchers used content analysis, cognitive mapping, and semi-structured interviews to assess the overall score. They found that a statistically significant impact was observed when the p-value was less than 0.050 [28]. Showing a positive impact through contact with nature, Yassein and Ebrahiem discussed contemporary biophilic interior design techniques used in residential and workplace settings based on a discussion with architecture students. This situated the notion of biophilic interior design and resulted in a conceptual model that strengthens and supports the human–nature relationship [29]. This further reinforces the connection between humans and nature.

2.6. The Impact of Biophilic Design on Children's Learning and Development

Incorporating biophilic design into educational facilities has various advantages, including higher test scores, greater health, and increased learning. The benefits extend to recreational spaces such as playgrounds, where biophilic design concepts can enhance children's behavior, attention span, and mental health. Biophilic design patterns have been demonstrated to improve stress reduction, cognitive abilities, sensitivity, atmosphere, and preference. According to the research findings, biophilic design boosts liking for the environment, decreases stress, increases happiness, and encourages focus [30]. McGee measured biophilic design attributes in Child Life settings to improve health and wellbeing. The study utilized a biophilic interior design matrix tool and an open-ended questionnaire to create 24 case studies. The results indicated that there is a correlation between biophilic environments and effective Child Life play areas, with higher scores representing greater effectiveness. Another study involved 90 Child Life specialists who used a photographic technique to identify elements that contribute to the design of ideal playrooms in different hospitals. The findings highlighted the crucial role of Child Life specialists in designing or modifying these settings [31,32]. Schools and educational institutions play a significant role in fostering a connection with nature. By providing a natural environment, these institutions can offer benefits to young people, such as promoting the use of nature, enhancing attention, cognitive performance, and mood. Moreover, a natural environment in schools improves concentration and self-discipline and reduces physiological stress [33]. The concept of biophilia might be used in the classroom to contribute to the creation of ecologically sound urban environments [20]. Fadda et al.'s research specifically demonstrated the beneficial effects of biophilic design on preschoolers' visual attention in indoor environments. Notably, even a brief exposure to plants on a desk for 48 seconds increased the duration of children's initial visual fixation on art displays [34].

There is a lack of research on the topic of integrating biophilic design elements into preschool interiors. By doing this, we can enhance children's cognitive abilities, lessen their stress levels, and foster their creativity. Additionally, it can encourage them to preserve nature by fostering a closer connection with it. But the Biophilic Design Matrix (BDM) offers a chance to advance design research while enhancing the health, security, and happiness of users. In an effort to address the difficulty of quantitatively evaluating biophilia, the BDM was created in 2011, with a focus on children's and adolescents' play areas. The BDM was successful in providing a visual inventory of biophilic features and the frequency of their use in hospital spaces intended for patient recreation and play. The current state of biophilic design research and practice in this context has also been explored in further research.

3. Materials and Methods

Biophilic design is a recognized methodology used to evaluate and incorporate nature-inspired elements in architectural settings, acknowledging the inherent need of children to be in environments with biophilic design patterns for their wellbeing [35]. In line with this, the objective of this study was to develop the biophilic interior design matrix (BID-M), specifically tailored for assessing biophilic contents in preschool interiors, utilizing Kellert's well-established design attributes [6]. These attributes, proposed by Kellert, encompass six biophilic design elements, including environmental features, natural shapes and forms, natural patterns and processes, light and space, place-based relationships, and human–nature relationships.

The BDM, initially developed by McGee and Marshall-Baker for identifying biophilic design elements in pediatric hospitals' play environments, served as the foundation for this study [8]. Notably, this is the first application of the BDM as a tool for interior design. Out of Kellert's 72 biophilic design elements, including those related to landscape and architecture, 53 were deemed appropriate for interior design purposes, while the remaining 19 were considered unsuitable [18]. This is because the 19 excluded are considered to be more relevant to landscape and architecture.

Accordingly, the present study aimed to expand the potential applications of the biophilic interior design matrix in different types of settings within Duhok, with a particular focus on preschool interiors. The BID-M technique employs a quantitative approach to evaluate biophilic features, providing a rating for each space and describing the presence of attributes both verbally and photographically. A total of 59 interior spaces from six preschools, consisting of four public and two private institutions, were selected for analysis (see Tables 1 and 2). Prior to data collection, the research team obtained the necessary approval letter from the Directorate of Education to ensure access to the preschools and the collection of relevant information for the study.

Table 1. The biophilic design elements and attributes.

Environmental Features		Light and Space	
1	Color	32	Natural light
2	Water	33	Filtered and diffused light
3	Air	34	Light and shadow
4	Sunlight	35	Reflected light
5	Plants	36	Light pools
6	Animals	37	Warm light
7	Natural material	38	Light as shape and form
8	Views and vistas	39	Spaciousness
9	Fire	40	Spatial variability
Natural Shapes and Forms		41	Space as shape and form
10	Botanical motifs	42	Spatial harmony
11	Tree and columnar supports	43	Inside/outside spaces
12	Animal	Place-Based Relationships	

Table 1. *Cont.*

13	Shells and spirals (invertebrates)	44	Geographic connection to place
14	Egg, oval, and tubular forms	45	Historic connection to place
15	Arches, vaults, domes	46	Ecological connection to place
16	Shapes resisting straight lines	47	Cultural connection to place
17	Simulation of natural features/biomorphy	48	Indigenous materials
18	Geomorphology	49	Landscape orientation/landscape features
19	Biomimicry		Human–Nature Relationships
	Natural Patterns and Processes	50	Prospect and refuge
20	Sensory variability/information richness	51	Order and complexity
21	Age, change, the patina of time	52	Curiosity and enticement
22	Central focal point	53	Change and metamorphosis
23	Patterned wholes		
24	Bounded spaces		
25	Transitional spaces		
26	Linked series and chains		
27	Integrations of parts to wholes		
28	Complementary contrasts		
29	Dynamic balance and tension		
30	Fractals		
31	Hierarchically organized ratios and scales		

Table 2. The biophilic interior design matrix scores of Chiman preschool settings.

	Chiman Preschool	**Multipurpose Hall**	**Cafeteria**	**Classroom 1**	**Classroom 2**	**Classroom 3**	**Classroom 4**	**Classroom 5**	**Classroom 6**	**Corridors**	**Total Subscores**
	(1) Environmental features										
1	Color	✓	✓	✓	✓	✓	✓	✓	✓	✓	9
2	Water	×	✓	×	×	×	×	×	×	×	1
3	Air	✓	✓	✓	✓	✓	✓	✓	✓	✓	9
4	Sunlight	✓	✓	✓	✓	✓	✓	✓	✓	✓	9
5	Plants	×	×	×	×	×	×	×	×	×	0
6	Animals	×	×	×	×	×	×	×	×	×	0
7	Natural material	×	×	×	×	×	×	×	×	×	0
8	Views and vistas	✓	✓	✓	✓	✓	✓	✓	✓	✓	9
9	Fire	×	×	×	×	×	×	×	×	×	0
	Subscores	4	5	4	4	4	4	4	4	4	37
	(2) Natural shapes and forms										
10	Botanical motifs	×	✓	✓	✓	✓	✓	✓	✓	✓	8
11	Tree and columnar supports	×	✓	×	×	×	×	×	×	×	1
12	Animal	✓	✓	✓	✓	×	✓	×	✓	×	6
13	Shells and spirals (invertebrates)	×	×	×	×	✓	×	✓	×	×	2

Table 2. *Cont.*

	Chiman Preschool	Multipurpose Hall	Cafeteria	Classroom 1	Classroom 2	Classroom 3	Classroom 4	Classroom 5	Classroom 6	Corridors	Total Subscores
14	Egg, oval, and tubular forms	×	✓	×	×	×	×	×	×	×	1
15	Arches, vaults, domes	×	×	×	×	×	×	×	×	×	0
16	Shapes resisting straight lines	×	×	×	×	×	×	×	×	×	0
17	Simulation of natural features/biomorphy	×	×	×	×	×	×	×	×	×	0
18	Geomorphology	×	×	×	×	×	×	×	×	×	0
19	Biomimicry	×	×	×	×	×	×	×	×	×	0
	Subscores	1	4	2	2	2	2	2	2	1	18
	(3) Natural patterns and processes										
20	Sensory variability/information richness	✓	✓	✓	✓	✓	✓	✓	✓	✓	9
21	Age, change, the patina of time	×	×	×	×	×	×	×	×	×	0
22	Central focal point	×	✓	×	×	×	×	×	×	×	1
23	Patterned wholes	×	×	×	×	×	×	×	×	×	0
24	Bounded spaces	✓	×	✓	✓	✓	✓	✓	✓	×	7
25	Transitional spaces	×	×	×	×	×	×	×	×	✓	1
26	Linked series and chains	×	✓	×	×	×	×	×	×	✓	2
27	Integrations of parts to wholes	✓	✓	✓	✓	✓	✓	✓	✓	✓	9
28	Complementary contrasts	×	×	×	×	×	×	×	×	×	0
29	Dynamic balance and tension	×	×	×	×	×	×	×	×	×	0
30	Fractals	×	×	×	×	×	×	×	×	×	0
31	Hierarchically organized ratios and scales	×	×	×	×	×	×	×	×	×	0
	Subscores	3	4	3	3	3	3	3	3	4	29
	(4) Light and space										
32	Natural light	✓	✓	✓	✓	✓	✓	✓	✓	✓	9
33	Filtered and diffused light	✓	✓	✓	✓	✓	✓	✓	✓	✓	9
34	Light and shadow	✓	✓	✓	✓	✓	✓	✓	✓	✓	9
35	Reflected light	×	×	×	×	×	×	×	×	×	0
36	Light pools	×	×	×	×	×	×	×	×	×	0
37	Warm light	×	×	×	×	×	×	×	×	×	0
38	Light as shape and form	×	✓	×	×	×	×	×	×	×	1
39	Spaciousness	✓	✓	✓	✓	✓	✓	✓	✓	×	8
40	Spatial variability	×	✓	×	×	×	×	×	×	×	1
41	Space as shape and form	×	×	×	×	×	×	×	×	×	0
42	Spatial harmony	×	×	×	×	×	×	×	×	×	0
43	Inside/outside spaces	×	✓	×	×	×	×	×	×	✓	2
	Subscores	4	7	4	4	4	4	4	4	4	39

Table 2. *Cont.*

	Chiman Preschool	Multipurpose Hall	Cafeteria	Classroom 1	Classroom 2	Classroom 3	Classroom 4	Classroom 5	Classroom 6	Corridors	Total Subscores
		(5) Place-based relationships									
44	Geographic connection to place	✓	✓	✓	✓	✓	✓	✓	✓	✓	9
45	Historic connection to place	×	×	×	×	×	×	×	×	×	0
46	Ecological connection to place	✓	×	×	×	×	×	×	×	×	1
47	Cultural connection to place	✓	✓	✓	✓	✓	✓	✓	✓	✓	9
48	Indigenous materials	×	×	×	×	×	×	×	×	×	0
49	Landscape orientation/landscape features	✓	×	✓	✓	✓	✓	✓	✓	✓	8
	Subscores	4	2	3	3	3	3	3	3	3	27
		(6) Human–nature relationships									
50	Prospect and refuge	×	×	×	×	×	×	×	×	×	0
51	Order and complexity	×	×	×	×	×	×	×	×	×	0
52	Curiosity and enticement	×	×	×	×	×	×	×	×	×	0
53	Change and metamorphosis	×	×	×	×	×	×	×	×	×	0
	Subscores	0	0	0	0	0	0	0	0	0	0
	Total subscores	16	22	16	16	16	16	16	16	16	150
	Average										16.66

In summary, the materials and methods of this study involved the development and application of the Biophilic Interior Design Matrix (BID-M), specifically tailored for preschool interiors in Duhok. The BID-M incorporates Kellert's design attributes and was used quantitatively to rate and describe the biophilic contents of each space within the selected preschools. Ethical considerations were taken into account, and approval was obtained from the Directorate of Education to conduct the research in the chosen preschools.

Methods

Purposive sampling was used to pick cases (preschools) in this research study. The researchers purposefully chose 6 preschools from a total of 21 accessible in Duhok that would best fit the research objectives. Purposeful sampling is a non-probability sampling strategy in which cases or participants are intentionally chosen based on certain qualities or attributes important to the research [36,37]. To guarantee variety in the sample, the researchers carefully picked preschools that represented several types of settings, including both public and private institutions. The researchers were able to guarantee that the selected instances were typical of the target population and could provide significant data for analysis by carefully selecting preschools that might provide important insights and information connected to the study questions. The researchers were able to focus on certain elements that were critical for the study, such as a variety of interior locations within the preschools, according to deliberate sampling. This systematic approach to case selection enabled a comprehensive assessment of biophilic design features in preschool environments, since the chosen cases provided the essential variance and depth of understanding.

The study was able to purposefully select preschools that reflected the required diversity and qualities essential to successfully examine and comprehend biophilic design aspects in preschool interiors by using purposeful sampling in this research. The researchers analyzed each location within the preschools using observation and photography, including the ceiling, walls, floor, and furniture. The objective was to see whether there were any special biophilic features in these places. This evaluation was led by Kellert's framework's six biophilic design principles and their related qualities. The researchers were able to analyze and score each space based on the identified biophilic features through meticulous observation and photographic recording. The visual attributes and features of the rooms were examined to see whether they displayed biophilic design principles. The results of this evaluation were utilized to build the Biophilic Interior Design Matrix, which provided insights into the presence and distribution of biophilic features throughout the preschools' various interior areas.

The use of observation and photography enabled a thorough analysis of the physical elements of the interior spaces. This method offered a visual record and documentation of the presence or absence of biophilic features within each space (see Figure 1), allowing for a more objective evaluation of the biophilic design elements present in the preschool environments.

(**a**) (**b**)

Figure 1. Class room of Shang preschool: (**a**) plan drawing from the directorate of the preschool; (**b**) photo by the author.

The biophilic design attributes of each preschool's spaces were determined separately by creating a matrix that included the number of spaces that would be allocated along the columns on the one hand and the set of biophilic features along the rows on the other. By indicating with a (✓) sign if present and a (×) sign if not present, these signs were converted to numbers (✓ = 1 and × = 0), in order to evaluate all of the listed attributes in a specific space. The maximum score for various biophilic features in the matrix was 53. In order to determine the overall biophilic inclusion of any space within the preschool, the biophilic attributes of each of the six biophilic elements were combined to create a subscore, as shown in Tables 1 and 2.

4. Results and Discussion

The quantitative analysis of the matrix was conducted to assess and compare the biophilic attributes among the preschools in Duhok. Scoring in the matrix was based on the presence or absence of specific attributes, enabling an evaluation of biophilia. The average overall score of the 59 analyzed interiors was determined to be 16.45. This score was then divided by the total number of features (53), resulting in a calculated percentage of 30% for the extent of biophilic presence, as presented in Table 3. Each attribute in the matrix contributed one score towards the assessment. The recorded scores ranged from 8 to 22, indicating the range of variation in the level of biophilic incorporation within each space.

These findings are visually represented and organized in Figure 2, offering a graphical depiction of the distribution of biophilic attributes across the analyzed preschool interiors.

Table 3. The biophilic interior design matrices of the whole preschools.

	Biophilic Interior Design Matrix	**Shang Preschool**	**Daveen Preschool**	**Avrocity Preschool**	**Zaryland Preschool**	**Kapir Preschool**	**Cihman Preschool**	**Total Scores**	**Average**
	(1) Environmental features								
1	Color	9	9	13	10	9	9	59	
2	Water	2	1	12	1	1	1	18	
3	Air	9	9	13	8	9	9	57	
4	Sunlight	9	9	13	8	9	9	57	
5	Plants	1	0	0	0	1	0	2	
6	Animals	0	0	0	0	0	0	0	
7	Natural material	0	0	0	0	2	0	2	
8	Views and vistas	9	9	12	8	9	9	56	
9	Fire	0	0	0	0	0	0	0	
	Subscores	39	37	63	35	40	37	251	4.25
	(2) Natural shapes and forms								
10	Botanical motifs	8	8	13	7	7	8	51	
11	Tree and columnar supports	0	1	0	0	0	1	2	
12	Animal	2	9	11	4	5	6	37	
13	Shells and spirals (invertebrates)	6	0	0	3	0	2	11	
14	Egg, oval, and tubular forms	0	1	0	0	0	1	2	
15	Arches, vaults, domes	0	0	9	0	0	0	9	
16	Shapes resisting straight lines	0	0	9	0	0	0	9	
17	Simulation of natural features/biomorphy	0	0	0	0	0	0	0	
18	Geomorphology	0	0	0	0	0	0	0	
19	Biomimicry	0	0	0	0	0	0	0	
	Subscores	16	19	42	14	12	18	121	2.05
	(3) Natural patterns and processes								
20	Sensory variability/information richness	9	9	13	10	9	9	59	
21	Age, change, the patina of time	0	0	0	0	0	0	0	
22	Central focal point	1	1	1	0	1	1	5	
23	Patterned wholes	0	0	0	0	0	0	0	
24	Bounded spaces	9	7	13	10	9	7	55	
25	Transitional spaces	1	1	0	1	1	1	5	
26	Linked series and chains	1	2	1	1	1	2	8	
27	Integrations of parts to wholes	9	9	13	10	9	9	59	
28	Complementary contrasts	0	0	11	0	0	0	11	
29	Dynamic balance and tension	0	0	0	0	0	0	0	

Table 3. *Cont.*

	Biophilic Interior Design Matrix	Shang Preschool	Daveen Preschool	Avrocity Preschool	Zaryland Preschool	Kapir Preschool	Cihman Preschool	Total Scores	Average
30	Fractals	0	0	0	0	0	0	0	
31	Hierarchically organized ratios and scales	0	0	0	0	0	0	0	
	Subscores	30	29	52	32	30	29	202	3.42
	(4) Light and space								
32	Natural light	9	9	13	9	9	9	58	
33	Filtered and diffused light	9	9	12	9	9	9	57	
34	Light and shadow	9	9	13	9	9	9	58	
35	Reflected light	0	0	0	0	0	0	0	
36	Light pools	0	0	0	0	0	0	0	
37	Warm light	0	0	0	0	0	0	0	
38	Light as shape and form	0	1	1	0	0	1	3	
39	Spaciousness	8	7	1	0	1	8	25	
40	Spatial variability	1	7	1	0	1	1	11	
41	Space as shape and form	0	0	0	0	0	0	0	
42	Spatial harmony	0	0	0	0	0	0	0	
43	Inside/outside spaces	0	1	1	0	0	2	4	
	Subscores	36	43	42	27	29	39	216	3.66
	(5) Place-based relationships								
44	Geographic connection to place	9	9	13	10	9	9	59	
45	Historic connection to place	0	0	0	0	0	0	0	
46	Ecological connection to place	1	1	13	0	0	1	16	
47	Cultural connection to place	9	9	13	10	9	9	59	
48	Indigenous materials	0	0	0	0	0	0	0	
49	Landscape orientation/landscape features	8	9	0	0	7	8	32	
	Subscores	27	28	39	20	25	27	166	2.81
	(6) Human–nature relationships								
50	Prospect and refuge	1	0	11	1	2	0	15	
51	Order and complexity	0	0	0	0	0	0	0	
52	Curiosity and enticement	0	0	0	0	0	0	0	
53	Change and metamorphosis	0	0	0	0	0	0	0	
	Subscores	1	0	11	1	2	0	15	0.25
	Total subscores	149	156	249	129	138	150	971	
	Average	16.55	17.33	19.15	12.9	15.33	16.66	16.45	16.45

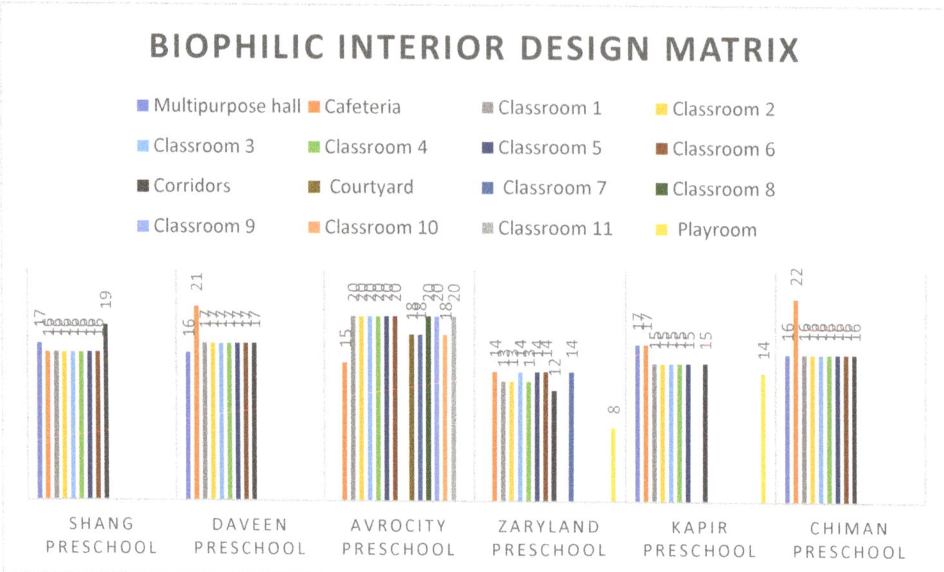

Figure 2. The biophilic interior design matrix of each space within the preschool.

The biophilic matrix of each preschool was organized into a separate table. Each table includes a set of preschool spaces distributed along the columns and a set of biophilic features distributed along the rows. Every space has subscores resulting from each existing attribute. These subscores were collected and the averages extracted to compare with one another in order to find the variances from the total average. The matrix results revealed that Avrocity preschool settings received a score of 19.15, higher than the average of the biophilic design of other preschools. The classrooms rated higher than average, while the cafeteria rated lower (Figures 2–4). The highest scoring was indicated through environmental features (4.25), followed by natural patterns and processes (3.42), as shown in Table 3.

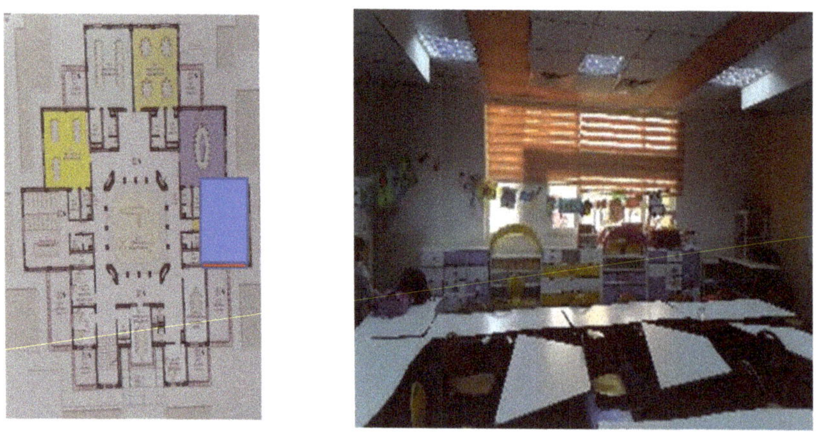

(**a**) (**b**)

Figure 3. Classroom of Avrocity preschool: (**a**) plan drawing from the directorate of the preschool; (**b**) photo by the author.

(**a**) (**b**)

Figure 4. Cafeteria of Avrocity preschool: (**a**) plan drawing from the directorate of the preschool; (**b**) photo by the author.

The next highest score of 17.33 was indicated in Daveen preschool settings through the light and space elements. The Chiman and Shang preschool spaces received scores of 16.66 and 16.55, respectively, which were nearly equal to the average of the biophilic design in the matrices. Kapir preschool spaces received a score of 15.33, below the average score, and Zaryland preschool settings scored 12.9 for biophilic design.

The highest scoring was presented through the environmental features, while the lowest was indicated through the human–nature relationships element. The cafeteria of Chiman preschool scored the highest in biophilia. The space was large and spacious and also provided direct and indirect biophilic features like color, water, air, and views. The space offered botanical motifs, animal representations, and a tubular form of column. The height of the celling and the skylight provided a central focal point for the space, as well as natural light (see Figure 5).

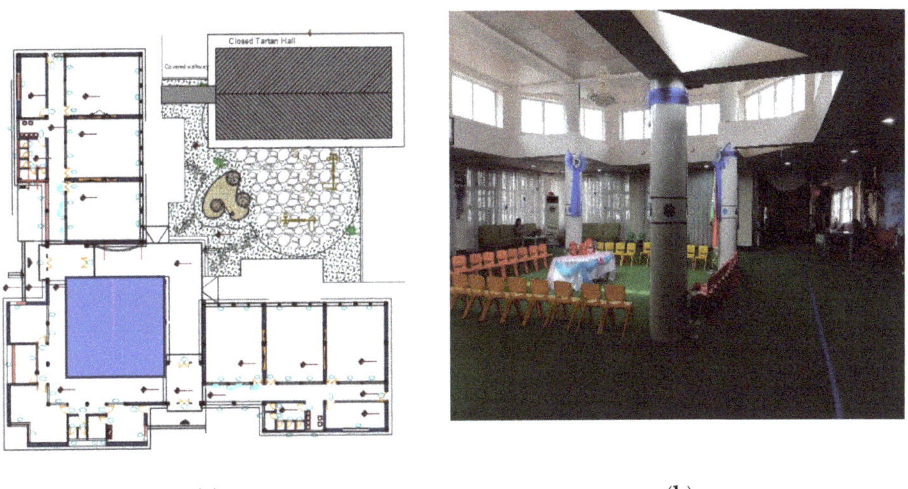

(**a**) (**b**)

Figure 5. Cafeteria of Chiman preschool: (**a**) plan drawing from the directorate of the preschool; (**b**) photo by the author.

The playroom in Zaryland preschool scored the lowest rating for biophilic design attributes in the matrix. Although it was a small space without windows, it offered the absence of natural light, air, and views. There are opportunities to improve the room if some adjustments could be made, such as expanding the area or adding a window. The ceiling adds to the space and the integration of parts into the whole, where each piece, when joined together, creates something greater. As well as adding sensory variability and information richness through various colors, toys added visual variety to the space (see Figure 6).

(**a**) (**b**)

Figure 6. Playroom of Zaryland preschool: (**a**) plan drawing from the directorate of the preschool; (**b**) photo by the author.

The environmental features element scored higher than the average, such as color, which could be found in all spaces. This element is classified as the organic dimension, which offers direct features (e.g., air, sunlight) and indirect features (e.g., views and vistas) that can be seen in most spaces. The highest ratings of the attributes among all elements were found in color, sensory variability, and information richness (e.g., sound from gaming devices, laptops, and TV), or various colors, which have sensory diversity, integration of parts into wholes (e.g., ceiling patterns), and geographical and cultural connections to place (e.g., near mountains, natural vegetation). The second-highest attributes were natural light and light and shadow (e.g., windows, skylights). Additionally, following in the sequence were some features such as air, sunlight, filtered and diffused light, and views and vistas that were present in most of the spaces. In contrast, the natural shapes and forms and human–nature relationships elements both displayed lower scores than the average, as indicated in Table 3. This may occasionally happen because designing a space with some direct and indirect biophilic features may be easier than incorporating another biophilic attribute, which may seem more challenging.

The features that received the lowest rankings among all elements included order and complexity, curiosity and enticement, change and metamorphosis, historic connection to place, indigenous materials, space as shape and form, spatial harmony, reflected light, light pools, warm light, age, change, the patina of time, patterned wholes, dynamic balance and tension, fractals, hierarchically organized ratios and scales, geomorphology, biomimicry, simulation of natural features, animals, and fire (fire is a common issue, particularly in preschool settings, and should be considered in safe spaces), as indicated in Table 3. None of these features were observed in any observed space. While more research may justify increasing features for safe, proper incorporation in children's spaces, the presence of biophilia was increased and supported by design elements with multiple attributes in the children's space. By integrating various attributes and undertaking additional research to determine their impact, this can assist in the development of future products [31]. It has been found that biophilia's engagement style in spaces is generally passive, whereas active engagement necessitates immersive experiences in nature.

4.1. Case Studies

Six preschools, both public and private, were selected to be analyzed in this study. These six preschools were selected from among many existing preschools within Duhok, due to the difficulty of obtaining permission to visit and photograph them for the collection of data required to conduct this research. As a result, 59 spaces were able to be included in the study.

4.1.1. Shang Preschool

One of the typical public preschools included nine spaces to be analyzed: six classrooms, a multipurpose hall, a cafeteria, and corridors. All spaces shared the presence of the same direct biophilic features (i.e., fresh air, sunlight) and indirect features (i.e., various colors, natural views for vegetation) through operable windows or doors. Additionally, the cafeteria and corridors added water (a sink) as another indirect biophilic feature. Botanical motifs, representations, or symbolic themes in paintings were found in most of the spaces, except for the multipurpose hall. In addition, animal representation features were found only in the multipurpose hall and corridors. Classrooms added shell and spiral features to the space through the paintings on the walls. Sensory variability and information richness created through visual sense (natural colors or toys), auditory sense (music), and tactile sense of touch (natural materials) were the common features in all of the spaces. Integration of parts into wholes could be seen through the ceiling, which was divided into parts that connected together to form the whole, and bounded space features were found in all spaces. All of the spaces offered natural light (through windows), filtered and diffused light (blinds), light and shadow, and artificial light. The classrooms, cafeteria, and multipurpose hall offered spaciousness and a sense of openness through their large areas. The classroom spaces were divided into two distinct areas: one area used chairs and tables, and the other area used pads stuffed for sitting or activities. The spatial variability feature produced a variety of play zones, with some toys found in the multipurpose hall. Transitional spaces and linked series and chains could be found in the corridors. A stage in the multipurpose hall used for play or participation activities provided a sense of separation from the main space and a sense of connection between nature and humans through the prospect and refuge attributes. Furthermore, the stage offered a central focal point for the space. All of the spaces offered geographic and cultural connections to places and landscape features. The corridors (Figure 7) and the multipurpose hall both scored higher than the average, whereas the classrooms and cafeteria were all given lower ratings than the biophilic matrix's overall average.

4.1.2. Daveen Preschool

In Daveen preschool, we analyzed six classrooms, a multipurpose hall, a cafeteria, and corridors. Direct environmental features (i.e., air, sunlight) and indirect features (i.e., color, views, and vistas) were found in all of the analyzed spaces, while water (a sink) was found in the cafeteria as an indirect feature. A lot of toys could be seen in spaces with animal representations. Except for the multipurpose hall, most of the spaces had paintings and posters with botanical motifs hanging on the walls. A lot of design elements could be found in the cafeteria (Figure 8), with multiple features like tree and columnar supports and egg, oval, and tubular forms. Sensory variability, information richness, and integration of parts into wholes were added to all spaces. Most of the spaces provided were bounded spaces, except for the cafeteria and corridors. Linked series and chains were offered for both corridors and cafeteria spaces, while transitional space features were found only in corridors. Outdoor access is one of the possibilities for enriching biophilic incorporation. For example, a courtyard layout can support other spaces by adding some natural themes and representational features through the visual connection. Natural, filtered, and diffused light and shadow attributes are provided through the windows in all spaces together, and they are accompanied by artificial light. Furthermore, the cafeteria added light in shape and form and a central focal point feature to the space through the skylight. Spaciousness features were found in most spaces except for the cafeteria and corridors. With the exception of the multipurpose hall and corridors, most of the spaces exhibited the spatial variability feature through zones of activities. Geographic and cultural connections to place, landscape orientation, and landscape features were found in all spaces. The human–nature relationship element was not present in all spaces. Generally, the majority of Daveen preschool settings (classrooms and corridors) scored a little higher than the average for the biophilic features in the matrix. The cafeteria had the second-highest score among all other preschool spaces, whereas the multipurpose hall scored lower than the average.

(**a**) (**b**)

Figure 7. Corridors of Shang preschool: (**a**) plan drawing from the directorate of the preschool; (**b**) photo by the author.

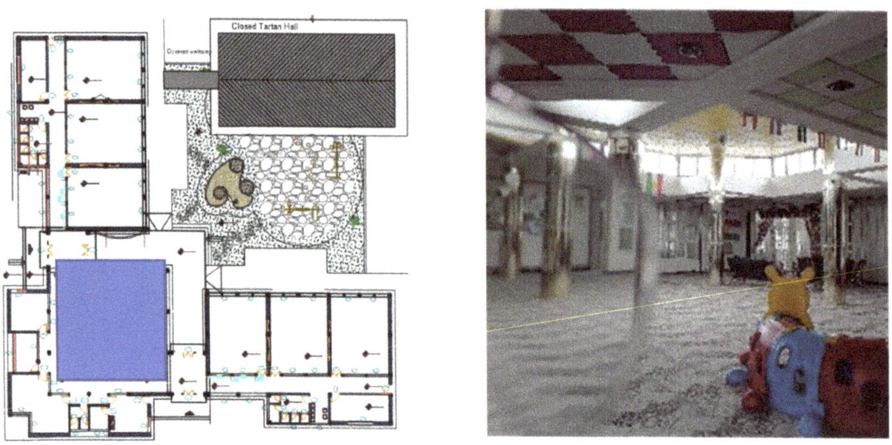

(**a**) (**b**)

Figure 8. Cafeteria of Daveen preschool: (**a**) plan drawing from the directorate of the preschool; (**b**) photo by the author.

4.1.3. Avrocity Preschool

One of the private preschools located within an apartment complex included 11 classrooms, a cafeteria, and a courtyard distributed over two floors. All of these spaces together displayed the highest average variance of biophilic features in the matrix. Color, water, air, sunlight, views, and vistas are examples of indirect environmental features that could be found in all spaces. Nevertheless, only the courtyard space (Figure 9) lacked water and views. A lot of photos or posters hanging on the walls could be seen in most spaces, with animal representations like fish or botanical motifs as natural shapes and forms. A variety of open and closed storage added complementary contrast features, and non-rectilinear desks or furniture in the classrooms added arches, vaults, domes, and shapes resisting straight lines. Sensory variability/information richness, integration of parts to wholes, and bounded space features were found in all spaces. Linked series and chains featured in courtyard spaces. Natural light, filtered and diffused light, and light and shadow attributes were provided through the windows in all spaces, and they were accompanied by artificial light. The courtyard added multiple features, such as spaciousness, light as shape and form, a central focal point, spatial variability, and inside/outside spaces through the skylight and the variance in the height. Geographic, ecological, and cultural connections to place features were found in all spaces. A prospect and refuge feature provided by an open small space separated from the classroom was one of the features of the human–nature relationship. Although Avrocity preschool's classrooms are generally smaller than those at Shang, Daveen, Chiman, and Kapir preschools, they are differentiated by a higher level of biophilic inclusion than those at other preschools. All of the spaces indicated were higher than the overall average of the biophilic design matrix.

Figure 9. Courtyard of Avrocity preschool.

4.1.4. Zaryland Preschool

Another private preschool located within an apartment complex included seven classrooms, a playroom, a cafeteria, and corridors. The classrooms (Figure 10) were the smallest compared to the classrooms of all other preschools. They include chairs and tables, a whiteboard on the wall, and some open and closed storage, some of which is high and not appropriate for children to use. Storage, particularly floor-level storage that children could access, was another frequent necessity. The incorporation of direct biophilic features (i.e., sunlight, air) and indirect features (i.e., views) was added to the classrooms and cafeteria, whereas the playroom was missing these features except for the indirect feature (color), since it lacked windows. In the cafeteria, corridors, and some classrooms, wall-mounted posters, artwork, and small toys were used. These added biophilic features, including botanical motifs, animals, shells, and spirals, but they were too high and challenging for the kids to reach, view, or touch. Sensory variability or information richness, integration of parts into wholes, and bounded space features were found in all spaces. Natural light (through windows), filtered and diffused light (blinds), and light and shadow, accompanied by artificial light features, were provided in all spaces except for the playroom, which had no windows. Geographic connection to place and cultural connection to place were found in all spaces. The lack of human–nature relationships in Zaryland preschool spaces revealed a lower level of biophilic inclusion than the overall average, especially in the playroom, which scored the worst rating for biophilia in the matrix.

(**a**)

(**b**)

Figure 10. Zaryland preschool settings: (**a**) classroom space; (**b**) corridor space.

4.1.5. Kapir Preschool

A typical public preschool located within a residential area included five classrooms, a multipurpose hall, a cafeteria, and corridors. All of the spaces added some indirect biophilic features, such as color, water (a sink), and natural views of vegetation, as well as direct natural features like windows (sunlight) and ventilation (natural air). The presence of plant and natural material features was limited to the cafeteria and multipurpose hall. Botanical motif features were found in the classrooms, corridors, and cafeteria, while animal representations were found only in the classrooms. Sensory variability and information richness, integration of parts into wholes, and bounded space features could be found in all spaces, with corridors providing transitional space as well as link series and chains. Natural light, filtered and diffused light, and light and shadow features were provided through the windows, combined with artificial lighting in all spaces. The windows have blinds in order to moderate the light, but the height of the windows restricts children from viewing the natural views. The stage in the multipurpose hall offers a sense of separation from the main space and a sense of connection between nature and humans through the prospect and refuge attributes; furthermore, features like focal points and spaciousness in the playroom space add prospect, refuge, and special variability features to the space (Figure 11). Most of the spaces offer geographic and cultural connections to places and landscape features. The majority of Kapir preschool's settings were ranked below the average of the biophilic matrix, whereas the multipurpose hall and the cafeteria ranked slightly higher.

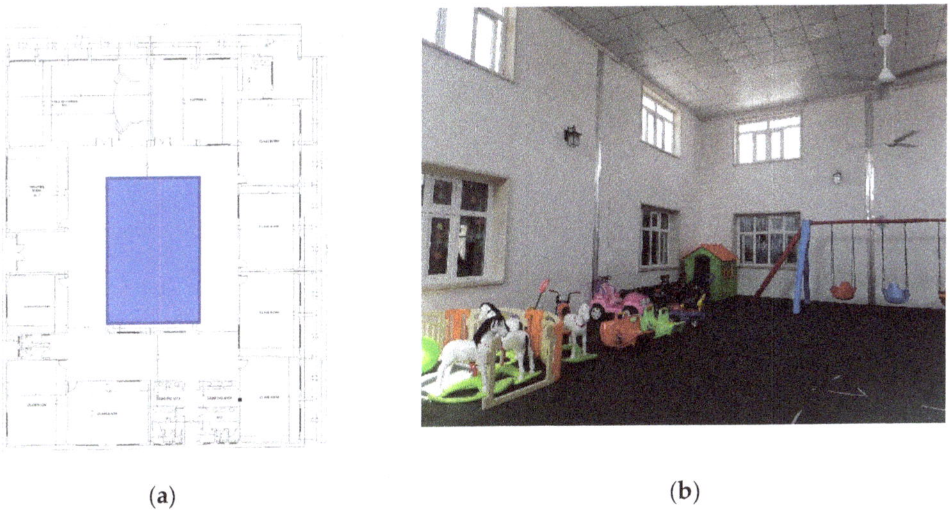

Figure 11. Play room of Kapir preschool: (**a**) plan drawing from the directorate of the preschool; (**b**) photo by the author.

4.1.6. Chiman Preschool

A typical public preschool located near the city center included six classrooms, a multipurpose hall, a cafeteria, and corridors. Direct environmental features (i.e., air, sunlight) and indirect features (i.e., color, views, and vistas) were observed in all spaces, while water (a sink) as an indirect feature was found in the cafeteria. Most of the spaces offered botanical motifs and animal representations through paintings and posters mounted on the walls. A few classrooms included features of shells and spirals in the space. A lot of design elements could be found in the cafeteria, with multiple features like tree and columnar supports, egg, oval, and tubular forms, and a central focal point. Sensory variability, information richness, and integration of parts into wholes were found in all spaces. Most of the spaces were provided with bounded spaces, except for the cafeteria and corridors. Links and chains were found in both the corridors and cafeteria spaces, while transitional space features were found only in corridors. Natural, filtered, and diffused light, with light and shadow attributes, is provided through the windows in all of the spaces, accompanied by artificial light. Furthermore, the cafeteria adds light as shape and form and a central focal point feature through the skylight, and attributes of both spatial variability and inside/outside spaces were found in the cafeteria as well (Figure 5). Spacious features were found in most of the spaces except for the corridors. Geographic and cultural connections to place, landscape orientation, and landscape features were found in all spaces. A lack of human–nature relationships featured in all spaces. The majority of Chiman preschool's spaces were rated lower than the average for the biophilic matrix. However, the cafeteria scored the highest among all other preschool spaces, as shown in Table 2.

The final scores of the whole matrix revealed the final averages of each preschool, which indicated Avrocity as the highest among the other preschools, as shown in Table 3.

5. Conclusions

The BID-M provides an innovative interior design perspective, enhancing the existing body of knowledge on biophilic design and its usefulness in the design process. It offers designers and practitioners a methodical and comprehensive vocabulary for the biophilic interior design of preschools, as well as a useful tool for incorporating and assessing biophilic features in interiors. The BID-M provides a variety of options that maximize the presence of biophilia in the space and can support multiple biophilic features through a single attribute.

The environmental features element was the highest-scoring in some preschools, such as Avrocity, Zaryland, Kapir, and Shang, through some features such as color, which could be seen in all spaces. This element is classified as the organic dimension, which offers direct features (e.g., air, sunlight) and indirect features (e.g., views and vistas), which can be seen in most spaces. On the other hand, the light and space element scored the highest in the Daveen and Chiman preschools, through natural light, filtered and diffused light, light and shadow, spaciousness, and spatial variability, which were found in most of the spaces.

According to the matrix results, Avrocity preschool recorded a higher inclusion of biophilic features through environmental features (e.g., color, water, air, sunlight, and views and vistas), natural shapes and forms, and natural patterns and processes in comparison with the other preschools, while Daveen preschool recorded a higher inclusion of biophilic features through light and space and place-based relationships. On the other hand, Zaryland preschool recorded a lower inclusion of biophilic features through environmental features, natural patterns and processes, light and space, and place-based relationships in comparison with the other preschools. In addition, Kapir preschool recorded a lower inclusion of biophilic features through natural shapes and forms, light and space, and place-based relationships.

The matrix revealed the levels of biophilic variability in the different spaces of each preschool, where the cafeteria at Chiman preschool showed the highest incorporation of biophilic features, followed by the cafeteria in Davin preschool and the classrooms in Avrocity preschool.

Remarkably, one of the most common issues discovered in most of the preschool spaces studied—especially in the Daveen and Chiman preschools—was a lack of biophilic features that strengthen the bonds between nature and humans, referred to as the human–nature relationships element, although this was minimally observed in Avrocity preschool. It can be supported by some small formations, such as independent play places that have privacy from the overall space and overlook the larger space in which they are located. Prospect and refuge refer to places where children enjoy playing when they feel they are in a special space. Another opportunity comes through the element of temptation and curiosity, perhaps through the creation of some features such as lighting and shadow or light and dark, and the presence of elements that stimulate the child's curiosity to discover order and complexity through furniture, materials, or colors.

The presence of some places with limited areas and the absence of windows resulted in a lack of daylight and natural air; thus, those spaces received the lowest ratings for biophilic design attributes. There are opportunities to make areas more biophilic by expanding them and adding some windows. Furthermore, the incorporation of biophilia through place-based relationships can be supported by outdoor access, which connects indoor spaces to the surrounding landscape.

Overall, some common issues were observed in these interiors. For example, the storage used by children in some spaces was high and difficult to reach. It would be better to design this furniture with consideration of children's scale and the variations of open or closed storage to make its use easier. Also, the posters and photos mounted on the walls are another issue to be considered when children cannot reach them to see or touch them for more interaction with these natural paintings. Additionally, the window height from the ground floor in some cases restricted children's ability to enjoy the natural views behind the windows, such as butterflies or feeling the breeze. Including natural themes and representations of the natural world in the spaces (e.g., Sun, plants, water, animals, or color) would make them more biophilic. Adding natural materials rather than artificial ones would be more biophilic and healthier. Another significant issue that was observed throughout the preschool environments was the unintentional random use of colors.

6. Further Study

Implications throughout time: Using biophilic design elements in preschool interiors may have long-term implications, which could be investigated in future studies. Researchers can measure the long-term effects on children's wellbeing, cognitive development, and academic achievement through longitudinal studies, offering important insights into the long-term advantages of biophilic environments.

Design interventions: More research is required to determine the efficacy of particular design solutions intended to improve biophilic features in preschool spaces. Researchers can investigate the use of naturally inspired components, including living walls, organic materials, and daylighting techniques, and evaluate their effects on students' engagement, concentration, and overall learning results.

Further research is needed to explore the health and wellbeing benefits of biophilic design in preschool environments. Factors such as stress reduction, immune system performance, creativity, and emotional wellbeing could be investigated. Cultural context and diversity are crucial for ensuring inclusion and meeting diverse community needs. Future studies should explore how cultural backgrounds, geographic regions, and socioeconomic issues affect the integration and effectiveness of biophilic features.

Understanding how users—children, instructors, and parents—perceive and behave in relation to biophilic design in preschool spaces is another potential research subject. Insights into user preferences, satisfaction, and the possibility for behavioral changes can be gained via qualitative research, surveys, and interviews about how various stakeholders interpret, value, and use biophilic features.

Researchers can improve their knowledge of biophilic design in preschool interiors by delving into these areas for future research, which will ultimately help to develop evidence-based recommendations and tactics for making learning environments for young children that are healthier, more interesting, and connected to nature.

Author Contributions: Conceptualization, I.M.; methodology, I.M.; software, I.M.; validation, I.M.; formal analysis, I.M.; investigation, I.M.; resources, I.M.; data curation, I.M.; writing—original draft preparation, I.M.; writing—review and editing, I.M.; visualization, I.M.; supervision, Z.O. and Ç.Ç. All authors have read and agreed to the published version of the manuscript.

Funding: This research received no external funding.

Institutional Review Board Statement: The study was conducted in accordance with the Declaration of Helsinki, and approved by the Institutional Review Board of the Directorate of Duhok Education (20 October 2022). All authors have read and agreed to the published version of the manuscript.

Informed Consent Statement: Informed consent was obtained from all subjects involved in the study.

Data Availability Statement: As stated in the full article, all data is accessible publicly.

Conflicts of Interest: The authors declare no conflict of interest.

References

1. Wells, N.M.; Evans, G.W. Nearby nature: A buffer of life stress among rural children. *Environ. Behav.* **2003**, *35*, 311–330. [CrossRef]
2. Flouri, E.; Midouhas, E.; Joshi, H. The role of urban neighbourhood green space in children's emotional and behavioural resilience. *J. Environ. Psychol.* **2014**, *40*, 179–186. [CrossRef]
3. Hand, K.L.; Freeman, C.; Seddon, P.J.; Recio, M.R.; Stein, A.; Van Heezik, Y. The importance of urban gardens in supporting children's biophilia. *Proc. Natl. Acad. Sci. USA* **2017**, *114*, 274–279. [CrossRef] [PubMed]
4. Gillis, K.; Gatersleben, B. A Review of Psychological Literature on the Health and Wellbeing Benefits of Biophilic Design. *Buildings* **2015**, *5*, 948–963. [CrossRef]
5. Beatley, T. *Handbook of Biophilic City Planning and Design*; Island Press: Washington, DC, USA, 2016.
6. Kellert, S.R. *The Theory of Biophilic Design*; John Wiley & Sons, Inc.: Hoboken, NJ, USA, 2008; pp. 3–19.
7. Marte, E.; Calumpit, A.; Bessa, B.D.S.; Toledo, A.; Fadda, R.; Skoler, T. Testing Reliability of Biophilic Design Matrix Within Urban Residential Playrooms. *Front. Psychol.* **2020**, *11*, 570099. [CrossRef] [PubMed]
8. McGee, B.; Marshall-Baker, A. Loving nature from the inside out: A biophilia matrix identification strategy for designers. *Health Environ. Res. Des. J.* **2015**, *8*, 115–130. [CrossRef] [PubMed]

9. MCGEE, B.L.S. *Designing Biophilia into Interior Environmental Practice: A Biophilic Design Assessment Tool Development*; University of Florida: Gainesville, FL, USA, 2018.
10. Kellert, S.R.; Heerwagen, J.; Mador, M. *Biophilic Design: The Theory, Science and Practice of Bringing Buildings to Life*; John Wiley & Sons: Hoboken, NJ, USA, 2011.
11. Browning, B.; Garvin, C.; Ryan, C.; Kallianpurkar, N.; Labruto, L.; Watson, S.; Knop, T. The Economics of Biophilia. In *The Economics of Biophilia*, 2nd ed.; Terrapin Bright Green LLC: New York, NY, USA, 2012; p. 40.
12. Heerwagen, J.; Hase, B. Building Biophilia: Connecting People to Nature in Building Design. *Environ. Des. Constr.* **2001**, *4*, 30–36. Available online: https://www.usgbc.org/Docs/Archive/External/Docs8543.pdf (accessed on 15 December 2022).
13. Beute, F.; de Kort, Y.A. Salutogenic Effects of the Environment: Review of Health Protective Effects of Nature and Daylight. *Appl. Psychol. Health Well-Being* **2014**, *6*, 67–95. [CrossRef]
14. Hartig, T.; Berg, A.E.V.D.; Hagerhall, C.M.; Tomalak, M.; Bauer, N.; Hansmann, R.; Ojala, A.; Syngollitou, E.; Carrus, G.; van Herzele, A.; et al. Health benefits of nature experience: Psychological, social and cultural processes. In *Forests, Trees and Human Health*; Springer: Dordrecht, The Netherlands, 2011; pp. 127–168. [CrossRef]
15. Kahn, P.H. Developmental Psychology and the Biophilia Hypothesis: Children's Affiliation with Nature. *Dev. Rev.* **1997**, *17*, 1–61. [CrossRef]
16. Matteson, D. *Learner Perceptions of Biophilia and the Learning Environment: A Phenomenological Study*; Walden University: Minneapolis, MN, USA, 2013.
17. Miller, H. *Nature-Based Design: The New Green Research Summary-Combine Leed with Essential Biophilic Elements*; Herman Miller, Inc.: Zeeland, MI, USA, 2013.
18. McGee, B.; Park, N.; Portillo, M.; Bosch, S.; Swisher, M. Diy Biophilia: Development of the Biophilic Interior Design Matrix as a Design Tool. *J. Inter. Des.* **2019**, *44*, 201–221. [CrossRef]
19. Cacique, M.; Ou, S.-J. Biophilic Design as a Strategy for Accomplishing the Idea of Healthy, Sustainable, and Resilient Environments. *Sustainability* **2022**, *14*, 5605. [CrossRef]
20. Downton, P.; Jones, D.; Zeunert, J.; Roös, P. Biophilic Design Applications: Putting Theory and Patterns into Built Environment Practice. *KnE Eng.* **2017**, *2*, 59–65. [CrossRef]
21. Walimbe, A.S.; Chitgopkar, A.S. Nurturing Children's Biophilia Through Nature Connectedness in School Buildings for a Sustainable Future. *Int. J. Civ. Eng. Technol.* **2018**, *9*, 187–192.
22. Ryan, C.O.; Browning, W.D.; Clancy, J.O.; Andrews, S.L.; Kallianpurkar, N.B. BIOPHILIC DESIGN PATTERNS: Emerging Nature-Based Parameters for Health and Well-Being in the Built Environment. *Archnet-IJAR Int. J. Arch. Res.* **2014**, *8*, 62–76. [CrossRef]
23. McGee, B.; Park, N.-K. Colour, Light, and Materiality: Biophilic Interior Design Presence in Research and Practice. *Interiority* **2022**, *5*, 27–52. [CrossRef]
24. McGee, B.; Jin, X.; Park, N.-K.; Ball, S.; Carr, A. Designers' perceptions of biophilia and testing of the biophilic interior design matrix in China. *Int. J. Archit. Res. Archnet-IJAR* **2022**, *16*, 517–535. [CrossRef]
25. Awad, N.A. *The Impact of Biophilia in Elementary Schools: A Study of Student Health and Well-Being*; University of Washington: Seattle, WA, USA, 2022.
26. Ondul, B.; Koparan, N.; Hasirci, D. Biophilic design in preschool furniture: A refuge pod design proposal as an educational tool. *EDULEARN21 Proc.* **2021**, *1*, 50–59. [CrossRef]
27. Yanez, R.E.; Fees, B.S.; Torquati, J. Preschool children's biophilia and attitudes toward nature: The effect of personal experiences. *Int. J. Early Child. Environ. Educ.* **2017**, *5*, 57.
28. Ünal, N.; Özen, E.S. Biophilic Approach to Design for Children. *Iconarp Int. J. Arch. Plan.* **2021**, *9*, 943–965. [CrossRef]
29. Yassein, G.; Ebrahiem, S. Biophilic Design in the Built Environment to Improve Well-Being: A Systematic Review of Practices. *J. Urban Res.* **2018**, *30*, 128–146. [CrossRef]
30. Ghaziani, R.; Lemon, M.; Atmodiwirjo, P. Biophilic Design Patterns for Primary Schools. *Sustainability* **2021**, *13*, 12207. [CrossRef]
31. McGee, B. An Inventory of Biophilic Design Attributes within Child Life Play Spaces. Master's Thesis, The University of North Carolina at Greensboro, Greensboro, NC, USA, 2012. Available online: https://www.researchgate.net/publication/268221979 (accessed on 10 March 2022).
32. Weinberger, N.; Butler, A.G.; McGee, B.; Schumacher, P.A.; Brown, R.L. Child Life Specialists' Evaluation of Hospital Playroom Design: A Mixed Method Inquiry. *J. Inter. Des.* **2017**, *42*, 71–91. Available online: https://digitalcommons.bryant.edu/apjou/54 (accessed on 15 April 2022). [CrossRef]
33. Cobreros, C.; Medoza-Ruvalcaba, N.; Flores-García, M.; Roggema, R. Improving Psychological Well-Being in Urban University Districts through Biophilic Design: Two Cases in Mexico. *Sustainability* **2023**, *15*, 5703. [CrossRef]
34. Fadda, R.; Congiu, S.; Roeyers, H.; Skoler, T. Elements of Biophilic Design Increase Visual Attention in Preschoolers. *Buildings* **2023**, *13*, 1160. [CrossRef]
35. Wardhani, D.K.; Susan, S. Strategy to Reduce the COVID-19 Transmission through Adaptation of Greenship Interior Space (IS) Criteria. *IOP Conf. Series Earth Environ. Sci.* **2021**, *738*, 012072. [CrossRef]

36. Palinkas, L.A.; Horwitz, S.M.; Green, C.A.; Wisdom, J.P.; Duan, N.; Hoagwood, K. Purposeful Sampling for Qualitative Data Collection and Analysis in Mixed Method Implementation Research. *Adm. Policy Ment. Health Ment. Health Serv. Res.* **2015**, *42*, 533–544. [CrossRef]
37. Campbell, S.; Greenwood, M.; Prior, S.; Shearer, T.; Walkem, K.; Young, S.; Bywaters, D.; Walker, K. Purposive sampling: Complex or simple? Research case examples. *J. Res. Nurs.* **2020**, *25*, 652–661. [CrossRef]

sustainability

MDPI

Article

Salutogenic Factors and Sustainable Development Criteria in Architectural and Interior Design: Analysis of Polish and EU Standards and Recommendations

Agnieszka Rek-Lipczyńska

Department of Interior Design, Faculty of Architecture, West Pomeranian University of Technology in Szczecin, Al. Piastów 17, 71-300 Szczecin, Poland; areklipczynska@zut.edu.pl

Abstract: This article examines the integration of salutogenic design principles into the legal and regulatory framework for buildings in Poland, situating the analysis within a broader European and international context. The study combines a systematic literature review, a comparative analysis of Polish building law and technical conditions with international standards and empirical evidence from case studies and Post-Occupancy Evaluations (POE) conducted in hospitals, offices, and schools. This triangulated approach allowed for both qualitative and quantitative assessment of key salutogenic factors, including daylight access, acoustic comfort, air quality, thermal conditions, aesthetics, ergonomics, and opportunities for social interaction. The results demonstrate that Polish regulations continue to focus on minimum hygienic and safety requirements, thereby reflecting a pathogenic rather than salutogenic approach. While some factors—such as daylight and ventilation—are partially included, critical dimensions of psychosocial well-being remain underrepresented. By contrast, both European EN standards and voluntary frameworks such as WELL systematically incorporate broader salutogenic dimensions. Evidence from Polish POE studies further highlights gaps between regulatory compliance and actual user experiences, e.g., excessive hospital noise, insufficient lighting in operating rooms, and the absence of biophilic or therapeutic features. The discussion highlights challenges of measurability, economic feasibility, and stakeholder acceptance, while also emphasizing long-term health and social benefits. The conclusions recommend incorporating salutogenic principles into Polish law and sustainability policies, promoting pilot projects, and strengthening interdisciplinary collaboration. In this way, buildings can evolve from merely preventing harm to actively supporting health, well-being, and resilience—aligning with global trends in human-centered sustainable design.

Keywords: salutogenesis; health-oriented architecture; building law; sustainable construction; WELL Building Standard; salutogenic factors; architecture; interior architecture

1. Introduction

Modern Approaches to Sustainable Development in the Context of Salutogenesis

Contemporary approaches to sustainable development encompass not only ecological and economic aspects but also human health and well-being, particularly that of building users. In academic literature, there are numerous competing definitions of health—ranging from a negative understanding as the mere absence of disease, through holistic concepts of biopsychosocial well-being, to dynamic models of equilibrium and adaptation. The most widespread remains the biomedical model, in which health is defined as the absence of illness, i.e., a state in which no identifiable pathological dysfunction occurs in the body [1].

https://doi.org/10.3390/su17219661

A milestone in this debate was Engel's proposition of the biopsychosocial model, which explicitly integrates biological, psychological, and social determinants of health and is now widely recognized as a more comprehensive framework [2].This model is frequently criticized for its biological reductionism and disregard for social and subjective dimensions of health. An alternative is offered by the World Health Organization (WHO), which defines health as "a state of complete physical, mental and social well-being and not merely the absence of disease or infirmity" [3]. While this definition introduces a positive framing of health, it has also been criticized for being utopian and impractical—since it is difficult to objectively define "complete well-being" and measure it in real-world settings [4]. There is growing recognition of dynamic health models that conceptualize health as a balance between individual resources and environmental demands. Socioecological models and processual approaches treat health as a variable potential, conditioned by factors such as the quality of the living environment [5]. In this context, the concept of salutogenesis, developed by Aaron Antonovsky, becomes highly relevant. Salutogenesis shifts the focus from pathogenic medicine to the origins of health. It defines health as a continuum between illness and full well-being, where the key factor in maintaining health is the Sense of Coherence (SOC)—a composite of comprehensibility, manageability, and meaningfulness in life experiences Individuals with a strong SOC cope more effectively with stress and make better use of available health resources [6]. The sense of coherence consists of three components: comprehensibility (belief, that stimuli from the environment are ordered and can be understood), resourcefulness (belief in accessibility resources to meet the requirements) and meaningfulness (the belief that life's requirements are worth the effort). These elements create the theoretical framework of the salutogenic model, in which health is the result of the ability to give meaning and cope with life's challenges, instead of just avoiding disease.

Salutogenesis provides a valuable complement to the paradigm of sustainable environmental design. Instead of merely minimizing pathogenic risks, it promotes the enhancement of health-supportive factors, including the design of environments that foster mental and social well-being. Antonovsky even described salutogenic design as a "measurable aspect of the design process that helps people function with maximum effectiveness and maintain physical and mental well-being" [7]. This means that the built environment can actively support health rather than just prevent disease. The key design question becomes: "How healthy is this place?"—a foundational consideration in salutogenic architectural thinking. Previous research on salutogenic architectural design has primarily focused on conceptual discussions, case studies, and applications within voluntary industry frameworks such as the WELL Building Standard or LEED [8].However, systematic analyses addressing the integration of salutogenic principles into binding legal frameworks remain scarce, particularly in Central and Eastern Europe. Poland, whose building law and technical requirements are still largely based on the so-called "hygienic minimum," illustrates a context where the gap between salutogenic theory and regulatory practice is particularly evident.

Against this background, the following research gap emerges: the lack of comprehensive understanding of the extent to which salutogenic principles have been (or could be) embedded in national legal regulations, and how the Polish situation compares to other European contexts.

Accordingly, this article is guided by the following research questions:

To what extent do current Polish building regulations and technical requirements incorporate salutogenic design principles?

How do these requirements compare with voluntary international standards for indoor environmental quality, such as the WELL Building Standard?

What lessons can be drawn from legislative changes in selected European countries that increasingly align with salutogenic objectives?

What recommendations can be formulated for the Polish legal system to more effectively promote user health and well-being in the built environment?

The salutogenic approach aligns with broader frameworks such as healthy architecture and healing environments, both of which seek to promote health through spatial and environmental quality. Numerous studies indicate that features of the built environment—particularly interior environmental conditions—significantly affect both the physical and mental health of occupants. From indoor air quality and lighting to acoustics, aesthetics, and spatial layout, such factors shape the comfort and condition of users. The environment can function as both a stressor (e.g., noise, crowding, toxins) and a supportive stimulus, reducing negative influences and enhancing positive ones that contribute to well-being. Recent empirical research confirms the positive impact of salutogenic design principles on the physical and mental health of architectural space users. These studies encompass a wide range of architectural interior types, most frequently focusing on public-use facilities such as workplaces, schools, and hospitals. Empirical evidence highlights improvements in user satisfaction and health outcomes within work environments designed in accordance with industry standards such as the WELL certification system. In order to evaluate the salutogenic potential of school buildings—defined as the capacity of the physical school environment to support health and well-being—et al. developed a transdisciplinary research model integrating environmental psychology, public health, and architecture. Within this framework, field studies were conducted in primary schools, analyzing key characteristics of the learning environment, including the quality of classroom acoustic insulation and the visual clarity of interiors, understood as access to daylight, exterior views, and spatial legibility. However, it is frequently noted that traditional building codes focus almost exclusively on the minimum technical requirements for safety and basic health in a pathogenic sense—i.e., avoiding harm (e.g., structural integrity, evacuation routes, hygiene). There is a notable lack of guidelines for salutogenic aspects of design, which would consider the emotional and cognitive impacts of space. In other words, current regulations are primarily concerned with physical parameters (e.g., temperature, ventilation, noise) and risk mitigation, rarely addressing how built environments can actively support health in a positive (psychosocial) sense. In recent years, Northern European countries have witnessed notable shifts in the legislative approach to the hygienic minimum requirements for daylight quality in residential buildings. The new European Committee for Standardization (CEN). (2018). EN 17037:2018 Daylight in buildings. Brussels: "Daylight in Buildings" represents a paradigm shift in the treatment of daylight. Countries such as Sweden and Norway are among the leaders in implementing these qualitative requirements. As noted by in the European comparative context, Northern states (e.g., Sweden) were quicker to adapt user-comfort-oriented elements in daylight regulations, moving beyond the previous "hygienic minimum". In many other European countries, simple quantitative indicators (such as the minimum daylight factor, DF) still dominate, whereas in Northern Europe, aspects such as required sunlight exposure duration, provision of views to the outside, and limitation of glare risk have begun to be incorporated. This shift of emphasis—from the mere quantity of daylight to the quality of spatial experience—is clearly evident in recent legislative changes in Sweden and Norway. Swedish law is moving away from a purely quantitative approach toward health- and well-being-oriented criteria: from the requirement of minimum daylight provision to ensuring that daylight genuinely enhances the comfort and well-being of building users.

The aim of this article is to examine the extent to which the principles of salutogenic design overlap with the legal requirements imposed on buildings and their interiors in

Poland, as well as with the recommendations of non-binding building quality standards. A secondary aim is to confront the Polish regulatory framework with that of selected European countries. Although architecture and interior design represent distinct domains of design practice—focusing respectively on the external form and structure of buildings and on the internal spatial experience—this article considers them jointly as part of a coherent regulatory construct. In the Polish legal and normative context, both are governed by the same legislative acts and technical standards (e.g., the Building Law and the Regulation on Technical Conditions), which establish requirements without differentiating between the architectural and interior dimensions of the built environment. For this reason, the comparative analysis presented here addresses architecture and interiors together, with an emphasis on the regulatory dimension of salutogenic factors. This approach allows for a consistent evaluation of how legal frameworks and voluntary standards (e.g., WELL) respond to salutogenic principles, while acknowledging the practical distinctions between architecture and interior design as professional disciplines.

In particular, it compares a set of key salutogenic design factors—identified through literature review—with the relevant provisions in Polish construction law and technical regulations, along with selected professional guidelines for health-promoting architecture. Such an analysis highlights the gaps and opportunities for improving legislation to better serve the health and well-being of building users—thus aligning more closely with the principles of socially sustainable development. Examples of legislation from selected European Union countries are also indicated, where recent changes show a stronger correlation with salutogenic objectives.

2. Materials and Methods

2.1. Identifying the Impact of Interior Building Features on Physical and Mental Health

In the first stage of the study, a review was conducted of research published over the last decade (2015–2025) concerning the impact of various interior design factors—such as air quality, lighting, acoustics, aesthetics (e.g., color schemes and the presence of artworks), ergonomic qualities of furnishings, and the spatial layout of rooms—on human health and well-being. Recent studies consistently indicate that the indoor environment of buildings significantly influences the physical and mental health of their occupants. Key features such as air quality, lighting conditions, acoustic comfort, aesthetic attributes (including color schemes and artistic elements), ergonomic design, and spatial arrangement all affect user comfort and overall condition. This section outlines the current state of scientific knowledge (primarily from 2015 to 2025) on the health-related effects of these factors, providing evidence-based outcomes and references to specific studies.

2.1.1. Indoor Air Quality (IAQ)

Indoor air quality is a key health determinant, particularly given that people spend the vast majority of their time indoors. Common indoor pollutants include carbon dioxide (CO_2), particulate matter ($PM_{2.5}$, PM_{10}), and volatile organic compounds (VOCs) emitted from building materials and furniture [1,2]. Numerous experimental studies have confirmed that deteriorated IAQ leads to reduced subjective comfort and productivity, increased fatigue, and symptoms of Sick Building Syndrome (SBS) in over 20% of exposed individuals [3]. For example, diminished air quality has been associated with decreased performance among students and office workers, alongside a higher incidence of complaints such as headaches and mucosal irritation [4].

Polluted indoor air affects not only subjective well-being but also objective physiological indicators. Elevated concentrations of typical pollutants (CO_2, particulates, VOCs) have been linked to increased heart rate and blood pressure, and reduced heart rate variability

(HRV), all of which suggest cardiovascular strain [5]. Exposure to CO_2 concentrations above ~1000 ppm has been shown to cause transient increases in heart rate and decreases in breathing rate and blood oxygen saturation [6]. Electroencephalogram (EEG) studies also suggest that moderate elevations in blood CO_2 reduce alertness and promote drowsiness—manifested by increased alpha and theta wave activity during cognitive tasks conducted in rooms with ~1200 ppm CO_2 [7,8]. Long-term exposure to poorly ventilated indoor environments may eventually lead to irreversible health problems, particularly in the respiratory and cardiovascular systems [9]. Vulnerable populations—such as the elderly, obese individuals, or those with chronic illnesses—are especially susceptible to the adverse effects of poor IAQ [10].

Importantly, there are positive implications: improving IAQ through enhanced ventilation and filtration can yield significant health benefits. Engineering interventions such as high-efficiency ventilation systems or HEPA filters reduce pollutant concentrations, resulting in improved lung function, decreased irritation symptoms, and enhanced cognitive performance among occupants [11]. Reviews of recent studies suggest that maintaining high indoor air quality may potentially lower the incidence of respiratory illnesses and even certain neurological disorders at the population level [12]. In summary, ensuring clean and well-ventilated interiors is a vital element of preventive health strategy in the built environment.

2.1.2. Natural and Artificial Lighting

Daylight plays a fundamental role in regulating human circadian rhythms, influencing sleep, mood, and overall hormonal balance. Modern societies spend the majority of the day indoors, often with limited access to natural sunlight. Field studies have confirmed that increased exposure to daylight indoors improves circadian health and psychological well-being. For instance, an experiment involving residents of apartments equipped with smart electrochromic windows showed that greater exposure to daylight during the day led to a marked improvement in circadian regulation: participants fell asleep approximately 22 min earlier, exhibited more regular sleep patterns, and reported higher levels of energy and positive mood throughout the day [1]. In weeks when the same individuals had reduced light exposure (e.g., covered traditional windows), their evening melatonin onset was delayed by an average of 15 min, and they reported lower vitality and more negative mood during the day [5]. Both interventional and observational studies support the association between interior lighting conditions and mental health. Individuals working in offices with windows and substantial daylight exposure tend to experience better sleep quality, longer nighttime rest, and higher levels of physical activity compared to those working in windowless spaces [3]. In one study, office workers with windows slept longer on average, reported better mood, and experienced less fatigue than their counterparts in artificially lit rooms [3]. Conversely, inadequate access to daylight has been linked to an increased risk of depressive and anxiety symptoms. For example, in a large survey involving 593 adults, those who spent more time outdoors and in brighter interiors were less likely to suffer from anxiety, depression, and insomnia [13]. Proper lighting in educational spaces also contributes to improved learning outcomes, as demonstrated in studies conducted across 12 European countries [14].

2.1.3. Interior Acoustics (Noise and Acoustic Comfort)

Acoustic comfort is another essential dimension of a healthy indoor environment. Excessive indoor noise—whether from loud equipment, reverberation, or conversations—can be a source of stress and reduced well-being. Research indicates that noise levels in many facilities, particularly hospitals and open-plan offices, often exceed the recommendations of the World Health Organization, and individuals exposed to such noise frequently report annoyance manifested as irritability, fatigue, and difficulty concentrating [15,16]. Noise is recognized as a significant environmental stressor; in hospital settings, for example, it is often cited as one of the main factors negatively impacting patients' anxiety levels and psychological comfort [17,18]. The adverse effects of noise on physiological health are well documented. Prolonged exposure to elevated noise levels has been linked to elevated heart rate, increased blood pressure, and dysregulation of the autonomic nervous system. Chronic exposure can lead to cardiovascular strain, reduced sleep quality, and impaired immune function. In educational and workplace settings, poor acoustics contribute to cognitive overload, reduced productivity, and learning difficulties [19].

In summary, noise is not only a nuisance but also a significant risk factor for both mental and physical health. Strategies to mitigate its impact include the use of sound-absorbing materials, zoning of noisy and quiet areas, proper spatial layout, and technological interventions such as active noise control systems. Ensuring acoustic comfort is a key requirement in designing spaces that support human well-being and health.

2.1.4. Interior Aesthetics and Color Design

The aesthetic qualities of interior spaces—particularly color, texture, shape, and the presence of artwork—have a significant impact on users' emotional and cognitive well-being. A growing body of research in environmental psychology and neuroarchitecture indicates that environments perceived as aesthetically pleasing can evoke positive emotions, reduce stress, and improve overall mental health [20]. Color in particular plays a crucial role in shaping mood and psychological response. Warm colors (such as reds and yellows) tend to stimulate and energize, while cool colors (such as blues and greens) generally have a calming and relaxing effect [21]. Studies have shown that blue and green interiors are associated with lower levels of perceived stress, greater feelings of safety, and increased satisfaction with the space [22]. In therapeutic or care settings, the use of calming color palettes has been linked to better emotional regulation among patients and residents. Furthermore, incorporating visual art into interior spaces—such as paintings, sculptures, or installations—can promote cognitive stimulation and emotional reflection. Several studies have demonstrated that exposure to visual art in hospitals, workplaces, or educational institutions is associated with decreased anxiety, improved mood, and a greater sense of identity and belonging [23,24]. In addition to passive viewing, participation in art-related activities (e.g., art therapy) has shown therapeutic benefits, particularly for individuals with mental health challenges or those at risk of social exclusion. Art therapy leverages creative expression as a tool for emotional regulation, social interaction, and self-reflection [25].

In sum, interior aesthetics—including color and the presence of art—should be regarded as more than decorative aspects. They constitute meaningful components of salutogenic design, supporting users' psychological well-being and offering sensory experiences that contribute to a restorative and engaging built environment.

2.1.5. Ergonomics and Spatial Layout

Ergonomics and the spatial arrangement of interior environments play a critical role in ensuring comfort, safety, and the functional efficiency of buildings. Well-designed spaces that account for human scale, natural movement patterns, and accessibility help reduce physical strain, improve navigation, and foster a sense of autonomy and control over the environment [26]. Poor ergonomic conditions—such as improperly designed furniture, awkward spatial layouts, or inadequate clearance—can contribute to musculoskeletal discomfort, fatigue, and frustration. For example, studies in office environments have shown that improperly adjusted seating and desk heights are associated with increased reports of back and neck pain [27]. Similarly, in residential and healthcare settings, poorly planned layouts can hinder daily functioning, reduce mobility, and elevate the risk of accidents, especially for elderly or disabled individuals [28]. Conversely, environments designed with ergonomic principles in mind improve users' functional independence and satisfaction. For example, clear spatial organization, intuitive circulation paths, and minimized physical barriers facilitate orientation and ease of movement. The concept of "universal design," which emphasizes accessibility for people of all ages and abilities, has been widely adopted in educational and healthcare architecture as a salutogenic approach that supports both physical and psychological well-being [29].

Additionally, spatial layout affects users' cognitive mapping and perception of control. Open, well-structured environments with clear sightlines and logical zoning help reduce spatial confusion and stress, while also promoting a sense of safety and empowerment. In hospital design, for instance, decentralized nursing stations and patient-centered layouts have been associated with reduced anxiety and improved satisfaction among patients [30].

Thus, ergonomic design and thoughtful spatial planning not only contribute to physical comfort and safety but also support cognitive clarity and emotional well-being, making them integral to the salutogenic quality of interior environments.

2.1.6. Spatial Layout and Interior Organization

The final aspect discussed here is the spatial configuration of interiors—including room arrangement, occupant density, and the level of privacy provided by the space. This domain influences health both physically and psychosocially. In office environments, numerous studies have compared open-plan offices with cellular (private) office layouts. Although open spaces promote interaction, they often prove less favorable for employee health. A 2022 systematic review and meta-analysis revealed that employees in open-plan offices were significantly more likely to take sick leave than those working in private rooms. The average risk of taking a sick day was estimated to be approximately 24% higher (OR $\approx$ 1.24) in open-plan offices compared to private offices [31]. Another study reported that individuals in open-plan environments missed 62% more workdays due to illness than their counterparts in enclosed offices [32]. Several factors may explain this phenomenon: increased transmission of infectious diseases in crowded shared spaces (e.g., easier spread of droplet-borne viruses), as well as heightened stress and mental strain caused by noise, lack of privacy, and frequent distractions [33,34]. Support for the infectious disease hypothesis comes from hospital settings—transitioning from multi-bed wards to single-patient rooms has been associated with reduced hospital-acquired infection rates (e.g., MRSA), and improved sleep and rest for patients due to fewer nighttime disruptions [35,36]. Privacy and environmental control are crucial factors influencing mental well-being in spatial contexts. Studies in workplace settings indicate that employees lacking privacy and quiet retreats (common in open-plan layouts) report lower job satisfaction, higher stress, and poorer social relationships than those working in enclosed offices [37,38]. The constant sense of being observed or at risk of distraction can lead to chronic psychological tension and fatigue. Conversely, excessive isolation is also detrimental—contemporary trends such as Activity-Based Working (ABW) seek to balance these needs by providing diverse zones within offices: open collaborative areas alongside small rooms for focused work or rest [39,40]. Preliminary studies suggest that relocating employees to ABW environments reduces perceived distractions (e.g., noise, movement) while preserving the benefits of enhanced inter-team collaboration [40]. Thus, balance is key—spaces should enable both social interaction and seclusion when needed. This principle also applies to schools (e.g., quiet corners for individual work) and homes (e.g., dedicated private areas within apartments). Another important element of spatial layout is legibility and accessibility—the ease of movement and orientation within a building. Disorganized, crowded layouts can cause disorientation and stress, particularly among older adults or individuals with cognitive impairments. In contrast, thoughtful design with clear sightlines, signage, and logical zoning improves users' sense of safety. For instance, in a Swedish emergency department redesign, strategic use of color schemes to delineate zones (neutral tones in public areas vs. bright colors in staff zones) significantly improved patients' and families' subjective sense of orientation and security [41,42].

In addition to spatial functionality and ergonomics, interior layout plays a vital role in shaping emotional experience and supporting psychosocial well-being. Thoughtful design strategies—such as creating quiet zones, providing restorative spaces, and supporting intuitive wayfinding—can reduce anxiety, promote calmness, and enhance a sense of autonomy and control, particularly in healthcare settings. Research on design for health emphasizes the need to accommodate the diverse needs of patients and visitors, especially in emotionally charged environments like hospitals or rehabilitation centers. Malchiodi points out that spaces enriched with elements of expression and symbolism can serve as emotional anchors and support healing, especially when art is used as a therapeutic medium [43]. Similarly, Carpman and Grant highlight that spatial legibility and supportive environments positively influence not only user orientation but also the quality of interpersonal interactions in healthcare facilities [44].

In summary, architectural layout influences health by modulating social interactions, stress levels, and exposure to harmful factors. Lower occupancy density and increased privacy generally support reduced disease transmission and stress (though potentially at the cost of less social integration). Evolving office concepts and evidence-based healthcare design aim to create environments that are both socially stimulating and psychologically comfortable. Research supports this direction—investments in healthy spatial organization translate into tangible health benefits, including reduced sick leave, lower levels of anxiety and depression, and higher satisfaction and efficiency among building occupants [31,33].

To deepen the analysis and provide it with a stronger empirical foundation, a comparative and triangulated methodological approach was applied, combining the review of literature and legal documents with selected case studies, post-occupancy evaluations (POE), and both quantitative and qualitative indicators.

Case studies. The study incorporated examples of buildings from Poland and other European countries (hospitals, schools, and office buildings) where health-promoting, salutogenic solutions had been implemented and documented in scientific literature or professional reports. The analysis of these cases made it possible to capture differences between minimum legal requirements and voluntary standards (e.g., WELL, BREEAM), as well as to highlight their practical implications for users.

Post-Occupancy Evaluations (POE). The research was further supplemented with qualitative and quantitative data concerning user experiences of buildings. Available surveys and reports assessing satisfaction with indoor air quality, acoustics, lighting, thermal comfort, and ergonomics were included. Such evaluations enabled the juxtaposition of legal regulations with the actual experiences and perceptions of building occupants.

Quantitative and qualitative indicators. The analysis also relied on empirical findings published between 2015 and 2025 that reported measurable environmental and health-related parameters. Examples include CO_2 and particulate matter concentrations (IAQ), noise levels in decibels, daylight and artificial lighting intensity (lux), absenteeism rates in open-plan offices, and self-reported well-being and stress among building users. By comparing these data with legal requirements and voluntary standards, it was possible to identify gaps and areas where legislation does not yet account for salutogenic factors, despite their proven significance for health.

2.2. Identification of Salutogenic Design Factors

In the second stage, a literature review was conducted to explore the concepts and guidelines of salutogenic design in architecture. This concept has been developed, among others, by Alan Dilani and Jan A. Golembiewski in the context of healthcare architecture. Based on the publications by Dilani (2008) and Golembiewski (2017), a set of the most commonly proposed salutogenic factors influencing building design was identified [8,9].

These factors include the following:

F1 Access to nature and restorative environments,
F2 Access to daylight,
F3 Thermal comfort and air quality,
F4 Space for social support,
F5 Control over the environment and a sense of privacy,
F6 Acoustic comfort,
F7 Sensory stimulation,
F8 Spatial legibility and navigation,
F9 Diversity and encouragement of physical activity,
F10 Sense of identity and meaning of place.

Access to nature and restorative environments: Incorporating elements of nature (vegetation, water, natural materials) and providing visual and physical contact with nature facilitate psychological regeneration. Studies show that people perceive natural environments as more conducive to relaxation and stress reduction than urban settings. This is confirmed by the theory of Restorative Environments, which states that natural settings allow for mental rest through the mechanism of "effortless attention" [45]. Contemporary research in environmental psychology, biophilic design theory, and neuroscience confirm that contact with nature—both direct and indirect—has a strong restorative effect on the human psyche, reduces stress levels, lowers blood pressure and heart rate, and improves concentration and general psychological well-being [46,47]. The Japanese practice of Shinrin-yoku, also known as forest bathing, has been shown to reduce stress and boost the immune system. According to researchers including these effects can be simulated in interior environments using biophilic stimuli [25].

Access to daylight and appropriate lighting: Maximizing natural light and ensuring adequate artificial lighting (in terms of spectrum and intensity) support circadian rhythms, well-being, and overall health. Access to daylight has been shown to improve mood and performance in office workers [48,49], while dynamic lighting supports night-time sleepiness and daytime alertness in older adults [50,51]. Systematic reviews also confirm the role of natural light exposure in accelerating patient recovery and improving well-being in hospital environments.

Thermal comfort and air quality: Maintaining optimal temperature, humidity, and air purity (through proper ventilation and pollution control) enhances physical comfort and, from the salutogenic perspective, is a prerequisite for long-term mental well-being [52]. Studies show that improving indoor air quality, especially in offices, increases productivity and reduces respiratory symptoms. Thermal comfort affects mood and cognitive functions, especially in educational settings, and occupants of naturally ventilated buildings report greater satisfaction with indoor conditions [11,12].

Spaces for social support: Designing environments that foster interpersonal interactions, integration, and the building of social bonds. This includes common areas (e.g., lounges, cafés, plazas) that encourage both formal and informal social encounters. Social support is a powerful moderator of environmental stress and significantly affects both physical and mental health. Shared spaces in residential areas increase social interaction and positively influence perceived quality of life. The presence of green spaces in communal areas has also been shown to foster social bonding and reduce loneliness in urban populations [53–55].

Control over the environment and a sense of privacy (crowding reduction): Design should enable users to choose whether to engage in social interaction or retreat to private zones. Avoiding overcrowding through spatial generosity and appropriate ceiling height reduces stress and feelings of claustrophobia. For example, research in student housing shows that the length of corridors and room layouts affect users' sense of control and perception of crowding [56,57]. The classic concept of the "behavioral sink," introduced by ethologist based on experiments with rodents, illustrates the detrimental effects of overcrowding and lack of social control [39]. Though conducted on animals, these findings are often cited in relation to human environments, particularly overcrowded and poorly designed institutional or residential spaces.

Acoustic comfort (low noise levels): Minimizing unwanted noise and vibration and ensuring good sound insulation. Excessive noise is linked to irritability, impaired communication, and hindered rest. As one of the key components of a salutogenic environment, noise reduction supports mental focus, regeneration, and general well-being [58].

Sensory stimulation: Color and art in the environment: Conscious use of colors, materials, and forms to evoke positive visual and emotional responses. Warm colors (reds, yellows) are stimulating, while cool colors (blues, greens) are calming. Art elements (paintings, sculptures, installations) enhance cognitive and emotional experiences and are used therapeutically (art therapy) to support mental health [25,46]. Aesthetic surroundings foster attention restoration [45] and reduce stress levels. Engaging with beauty activates brain areas associated with pleasure, reflection, and meaning-making. Spatial legibility and wayfinding (landmarks): Use of distinct visual cues (e.g., architectural elements, artworks, color-coded signage) to facilitate navigation and orientation [59]. A well-organized layout reduces stress and boosts autonomy. In hospitals and elder care settings, accessible navigation systems—based on color, texture, and light—are essential for safe, independent functioning [60,61].

Diversity and promotion of physical activity: Providing varied spatial experiences (e.g., surprise elements, spatial discovery) and encouraging movement through well-designed staircases, walking paths, or public spaces. Physically active environments improve mood, reduce stress, and support physical health. According to spatial variety and opportunities for movement contribute to both physical and cognitive well-being [48,49].

A sense of identity and meaning of place: Creating environments that evoke meaningfulness—by referencing local culture, history, and values. Spaces that tell a story or incorporate symbolic architecture enhance users' sense of belonging and purpose. Antonovsky (1988) emphasized meaningfulness as a pillar of the sense of coherence, and architectural spaces reflecting cultural identity can reduce stress and strengthen psychological resilience [7]. He above list formed the basis for further analysis.

It is important to emphasize that salutogenic factors are complementary in nature—only their holistic integration enables the creation of environments that comprehensively support health. In practice, implementing all of these elements simultaneously can be challenging. However, it is increasingly common to include at least some of them—especially in buildings such as hospitals, schools, and offices, where the health and well-being of users are critical.

2.3. Analysis of Legal Regulations and Industry Guidelines

In the next stage, the identified set of salutogenic factors was compared with the requirements of Polish law and selected industry standards.

National legal framework: The analysis included the Act of 7 July 1994—Construction Law, which constitutes the primary legal act regulating the design, construction, maintenance, and demolition of buildings. It provides general requirements regarding safety, hygiene, health, functionality, and accessibility. Particular attention was paid to the related executive regulation: Regulation of the Minister of Infrastructure of 12 April 2002 on the technical conditions to be met by buildings and their location, which contains detailed technical and construction requirements related to daylight access, ventilation, usable space, noise, temperature, accessibility, etc. Another key regulation was the Regulation of the Minister of Health of 26 June 2012 on the detailed requirements for premises and equipment of healthcare entities (Dz.U. 2012 poz. 739, as amended), which is especially relevant in the context of healthcare facilities, as it indirectly addresses issues of comfort, functionality, and hygiene.

Also included was the Regulation of the Minister of Labour and Social Policy of 26 September 1997 on general provisions of occupational health and safety (Dz.U. 2003 No. 169, item 1650, as amended), which applies to public-use and office buildings, including requirements concerning lighting conditions, noise, ventilation, and ergonomics.

The primary focus of this legal analysis was on regulations concerning the hygienic and health conditions of building use, protection against noise, daylight access, and other aspects of user experience and functionality.

Industry standards and guidelines: As a representative example of non-legally binding initiatives promoting healthy buildings, the WELL Building Standard (WELL v2 version) was selected. This international certification system, developed by the International WELL Building Institute, emphasizes the health, comfort, and well-being of building users. In addition, guidelines from related certification systems (e.g., LEED, BREEAM) were considered in terms of user well-being, along with relevant World Health Organization (WHO) recommendations for healthy residential environments where applicable.

The following diagram presents the research process used in this article. To clarify the research method adopted, a step-by-step flowchart was created [Figure 1].

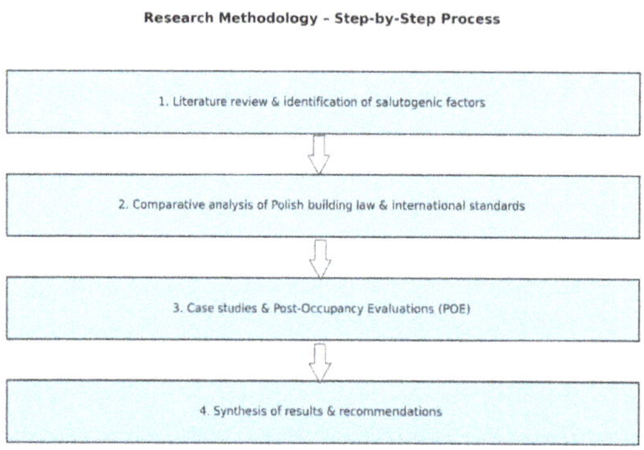

Figure 1. Research methodology applied in the study. The process consisted of four stages: (1) literature review and identification of salutogenic factors; (2) comparative analysis of Polish building law and international standards; (3) case studies and post-occupancy evaluations (POE); and (4) synthesis of results and formulation of recommendations.

3. Results

3.1. The Impact of Building Interior Features on Physical and Mental Health

Across the reviewed evidence (2015–2025), interior environmental quality emerges as a robust determinant of both physical and mental health. Deteriorated indoor air quality (elevated CO2, PM, VOCs) consistently predicts reduced cognitive performance, increased SBS symptoms, and adverse physiological responses, while improvements via ventilation/filtration yield measurable benefits. Adequate daylight and lighting support circadian alignment, sleep quality, mood, and learning outcomes; insufficient exposure is associated with higher risks of anxiety, depressive symptoms, and fatigue. Excess noise is a pervasive stressor in healthcare, education, and offices, impairing sleep, autonomic balance, concentration, and productivity; targeted acoustic strategies are therefore health-critical.

Beyond physical parameters, aesthetics and color (including the presence of art) contribute to reduced stress, improved affect, and greater spatial satisfaction, underscoring their role as nontrivial salutogenic inputs. Ergonomics and spatial layout affect safety, autonomy, wayfinding, and perceived control; user-centered layouts (clear sightlines, intuitive circulation, minimized barriers, universal design) enhance comfort and functional independence. Finally, interior organization and privacy—including density, zoning, and availability of retreat spaces—modulate infection risk, stress, and social functioning; evidence favors configurations that balance interaction with access to seclusion (e.g., ABW in offices; single-patient rooms in hospitals).

In sum, health-supportive interiors require an integrated approach that couples baseline hygienic/safety provisions with psychosocial and experiential qualities. The literature provides both objective metrics (e.g., CO2 ppm, dB, lux) and subjective indicators (well-being, stress, satisfaction) suitable for post-occupancy evaluation and for benchmarking salutogenic performance in subsequent analyses.

3.1.1. Case Studies and Post-Occupancy Evaluations (POE)

In Poland, there is a growing interest in evaluating hospital buildings after their commissioning from the perspective of users, in line with the Post-Occupancy Evaluation (POE) approach and Evidence-Based Design (EBD). Researchers emphasize that architectural solutions in hospitals—including functional and spatial layout—have a significant impact on patients' healing processes and comfort, as well as on the well-being and efficiency of medical staff. In recent years, Polish research centers have undertaken studies addressing the needs of three main user groups: patients, medical (and auxiliary) staff, and visitors. The following section presents selected studies and reports published after 2015 that focus on post-occupancy assessments or related evaluations of the hospital environment in Poland, with particular attention to environmental factors (acoustics, lighting, indoor climate, ergonomics, and contact with nature) and to public and university hospitals (including those located in Kraków).

Lighting comfort is another important aspect of the hospital environment assessed in post-occupancy studies. Pilot measurements conducted by CIOP-PIB in selected hospital rooms demonstrated that illuminance levels in key spaces were often insufficient. In the examined operating theatre, sterilization room, and histopathology laboratory, under-illumination was recorded—lighting levels did not meet the required standards and required improvement [62].

Increasing attention is being paid to thermal comfort and indoor air quality in hospitals, as these factors affect both patients' well-being and staff performance. A recent example is a comprehensive survey published in Building and Environment in 2023, which investigated patients' perceptions of inpatient rooms in Polish [63]. The aim of this study was to determine how hospitalized patients perceive their room conditions—particularly thermal aspects (temperature, drafts) and air quality—and whether individual characteristics (e.g., age, gender) influence these perceptions. The research also examined patients' subjective evaluation of how environmental conditions affect the healing process. The findings demonstrated that patients place significant importance on microclimatic parameters, with thermal comfort emerging as a key factor for overall well-being. At the same time, it was noted that in multi-bed hospital rooms it is very difficult to ensure ideal thermal conditions for every patient due to individual differences (e.g., health status, medication. The authors suggest developing design solutions that enable more individualized control of the environment around each patient (e.g., regulating temperature/airflow at the bedside), which would support recovery. Similarly, the study emphasized the need to update existing regulations, as current standards often remain limited to general ranges of temperature and humidity without fully accounting for the impact of the indoor environment on the healing process.

The body of research indicates that favorable thermal–humidity conditions and clean air are crucial both for patients—accelerating recovery and increasing satisfaction with care and for staff, as they enhance work performance and reduce fatigue. In Poland, work is underway on new design guidelines for ventilation and air-conditioning systems in healthcare facilities, since the currently applicable national recommendations are outdated (some originating as far back as the 1980s) and do not reflect current scientific knowledge. Recent studies (such as the aforementioned patient and staff surveys) provide data that will inform the development of standards ensuring a safe and comfortable indoor environment in hospital wards.

Inadequate lighting at workstations can deteriorate staff comfort and compromise patient safety. The aforementioned survey among surgeons confirmed this issue—nearly two-thirds of operators considered the lighting in operating rooms insufficient and potentially detrimental to the precision of surgical procedures. In response to these problems, many Polish hospitals are modernizing their lighting systems, introducing, for example, LED lighting with adjustable intensity, aimed at improving patients' visual comfort and reducing staff fatigue (as indicated in professional practice reports).

Serious problems with excessive noise, negatively affecting both patients and staff, have been identified in Polish hospitals. A 2017 study conducted by the Regional Sanitary and Epidemiological Station in Rzeszów covered 26 patient rooms in four district hospitals (Podkarpackie Voivodeship) [64].

Measurements of sound levels (day/night) revealed that permissible noise thresholds were exceeded in as many as 58% of the rooms. Three primary sources of noise were identified: medical staff (activities and communication), technical building equipment (e.g., elevators, doors, air-conditioning systems), and the external environment (traffic, parking areas). These findings confirm that patients perceive silence as a key factor of comfort, crucial for recovery. The authors recommended implementing organizational and technical changes in hospitals to reduce noise—often feasible at relatively low cost—since ensuring quiet and calm conditions is conducive to more effective treatment.

The issue of acoustics also extends to the working environment of hospital staff. A survey conducted among surgeons working in operating theatres revealed that more than 86% of surgeons rated the environmental conditions in operating rooms as uncomfortable, and as many as 76% demanded improvements. The most burdensome factors were reported to be the indoor climate and environmental conditions such as noise and lighting—both of which over 62% of surgeons assessed as uncomfortable during their work. Insufficient lighting in operating rooms was noted as a factor that can directly affect the quality of surgical procedures. These conclusions are consistent with international research findings and with WHO guidelines, which recommend very low noise levels in hospital environments (around 30–40 dB) in order to safeguard patient well-being and support staff concentration [65].

Subjective assessment of indoor air quality and thermal comfort in hospital rooms plays a key role in patients' perceived well-being and may influence the healing process. A recent survey conducted in Polish hospitals revealed that patients are particularly sensitive to factors such as air quality, drafts, and difficulties in maintaining individualized thermal comfort—especially in multi-bed rooms. Diverse needs stemming from patients' health status, age, or medication use mean that standardized parameters often fail to accommodate everyone. The authors suggest that hospital environments should be designed to enable more personalized microclimate control at the bedside [66].

In 2020, an analysis of hospital functionality and workplace ergonomics was published in the context of an aging medical workforce [67]. The author argued that the traditional approach to hospital design as merely a "building with infrastructure for carrying out medical procedures" is insufficient; instead, a more humanistic approach is required, one that takes into account the needs and limitations of users (physicians and nurses). It was emphasized that improving working conditions (ergonomics) can help prevent professional burnout and positively influence the quality of patient care. Recommendations include optimizing the work environment by improving hospital circulation systems, eliminating unnecessary noise and stressors, providing rest areas for staff, and adapting workstations to ergonomic requirements—all of which, in turn, enhance patient safety.

Contact with nature and the therapeutic hospital environment promotes stress reduction, alleviates pain, and may shorten recovery time. In Poland, the idea of creating hospital gardens with therapeutic functions is also gaining importance. A review of the literature and case studies conducted by Czałczyńska-Podolska and Rzeszotarska-Pałka (West Pomeranian University of Technology in Szczecin) illustrated how hospital gardens can become spaces of therapy and convalescence for patients [68]. The authors identified features that a hospital garden with high therapeutic potential should fulfill: a sense of familiarity and safety, provision of diverse stimuli and environmental variety, and rich sensory qualities (visual, olfactory, and auditory stimuli). These three components—integrated into appropriate spatial arrangements—determine the friendliness of a garden for different user groups (chronically ill patients, rehabilitation patients, visitors, as well as staff seeking respite). The publication also presented specific examples of such gardens, emphasizing that fulfilling these criteria makes a garden an integral part of the therapeutic process, enhancing patients' well-being and psychological comfort [69].

In clinical practice in Poland, these recommendations are increasingly being implemented. A recent example is the sensory garden opened in 2025 at the Małopolska Rehabilitation Center in Radziszów near Kraków. This project—designed for children and adolescents undergoing long-term rehabilitation—aims to provide a safe green space offering relief from hospital routines while stimulating the senses through horticultural therapy. As the hospital's deputy director stressed, "Nature has an extraordinary power to support healing and reduce stress... Contact with nature is not only a moment of respite, but also an important element of rehabilitation". The establishment of such gardens responds to the needs expressed by specialists and users alike, demonstrating an understanding that the quality of the healing environment extends beyond the walls of the building, encompassing outdoor spaces that provide patients with psychological comfort.

Although formal POE projects in Polish hospitals are only beginning to gain momentum, several completed evaluation studies of this type can already be identified that address the overall assessment of healthcare facilities. Between 2015 and 2021, a research team from Poznań University of Technology, led by Professor Ewa Pruszewicz-Sipińska, conducted a multi-stage study on contemporary trends in healthcare architecture, incorporating the philosophy of Evidence-Based Design (EBD) As part of this work, new hospital projects were analyzed with regard to user comfort and functionality [70]. The 2021 report summarizing this project highlighted the strong potential of applying POE/EBD in the design of hospitals in Poland, as well as the need for broader use of scientific knowledge by investors and decision-makers. The authors emphasized that the conscious use of research findings (e.g., on the impact of noise, interior color schemes, or access to daylight) can significantly improve both patients' experiences and staff working conditions—an approach reinforced by increasing social pressure to raise the quality of healthcare services.

3.1.2. Results of Empirical Research Published in 2015–2025

The study by (Ildiri et al. 2022) examined the effectiveness of WELL certification in improving working conditions from the perspective of office employees, while the study by (Marzban et al. 2023) compared 14 open-plan offices in the Asia–Pacific region (Australia, New Zealand, Hong Kong) [71,72]. The authors of the first study analyzed more than 1300 survey responses collected before and after relocating the same group of employees (from six North American companies) from uncertified offices to WELL-certified offices. The results demonstrated a significant improvement in workplace satisfaction, self-reported health, well-being, and productivity after the transition to WELL offices. For example, the percentage of employees satisfied with their workplace increased from 42% before relocation to 70% after relocation (an increase of about 28%); the overall well-being index increased by 26%, while mental health and self-rated productivity indicators improved by about 10% each. Statistical analyses indicated that most aspects of workplace satisfaction improved significantly—the mean ratings increased by about 1.1 points on a 7-point scale, and large effects (practically meaningful changes) were observed for 8 out of 12 satisfaction parameters. In the second study cited above, 1403 survey responses from employees in the examined offices were analyzed, focusing on factors affecting work efficiency and well-being. The objective was to identify the key drivers of productivity, creativity, and health in high-performance office environments and to compare satisfaction and perceived health among employees in WELL-certified versus uncertified offices, in order to assess whether WELL certification translates into improved user outcomes. The findings revealed that the primary factors supporting productivity were contact with the external environment (e.g., views of greenery, access to nature) and visual privacy in the workplace. For creativity, the most critical factor was the layout and interior design of the office; for mental health, organizational factors (e.g., management style and workplace culture) played the key role; for physical health, privacy and indoor air quality were most important; and for overall health, again, contact with the external environment was the most important factor. The key results of the WELL certification studies discussed here are presented in the diagram below [Figure 2].

Research by developed a transdisciplinary research model combining environmental psychology, public health, and architecture [15]. Within this framework, field studies were conducted in primary schools to analyze features of the learning environment (e.g., the quality of classroom acoustic insulation and visual clarity, understood as access to daylight, outdoor views, and spatial legibility). At the same time, children's stress levels were assessed via cortisol samples (saliva tests), and their attention capacity was measured using standardized tests and behavioral observation during lessons. Correlational analyses showed that better acoustics (lower noise, good sound insulation) and higher quality of visual conditions (abundant daylight, views of greenery, orderly space) were associated with lower cortisol levels during classes and with better concentration and reduced impulsivity. In other words, children learning in classrooms with quieter, calmer surroundings and better natural lighting exp—such as ensuring acoustic comfort and high-quality visual conditions in classrooms—can help reduce stress in children (as reflected by lower cortisol) and enhance their ability to focus and learn. The findings provide policymakers and school designers with scientific evidence to treat the school building as an active ally in promoting health: through appropriate architecture, it is possible to support children's well-being and cognitive functions, which in the long term may contribute to better educational and health outcomes in younger generations.

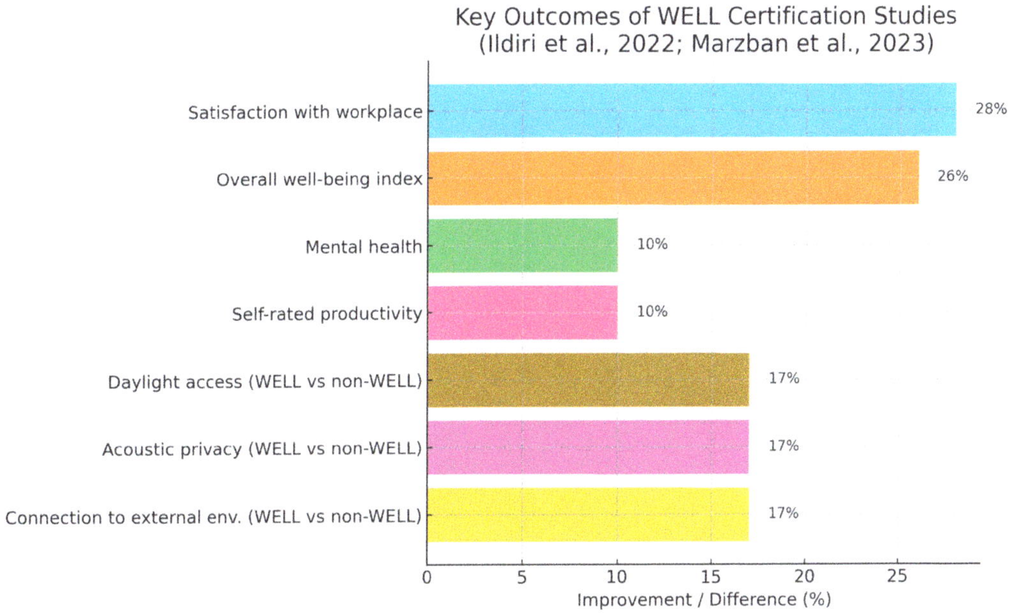

Figure 2. Key outcomes of WELL certification studies. Comparison of user-reported improvements in workplace satisfaction, overall well-being, mental health, and productivity after relocation to WELL-certified offices, as well as differences in daylight access, acoustic privacy, and connection to the external environment between WELL-certified and non-certified offices in the Asia–Pacific region [71,72]. Author: Agnieszka Rek-Lipczyńska.

3.2. Comparison of Legal Requirements and Salutogenic Design Factors

The analysis followed a qualitative comparison methodology, examining whether each salutogenic design factor:

(a) is strictly required by national legislation (or normative acts with legal force),
(b) is recommended or indirectly required by industry standards,
(c) is not present in any official legal or professional requirements.

In Poland, voluntary certification systems such as the WELL Building Standard are currently the most commonly applied benchmarks for health-oriented building design. Unlike in some other European countries, Poland does not yet have widely adopted national standards addressing salutogenic factors beyond the minimum building law requirements. For this reason, WELL is used here as a representative international framework. To provide a broader regulatory context, an additional comparative table (Table 1) has been included, juxtaposing Polish regulations with selected European EN standards relevant to salutogenic design.

Table 1. Inclusion of Salutogenic Design Factors in Legal Regulations and Industry Standards. Author: Agnieszka Rek-Lipczyńska.

No.	Salutogenic Factors	Inclusion in Polish Mandatory Regulations (Construction Law, Technical Conditions)	Inclusion in Industry Standards (e.g., WELL Building Standard)
1	Access to nature and restorative environments	No—direct requirements (there is no obligation to have internal greenery or a view to the outside, apart from the standards regarding plot sunlight).	Yes—the standards recommend biophilic elements (e.g., WELL Feature Biophilia).
2	Access to daylight	Yes—requirement for daylighting of rooms intended for people (condition: minimum window area and sunlight).	Yes—emphasis placed on access to daylight (WELL Light category)
3	Air quality	Yes—mandatory requirements regarding natural or mechanical ventilation, pollution limits (CO_2, other) in rooms.	Yes—very strongly emphasized (WELL Air category—air quality control, filtration, etc.).
4	Thermal comfort	Yes—requirements regarding thermal insulation and heating; minimum temperature in utility rooms (e.g., 20 °C) specified in the regulations.	Yes—recommendations regarding thermal comfort (WELL Thermal Comfort—compliance with comfort standards).
5	Acoustic comfort	Yes—standards for acoustic insulation of partitions and permissible noise levels in residential and public buildings.	Partially—the standards recommend good acoustics (e.g., WELL Sound), although not always obligatory.
6	Common and social space	No—the regulations do not require the creation of additional integration spaces (apart from, for example, the requirement for a common room in boarding schools or a playground in larger housing estates).	Yes—WELL and other recommendations promote recreational spaces and common areas for users (impact on the Community).
7	Possibility of privacy, avoiding crowds	Partially—indirectly through the standards of minimum living space per person (e.g., at least 8 m^2 per person in a living room), but the lack of regulations regarding the subjective sense of privacy.	Yes—the standards suggest designing taking into account various zones (public, semi-private, private) for the comfort of users.
8	Readability of orientation (wayfinding)	No—apart from the requirements regarding marking escape routes, there are no regulations regarding orientation facilities (e.g., there is no requirement for "landmarks").	Yes—good layout design and space marking recommended (e.g., Design for mind in WELL).
9	Aesthetics, colors, art	No—aesthetic and color issues are not subject to legal standards (except, for example, safety colors for occupational health and safety markings).	Yes—e.g., WELL in the Mind category encourages you to include art, biophilic design, and elements that improve the mood of user.
10	Physical activity and movement	Partially—the regulations require infrastructure for activities only in specific cases (e.g., playgrounds at schools). Generally, there is no requirement, e.g., to encourage the use of stairs.	Yes—the standards (WELL Fitness) promote solutions that encourage exercise: availability of stairs, physical activity zones in buildings.

To organize the results, a summary comparison table was developed (Table 1), (Figure 3), presenting the classification of each design factor. Table 1 presents a comparative summary of key salutogenic design factors in relation to the corresponding requirements of Polish construction law (including relevant executive regulations) and the recommendations found in professional industry standards, using the WELL Building Standard as a representative example.

Figure 3. Showing the degree of consideration of the postulated salutogenic factors in Polish construction law and industry standards. Author: Agnieszka Rek-Lipczyńska.

The comparison presented in the table and diagrams clearly indicates that Polish legal regulations concerning building design primarily focus on the physical and hygienic aspects of the environment (lighting, ventilation, heating, noise) and on user safety, while they only marginally address psychological and social salutogenic factors. Certain elements—such as the required minimum daylighting or ventilation—can be considered consistent with the salutogenic approach. However, these are minimum requirements aimed more at preventing harm (e.g., hypoxia, darkness) than at maximizing positive health outcomes. Other factors, such as the integration of nature into interiors, the provision of social spaces, or the use of decoration and color schemes to enhance well-being, are not covered by any mandatory regulations. As a result, their implementation depends largely on the awareness and initiative of the investor or designer. In contrast, industry standards such as the WELL Building Standard cover nearly all of the identified salutogenic factors (divided into categories such as Air, Light, Thermal Comfort, Sound, Community, and Mind), though their application is voluntary. The WELL standard explicitly places user health and comfort at the center of its framework: it is a certification system specifically designed to improve the health and well-being of people spending time in buildings.

In Northern European countries, the integration of salutogenic design principles and WELL certification requirements is increasingly evident. Countries such as Sweden have long required, through their building regulations (Boverkets Byggregler, BBR), that every habitable room provide access to daylight through windows, with glazing corresponding to at least 10% of the floor area (equivalent to a daylight factor of approximately 1% [1]). However, the current approach goes further. In recent years, Sweden has updated the BBR in line with EN 17037:2018 Daylight in buildings. Brussels: CEN., placing greater emphasis on the quality of daylight. References to EN 17037:2018 Daylight in buildings. Brussels: CEN [73].now appear as both an assessment method and as design guidance—for example, Boverket documents highlight new recommended daylight illuminance levels (lux) instead of relying solely on a fixed daylight factor value [3]. Importantly, Swedish regulations now also incorporate the concept of "utblick"—the requirement to ensure a view to the outside. Every dwelling room must have a window with a view of an open outdoor space, reinforcing a sense of connection with the environment and supporting the salutogenic aspects of comprehensibility and meaningfulness of space. Although the BBR has not yet established criteria for glare protection from daylight, regulatory trends indicate a growing interest in this issue.

To complement the tabular comparison, Figure 4 illustrates the extent to which salutogenic factors are reflected in Polish regulations compared with European EN standards. The results indicate that Polish technical requirements cover mainly basic hygienic parameters—such as minimum daylight factors, ventilation rates, or acoustic insulation thresholds—without addressing broader qualitative aspects of user well-being. In contrast, European EN standards (e.g., EN 17037, EN 15251/EN 16798, EN ISO 3382) provide more detailed and holistic criteria, encompassing daylight quality, access to external views, acoustic comfort, thermal categories, and indoor air quality thresholds [74–76]. This divergence highlights a regulatory gap: while European frameworks increasingly align with salutogenic principles, Polish regulations remain limited to pathogenic protection.

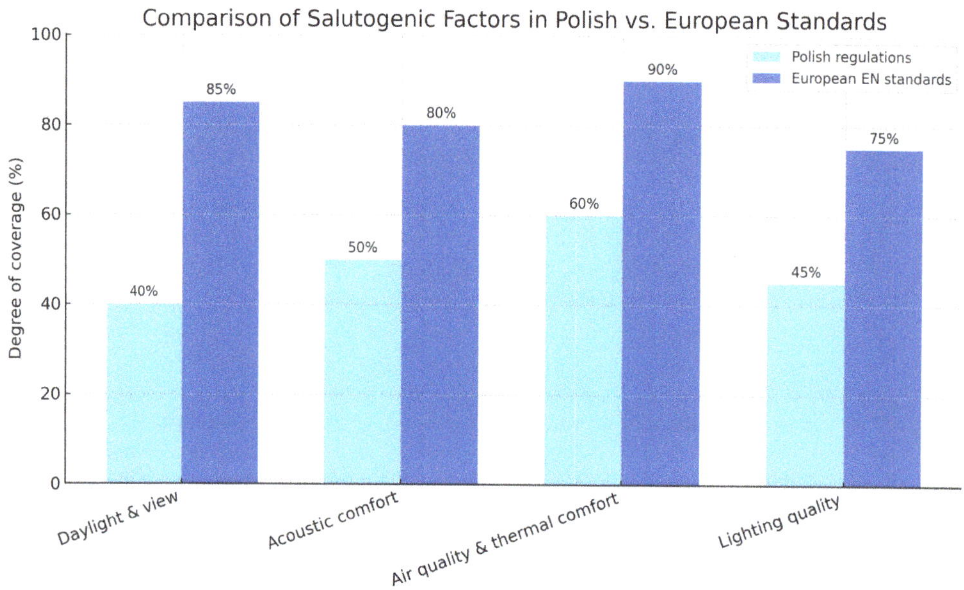

Figure 4. Comparison of salutogenic factors in Polish regulations versus European EN standards. Author: Agnieszka Rek-Lipczyńska.

The diagram illustrates the relative degree to which selected factors are addressed: daylight and access to view, acoustic comfort, indoor air quality and thermal comfort, and lighting quality. Polish regulations reflect mainly basic hygienic and technical requirements, while European EN standards provide broader, more detailed coverage, aligning more closely with salutogenic design principles.

In practice, this means that buildings aspiring to WELL certification must meet a number of criteria that go beyond legal requirements—for instance, ensuring access to relaxation areas, providing views of greenery from workspaces, using low-toxicity finishing materials, introducing biophilic design (nature-based elements within interiors), and implementing programs that promote physical activity among users. Comparable requirements can be found in other green certification systems such as LEED or BREEAM, particularly in the category of Indoor Environmental Quality (IEQ), although their emphasis on salutogenic aspects is generally less pronounced than in the WELL framework. In Poland, similar approaches are occasionally observed in the design of modern, ecologically certified hospitals and office buildings, but they have not yet become common practice in mainstream building standards. It is worth emphasizing that certifications such as WELL, LEED, or BREEAM are voluntary—meaning their application is not legally mandated and depends entirely on stakeholder initiative.

Limitations and criticism of industry standards:

The adoption of such standards depends largely on investor decisions—in the public sector, certification is less frequently pursued due to cost considerations.

In Poland, the use of WELL or BREEAM is still gaining traction, primarily within the office and hospitality sectors.

There is also a lack of mandatory post-occupancy evaluation mechanisms unless the certification process was commissioned from the outset.

As noted in an article published in Building Research & Information, many building certification systems (e.g., LEED, BREEAM) have been criticized for focusing too heavily on measurable indicators and procedural compliance, while paying insufficient attention to actual user experience or long-term health and environmental outcomes [50]. In a literature review by Zuo and Zhao (2014), the authors highlight that the effectiveness of implementing standards such as LEED or BREEAM is often limited in developing countries or localized contexts where resources, expertise, or regulatory frameworks may be insufficient and sustainable development goals tend to be secondary to cost constraints [53].

3.3. Identification of Gaps and Opportunities for Improvement

The comparative analysis outlined above reveals several concrete gaps in current Polish regulations in the context of salutogenic design: Lack of requirements related to nature and biophilic design: Polish law does not mandate the inclusion of greenery or nature-based elements inside or around buildings, beyond general land-use planning requirements concerning biologically active surface area on a plot. Meanwhile, research strongly emphasizes the health benefits of contact with nature in living or working environments—including stress reduction and improved cognitive functioning.

Opportunity for improvement: Introducing recommendations (even if non-binding) regarding the inclusion of green elements—such as green courtyards, plant walls, or visual access to trees from windows—especially in public-use buildings (e.g., schools, hospitals). This approach aligns with the strategic direction of the European Union, which promotes holistic indoor environmental quality through funding research and implementation projects.

One of the key initiatives in this field is COST Action CA16114 "RESTORE—Rethinking Sustainability Towards a Regenerative Economy," funded under the Horizon 2020 programme. The initiative aimed to establish a regenerative design framework for indoor environments, covering not only physical aspects (thermal comfort, air quality, acoustics, lighting), but also psychosocial values such as sense of place, aesthetics, and spaces for rest and social integration [64]. The project developed a set of Key Performance Indicators (KPIs) for health-promoting architecture, including visual access to nature, use of biophilic materials, spatial diversity, and inclusion of art. Insufficient consideration of social spaces: Polish Building Law does not require the design of shared spaces for social interaction beyond circulation areas. Multifamily housing and office buildings can meet legal standards with just corridors and staircases, lacking any spaces that encourage interaction. Suggested improvement: Introducing incentives for designing communal spaces (e.g., lounges, terraces, courtyards) in residential and office buildings, to support social capital and indirectly enhance mental well-being.

Lack of criteria addressing aesthetics and psychological comfort: Interior color schemes, presence of art, or efforts to create a welcoming atmosphere are not addressed in legal requirements. Although most legal systems do not regulate such aspects, an increasing number of researchers argue for their inclusion in planning regulations or spatial policy frameworks. Rezafar and Sence analyze the need to formally integrate aesthetics into planning law [55]. Vartanian et al. demonstrate that architectural form directly influences users' emotional and behavioral responses to space [56]. Similarly, Ionescu emphasizes that emotional and aesthetic dimensions of interior environments are underrepresented in formal standards, despite their relevance for mental [57] Ward et al. explores links between aesthetics, sensory and synesthetic perception, reinforcing the argument for multisensory design standards [58]. These conclusions are consistent with WHO guidelines, which acknowledge the impact of environmental quality—including aesthetics—on mental and physical health, particularly in vulnerable populations.

Opportunity for improvement: Development of soft guidelines (e.g., in the form of industry standards or universal design manuals) that highlight the importance of aesthetic factors for well-being, intended for voluntary application by designers.

Minimal compliance vs. health optimization: Current legal requirements establish only the minimum acceptable conditions (e.g., minimum daylighting or minimum apartment area). They do not encourage exceeding these thresholds. As a result, many designs merely meet the legal minimum without necessarily offering high-quality living environments. For instance, current regulations mandate only 3 h of direct sunlight in a single room per day around the equinox, effectively allowing most apartment spaces to remain sunless for much of the year.

Recommendation: Introducing motivational mechanisms (e.g., tax incentives or certification schemes) for developers who exceed minimum standards in favor of improving residents' health—e.g., by designing homes with better access to light, quieter interiors, or views of nature.

The literature indicates that effective implementation of salutogenic design strategies requires broad stakeholder involvement—from architects and developers to policymakers and regulatory authorities. An interdisciplinary approach is needed, combining knowledge from architecture, environmental psychology, medicine, and environmental sciences. Although the number of related initiatives and studies is growing internationally, comprehensive legal frameworks promoting health-oriented architecture remain rare. Poland is no exception: there are some isolated recommendations (e.g., by the State Sanitary Inspection regarding lighting and air quality), but no integrated regulatory approach. Standards such as WELL represent progress in the right direction, yet their impact depends on voluntary market adoption.

4. Discussion

The results of the analysis clearly show that current building regulations in Poland provide only a basic level of health protection, primarily focusing on eliminating hazards and ensuring minimal comfort. This approach aligns with the traditional pathogenic model—protecting users from disease, injury, or discomfort. In contrast, the salutogenic concept asks what more can be done to actively enhance people's health and well-being through space. In this perspective, the building is no longer seen as a neutral backdrop but as a tool for health promotion. From a sustainable development standpoint, health should be regarded as a pillar equal to environmental and economic concerns. Sustainable buildings are not only those that consume little energy and have a low carbon footprint, but also those that positively affect their users—ensuring comfort, well-being, and even supporting community development. This approach represents a holistic vision of sustainable architecture, combining environmental goals with health-promoting strategies. Integrating the salutogenic dimension complements the concept of green building by adding a social (well-being) aspect, consistent with the idea of the well-being economy and healthy cities.

Feasibility and measurability. One of the key challenges in integrating salutogenic principles into legal frameworks is their quantification. While it is relatively straightforward to define a minimum daylight factor or ventilation rate, it is much more complex to legislate for qualities such as "a sense of calm" or "aesthetic conditions that foster reflection." Excessive rigidity in regulating these aspects could even be counterproductive. Therefore, qualitative guidelines rather than rigid numerical standards are recommended. Tools such as the Design for Wellbeing Toolkit or the Perceived Restorativeness Scale (PRS) can serve as practical methods of translating qualitative aspects into design guidance.

Costs and economic feasibility. Another crucial dimension relates to costs. Many health-promoting design solutions—such as indoor gardens, more generous spatial layouts, or higher-quality materials—are associated with increased upfront investments. Developers and hospital managers may perceive these as additional burdens. On the other hand, evidence shows that salutogenic interventions can generate long-term savings by reducing disease burdens, lowering healthcare expenditures, and improving staff retention and productivity. Consequently, policy frameworks could combine requirements with incentive mechanisms (e.g., tax relief, subsidies, or preferential loans for projects that incorporate salutogenic design).

Stakeholder acceptance. Successful implementation also depends on broad stakeholder buy-in. Participatory design and Post-Occupancy Evaluation (POE) methods allow patients, staff, and visitors to directly assess environments and provide feedback for improvements. Such approaches enhance acceptance, ensure that solutions are user-centered, and provide valuable data for the further development of regulatory frameworks. Involving different stakeholders early in the design process reduces resistance and increases the legitimacy of salutogenic strategies.

Implementation challenges and cultural context. Salutogenic solutions must be adapted to local conditions and cultural expectations. What is considered restorative or privacy-enhancing in one context may not translate directly to another. For example, daylight access and greenery are universally valued, but aesthetic preferences, privacy norms, or the balance between individual and social needs differ across societies. Polish regulations should therefore be based on universal principles but implemented in ways tailored to national needs and expectations.

Pathways for integration. In conclusion, the findings highlight the importance of balancing mandatory and voluntary approaches. An optimal solution may involve integrating general health-promoting objectives into the Building Law (e.g., explicitly including user well-being alongside safety), while leaving detailed design strategies to voluntary standards and certifications such as WELL or LEED. In this way, the market can gradually adapt salutogenic innovations, while legal frameworks provide guidance and direction without imposing rigid mandates.

5. Conclusions

The conducted analysis reveals a significant gap between the principles of salutogenic design and the minimum requirements imposed by current Polish building regulations. The law focuses primarily on ensuring structural safety and meeting basic hygienic and health-related conditions (in the traditional sense: lighting, ventilation, temperature, noise), while key factors supporting broader well-being—such as contact with nature, sensory stimulation, social support, and the creation of friendly and meaningful spaces—remain largely unregulated. On the other hand, the dynamic development of voluntary building standards (e.g., the WELL Building Standard) and the growing number of health-oriented architectural projects demonstrate both the need and the potential for creating buildings that are more user-friendly than what current legal frameworks require. Such buildings yield measurable benefits: increased productivity in workplaces, improved treatment outcomes in hospitals, and higher residential comfort. This aligns with global trends advocating for human-centered design in buildings and cities—an approach that complements the goals of sustainable development.

The potential costs of implementing salutogenic solutions may be offset by lower expenditures on healthcare and by the broader social benefits of improved well-being. In this context, the principle of "prevention is better than cure" gains special significance—health-supportive buildings represent an investment not only in users but in the economy as a whole. The inclusion of the principles of salutogenesis into construction law in Poland is not only possible but also advisable in the face of the health challenges of modern society. International studies show that a well-designed environment can become a protective factor for health at the population level, reducing the burden of lifestyle diseases and improving the quality and length of life. The 2024 Polish reform of technical conditions represents a promising step in this direction, although it currently addresses only selected aspects.

The results of the author's comparative analysis—including a detailed matrix of regulatory coverage of salutogenic factors—demonstrate that while Polish technical standards address some physical determinants of comfort (light, air, thermal conditions), they fail to incorporate essential psychological, social, and aesthetic components. The model-based study further shows that the integration of salutogenic features (e.g., biophilic elements, clear spatial orientation, zones of rest and privacy, sensory stimulation) enhances users' perceived sense of coherence (SOC)—including comprehensibility, manageability, and meaningfulness—in architectural spaces.

As confirmed by qualitative pilot studies carried out by the author in immersive environments simulating different design parameters, even subtle changes in layout logic, access to nature, acoustic climate, and material expression can significantly affect the perception of space as understandable, supportive, and valuable. These findings support the hypothesis that salutogenic design contributes not only to well-being but also to users' psychological resilience and cognitive orientation in space.

This study demonstrates that a significant gap remains between the principles of salutogenic design and the minimum requirements of current Polish building regulations. While the law ensures structural safety and covers basic hygienic conditions (lighting, ventilation, temperature, noise), it does not adequately address psychosocial and experiential dimensions of health, such as contact with nature, sensory stimulation, social support, or restorative aesthetics.

By integrating findings from post-occupancy evaluations (POE), case studies of hospitals, schools, and offices, as well as international benchmarking, the analysis provides evidence that salutogenic environments generate measurable health, well-being, and productivity benefits. For example, improved acoustic and lighting conditions reduce stress and enhance surgical precision, while therapeutic hospital gardens or WELL-certified office spaces demonstrate tangible psychosocial and economic advantages.

The results highlight three main implications:

Regulatory innovation: Polish building law should move beyond a pathogenic focus and include explicit well-being objectives, aligning with European trends such as the adoption of EN 17037:2018 Daylight in buildings. CEN.daylight standards and user-centered healthcare design [73].

Economic and social value: While salutogenic features may increase upfront investment, they can yield long-term savings in healthcare costs, enhance staff retention, and improve societal well-being—making preventive architecture a viable public health strategy.

Implementation pathways: A balanced approach is required, combining general legal provisions for user well-being with voluntary standards (e.g., WELL, BREEAM) and soft guidelines to ensure flexibility, stakeholder acceptance, and adaptation to local cultural contexts.

Overall, integrating salutogenic thinking into mainstream architectural practice and legislation represents the next step in sustainable development. Future Polish reforms should seize this opportunity by embedding salutogenic principles in technical standards, promoting pilot projects, and fostering interdisciplinary collaboration. Buildings of the future should not merely prevent harm but actively promote health, resilience, and meaningful human experience—thus strengthening both individual well-being and collective societal sustainability.

Funding: This research received no external funding.

Institutional Review Board Statement: Not applicable.

Informed Consent Statement: Not applicable.

Data Availability Statement: The original work presented in this study can be found in the article. If you have any further questions, please contact the corresponding author.

Acknowledgments: While preparing this manuscript, the author used the Chat GPT 4 tool for data collection, generating a diagram and language translation. The author reviewed and edited the results and takes full responsibility for the content of this publication.

Conflicts of Interest: The author declares no conflicts of interest.

References

1. Domaradzki, J. O skrytości zdrowia. O problemach z koncepcją zdrowia. *Hygeia Public Health* **2015**, *50*, 441–447.
2. Engel, G.L. The need for a new medical model: A challenge for biomedicine. *Science* **1977**, *196*, 129–136. [CrossRef]
3. WHO. Constitution of the World Health Organization. 1946. Available online: https://www.globalhealthrights.org/instrument/constitution-of-the-world-health-organization-who/ (accessed on 21 July 2025).
4. Jadczyk, M. *Zdrowie Jako Kategoria Filozoficzna i Pedagogiczna*; Oficyna Wydawnicza Uniwersytetu Zielonogórskiego Publishing: Zielona Góra, Poland, 2009.
5. Antonovsky, A. The salutogenic model as a theory to guide health promotion. *Health Promot. Int.* **1996**, *11*, 11–18. [CrossRef]
6. Eriksson, M.; Lindström, B. Antonovsky's sense of coherence scale and the relation with health: A systematic review. *J. Epidemiol. Community Health* **2006**, *60*, 376–381. [CrossRef]
7. Antonovsky, A. *Unraveling the Mystery of Health. How People Manage Stress and Stay Well*; Jossey-Bass: San Francisco, CA, USA, 1988.
8. Golembiewski, J.A. Salutogenic Architecture in Healthcare Settings. In *The Handbook of Salutogenesis*; Mittelmark, M.B., Sagy, S., Eriksson, M., Bauer, G.F., Pelikan, J.M., Lindström, B., Espnes, G.A., Eds.; Springer Publishing: Cham, Switzerland, 2017; pp. 267–276.
9. Dilani, A. Psychosocially Supportive Design: A Salutogenic Approach to the Design of the Physical Environment. *Des. Health Sci. Rev.* **2008**, *1*, 47–55.
10. Izadpanah, S.; Majedi, H.; Zabihi, H. Wprowadzenie zmiennych architektonicznych zdrowych szkół, zgodnych z teorią salutogeniczną w poprawie jakości umysłowej uczennic szkół średnich w mieście Gorgan. *J. Conserv. Archit. Iran* **2024**, *14*, 103–124.
11. Nie, T.; Zhang, G.; Sun, Y.; Wang, W.; Wang, T.; Duan, H. Effects of Indoor Air Quality on Human Physiological Impact: A Review. *Buildings.* **2025**, *15*, 1296. [CrossRef]
12. Nagare, R.; Woo, M.; MacNaughton, P.; Plitnick, B.; Tinianov, B.; Figueiro, M. Access to Daylight at Home Improves Circadian Alignment, Sleep, and Mental Health in Healthy Adults: A Crossover Study. *Int. J. Environ. Res. Public Health* **2021**, *18*, 9980. [CrossRef] [PubMed]
13. Landvreugd, A.; Nivard, M.G.; Bartels, M. The Effect of Light on Wellbeing: A Systematic Review and Meta-analysis. *J. Happiness Stud.* **2025**, *26*, 1–24. [CrossRef]

14. Zhou, T.; Wu, Y.; Meng, Q.; Kang, J. Influence of the Acoustic Environment in Hospital Wards on Patient Physiological and Psychological Indices. *Front. Psychol.* **2020**, *11*, 1600. [CrossRef]
15. Sander, E.J.; Marques, C.; Birt, J.; Stead, M.; Baumann, O. Open-Plan Office Noise Is Stressful: Multimodal Stress Detection in a Simulated Work Environment. *J. Manag. Organization.* **2021**, *27*, 1358–1376. [CrossRef]
16. Lee, H.; Lee, Y. Effects of Postural Changes Using a Standing Desk on the Craniovertebral Angle, Muscle Fatigue, Work Performance, and Discomfort in Individuals with Forward Head Posture. *Healthcare* **2024**, *12*, 2436. [CrossRef] [PubMed]
17. Rodriguez, T. Color in Health Care: How Mindful Design Can Improve Clinical Spaces. Psychiatry Advisor. 2024. Available online: https://www.psychiatryadvisor.com/features/color-in-health-care-design-spaces/ (accessed on 13 July 2025).
18. Dawood, S. Filling Hospitals with Art Reduces Patient Stress, Anxiety and Pain. 2019. Available online: https://www.designweek.co.uk/issues/22-28-july-2019/chelsea-westminster-hospital-arts-research// (accessed on 11 July 2025).
19. Ustawa z Dnia 7 Lipca 1994 r.–Prawo Budowlane (Tekst Jednolity: Dz.U. 2023 poz. 682 z późn. zm.). 2023. Available online: https://isap.sejm.gov.pl/isap.nsf/DocDetails.xsp?id=WDU20230000682 (accessed on 11 July 2025).
20. Rozporządzenie Ministra Infrastruktury z dnia 12 Kwietnia 2002 r. w Sprawie Warunków Technicznych, Jakim Powinny Odpowiadać Budynki I Ich Usytuowanie (Tekst Jednolity: Dz.U. 2022 poz. 1225 z późn. zm.). Available online: https://isap.sejm.gov.pl/isap.nsf/DocDetails.xsp?id=WDU20220001225 (accessed on 11 July 2025).
21. International WELL Building Institute (IWBI). WELL Building Standard v2. IWBI: Nowy Jork. 2020. Available online: https://www.wellcertified.com/ (accessed on 15 July 2025).
22. Kaplan, R.; Kaplan, S. *The Experience of Nature: A Psychological Perspective*; Cambridge University Press Publishing: Cambridge, UK, 1989.
23. Ulrich, R.S.; Simons, R.F.; Losito, B.D.; Fiorito, E.; Miles, M.A.; Zelson, M. Stress recovery during exposure to natural and urban environments. *J. Environ. Psychol.* **1991**, *11*, 201–230. [CrossRef]
24. Berto, R. Exposure to restorative environments helps restore attentional capacity. *J. Environ. Psychol.* **2005**, *25*, 249–259. [CrossRef]
25. Li, Q. Effect of forest bathing trips on human immune function. *Environ. Health Prev. Med.* **2010**, *15*, 9–17. [CrossRef]
26. Viola, A.U.; James, L.M.; Archer, S.N.; Dijk, D.J. Blue-enriched white light in the workplace improves self-reported alertness, performance and sleep quality. *Scandinavian. J. Work. Environ. Health* **2008**, *34*, 297–306. [CrossRef]
27. Cook, M.B.; Smallman, H.S.; Lacson, F.C.; Manes, D.I. Situation displays for dynamic UAV replanning: Intuitions and performance for display formats. *Proc. Hum. Factors Ergon. Soc.* **2010**, *54*, 492–496. [CrossRef]
28. Figueiro, M.G.; Plitnick, B.; Lok, A.; Steverson, B. Tailored lighting intervention improves sleep and mood in older adults with Alzheimer's disease and related dementia. *Sleep Health* **2017**, *3*, 214–221.
29. Czeisler, C.A. Perspective: Casting light on sleep deficiency. *Nature* **2013**, *497*, S13. [CrossRef]
30. Wargocki, P.; Wyon, D.P.; Sundell, J.; Clausen, G.; Fanger, P.O. The effects of outdoor air supply rate in an office on perceived air quality, sick building syndrome (SBS) symptoms and productivity. *Indoor Air* **2000**, *10*, 222–236. [CrossRef] [PubMed]
31. Zhang, H.; Arens, E.; Kim, D.; Buchberger, E.; Bauman, F.; Huizenga, C. Comfort, perceived air quality, and work performance in a low-power task–ambient conditioning system. *Build. Environ.* **2011**, *46*, 1001–1009. [CrossRef]
32. Frontczak, M.; Wargocki, P. Literature survey on how different factors influence human comfort in indoor environments. *Build. Environ.* **2011**, *46*, 922–937. [CrossRef]
33. Evans, G.W. The built environment and mental health. *J. Urban Health* **2003**, *80*, 536–555. [CrossRef]
34. Kim, J.; Kaplan, R. Physical and psychological factors in sense of community: New urbanist Kentlands and nearby Orchard Village. *Environ. Behav.* **2004**, *36*, 313–340. [CrossRef]
35. Kweon, B.-S.; Sullivan, W.C.; Wiley, A.R. Green common spaces and the social integration of inner-city older adults. *Environ. Behav.* **1998**, *30*, 832–858. [CrossRef]
36. Altman, I. *The Environment and Social Behavior: Privacy, Personal Space, Territory, Crowding*; Brooks/Cole Publishing: Monterey, CA, USA, 1975.
37. Evans, G.W.; Wener, R.E. Crowding and personal space invasion on the train: Please don't make me sit in the middle. *J. Environ. Psychol.* **2007**, *27*, 90–94. [CrossRef]
38. Sundstrom, E. Privacy protection in a behavioral world. *J. Soc. Issues* **1978**, *34*, 47–65. [CrossRef]
39. Calhoun, J.B. Gęstość zaludnienia i patologia społeczna. *Sci. Am.* **1962**, *206*, 139–149. [PubMed]
40. Haapakangas, A.; Hongisto, V.; Varjo, J.; Lahtinen, M. Benefits of quiet workspaces in open-plan offices–Evidence from two office relocations. *J. Environ. Psychol.* **2014**, *38*, 38–44. [CrossRef]
41. Burry, M. Wrażenie Barwne–Wpływ Kolorów Na Ruch I Emocje. In *Konsekwencje Pracy Nad Świadomością Ciała: Podejścia, Doświadczenia, Praktyki*; Tatar, M., Ed.; Wydawnictwo Uniwersytetu Śląskiego Publishing: Katowice, Poland, 2020; pp. 114–116.
42. Ziat, M.; Balcer, C.A.; Shirtz, A.; Rolison, T. The hue-heat hypothesis: Does color truly affect temperature perception? *Sci. Rep.* **2014**, *4*, 5527. [CrossRef]
43. Malchiodi, C.A. *Art Therapy and Health Care*; Guilford Press Publishing: New York, NY, USA, 2012.

44. Carpman, J.R.; Grant, M.A. *Design That Cares: Planning Health Facilities for Patients and Visitors*; John Wiley & Sons: Hoboken, NJ, USA, 2001.
45. Lynch, K. *The Image of the City*; MIT Press Publishing: Cambridge, MA, USA, 1960.
46. Passini, R. Wayfinding design: Logic, application and some thoughts on universality. *Des. Stud.* **1996**, *17*, 319–331. [CrossRef]
47. Carpman, J.R.; Grant, M.A. *Design That Cares: Planning Health Facilities for Patients and Visitors*, 3rd ed.; John Wiley & Sons: Hoboken, NJ, USA, 2016.
48. Frumkin, H.; Bratman, G.N.; Breslow, S.J.; Cochran, B.; Kahn, P.H.; Lawler, J.J.; Levin, P.S.; Tandon, P.S.; Varanasi, U.; Wolf, K.L.; et al. Nature contact and human health: A research agenda. *Environ. Health Perspect.* **2017**, *125*, 075001. [CrossRef] [PubMed]
49. Sallis, J.F.; Cerin, E.; Conway, T.L.; Adams, M.A.; Frank, L.D.; Pratt, M.; Salvo, D.; Schipperijn, J.; Smith, G.; Cain, K.L.; et al. Physical activity in relation to urban environments in 14 cities worldwide: A cross-sectional study. *Lancet* **2016**, *387*, 2207–2217. [CrossRef]
50. Lengen, C.; Kistemann, T. Sense of place and place identity: Review of neuroscience evidence. *Health Place* **2012**, *18*, 1162–1171. [CrossRef]
51. Manzo, L.C. For better or worse: Exploring multiple dimensions of place meaning. *J. Environ. Psychol.* **2005**, *25*, 67–86. [CrossRef]
52. Cole, R.J.; Valdebenito, M.J. The importation of building environmental certification systems: International usages of LEED. *Build. Res. Inf.* **2013**, *41*, 662–676. [CrossRef]
53. Zuo, J.; Zhao, Z.Y. Green building research–current status and future agenda: A review. *Renew. Sustain. Energy Rev.* **2014**, *30*, 271–281. [CrossRef]
54. Mfon, I.E. Aesthetic considerations in architectural design: Exploring pleasure, arousal, and dominance. *Int. J. Res. Publ. Rev.* **2023**, *4*, 923–935.
55. Rezafar, A.; Turk, S.S. Czynniki urbanistyczne wpływające na ocenę estetyczną nowo wybudowanych środowisk i ich uwzględnienie w ustawodawstwie: Przykład Stambułu. *Urbani Izziv* **2018**, *29*, 83–95. [CrossRef]
56. Vartanian, O.; Navarrete, G.; Chatterjee, A.; Fich, L.B.; Leder, H.; Modroño, C.; Skov, M. Wpływ konturu na osądy estetyczne i decyzje o podejściu lub unikaniu w architekturze. *Proc. Natl. Acad. Sci. USA* **2013**, *110* (Suppl. S2), 10446–10453. [CrossRef] [PubMed]
57. Ionescu, A.M. The interior as interiority: Architecture, affect, and the aesthetics of experience. *Palgrave Commun.* **2018**, *4*, 1–9. [CrossRef]
58. Ward, J.; Moore, S.; Thompson-Lake, D.; Salih, S.; Beck, B. Atrakcyjność estetyczna percepcji synestetycznych słuchowo-wzrokowych u osób bez synestezji. *Perception* **2008**, *37*, 1285–1296. [CrossRef] [PubMed]
59. World Health Organization. Environment and Health Risks: A Review of the Influence and Effects of Social Inequalities. WHO Regional Office for Europe. 2010. Available online: https://iris.who.int/server/api/core/bitstreams/ca617634-4487-40f0-a5b9-7babeb66bc82/content (accessed on 15 July 2025).
60. Deuble, M.P.; de Dear, R.J. Czy tu jest gorąco, czy tylko ja tak mam? Walidacja oceny poeksploatacyjnej. *Intell. Build. Int.* **2014**, *6*, 112–134. [CrossRef]
61. Hartig, T.; Korpela, K.; Evans, G.W.; Gärling, T. A measure of restorative quality in environments. *Scand. Hous. Plan. Research.* **1997**, *14*, 175–194. [CrossRef]
62. Smagowska, B. Wyniki Badań i Oceny Hałasu, Oświetlenia i Mikroklimatu na Wybranych Stanowiskach Pracy. Cent. Inst. Ochr. Pr.–Państwowy Inst. Badaw. (CIOP-PIB). 2018. Available online: https://m.ciop.pl/CIOPPortalWAR/file/91827/NA_hlalas_oswietlenie_mikroklimat_na_wybranych_stanowiskach_pracy_wyniki_badan.pdf (accessed on 15 July 2025).
63. Carpman, J.R.; Grant, M.A. *Design That Cares: Planning Health Facilities for Patients and Visitors*, 2nd ed.; American Hospital: Chicago, IL, USA, 1993.
64. Norberg-Schulz, C. *Genius Loci: Towards a Phenomenology of Architecture*; Rizzoli: New York, NY, USA, 1980.
65. Joarder, M.A.R.; Price, A.D.F.; Mourshed, M. Systematic Study of the Therapeutic Impact of Daylight Associated with Clinical Recovery. In Proceedings of the PhD Workshop of HaCIRIC's International Conference 2009: Improving Healthcare Infrastructures Through Innovation, HaCIRIC, Brighton, UK, 2 April 2009; pp. 25–31.
66. Uścinowicz, P.; Bogdan, A.; Szyłak-Szydłowski, M.; Młynarczyk, M.; Ćwiklińska, D. Subjective assessment of indoor air quality and thermal environment in patient rooms: A survey study of Polish hospitals. *Build. Environ.* **2023**, *228*, 109840. [CrossRef]
67. Szuba, A.; Martynowicz, H.; Zatońska, K.; Iłów, R.; Regulska-Iłów, B.; Różańska, D.; Zatoński, W. Częstość występowania nadciśnienia tętniczego w populacji polskiej badania PURE Poland. *J. Health Inequalities* **2017**, *2*, 157–162. [CrossRef]
68. Czałczyńska-Podolska, M.; Rzeszotarska-Pałka, M. Ogród szpitalny jako miejsce terapii i rekonwalescencji. Kosmos. *Probl. Nauk. Biol.* **2016**, *65*, 609–619.
69. Carroll, C.; Higgs, J.; McCray, S.; Utter, J. Implementation and Impact of Health Care Gardens. *J. Integr. Med.* **2024**, *30*, 431–449. [CrossRef]

70. RESTORE COST Action CA16114. *Rethinking Sustainability Towards a Regenerative Economy: Report on Regenerative Design and Assessment*; European Cooperation in Science and Technology: Brussels, Belgium, 2020. Available online: https://www.eurestore.eu (accessed on 11 July 2025).
71. Ildiri, N.; Bazille, H.; Lou, Y.; Hinkelman, K.; Gray, W.A.; Zuo, W. Impact of WELL certification on occupant satisfaction and perceived health, well-being, and productivity: A multi-office pre-versus post-occupancy evaluation. *Build. Environ.* **2022**, *224*, 109539. [CrossRef]
72. Marzban, S.; Candido, C.; Avazpour, B.; Mackey, M.; Zhang, F. The potential of high-performance workplaces for boosting worker productivity, health, and creativity: A comparison between WELL and non-WELL certified environments. *Build. Environ.* **2023**, *243*, 110708. [CrossRef]
73. *EN 17037:2018*; Daylight in Buildings. European Committee for Standardization: Brussels, Belgium, 2018.
74. *EN 16798-1:2019*; Energy Performance of Buildings–Ventilation for Buildings–Part 1: Indoor Environmental Input Parameters for Design and Assessment of Energy Performance of Buildings Addressing Indoor Air Quality, Thermal Environment, Lighting and Acoustics. European Committee for Standardization: Brussels, Belgium, 2019.
75. *EN ISO 3382-1:2009*; Acoustics–Measurement of Room Acoustic Parameters–Part 1: Performance Spaces. International Organization for Standardization: Geneva, Switzerland, 2009.
76. *EN ISO 3382-2:2008*; Acoustics–Measurement of Room Acoustic Parameters–Part 2: Reverberation Time in Ordinary Rooms. International Organization for Standardization: Geneva, Switzerland, 2008.

sustainability

MDPI

Article

Multisensory Coding of Red and Blue in Interior Design for Older Adults and Visually Impaired Users: An Inclusive Design Perspective

Agnieszka Rek-Lipczyńska * and Aleksandra Kowalska

Department of Interior Design, Faculty of Architecture, West Pomeranian University of Technology in Szczecin, Al. Piastów 17, 71-300 Szczecin, Poland; aleksandra.kowalska01@wp.pl
* Correspondence: areklipczynska@zut.edu.pl

Abstract: This article analyses the design aspects of multisensory colour experience in interior architecture. It presents the results of comparative studies on the perception of red and blue spatial environments through various senses (vision, hearing, smell, touch, and thermal sensation). The phenomenon of synaesthesia and its potential application in design are discussed, demonstrating how colors can be encoded and perceived through senses other than sight. Spatial solutions are proposed to enable blind individuals to experience color through touch (e.g., differentiated surface textures), sound (appropriate tones or rhythms), temperature (warm vs. cool lighting), and other environmental properties. The text includes references to the cultural and symbolic meanings of red and blue—covering cross-cultural differences and spatial characteristics attributed to these colors—based on an analysis of Stanisław Lem's novel Solaris. Furthermore, the article formulates design recommendations from the perspective of sustainable development and inclusivity, taking into account European accessibility guidelines and universal design principles. The aim is to present an integrated approach to interior design, in which color becomes an experience accessible to all users, in line with the concept of "Design for All" and the principles of sustainable design.

Keywords: architecture; interior architecture; Design for All; multisensory design

1. Introduction

Color is one of the most powerful factors shaping the perception of space in interior architecture. Traditionally, colors have been considered mainly in terms of visual composition and aesthetics. Today, however, there is growing interest in multisensory design, which engages not only vision but also other senses, enabling the built environment to provide holistic experiences for its users. Research in environmental psychology suggests that the color scheme of an interior influences tactile sensations, thermal comfort, and even the perception of sounds and smells [1,2]. For designers, this implies a need for a broader perspective: surface color not only defines the aesthetic character of a space but can also modulate mood, behaviour, and thermal and acoustic comfort for its occupants.

At the same time, the inclusivity of color design—ensuring that people with sensory impairments (e.g., blind or visually impaired individuals) can also "experience" the colors used in an interior—has become an increasingly important challenge.

In an era of ageing populations and legal regulations such as the European Accessibility Act, there is growing emphasis on universal design, according to which spaces should be usable by all, regardless of limitations.

This article addresses these issues by focusing on two contrasting colors—red and blue—as examples for analysing multisensory color perception. It compares how red and blue are perceived through different senses, discusses the phenomenon of synaesthesia and its potential applications in design, and proposes strategies for enabling blind and partially sighted individuals to perceive color information through other sensory channels. Section 5 offers recommendations for sustainable and inclusive color design in interior spaces, drawing on best practices and European accessibility guidelines.

2. Materials and Methods

2.1. Identifying the Impact of Color on Emotions

In the first stage, a review was conducted of studies from the last decade (2015–2025) concerning emotional branding and sensory design.

The most recent systematic review by Jonauskaite & Mohr covering 132 studies conducted between 1895 and 2022 (more than 42,000 participants from 64 countries) on the associations between colors and emotions, confirms that people across the globe consistently link certain colors with particular emotions [3]. In earlier cross-cultural research, involving 4598 participants from 30 countries, demonstrated the existence of robust, universal color–emotion associations while also highlighting the role of culture in modulating these associations. Significant local differences were observed: machine-learning analysis revealed that the profile of color–emotion associations additionally depends on the country and language of the respondents [4].

Subsequent research by Takei & Imaizumi demonstrated that certain colors can perceptually enhance the recognition of corresponding emotions (e.g., yellow facilitates the detection of happiness). This was shown in a perceptual experiment on the influence of background color on the speed of recognising facial expressions associated with culturally linked emotions (yellow for happiness, blue/grey for sadness) [5].

Equally important are the findings of Alvarado, showing that specific combinations of color properties (hue, lightness, saturation) convey emotional content more effectively than single colors. This innovative study, at the intersection of perception psychology and artificial intelligence, used generative text-to-image models to analyse color–emotion associations. Neural networks generated images based on descriptions of emotions defined by valence, arousal, and dominance, and the resulting images were analysed for color features (e.g., brightness, saturation, tonality) to determine which combinations the model "learned" for specific emotions [6].

In a psychophysiological study examining the effect of color range in virtual reality on a mobile device, Bower and colleagues exposed eighteen participants to different color conditions within an immersive VR environment: a room with neutral white walls (control), a room with intensely blue walls, and a baseline condition (total darkness–black surroundings). Compared to the white (and black) conditions, the blue environment caused a significant increase in breathing variability and skin conductance responses, as well as noticeable changes in brain activity—including increased frontal alpha power (particularly the frontal-midline rhythm), greater interhemispheric asymmetry, and elevated theta, alpha, and beta power in multiple regions while in the blue room. In other words, an environment with a cool, saturated color induced stronger physiological arousal and a distinct neural response compared to an achromatic environment [7].

As Wang, Ku, and Lin, who studied changes in color preferences and emotional responses among young people during the COVID-19 pandemic in 2024, showed, color perception preferences at the population level are not constant [8]. Overall, the pandemic reduced the preference for vivid, warm colors and increased interest in muted, cool tones, reflecting a greater need for emotional calm during uncertain times. The study illustrates how a global stressor can reprogram our color associations [8].

Rui and Firzan's latest study, 2025, examines the emotional dimension of interior design, focusing on the integration of cultural and emotional elements in Chinese living spaces [9]. Using qualitative methods—interviews with five interior designers and a focus group with six experts—the authors found that culturally inspired decorative motifs strongly influence the emotional reception of a space. Interiors rooted in tradition (in this case, the "new Chinese style") can evoke a sense of identity and belonging, which positively affects users' emotions [9].

2.2. Interior Design and Multisensory Design in Research

In his latest review article, Spence, a leading experimental psychologist, highlights how, amidst growing interest in neuroscience and sensory psychology, a trend toward synesthetic design—engaging multiple senses simultaneously—has emerged. The author discusses sensory interactions in the environment, for example, how the color of lighting can influence the perceived temperature. A key concept examined in his work is crossmodal correspondences–natural associations between stimuli from different modalities (e.g., cool colors are associated with cool temperatures, high-pitched sounds with bright colors). Spence argues that architects should consciously use these relationships to create coherent experiences. He concludes that a multisensory approach in design can enhance user well-being–positively influencing social, emotional, and cognitive development–in contrast to monofunctional, dehumanized spaces that often impair such outcomes [10].

2.2.1. Links Between Color and Sound Perception

Recent studies have shown that in deaf individuals fitted with a cochlear implant (CI), auditory stimulation can activate areas of the visual cortex. This is the result of a reorganization of sensory pathways—signals from the cochlear implant, rather than being processed solely in the auditory cortex, partially engage regions originally dedicated to processing visual input (crossmodal plasticity). In practice, this means that the brain can recruit previously unused resources to enhance auditory perception, improve sound localization, and increase discrimination of acoustic features [11].

In a series of three Stroop-type experiments, Miyamoto et al. in 2023 investigated how correlations between colors and sounds influence automatic perceptual processes in non-synesthetic individuals [12]. They confirmed the existence of bidirectional associations: the presentation of a given color could either facilitate or interfere with recognition of a simultaneously presented sound, and vice versa, provided that the stimuli matched known associations. For example, combinations of blue with the sound of a "falling water droplet" and yellow with a "bright, sparkling" sound produced the strongest interference or facilitation effects. Importantly, in the third experiment it was shown that the effect depended on semantic categorization—when an intermediate color was presented, the Stroop effect only occurred after participants consciously classified it as, for example, "yellow", and then heard a congruent/incongruent sound. This finding indicates that verbal labels and the semantic meaning of colors play a role in perceptual crossmodal interactions [12].

In 2025, Bartulienė et al. described a unique case of sound → color synesthesia in a visually impaired individual [13]. Using machine learning, they analyzed the acoustic features of sounds that triggered specific color perceptions. After confirming the synaesthesia (the participant "saw" colors when hearing voices), recordings of speech were collected and labelled according to the color perceived (e.g., blue, pink). A deep neural network was then trained to predict the synaesthetic color evoked by a given voice based on its acoustic parameters (including MFCC coefficients, timbre, and signal energy). The model achieved high classification accuracy (~84%), and feature analysis revealed that pitch, spectral characteristics, and loudness/energy were the most influential factors shaping color perception in this synaesthete [13].

2.2.2. Links Between Color and Aroma Perception

Human senses do not operate in isolation—visual, olfactory, tactile, and auditory information is integrated by the brain to create a unified perceptual experience. In this context, color plays not only an aesthetic role but can also modulate the perception of stimuli from other modalities, including scent. Numerous studies in perception psychology, neurogastronomy, and multisensory design demonstrate that color can influence expectations and the perceived intensity of scents—sometimes even before the odour reaches the olfactory receptors [14].

This is evidenced by a psychophysical study conducted by Ward et al. in 2023, which showed that color perception can be subtly altered by olfactory signals [15]. During experiments, participants were presented with various scents and asked to adjust a visual stimulus to match an achromatic (neutral grey) reference. A systematic shift in the neutral point toward warmer tones was observed in the presence of certain odours. For example, the scent of cherries caused the neutral stimulus to be perceived as slightly reddish-brown—consistent with the expected association between cherry aroma and the color red. Four of the five tested scents induced such color biases aligned with their "color" connotations, demonstrating a crossmodal effect: olfactory information influences visual color perception [15].

2.2.3. Links Between Color and Taste/Flavour Perception

Taste and aroma perception is a multisensory process in which visual cues play a key role alongside the chemical stimuli detected by taste and smell receptors. The color of food, tableware, packaging, and lighting can significantly shape consumer expectations and subjective assessments of taste and aroma. This phenomenon is explained by crossmodal correspondences—natural, though largely culturally mediated, associations between sensory modalities.

In their 2021 study, Spence and Levitan examined how food color, tableware, and lighting affect taste perception. They found that red and orange increased perceived sweetness, while blue and green reduced taste intensity. Packaging color also modulates flavour expectations [16]. Similarly, Reinoso Carvalho et al. found that beer color (light vs. dark), as well as bottle and label color, influenced both taste expectations and actual perception. Darker beers were perceived as stronger and more bitter, regardless of their actual flavour profile [17].

Environmental cues may also play a role: Tu et al. in 2020 showed that red lighting increased the perceived sweetness and pleasure of drinking a beverage, while blue lighting re-evoked the perceived flavor intensity and elicited cooler emotional responses [18].

Wang & Spence demonstrated that a "sensory language" develops from early childhood—both children and adults associate specific colors with tastes (e.g., red = sweet, green = sour, brown = umami) [19]. In 2023, Chen and Spence confirmed strong associations between colors and basic tastes: yellow with sourness, brown with bitterness, and red with sweetness [20]. Their study also used three-dimensional stimuli (beverages, food) [20]. Warm hues such as red, orange, and yellow frequently enhance the perception of sweetness and fruity notes, whereas cool shades of blue and green tend to evoke impressions of freshness, lightness, or acidity.

2.2.4. Links Between Color and Touch Perception

Touch, like other senses, does not function in isolation—haptic sensations are often influenced by stimuli from other modalities, including visual signals. One of the best-documented examples of such interaction is the influence of color on how we interpret the tactile properties of objects.

In a series of experiments in 2024, Ho et al. asked participants to touch objects of different colors and temperatures [21]. Paradoxically, blue objects of a given temperature were more likely to be rated as warmer than red ones—an effect that contradicts common cultural associations [21].

Ludden et al.'s 2020 study examined how the consistency or inconsistency of a product's color and texture affects tactile perception [22]. Products with congruent combinations (e.g., red + rough texture) were rated as more natural and pleasant to touch. Reanalysis of these findings confirmed the existence of the crossmodal congruence effect—alignment between visual and tactile cues increases perceptual comfort [22].

Similarly, Wastiels et al. investigated how color and surface gloss influence tactile expectations. Light, matte colors were associated with softness, while dark, glossy colors were linked to hardness and coolness [23]. These findings confirm that color shapes expectations about texture before the object is even touched, which is significant for interior and product design.

In a 2022 marketing study, Hagtvedt and Brasel showed that color can strengthen or weaken textural associations—for example, brown and beige enhance the perception of fabric softness, while gray and black are associated with smoothness or coolness [24].

Neurological evidence comes from a study by Ward and Simner in 2013, who described a case of tactile-visual synesthesia: a study participant consistently experienced specific colors when touching specific textures (e.g., soft = yellow, rough = dark green) [25].

These examples illustrate that crossmodal correspondences—natural or culturally reinforced associations between different sensory modalities—can shape our tactile expectations regarding an object's texture, temperature, weight, or hardness. Warm colors such as red and orange often evoke associations with rough, heavy, or warm surfaces, whereas cool shades of blue or green tend to suggest smoothness, lightness, or coolness. Understanding these links is crucial for multisensory design—both in architecture and product design—where the coherence between color and tactile characteristics can enhance comfort, intuitiveness, and emotional resonance.

2.3. *Multisensory Color Coding for Blind, Visually Impaired, and Elderly Users*

A review was conducted of studies from the last decade (2015–2025) addressing multisensory color-coding strategies for individuals with various forms of color perception impairment.

Color is one of the most important elements of visual communication. However, for blind or partially sighted individuals, as well as for many older adults—whose ability to distinguish colors naturally deteriorates with age—color perception is significantly limited. This phenomenon affects not only comfort in navigating spaces but also orientation, safety, and the ability to fully participate in social and cultural life.

Contemporary research in psychophysics, neuroscience, and inclusive design suggests that the loss or weakening of color perception can be partially compensated for by implementing multisensory color coding. This involves using auditory, tactile, thermal, olfactory, or acoustic cues as alternative carriers of color information. Such systems and environments allow color information to be perceived and interpreted through other sensory modalities, often in ways that are intuitive and emotionally engaging.

These solutions have practical applications in interior design, public spaces, educational aids, and digital interfaces. They are particularly important in the context of ageing populations and growing requirements for universal design, which aims to ensure environmental accessibility and clarity for all users, regardless of sensory abilities.

Cavazos Quero, Lee & Cho developed a color-coding system using sound and scent to represent colors, aimed at helping blind individuals access information about colors in works of art. Melodies were mapped to specific hues (melodic tonality reflecting hue), while scent intensity corresponded to brightness and saturation. In an experiment involving 18 blind or partially sighted participants, the new system was compared with an audio-only approach. Color identification accuracy improved for 39% of participants, remained similar for 28%, and decreased for 33%. Despite mixed results, the prototype using multisensory coding was rated higher in terms of usability (mean 78.6 points vs. 61.5 on the SUS scale, compared to traditional tactile graphics). Participants reported that the multisensory color code increased comfort and confidence when exploring artworks.

A different multisensory approach was proposed by Iranzo Bartolomé, Cho & Cho combining sound and thermal cues to convey color information to blind users. The authors developed a mapping in which musical sounds and warm/cool sensations corresponded to specific colors and their attributes (e.g., warm colors such as red represented by higher temperatures, cool colors such as blue by lower ones). In experiments with 18 participants, the system enabled identification of 24 colors with significantly better-than-chance accuracy. Detailed semantic surveys were also conducted to determine the most intuitive associations—e.g., certain sounds were naturally associated with "warm" or "cool" colors. The final algorithm was validated on a group of 12 new participants, confirming its usability and effectiveness [26]. Recent studies emphasize that multisensory spatial design can significantly enhance accessibility and emotional engagement for users with visual or sensory limitations. Xu et al. (2025) demonstrated that the integration of visual, auditory, tactile, and olfactory cues within an interior environment can effectively guide perception and improve spatial orientation [27]. Their case study of the Suzhou Coffee Roasting Factory exemplifies how coordinated sensory stimuli—such as lighting, soundscapes, material textures, and ambient aromas—can transform interior spaces into inclusive and emotionally resonant environments, aligning with the principles of multisensory color coding proposed in this paper [27].

2.4. Color Perception in Older Adults: Seeing Red and Blue in the Built Environment

Literature Review: 2015–2025

Ageing significantly alters color perception due to changes in the visual system. Natural processes such as lens yellowing, reduced transmission of blue light, decreased contrast sensitivity, and slower adaptation to lighting changes influence color discrimination. Older adults often distinguish warm colors (red, orange, yellow) more easily, whereas cool tones (blue, green) may appear less saturated or be harder to differentiate. These changes affect visual comfort, spatial orientation, safety, and overall well-being. Understanding these shifts is essential for designing legible, friendly, and inclusive environments that support functionality and quality of life for older adults.

Wang & Cho analyzed how colors in residential interiors attract the attention of older adults using eye-tracking technology. In an experiment, seniors were shown visualizations of interiors with different door and wall colors, including combinations of so-called "safety colors" (e.g., red) with black or white. Red doors, particularly with a black frame (high red–black contrast), attracted the strongest and fastest visual attention and sustained gaze for longer periods compared to white doors in the same position. In contrast, combinations with blue or green did not elicit as strong a response. Even in participants with lower cognitive scores (MMSE), the preference for red persisted despite more dispersed and less focused scanning patterns. The findings suggest that appropriate color selection for environmental elements critical to safety can improve their visibility and enhance safety in the home [19].

In 2024, Jaglarz also raised the issue of the importance of interior colors for the well-being of older people [28]. The author reviewed research on age-related changes in color perception and conducted a survey among seniors. Results indicated that older adults strongly rejected monochromatic interiors devoid of color accents, perceiving them as sad and depressing. Preferred schemes drew from nature-inspired palettes—earth tones, greens, blues, muted browns—creating a harmonious, welcoming atmosphere. Interestingly, despite reduced sensitivity to cool colors with age, respondents most often cited blue and green as their favourite environmental colors. These findings inform design recommendations to use color strategically (nature inspiration, avoiding monotony) to improve comfort, orientation, and mood in senior living environments [28].

2.5. Symbolism and Perception of Red and Blue

Red and blue have long been associated with rich cultural symbolism, evoking distinct emotional reactions and associations. Red is widely linked with energy, dominance, passion, but also danger. It is often perceived ambivalently—symbolizing courage, struggle, and blood on one hand, and love and sensuality on the other. In many cultures, red holds special status: in China and Japan it is a color of luck and prosperity, while in Buddhist tradition red ink is used to write the name of the deceased, lending it funerary associations.

Blue carries associations with calmness, harmony, and spirituality. In Western culture, blue is connected with wisdom, trust (e.g., widely used in business and healthcare as a professional color), and relaxation. However, cultural variations exist: in Iran, blue may be associated with mourning, while in Hinduism it can symbolize war. Generally, warm red is perceived as stimulating and energizing, whereas cool blue is calming and contemplative.

Color psychology confirms that warm tones (reds, oranges, yellows) tend to increase activity and emotional arousal, while cool tones (blues, greens, purples) promote relaxation and reflection.

3. Results

The results presented in this section highlight the multidimensional nature of color perception in the context of multisensory design, with particular emphasis on individuals who are blind, visually impaired, and elderly. The analysis is based both on empirical research findings and on a review of scientific literature from the past decade (2015–2025), encompassing psychophysics, neuroscience, environmental psychology, and inclusive design practices.

In particular, data concerning the perception of two contrasting colors—red and blue—were examined in relation to various sensory modalities: vision, touch, smell, taste/aroma, hearing, and thermal sensations. The results reveal characteristic patterns of how these colors are perceived, their associations with emotional and physiological responses, and their cultural symbolism. Mechanisms of crossmodal correspondences and the phenomenon of synesthesia were also analyzed, as they may serve as an inspiration for the development of color codes accessible beyond the visual modality.

The findings not only document existing interrelations between color and multisensory perception but also point to practical opportunities for their application in interior design—so that color may become a universal spatial language, "readable" by all users, regardless of their perceptual abilities.

3.1. General Emotional Patterns Associated with Color

An analysis of a systematic review conducted by Jonauskaite and Mohr in 2025 found consistent cross-cultural patterns linking color with emotional responses [3]. A key factor determining emotional appraisal is not the color name itself but rather its brightness, saturation, and color temperature. Light colors are globally associated with positive emotions, while dark colors are linked to negative ones. Moreover, saturated warm colors elicit emotions characterized by higher levels of arousal compared to cool and muted tones. Below, we present a summary of emotions associated with specific colors, based on the analysis by Jonauskaite and Mohr [3] [Table 1].

Table 1. Summary of color emotions based on the research of Jonauskaite, D. and Mohr, C. from 2025 [3]. Author: Agnieszka Rek-Lipczyńska.

	Colors	Emotions Positive	Emotions Negative	Level of Arousal	Comments
1.	red	X love	X anger	TALL	Strong emotional reaction, ambiguous
2.	yellow	X joy		TALL	Associated with optimism and light
3.	orange	X enthusiasm		TALL	Energizing, but not as aggressive as red
4.	blue	X composure		SHORT	It creates a feeling of security
5.	green	X relaxation		SHORT	It is associated with nature and harmony
6.	white	X cleanliness, openness		SHORT	Neutral, clean, soothing
7.	pink	X sensitivity		SHORT	Delicate, gentle, feminine
8	purple	X strength, spirituality		TALL	It represents dignity, prestige, spirituality
9	gray		X sadness	SHORT	Most often associated with melancholia
10	black		X fear, sadness	TALL	The color of strong negative emotions

Across all studied cultures, red was typically associated with high-arousal emotions, yellow with joy, and black with sadness; however, the strength and nuances of these associations varied depending on the geographical and linguistic proximity of nations. Jonauskaite et al. (2020) [4]. Such knowledge can be applied in the design of spaces and products tailored to different cultural contexts, as well as in well-being–oriented practices (e.g., chromotherapy, cross-cultural marketing), where accounting for local variations in color perception may increase the effectiveness of interventions.

Takei and Imaizumi demonstrated that specific background colors can modulate the speed and accuracy of recognizing emotions from facial expressions. A yellow background significantly facilitated the identification of happy facial expressions, whereas the traditionally assumed association "sadness = blue" was not confirmed in this study. In practice, this suggests that background colors may be strategically used in interfaces, visual signals, or therapeutic settings to enhance positive emotions. At the same time, it emphasizes that not all colloquial color–emotion metaphors correspond to real perceptual mechanisms [5].

Other analyses confirm that specific combinations of color properties—hue, brightness, saturation—convey emotional meaning more effectively than single parameters [6]. Environmental studies further demonstrate that color embedded in architectural surroundings can modulate emotional and physiological responses. For example, blue environments increase autonomic arousal and affect brain activity patterns linked to emotional processing. This has direct implications for the design of public spaces (e.g., hospitals, offices, schools) in terms of psychological well-being: a conscious selection of interior colors may support desired states (e.g., calming vs. stimulating), which translates into measurable health and social benefits at the population level [7].

Research conducted during the COVID-19 pandemic also showed that global stressors can recalibrate color associations. Under conditions of elevated stress, preference for cool colors (blue, green)—associated with safety and calm—increased, while preference for warm colors declined [8].

Spence's review (2020) provides a theoretical foundation for the concept of sensory design and, indirectly, emotional branding [10]. It demonstrates that human perception of space is holistic—for example, visually impaired individuals compensate for the lack of visual information through hearing, touch, and smell. Multisensory interior design for seniors or visually impaired people is therefore not only a matter of accessibility but also of shaping mood and emotion. In this context, red and blue do not need to operate exclusively through the visual channel—their perception may be reinforced by correlated stimuli: red may "warm" a space through higher temperature and soft textures, while blue may be conveyed through gentle sounds and airy materials.

Such synergy of stimuli creates atmospheres more easily perceivable by older adults and visually impaired users, even if they cannot directly see the color. Emotional branding, in this sense, means building an emotional bond between the user and the place through multisensory impressions—for example, evoking memories through scent, a sense of safety through sound, or energy through colored light [9].

3.2. Interior Design and Multisensory Design

The study by Xu, H., Zhao, J., Jin, C., Zhu, N., & Chai, Y. proposed a set of multisensory strategies for interior design, including:

The selection of colors and illumination to create the desired emotional climate;

The design of soundscapes that enhance users' immersion in space;

Tactile elements providing physical comfort and textural diversity;

Scents that evoke associations and memories, supporting a coherent sensory narrative [27].

These strategies were implemented in a conceptual design of a coffee-smoking lounge, where coordinated stimuli—colors, lighting, coffee aroma, and ambient sounds—created the impression of a coherent "scenography" of space. This approach illustrates that the conscious interplay of color with other sensory modalities can significantly enhance user experience and generate a unique emotional climate within interiors.

Scope of application: commercial and public interiors where a high level of immersion and sensory engagement is expected—e.g., cafés, restaurants, boutique hotels, exhibition spaces.

In Spence's analysis, three levels of emotional design were identified:

(1) Visual–aesthetic—including novelty, appropriateness of color schemes, order, and richness of form;
(2) Functional—referring to usability, comfort, and privacy;
(3) Individual—related to personalization, fashion, and the incorporation of cultural elements.

Importantly, color appears here both as a key factor at the aesthetic level (e.g., color coherence with cultural expectations) and as an element influencing comfort (e.g., a suitable color can calm or stimulate). The authors argue that the integration of emotion and culture can help overcome many design challenges—for example, making interiors more creative and user-friendly [10].

3.3. *Multisensory Color Coding for the Blind, Visually Impaired, and Elderly*

The study by Cavazos Quero, Lee, & Cho directly pursued the idea of "colors beyond vision," proposing alternative sensory channels—auditory and olfactory—for conveying color information to individuals who are blind or have severe visual impairments. In the context of interior design, their findings suggest that essential colors such as red and blue may be communicated through sounds or aromas associated with these hues, enabling their experience through non-visual modalities, in line with inclusive design principles [14].

A similar approach is found in the work of Iranzo Bartolomé et al. who combined thermal sensations with color perception. Interior elements may emit subtle warmth or coolness, allowing older adults and visually impaired individuals to differentiate colors: a heated surface may signal red (warm), while a cooler one may indicate blue (cold). Combining temperature with sound adds an emotional dimension and improves memorability, confirming that multisensory experiences are more engaging and durable than single-channel ones [26].

The findings of Wang & Cho highlight the specific role of red in the visual perception of seniors. Red—a long-wavelength color—is most visible and attracts attention fastest, whereas cool colors (e.g., blue) are perceived less effectively. This phenomenon results from age-related ocular changes, such as lens yellowing and reduced transmission of blue light. For interior design, this implies that warm colors (red, orange, yellow) should be prioritized in marking functionally important elements, such as doors, handrails, or emergency buttons [19].

Jaglarz provides a theoretical basis for inclusive and sustainable color design. The study confirms that red and other warm colors are more legible to seniors than blues, while also stressing that design should not be limited to function—colors affect emotions and well-being. Overly monotonous, colorless interiors may exacerbate apathy, whereas carefully designed color accents stimulate activity and improve mood. In the spirit of emotional branding, colors may be "felt" not only visually but also through the mood they bring into a space [28].

The issue of multisensory color coding also includes associations between color and sound perception. Studies on crossmodal correspondences show that colors have consistent acoustic associations: for example, red is linked to full and low-pitched sounds, while blue is associated with bright and high-pitched tones [17]. In multisensory interface design, this may allow the adaptation of auditory signals to color codes, also in the context of spatial orientation for blind users.

Research on synesthesia and the use of artificial intelligence to simulate synesthetic experiences provides further contributions to this issue [13]. Findings show that it is possible to predict which sound properties correspond to a given color, which may, in the future, support the development of tools translating speech sounds or acoustic signals into color information.

Evidence of the brain's adaptive capacity in the context of multisensory coding comes from studies of crossmodal plasticity observed in individuals with cochlear implants. Research shows that auditory stimuli can activate visual cortex areas, demonstrating that the brain can reorganize its pathways and transfer information from one modality to another. In terms of interior design, this means there is a real possibility of creating systems that translate color into auditory, tactile, or thermal stimuli, which users—thanks to neural plasticity—can learn to interpret as naturally as visually perceived colors.

3.4. Links Between Color and Aroma Perception

Research has shown that the integration of olfactory and visual stimuli can influence color perception, although the effect is small yet consistent. The findings suggest that the brain integrates olfactory information with visual input at an early stage of color perception, confirming the existence of multisensory integration mechanisms. While the study was fundamental (laboratory-based), its implications may be applied in areas such as VR/AR experience design, museum and exhibition installations, and sensory marketing, where scent is used to modulate visual perception. This points to the potential of employing fragrance as a subtle tool for enhancing or modifying color perception in the context of shaping mood and user experience.

3.5. Links Between Color and Tactile Perception

One of the first well-documented studies on tactile-visual synesthesia was conducted by Ward and Simner in 2013 [25]. It demonstrated that connections between color and touch may have a neurological basis, even if for most people they remain nonverbal and unconscious. This may explain why, in interior design, certain colors are "intuitively" associated with specific materials—for example, white with softness, black with smoothness, and red with roughness. The authors suggest that crossmodal correspondences (natural associations between senses) may originate from neurological, cultural, or learned factors. Such research provides valuable insights into how sensory integration operates even in neurotypical individuals, which may inspire more coherent multisensory design [Table 2].

Table 2. Experiencing Color without Vision: Design for the Visually Impaired and Blind. Author: Agnieszka Rek-Lipczyńska.

Colors	Correlated Tactile Property	Examples of Materials/Surfaces	Notes/Source
yellow	Soft, fluffy, delicate	Blanket, wool, velvet, MW foam	felt yellow when touched by a soft blanket/Ward & Simner, 2013 [25]
red	Rough, grainy, slightly rough	Sandpaper, raw wood, string	Strong color reaction of MW to rough cardboards/Ward & Simner, 2013 [25]
dark green	Rough, fibrous	Carded fabrics, jute, natural fabrics	Subjectively perceived as a "grainy" color of touch
blue	Smooth, cool, wet or slippery	Glazed ceramics, metal, glass	Color often associated with smoothness and coolness/Wastiels et al., 2021 [23]
white	Loose, dry, light	Flour, powder, chalk	Fine-grained feeling, "neutral" touch/Ward & Simner, 2013 [25]
brown	Hard, stiff, massive	Wood, leather, cork	Associated with weight and stability
black	Hard, slippery, cold or unfriendly	Marble, polished stone, metal	Often correlated with hardness and cold/Ludden et al., 2020 [22]
pink	Smooth, soft, warm	Velvet, latex, silicone	Associated with pleasure and tactile comfort

For a person who is completely blind, the colors of the external world do not exist in direct perception but can be indirectly experienced through other sensory impressions. Therefore, interior designers aiming to create truly inclusive spaces should consider solutions that enable information encoded in color to be conveyed through other senses. The idea of "design for the blind" assumes that every element available visually (e.g., a wall color, a zone marking, or an on-screen message) should have a tactile, auditory, or olfactory equivalent understandable to people with visual impairments.

The simplest carrier of information for the blind is touch. Interiors can employ diverse textures and shapes corresponding to specific colors. For example, a wall or handrail marked in red might have a distinct tactile pattern (e.g., raised stripes, rough texture), while blue surfaces might have a contrasting pattern (e.g., smooth, cooler-to-the-touch panels). Existing systems of tactile color codes—such as the Scripor Alphabet—assign each color a unique, easily recognizable tactile symbol. Originally developed for clothing and color identification of garments by the blind, this system can be successfully adapted to architecture, e.g., by marking furniture edges or picture frames in museums with the appropriate tactile color symbols. The Scripor standard effectively translates the concept of color into the language of embossed signs, transforming color from a purely visual category into a tactile experience. Similar approaches have been tested with tactile maps and building plans, where different embossed patterns represent colors on printed maps, enabling blind individuals to "read" the color-coded legend with their fingers.

Another important channel is sound. Although sound has traditionally been rarely used for color coding in space, new technologies open wide possibilities. Blind people often use smartphone applications that recognize the color of an object via the camera and deliver the information verbally ("red shirt," "blue door"). Embedding auditory cues directly into the environment, however, may create a more seamless experience. For instance, one can imagine sound installations reacting to proximity to a color-marked zone—upon approaching a red-coded room, a high-pitched or fast sound sequence is heard, while a blue-coded room may be accompanied by lower or slower tones. These would act as sonification of color, i.e., translating color information into acoustic signals. In practice, such systems would require discreet sensors and directional speakers to avoid excessive noise—a technical challenge, but feasible in specific applications (e.g., educational spaces, multimedia exhibitions).

A particularly inspiring example is Neil Harbisson, an artist recognized as the world's first cyborg, who perceives colors through a permanently mounted camera-antenna that converts surrounding colors into sound vibrations transmitted via bone conduction. Harbisson, who has complete achromatopsia, learned to distinguish colors through different sound frequencies generated by his device—each color became a note or chord. While an individualized high-tech solution, it demonstrates the potential of translating visual signals into auditory ones for blind users. In architecture, instead of implants, color sensors and sound emitters could be used: for instance, a cane equipped with an RGB sensor could detect the color of a room and emit a corresponding tone (such prototypes already exist). However, universal design principles recommend that solutions be embedded in the environment itself rather than requiring special devices from users. Consequently, architects increasingly envision intelligent, speaking environments that autonomously provide voice or sound cues (e.g., elevators announcing floors, acoustic traffic signals). Extending this concept, auditory cues could be directly coded to the color of the environment.

The thermal sense can also be engaged. Although less intuitive, in certain contexts ambient temperature can be modulated according to color, allowing blind individuals to perceive differences physically. For example, in an art gallery presenting "red" and "blue" installations, heating or cooling could be independently controlled: the "red" room slightly warmer (e.g., halogen lights emitting heat), and the "blue" room cooler (e.g., LED lighting and air conditioning). The difference need not be large—just a few degrees—but perceivable on the skin, providing an additional clue to the "climate" of the space. Another approach involves heated or cooled tactile surfaces—for instance, a handrail at the entrance to a red-coded zone could be warm, while one leading to a blue-coded zone could be cool (or made from naturally cooler materials). Although such measures are rarely used due to cost and complexity, they may be valuable in specific settings, such as multisensory experience rooms, therapeutic centers, or special exhibitions.

Scent may also assist blind individuals in identifying the character of a color-coded space. If a given room has a distinctive fragrance, users learn to associate it with that place or function (a process that occurs naturally, as every building has its own "smell"). Designers can amplify this by introducing subtle olfactory indicators—for example, a floral fragrance for a red room and a forest scent for a blue one. Importantly, these scents must be subtle and non-intrusive to maintain their informational function. Increasingly, signature scents—unique brand fragrances—are used in modern offices and hotels, which can also aid orientation for blind individuals (recognizing a building by its smell). In this context, scent may serve as a complementary layer in conveying the "color atmosphere" of a space: for instance, a red, energetic sales zone could be emphasized with a stimulating citrus aroma, while a blue relaxation lounge with a calming lavender fragrance.

In summary, enabling blind individuals to experience color involves applying redundancy of information—each color-coded message should be accompanied by an alternative cue in tactile, auditory, olfactory, or textual form. For example, if offices are color-coded, each room should also have Braille inscriptions with the color name or embossed symbols, or an automatic voice message at the entrance ("This is the Blue Room"). Such practices are increasingly becoming standard within universal design. European accessibility guidelines emphasize that information should not rely on color alone—any color-coded information (e.g., evacuation plans, door markings) must have an alternative in the form of text, shape, or texture, accessible to individuals with visual impairments. This ensures that color functions as an additional layer of information rather than the sole carrier of meaning.

For the visually impaired (low vision), the key principle is to provide strong color contrasts between elements (floor–wall, door–wall, text–background), facilitating form recognition. For color-blind individuals, color-blind-friendly schemes are recommended, avoiding the most frequently confused pairs (e.g., red and green) and supplementing colors with descriptions or pictograms. This, too, belongs to multisensory design, as it engages the sense of shape (visual or tactile) rather than relying exclusively on color information.

3.6. Survey on Multisensory Associations of Red and Blue

To complement the theoretical analysis, an exploratory survey was conducted by the co-author of this article, Aleksandra Kowalska, a fifth-year architecture student at the West Pomeranian University of Technology in Szczecin. The study involved 33 student participants (aged 21–26), recruited from the Faculty of Architecture and Design. The purpose of the survey was to investigate how individuals intuitively associate two contrasting colors—red and blue—with different sensory modalities.

Participants were asked to freely assign their first associations with the colors red and blue across four categories: (1) emotions, (2) sensory impressions (sound, touch, temperature), (3) scents, and (4) spatial forms. The study was conducted in a controlled classroom environment with neutral lighting to avoid external color bias. Responses were collected through an open-ended questionnaire and subsequently categorized into thematic groups for analysis.

Red was most frequently associated with high-arousal emotions such as energy, passion, and aggression (Figure 1).

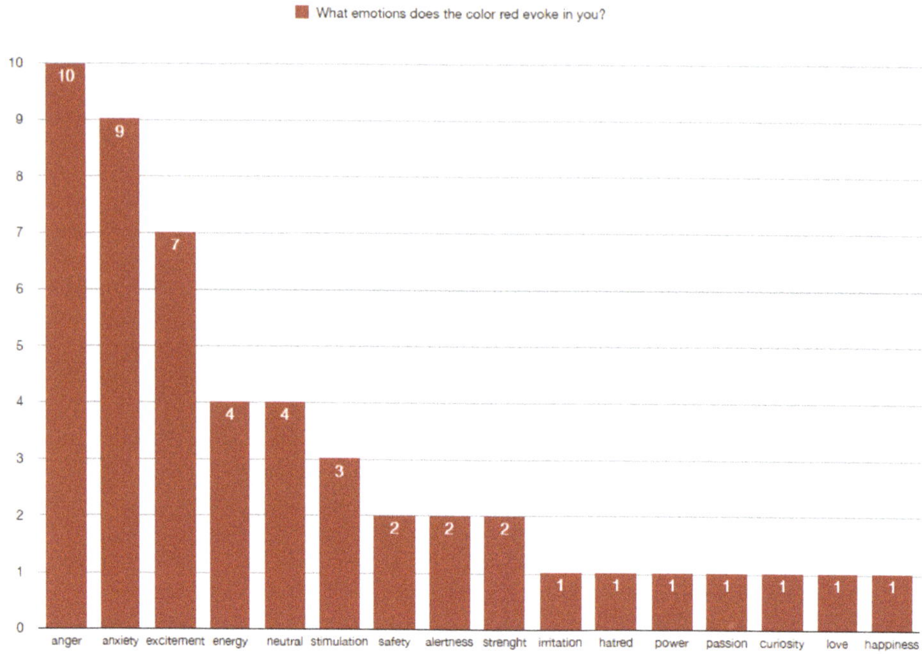

Figure 1. Diagram showing the results of a survey on the "first association" of the color red with an emotion. Author: Aleksandra Kowalska.

Blue, in contrast, was associated with calmness, relaxation, and melancholy (Figure 2). Sensory links included coolness, smooth textures, and soft or muted sounds. Respondents rarely indicated a strong scent association, though freshness (e.g., water, air) appeared in several answers. In spatial terms, blue was linked with open, flowing, and light forms, perceived as ordered and expansive [Figure 2].

In sensory terms, respondents linked red with warmth, sharp or rough textures, and loud or dynamic sounds. Olfactory associations included sweet and intense fragrances (e.g., flowers, spices). Spatially, red was connected with angular, dense, and enclosing forms, often perceived as heavy or overwhelming.

The most frequent spontaneous association with red was blood, followed by love, anger, fruit, and danger. Other responses included warmth, fire, and rose and less common associations such as lust, strength, heart, and warning. These results indicate that red is strongly linked to biological and emotional activation, often connected with life, passion, danger, and intensity. The semantic field of red thus emphasizes energy, alertness, and affective arousal, in line with previous studies in color psychology [Figure 3].

By contrast, the color blue was most commonly associated with sky (27 indications), followed by water (8) and peace (6). Less frequent were coldness (3), rest (2), and individual responses such as ice, car, flowers, and blueberry. These findings demonstrate that blue evokes a strong sense of calmness, spaciousness, and stability, associated with nature (sky, water) and tranquil affective states [Figure 4].

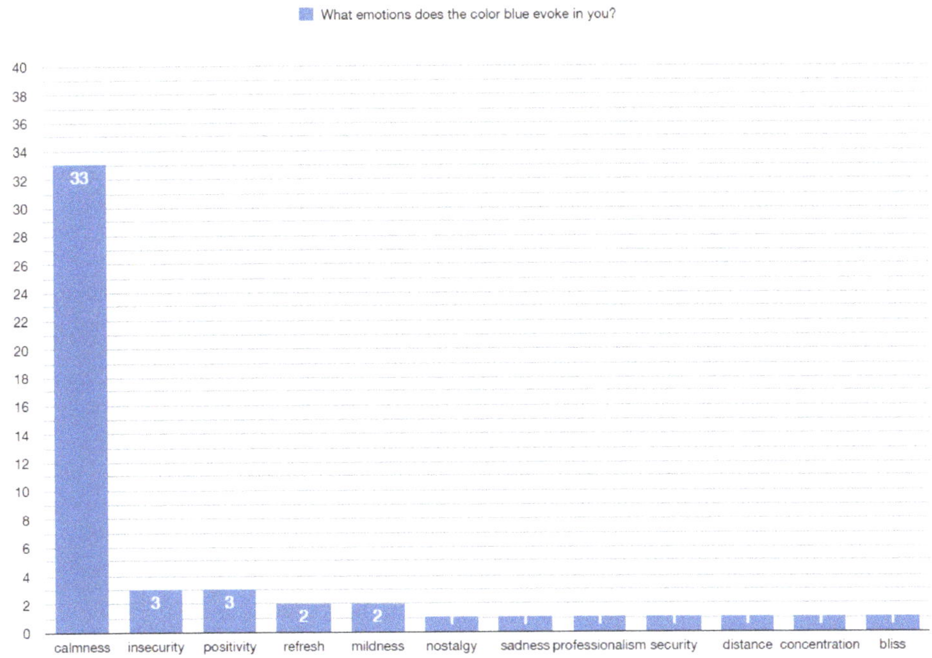

Figure 2. Diagram showing the results of a survey on the "first association" of the color blue with an emotion. Author: Aleksandra Kowalska.

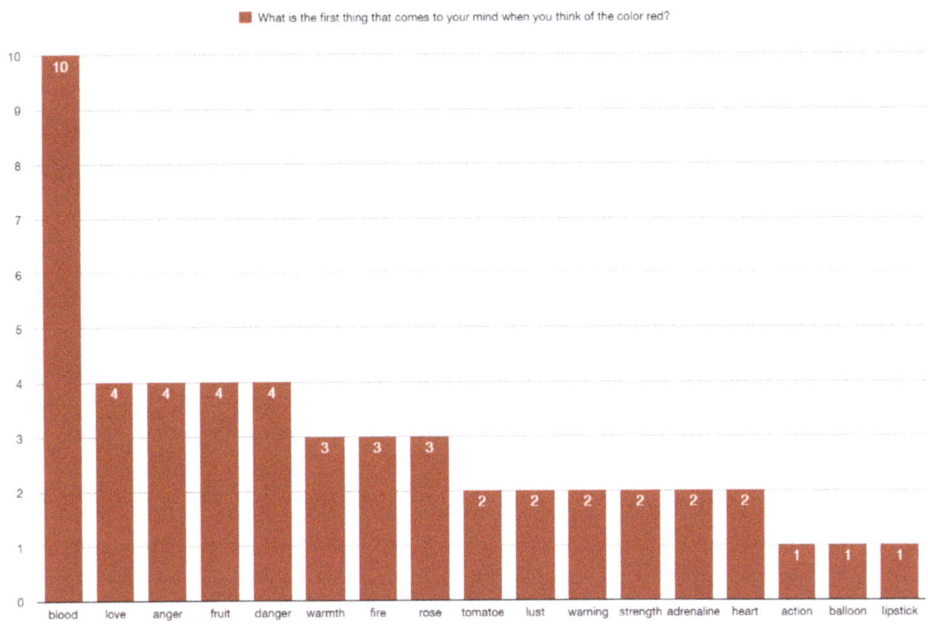

Figure 3. Diagram showing the results of the "first association" red survey. Author: Aleksandra Kowalska.

The clear opposition between the associative domains of red and blue—activation versus relaxation, heat versus coolness, closeness versus distance—supports the theoretical framework of cross-modal emotional coding of color discussed in the article. These associative patterns also align with established perceptual theories, confirming that even in free-association tasks, color stimuli activate multisensory and emotional representations rather than purely visual ones.

The survey results align with crossmodal correspondence literature, which shows systematic links between warm colors (such as red) and high-arousal sensory associations, versus cool colors (such as blue) and low-arousal, expansive impressions. Importantly, the findings confirm that participants tend to perceive colors not only visually but as part of a multisensory experience, linking hue with emotion, materiality, soundscape, and spatial form. These associations provide a practical basis for applying color strategies in interior design, particularly in contexts where multisensory cues are needed for inclusivity. To enhance the reproducibility and credibility of the study, it is essential to provide more detailed information about the applied methodology. In particular, the following aspects should be specified: sample selection criteria (e.g., number of participants, age, student status, presence or absence of sensory impairments), experimental environment settings (e.g., type of lighting, ambient temperature, background noise during the test), and data collection tools (e.g., survey questionnaires, rating scales, sensory testing devices). Presenting these details allows for greater comparability of results across studies and strengthens the empirical validity of the findings.

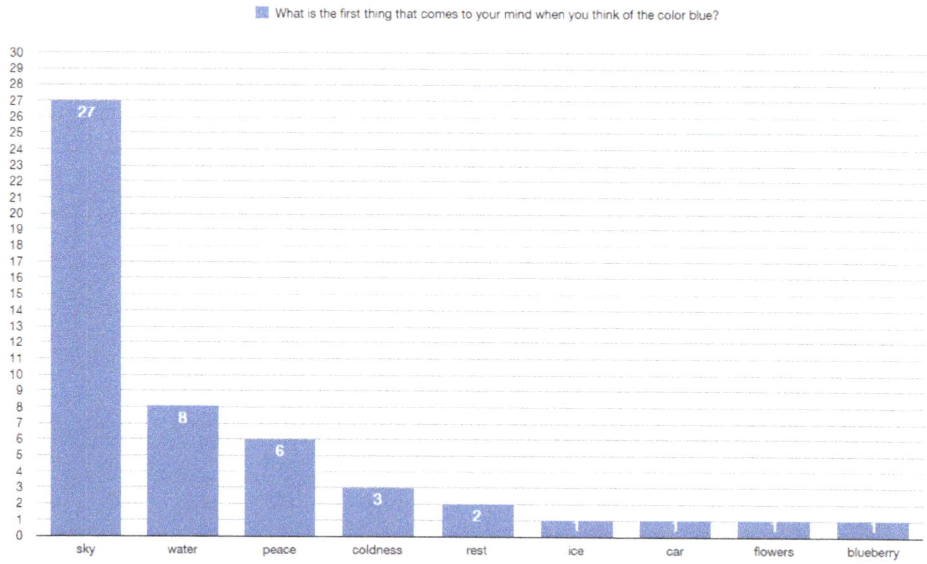

Figure 4. Diagram showing the results of the "first association" blue survey. Author: Aleksandra Kowalska.

The spatial-form associations revealed clear contrasts between red and blue. Participants most frequently linked red with closed, compact, and sharply defined forms, often describing it as intense, energetic, or contained [Figure 5]. By contrast, blue was associated with open, wavy, and flowing shapes, evoking impressions of spaciousness and continuity. These findings support the notion that warm and cool colors elicit distinct perceptual and spatial schemas—red being perceived as dense and confined, while blue conveys openness and fluidity [Figure 6].

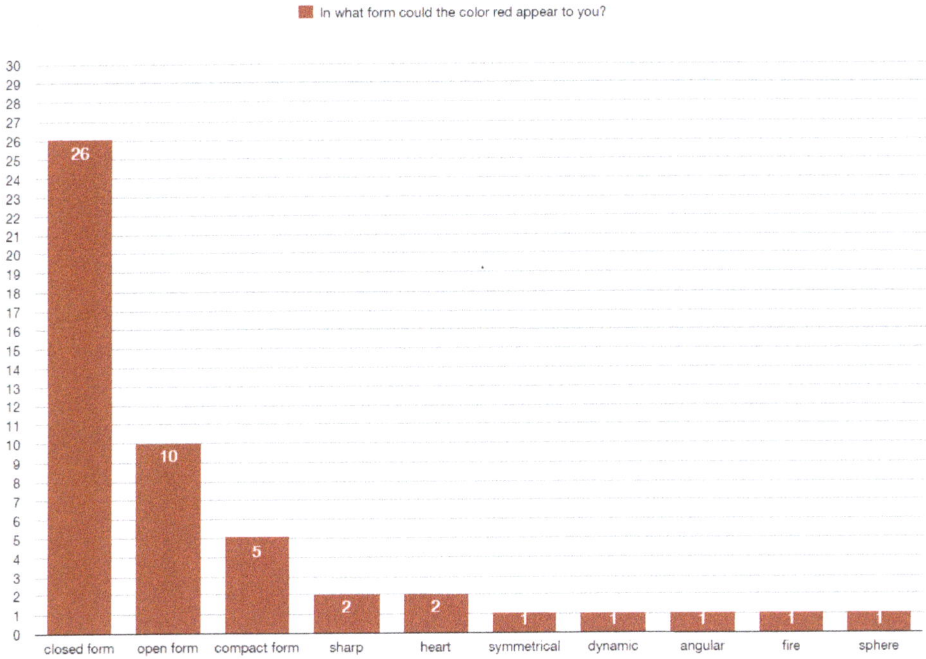

Figure 5. Diagram showing the survey results of the "first association" of red with spatial form. Author: Aleksandra Kowalska.

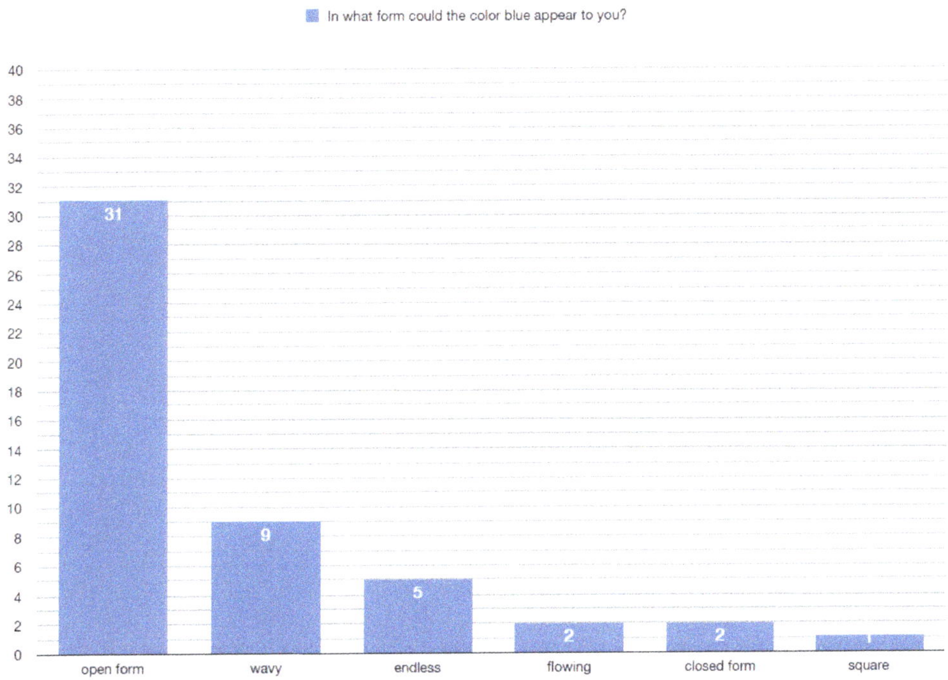

Figure 6. Diagram showing the survey results of the "first association" of blue with spatial form. Author: Aleksandra Kowalska.

3.7. The Contrast Between Red and Blue as a Literary Paraphrase of Narrative Tension in Solaris

The authors of this article conducted an analysis of the literary vision of the colors red and blue in Stanisław Lem's novel Solaris. The purpose of this literary analysis was to demonstrate how language and the symbolism of color can be interpreted in the context of multisensory perception of space. In the novel, colors do not function merely as descriptive attributes of the fictional world; rather, they serve as active carriers of meaning, emotion, and narrative tension. The ocean and the sky—the central elements of the setting—change color in ways that reflect the dynamics of the plot and the psychological states of the characters, becoming a metaphorical "screen" for existential experiences and confrontations.

Red and blue in Solaris create a form of dualism: the former symbolizes life, energy, and emotional intensity, but also chaos, dread, and unease; the latter conveys cold distance, otherness, and apparent tranquility, concealing silence and lack of response. This contrast, embedded in Lem's literary descriptions, resonates with the psychology of color and may be interpreted within the framework of multisensory design—as a model example of how colors can be used to construct atmosphere, mood, and meaning in space.

The ocean is often described as shifting in color—from blue to blood-red, sometimes gray, silvery, or almost lifeless. Symbolically, these two hues may be associated with:

Red—life, pulsating matter, threatening force, anxiety.

Blue—cold intelligence, otherness, inscrutable mystery, cosmic chill.

The juxtaposition of these colors expresses the dual nature of the ocean: something organic and alive, yet simultaneously inhuman, alien, and impersonal. It seems to react to humans, but in ways that remain beyond human comprehension.

In Solaris, the sky does not serve as a source of hope or transcendence, as it often does in classical literature. On the contrary, it appears as:

empty, cold, and silent,

devoid of spiritual presence,

stripped of divine meaning—humanity has not found God but an alien, incomprehensible ocean.

The sky thus becomes a metaphor for the psychological states of the novel's characters.

Blue sky—symbolizes the cold, alien calm of the planet; it may be read as an illusion of normality to which the characters cling; it conveys cosmic indifference that provides no answers.

Red sky—appears during moments of emotional intensity (e.g., when Rheya manifests); it symbolizes human interference with what is natural, as well as unease or the overwhelming force of the ocean. It is a "wounded" sky, marked by human emotion, fear, shame, and guilt.

The sky in Solaris is not merely a backdrop—it reacts to the presence of humans, much like the ocean itself. It forms part of what may be described as the "planet's consciousness." Just as the ocean creates "visitors," the sky reflects psychological moods and existential tension.

Red sky = moments when emotions dominate, when the protagonist confronts his past.

When the ocean was calm or "silent," the sky appeared in cool tones—often blue, pearly, or whitish. When the ocean was active—pulsating, generating structures (symmetriads, mimoids)—the sky shifted to red, orange, or even luminous, as though internally illuminated. The planet's atmosphere was dependent on the ocean, which functions as a living, conscious organism.

This is further underscored by Solaris's orbit around two suns—one red and one blue. Depending on which sun dominated the horizon, the sky acquired corresponding hues:

Under the blue sun—the sky appeared cold, gray, and distant.

Under the red sun—the sky became warm, intense, and almost unnatural in its coloration.

These were not ordinary day-night cycles; transitions between "days" carried symbolic and emotional weight. One may also interpret the changing sky as reflecting the inner states of the protagonist, especially Kelvin:

In moments of calm, detached observation, he described the sky as clear, ashen, or emotionless.

In moments of fear, memories or confrontation with the past, the sky glowed crimson, suffused with oppressive, suffocating colors. The sky thus acts as an expression of emotion—not literal, but embedded in the poetics of the fictional world: "The sky was orange-gray, and the ocean gleamed faintly under the fading light of the red sun. Soon the blue one would rise—and everything would once again become cold, dead."

The juxtaposition of red and blue in Solaris demonstrates how literary symbolism of color can mirror psychological states and existential tension. As shown in Table 3, the dualism between red (life, chaos, emotional intensity) and blue (distance, calmness, cosmic indifference) provides a model for understanding how colors may shape atmosphere and meaning in multisensory design. This highlights the potential of narrative-inspired color strategies to enrich architectural spaces with layered emotional and symbolic depth.

Table 3. Analysis of Colors in the Context of Solaris. Author: Agnieszka Rek-Lipczyńska.

Narrative Aspect	Red	Blue
Meaning in the description of the ocean—Life	pulsating matter, threatening force, emotional tension	Cold intelligence, otherness, mystery, cosmic chill
Emotional context of the characters	Moments of confrontation with the past, surges of emotion, fear, guilt	Pseudo-calm, superficial order, cold distance
Connection with the day cycle on Solaris	Red sun—warm, intense, dramatic light	Blue sun—cold, pale, indifferent sky
Symbolic space—Chaos	claustrophobia, dominance, heaviness	Order, openness, submission, lightness

Blue sky = pseudo-calm, surface order masking emptiness or existential dread.

The spatial associations of red and blue emphasize the multisensory dimension of color perception, extending beyond the visual domain to include tactile, acoustic, and olfactory cues. As illustrated in Table 4, red tends to evoke intensity, density, and stimulation, while blue is linked to calmness, openness, and restraint. These findings underscore the importance of integrating color with other sensory modalities in interior design to achieve coherent and immersive user experiences.

Table 4. Spatial Impressions and Multisensory Perception of Colors. Author: Agnieszka Rek-Lipczyńska.

Spatial Feature	Red	Blue
Lighting	Warm light, strong shadow contrasts	Cold light, hidden sources
Form	Sharp angles, chaotic arrangement, crowding	Fluid forms, rounded shapes, few elements
Material	Rough, coarse surfaces	Smooth, light-reflecting surfaces
Acoustics	Noise, echo, loud effects	Silence, muffled sounds
Olfactory perception	Sweet fragrance	No distinct scent, freshness
Tactile perception	Oppressiveness, warmth, narrowness	Lightness, spaciousness, coolness

The analysis of the meaning of red and blue in the authors' study is consistent with international research in psychophysics, environmental psychology, and neuroscience.

Red emerges as a symbol of passion, strength, and emotional intensity, but also of chaos, claustrophobia, and confrontation with the past. It is a high-arousal color: it increases heart rate and blood pressure, activates the nervous system, and stimulates emotions (both positive and negative). This dualism corresponds closely to the findings of Jonauskaite and Mohr. In contrast, blue appears as a symbol of distance, cold intelligence, and cosmic indifference. The spatial dimension associated with blue is ordered, bright, and light [3].

The analyzed contrast between red and blue confirms the classical dualism in color psychology: emotion ↔ rationality, energy ↔ calmness, closeness ↔ distance. In the authors' study, both colors create a complementary emotional code—their juxtaposition in design can guide the user's mood and shape the rhythm of spatial experience.

The author made a preliminary attempt to determine how an architectural interior could be designed to be dedicated specifically to the multisensory perception of a single color. The comparison includes fundamental spatial factors. The data are presented in the table below (Table 5).

Table 5. Summary of core factors for multisensory perception and model interior in "red" and "blue". Author: Agnieszka Rek-Lipczyńska.

Red Color	Blue Color
Warm light	Cold light
Chaotic, disorderly arrangement	Orderly layout
Geometric, sharp forms, angles	Liquid forms, waves, curves
Dense arrangement of elements, tightening space	Open space, few elements
The layering of elements makes the space tighten	Smooth, light-reflecting materials. Use of airy materials
Overwhelm effect, Shadow play—strong chiaroscuro contrasts	High room. Hidden light sources
Feeling warm	Feeling cold
Dynamic sound effects	Silence
Rough, rough surfaces	Cleanliness
Sweet scent	The smell of freshness

In the context of interior architecture, red and blue carry a number of predictable perceptual effects. It has long been recognized that color schemes influence the subjective perception of a room's dimensions and atmosphere. Walls painted in intense red may create the impression that the space is "closing in" on the observer—red, as a warm color, visually reduces the interior, adding a sense of coziness but also of confinement when used in excess. The opposite effect is produced by cool blue: hues from the blue–green palette visually enlarge and "push back" the boundaries of a room, creating a sense of spaciousness [Table 5].

This impression is strongly linked to the psychological perception of color temperature—blue tones are experienced as cold, whereas reds are perceived as warm and advancing, both of which affect the way we judge depth and scale in space. Studies confirm that people generally experience light blue as "cold" in perception, while red or orange are felt as "warm." Moreover, color selection may influence the subjective sense of thermal comfort—for instance, interiors dominated by red appear warmer, while those characterized by blue tones are perceived as cooler, even at the same ambient air temperature. As a result, color schemes may directly impact user decisions: in a "cool" blue-toned room, occupants may feel a need to increase heating, whereas in a warm red environment, the temperature may feel higher than it actually is.

In addition to influencing the visual appraisal of space, colors also affect other sensory experiences in an indirect manner. For example, color schemes can modulate gustatory–olfactory impressions and appetite. Red is known to have a stimulating effect—in gastronomy and dining-room design, red accents (such as tablecloths, walls, or restaurant logos) are often used, as warm colors tend to increase appetite and encourage social interaction at the table [Tables 4 and 5]. Conversely, cool blue is often considered an appetite-suppressing color—it is rarely used in dining environments, and studies suggest that meals served on blue plates appear less appetizing and are consumed in smaller quantities. Blue, associated with cleanliness and freshness, is instead well suited to bathrooms and kitchens, where it enhances the impression of hygiene. Red, on the other hand, is often avoided in bedrooms, as its energetic character may hinder relaxation before sleep.

Colors also influence sound perception—an effect that is less intuitive. Experimental studies have shown that the visual environment can alter the perceived loudness of noise. For example, the noise of a household appliance was subjectively rated as louder and more unpleasant when accompanied by a red visual stimulus, compared to the identical sound presented with green–blue stimuli. In other words, red can "amplify" the perception of sound (similarly to white and other bright colors), while cool greens and blues exert a calming effect, making noise seem somewhat more tolerable.

Below is a summary of the key features and effects attributed to red and blue in the context of the multi-sensory experience of interior spaces. [Table 6.]

Table 6. Summary of the key features and effects attributed to red and blue in the context of the multi-sensory experience of interior spaces. Author: Agnieszka Rek-Lipczyńska.

Aspekt	Red Color	Blue Color
Cultural symbolism	Energy, power, passion, insecurity; happiness (China, Japan); also mourning (e.g., name of the deceased in Buddhism)	Peace, wisdom, harmony, spirituality; mourning (Iran); professionalism (business, medicine in Western culture)
Effect emotional	Stimulation, excitement, increased dynamics. Increases energy and activity levels. May cause anxiety if consumed in excess.	Calming, relaxation, contemplation. Reduces the level of stress and tension. It may contribute to a slight depression of mood (symbolism of sadness).
Perception of space	It optically reduces and zooms in on the walls ("pressing" color). It gives coziness, but too much of a feeling of claustrophobia	Optically, it enlarges and recedes the borders ("receding" color). It gives the impression of spaciousness and depth. It eliminates claustrophobia in small interiors.
Thermal comfort-feeling	"Warm"—subjectively increases the feeling of ambient temperature. The interior feels hotter; a color associated with fire and the sun	"Cold"—subjectively reduces the feeling of temperature. The interior feels cooler; a color associated with water and ice
Acoustics-noise	It enhances the perception of loudness—sounds seem louder and more insistent in a red environment	Softens the perception of loudness—noise is perceived as slightly quieter and less bothersome in a blue-green environment
Taste-smell	Sweet, intense—associated with ripe fruit, spices, flowers (rose). Red in the environment can stimulate appetite	Fresh, clean—associated with mint, air, water (sea breeze). Blue in the environment often suppresses appetite
Touch, texture-associations	Warm, rough—in synesthetic sensations, it may be associated with the impression of warmth, roughness or greater weight.	Cool, smooth—combined with the impression of coolness, smoothness and lightness. Often described as "lighter" and softer than red.

Moodboard—Red

The red moodboard illustrates the multisensory associations of red as a high-arousal, "warm" color. Coarse brick textures, matte painted walls, and fibrous fabrics evoke tactility linked with warmth, density, and a sense of enclosure. These material and chromatic choices confirm the findings of psychophysical studies in which red is consistently associated with stimulation, heightened attention, and increased perceived temperature. In spatial perception, such surfaces may foster feelings of coziness, but also claustrophobia when applied excessively. This aligns with experimental evidence demonstrating that red environments enhance arousal and can amplify the perception of noise, thereby intensifying the sensory climate of interiors [Figure 7].

Figure 7. Sensory mood board illustrating the multisensory associations of red and blue in interior design (author's concept).

Moodboard—Blue

The blue moodboard conveys blue as a "cold," low-arousal color associated with calmness, expansion, and harmony. Smooth, glossy, and water-inspired textures (e.g., satin, polished stone) generate tactile sensations of coolness and lightness, reinforcing the perception of depth and spaciousness. In alignment with environmental psychology research, blue-dominant environments are reported to reduce stress levels and attenuate the subjective perception of noise, thereby supporting restorative experiences. From a design perspective, these qualities position blue as an optimal color strategy for interiors intended to promote relaxation, contemplation, and a sense of openness [Figure 7].

Below is a conceptual, original mood board illustrating the multisensory impact of red on interior design. Raw textures, warm lighting accents, and dynamic forms emphasize energy, warmth, and stimulation. The arrangement demonstrates how red, as an "unfolding" color, creates intimacy and intensity while simultaneously enhancing sensory activation through tactile and thermal associations [Figure 8].

Below is a conceptual mood board illustrating the multisensory impact of blue in interior design. Smooth, reflective surfaces, metallic and satin finishes, and fluid forms emphasize coolness, spaciousness, and tranquility. The composition highlights how blue, as a "receding" color, enhances the perception of openness and relaxation through visual, tactile, and acoustic stimuli [Figure 9].

Figure 8. Moodboard—Red Interior Concept. Author: Agnieszka Rek-Lipczyńska.

Figure 9. Moodboard—Blue Interior Concept. Author: Agnieszka Rek-Lipczyńska.

4. Discussion

4.1. Multisensory Perception of Color: Vision, Sound, Smell, Touch, and Temperature

The experience of color in space usually takes place through the dominant sense of vision, yet other senses also participate in perceiving indirect stimuli associated with color. In a well-designed multisensory interior, visual impressions harmonize with acoustic, thermal, tactile, and even olfactory sensations, creating a coherent atmosphere. As discussed in the previous section, red and blue can trigger predictable reactions beyond the visual modality—for instance, influencing the perceived loudness of noise or modulating sensory arousal levels related to thermal comfort or appetite. The mechanisms underlying these phenomena often involve cross-sensory associations (crossmodal correspondences), shaped both biologically and culturally. For example, the brain integrates inputs from multiple senses into a single interpretation of the environment: higher temperatures are associated with "warm" colors, while lower temperatures are linked with "cool" colors, even if the wall paint does not alter the actual air temperature. Similarly, lower-pitched sounds (bass) are intuitively matched with darker or visually "lower" colors, while brighter colors correspond with higher tones.

Taste and aroma. Strong associations exist between color and the taste or aroma of food—for instance, the taste of an orange is tightly bound to its color, so a green-colored beverage with orange flavor may be perceived as less authentically "orange." Studies have shown that the presence of specific scents in the environment can distort color perception—for example, the smell of lemon or mint may make a hue appear more yellowish or greenish than it actually is. This occurs because the mind generates unconscious associations (lemon = yellow, mint = green) that influence the perception of background color. Conversely, color can also modify flavor and aroma perception. In one experiment, the same wine was judged as sweeter and more aromatic when served under red lighting compared to blue or neutral light [21]. Such phenomena are known as crossmodal effects, in which stimuli from one sense influence perceptions in another. Experimental models of multisensory color perception often manipulate the hue of a room and measure changes in taste or smell, confirming that visual cues can systematically bias gustatory experiences.

Auditory perception of color. Colors also affect the acoustic mood of interiors. In rooms dominated by red tones, users may unconsciously speak louder and faster, and background sounds (e.g., street noise or air conditioning) may be judged as more disturbing. In blue-toned interiors, people tend to speak more quietly, and background noise is perceived as less intrusive. In rare cases of synesthesia, sounds trigger color sensations or vice versa—listeners "see" colors while hearing music. Composers such as Scriabin experimented with assigning musical keys to specific colors [29]. In general, high-pitched sounds are more frequently associated with bright, cool colors (yellow, light blue), while low-pitched tones are linked with dark, warm hues (burgundy, brown).

Olfactory and gustatory perception of color. Olfactory descriptions often employ "color terms" (e.g., "green" scent of grass, "warm" spicy aroma). When using aroma marketing in interiors, designers should ensure that scents are consistent with the color palette. For example, the intense fragrance of cinnamon or vanilla is more congruent with warm-colored environments (oranges, reds) than with cool blue ones, where it may appear incongruent or unpleasant. Research shows that sensory congruence between color and aroma enhances overall evaluations of spaces by users. In practice, this may be applied by thematically matching scents to interior zones: a "red" energy zone in a spa might be reinforced with clove or citrus notes, while a "blue" relaxation area could be complemented with eucalyptus or marine aromas. The influence of color on taste is also documented: in experiments, beverages of identical composition tasted different depending on the color of packaging or lighting—red cues enhanced perceptions of measure changes in taste or smell, confirming that visual cues can systematically bias gustatory experiences. Tactile perception of color. Color is also perceived through materials and textures, which people intuitively associate with particular hues. Even sighted individuals, when closing their eyes, can "imagine" that a warm, coarse textile (e.g., wool) corresponds with red or orange, whereas smooth, cool silk aligns with blue or silver. In some cases of synesthesia, tactile stimuli can directly evoke color sensations (e.g., a given texture "feels green" even without visual input). In design practice, warm, soft, and textured materials (wood, velvet) amplify the warm character of red interiors, while glass, metal, or polished stone emphasize the visual coolness of blue spaces. Interestingly, architectural theorists also speak of the "weight of colors"—subjectively, colors differ in perceived heaviness: black and red appear "heavy," while blue and white are perceived as "light." This has implications for how interiors are experienced—for example, a dark red floor may be perceived as more solid (although visually lighter) than a pale blue one [30].

These complex phenomena illustrate that a designer consciously employing color can simultaneously influence multiple senses—even if the user remains unaware of such effects.

4.2. *Challenging Assumptions: Color, Temperature, and Synesthesia*

Interestingly, recent experiments have questioned some of the traditional assumptions. For instance, Ho et al. found that when participants touched colored objects, the effect was opposite to expectations: a blue object at a given temperature was more often perceived as warm than a red object at the same temperature. This suggests that the brain integrates direct thermal sensations with prior associations of color and warmth in a contrasting manner—possibly as a corrective mechanism to the anticipated "color effect."

4.2.1. Synesthesia and Applications in Design

The study also refers to synesthesia, a special case of multisensory perception. The findings are scientifically significant—revealing which properties of sound correspond to color experiences—and practically relevant, as they point to opportunities for using AI to predict or simulate synesthetic sensations. In the future, such knowledge may support the design of tools for blind individuals (e.g., translating speech into color cues) or inspire multisensory art and interfaces that connect sound and color based on actual cross-sensory correlations [13].

Synesthesia occurs when stimulation of one sense simultaneously triggers sensations characteristic of another. Individuals with synesthesia may, for example, "hear" colors (each hue evoking a specific sound), "taste" words, or "see" colors in response to numbers or music. Although synesthesia is rare as a congenital neurological trait, it has long inspired artists and designers to create multisensory experiences accessible to all. In architecture, one speaks of "synesthetic architecture", in which spaces are designed so that stimuli from different senses converge into a unified perceptual whole. A well-known example is the Jewish Museum in Berlin by Daniel Libeskind, where slanted floors, sharp angles, and cold lighting deliberately disrupt conventional stimuli, producing not only visual disorientation but also bodily sensations of imbalance and confusion—thus evoking deeper emotions aligned with the museum's theme.

In synesthetic design, a stimulus intended for one sense is deliberately crafted to suggest another. For instance, a specific texture may evoke a thermal impression, or a building material may suggest a sound association: rough brick may "feel" warm and dry, while smooth chrome-plated metal may appear cold and "resonant" (as we imagine the sound it might produce). Designers can harness these associations: to make a space feel cozy and warm, one may use not only warm colors but also soft, matte textures that absorb sound; to create a clinical, laboratory-like atmosphere, one may combine cool hues with smooth, glossy materials that reflect both light and sound, producing an echoing sense of "sterility".

Synesthesia also has direct applications in multisensory information design. Since in some individuals the senses are "cross-wired," this principle can be adapted as a universal design language—for example, assigning specific sounds, shapes, or scents to particular colors, so that color-coded information is accessible across multiple modalities. Historically, attempts at "color instruments" such as the color organ sought to match sounds with light hues. Today, designers experiment with multimedia installations that combine color, sound, image, and touch (e.g., vibrations) into a single immersive experience. One example is an interactive bridge illumination project, where a pedestrian's sound (e.g., a clap) triggers a wave of light in a corresponding color. Another example involves therapeutic sensory rooms for children with disabilities, where wall colors respond to sound or touch, allowing children to "see music" or "hear touch." Although these ideas once seemed futuristic, LED, sensor, and synthesizer technology has advanced such that these multisensory concepts are increasingly feasible. Thus, in multisensory navigation systems for buildings, these principles can be applied: a red path could be marked not only by visual color but also by a high-pitched, fast rhythm sound, while a blue path could be accompanied by a lower tone and slower pulse, ensuring that both visual and auditory cues are congruent and intuitive. Similarly, functional zones may be paired with olfactory motifs—for instance, lavender in a blue relaxation area, and citrus notes in a red activity zone. In this sense, synesthetic design aims at the harmonious integration of stimuli, enabling them to reinforce each other and convey meaning even when one sensory channel is limited.

Sustainable and Inclusive Color Design

Multisensory and inclusive design are closely aligned with the principles of sustainable development—particularly in the social dimension (ensuring equal access) and in promoting user well-being. According to the UN 2030 Agenda, the Sustainable Development Goals include building inclusive, safe urban spaces (SDG 11) and promoting health and well-being (SDG 3). Considering the needs of people with different disabilities when designing interior color schemes directly supports these goals. European legislation and standards—such as the European Accessibility Act of 2019—emphasize the application of universal design principles in all new projects.

This means that multisensory perception of space should be planned already at the conceptual stage of architectural design: from providing adequate lighting (for the visually impaired), to acoustics (important for those using hearing aids in noisy interiors), to clear tactile and contrast-based markings. Publicly funded projects in the EU increasingly require accessibility audits, which also assess elements such as the use of colors that meet contrast criteria (e.g., EN 17161 standards on visual accessibility and color perception in the built environment).

It is important to stress that inclusive design does not mean eliminating intense colors but rather using them consciously. For example, UK Design Council guidelines on design for dementia recommend avoiding patterned, distracting surfaces but at the same time employing strong color accents to mark key elements (e.g., toilet doors in contrasting hues, handrails in bright yellow). In design for individuals with autism, by contrast, muted, cool palettes are preferred to prevent sensory overload—here blues and greens are more suitable than vivid reds. The context of the user and the function of the space thus determine color choices from a sustainable design perspective: the aim is to create environments that support health and well-being for all.

Interestingly, despite the decline in sensitivity to cool colors with age, surveys show that elderly respondents most often choose blue and green as their favorite environmental colors. This apparent paradox (reduced perception of blues vs. preference for blues) continues to puzzle researchers. Age-related visual changes include:

Yellowing of the lens, acting as a "warm filter" that reduces blue light transmission;

Decline in cone sensitivity, particularly S-cones responsible for blue perception;

Increased need for lighting—older adults require more light to recognize colors;

Greater reliance on mesopic adaptation, i.e., transitional conditions between day and twilight [28].

How then should we interpret the dominant color preferences of older adults in light of the Purkinje phenomenon?

4.3. Conceptual Prototypes of Multisensory Light Switches

To further illustrate how multisensory coding of color can be translated into practical design applications, two conceptual prototypes of light switches were developed. These design studies are based on the analyses presented earlier in this article, which highlighted how red and blue evoke contrasting yet predictable sensory associations across vision, touch, hearing, thermal perception, and even olfactory imagination. By integrating these findings into product-scale design, the prototypes demonstrate how inclusive design strategies may extend beyond spatial environments to everyday interfaces (Table 7).

Table 7. Comparative multisensory features of the conceptual light switches "Red Code" and "Blue Flow." Author: Agnieszka Rek-Lipczyńska.

Sensory Modality	Red Code (Stimulation and Intensity)	Blue Flow (Calmness and Balance)
Form and layout	Sharp, geometric, layered buttons (triangles, rhombuses, jagged edges)	Smooth, flowing, wavy shapes reminiscent of water ripples
Surface texture	Rough, tactile, sandblasted or ceramic with grooves	Glossy, sleek, polished glass/ceramic with seamless finish
Thermal perception	Warm to the touch (wood, ceramic with micro-heating, warm polymers)	Cool to the touch (glass, stone, polished ceramic)
Color and lighting	Dark red casing, warm LED backlight (2700 K) highlighting contrasts	Light-to-medium blue casing, cool LED backlight (5000–6500 K) accentuating softness
Touch	Distinct roughness and warmth	Smoothness, gloss, coolness
Sight	Strong light–shadow contrasts, intense presence	Gentle reflections, soft contours, calming openness
Sound	Sharp, mechanical click sound	Muted, low-pitched click sound

The design emphasizes sharp, geometric button shapes with rough, textured surfaces and warm tactile qualities, reflecting the stimulating and dynamic sensory profile of red [Figure 10].

Figure 10. Conceptual rendering of a multisensory light switch inspired by the color red. Generated by ChatGPT-5 based on the developed design assumptions.

Concept Proposal—Multisensory Light Switch "Red Code"
Form and Layout:

The panel consists of several sharp, geometric buttons (e.g., triangles, rhombuses, jagged edges).

The buttons are arranged in a multi-element, layered composition, visually and tactilely resembling a dynamic structure (e.g., a crystal or a fragment of a wall).

Surface:

A rough texture—for example, cast from sandblasted aluminum, ceramic, or imprinted architectural concrete.

Small grooves and protrusions are clearly perceptible to the touch.

Thermal Impression:

The buttons are made of a "warm-touch" material—e.g., impregnated wood, ceramic with light surface heating (a low-energy micro-heater).

Alternatively, a polymer composite can be used, which naturally warms quickly to the hand's temperature compared to metal.

Color and Lighting:

A dark red casing with subtle warm LED backlighting (2700 K) glowing from the gaps between the buttons, highlighting the dynamism of the form and signaling activation.

Multisensory Qualities:

Touch: a distinctly rough surface and a warm tactile impression.

Sight: sharp shapes, contrasts of light and shadow.

Sound: each click produces a clear, "sharp" mechanical sound that matches the geometric form.

This concept was developed on the basis of the earlier analyses presented in this article, which explored how the color red can be encoded and experienced across multiple sensory modalities in interior design.

The design features smooth, wavy button forms with glossy, cool-to-touch surfaces and subtle cool backlighting, reflecting the calming and spacious sensory profile of blue [Figure 11].

Figure 11. Conceptual rendering of a multisensory light switch inspired by the color blue. Generated by ChatGPT-5 based on the developed design assumptions.

Concept Proposal—Multisensory Light Switch "Blue Flow"

Form and Layout:

The panel is composed of smooth, flowing, and wavy-shaped buttons, arranged in a minimalistic and harmonious composition.

The design visually and tactilely evokes the impression of water ripples or fluid architectural forms, reinforcing associations of openness and calmness.

Surface:

A glossy, sleek texture—for example, polished glass, glazed ceramic, or lacquered composite materials.

The buttons are intentionally smooth to the touch, emphasizing a cool and refined tactile quality.

Thermal Impression:

The material selection ensures a "cool-touch" sensation—e.g., glass, stone, or polished ceramic.

The surfaces retain a slightly lower temperature than the surrounding air, providing a refreshing tactile experience.

Color and Lighting:

A light-to-medium blue casing with subtle cool LED backlighting (5000–6500 K) shining gently around the edges of the buttons.

This lighting effect highlights the fluid contours and underscores the calming atmosphere of the design.

Multisensory Qualities:

Touch: smoothness, gloss, and cool tactile sensations.

Sight: soft, flowing shapes with gentle contrasts and reflections.

Sound: each click produces a muted, low-pitched sound consistent with the tranquil character of the form.

This concept was developed on the basis of the earlier analyses presented in this article, which demonstrated how the color blue can be experienced multisensorially in interior design—as a symbol of spaciousness, serenity, and sensory balance.

The two concepts—Red Code and Blue Flow—represent contrasting sensory strategies. The red prototype emphasizes stimulation and intensity: sharp geometries, rough textures, warmth to the touch, and a sharp mechanical sound upon activation, all of which reflect red's energetic and activating qualities. By contrast, the blue prototype highlights calmness and balance: smooth, flowing forms, glossy cool-touch surfaces, and a muted, low-pitched click that resonates with blue's associations of serenity and spaciousness. Together, these designs showcase how theoretical insights into multisensory perception of color can be materialized into tangible objects. They also provide a practical framework for inclusive design, where everyday elements such as light switches can be enriched with multisensory cues, ensuring accessibility and enhancing the user experience for both sighted and visually impaired individuals.

4.4. Color, Perception, and the Purkinje Effect

Considering the lower lighting conditions that often prevail in the homes of older adults (diffused lighting, local lamps, lack of strong daylight), the Purkinje effect naturally enhances the perception of blue–green tones relative to red. Although objective visual sensitivity declines, in subjective perceptual experience (i.e., how one feels in a given space), blues and greens appear brighter, calmer, and more present. It can therefore be assumed that seniors subconsciously prefer colors that harmonize with their perceptual lighting conditions—for instance, hues that remain "visible" under dim, natural, or evening light—consistent with the Purkinje mechanism. In summary, under reduced lighting conditions typical of domestic environments for the elderly, the visual system becomes more sensitive to wavelengths corresponding to blues and greens, which may unconsciously influence the sense of their brightness and presence [31].

Sustainable and inclusive design benefits from a multisensory approach to color. Biophilic design, recognized as a component of sustainable architecture, emphasizes the need to engage multiple senses through natural elements—not only the green of plants, but also their scent, the sound of wind, or the texture of wood. The introduction of natural materials and colors into interiors simultaneously fulfills several goals: it is ecological (local materials, fewer synthetics), health-promoting (improved air quality, humidity regulation by wood), and provides rich multisensory stimuli. For example, wooden finishes not only add a warm, honey-like hue to a space but also produce a subtle resin scent, are pleasant to the touch, and improve room acoustics (wood reduces echo). Plants enrich interiors with soothing greenery for the eyes, filter the air while providing fresh fragrance, and introduce natural sounds (the rustling of leaves). Multisensory design thus draws upon the principles of sustainable design to create spaces that are both welcoming and health-promoting.

Based on the studies reviewed, a set of key design recommendations can be formulated:

Ensure multisensory coherence. When designing for all senses, stimuli should be aligned—for example, if a particular color palette is applied, it should be accompanied by a congruent soundscape and fragrance, avoiding dissonance (e.g., bright red décor paired with a cold, acoustically harsh space). Non-material elements such as acoustics, background music, scents (derived from natural materials or tailored to the room's function), and tactile features (textures, surface temperatures) should complement visual design.

Use redundant multisensory coding. Color-coded information can be supported by other modalities. For instance, a specific sound or scent can signal the presence of red or blue in the environment. Inclusive designers should consider hybrid sensory strategies—combining stimuli (e.g., sound with scent)—to increase accessibility for visually impaired users. Such systems must remain intuitive, easy to learn, and simple to perceive.

Leverage natural crossmodal correspondences. For example, to encode red in space, designers might employ warm materials or heated elements paired with energetic sound cues; conversely, blue may be conveyed through cooler tactile materials and calmer sounds. Balanced, correlated stimuli (aligned with psychological crossmodal correspondences) enhance perception, provided no single sense overwhelms the others.

Design senior-friendly interiors. Residential environments for older adults should emphasize high-contrast warm colors. Key elements (doors, handrails, stair edges) should be marked with colors such as red against a dark background (e.g., black) to maximize visibility. Blue and other cool hues may be included decoratively but should not be the sole identifiers of functionally critical features, as they may be overlooked by users with weakened vision. For seniors with cognitive decline, color cues should be even simpler and more distinct—for example, universal color codes (red = caution/danger, green = safe) applied consistently throughout the interior.

Apply practical guidelines. Warm, highly saturated colors (reds, oranges, yellows) should highlight important features, while problematic color combinations for seniors (e.g., blue text on green background, dark navy details on black surfaces) should be avoided. High brightness contrasts are essential (e.g., light-colored doors on dark walls). Excessive use of dark hues in small rooms should be avoided, as they can increase feelings of confinement and reduce visibility in low light. Instead, bright, light-reflecting backgrounds complemented with strong color accents are recommended. Functionality should be balanced with aesthetics—for instance, yellow can serve as a warning color in communal spaces, while in living areas it should be used more sparingly to prevent visual fatigue.

Ensure redundancy in information design. Where color is used to convey functional meaning (e.g., zones of a building, evacuation routes, wayfinding systems), it should always be accompanied by redundant cues such as text, pictograms, or tactile markers, ensuring accessibility for people with low vision or color blindness.

Test with users and follow standards. Participatory approaches should be applied—consulting with blind users to evaluate tactile signage, or with seniors to ensure color schemes are not visually tiring. Designers should also reference established standards and guidelines (e.g., Guide to Universal Design, BS 8300 in the UK, DIN 32975 in Germany on visual contrast, or Polish Universal Design Guidelines).

All of these measures reflect the principles of human-centered design, where the ultimate goal is user well-being and a complete sensory experience of space. The result is interiors that are more inclusive, comfortable, and safe—an inherent part of sustainable social development. Moreover, such spaces are often aesthetically richer, as multisensory design adds depth to the experience. In today's competitive environments (retail, museums, hotels), design that appeals to multiple senses creates memorable, distinctive impressions—supporting both marketing value and health-promoting benefits by fostering stronger emotional bonds with place and a deeper sense of place.

5. Conclusions

The conducted study demonstrates the potential of developing theoretical interior models as a methodological tool for architectural and design research. By systematically translating sets of perceptual and sensory features into spatial, material, and atmospheric characteristics, such models enable a deeper understanding of how interiors can be shaped to evoke specific experiences. They provide an experimental framework that links design parameters (light, geometry, materiality, acoustic and olfactory qualities) with multisensory responses, thereby enriching traditional approaches to architectural analysis.

Equally significant is the role of conceptual visualizations, which allow abstract qualities—such as warmth, spaciousness, tension, or ethereality—to be expressed in a tangible form. These visual representations facilitate not only the communication of design intentions but also the validation of theoretical assumptions through experiential simulation. In this way, visualizations act as a bridge between theoretical constructs and practical design applications.

The results indicate that creating theoretical models and their visualizations offers valuable opportunities for further interdisciplinary research, particularly in the fields of sensory studies, environmental psychology, and sustainable design. By combining conceptual rigor with perceptual exploration, this approach supports the development of more human-centered and experientially rich architectural environments [Table 8].

Table 8. Conceptual interior model; "Chaotic and overwhelming" and "Cold and Ethereal" Variant. AI-generated models. Author: Agnieszka Rek-Lipczyńska.

Aspekt	Interior Model—"Cold and Ethereal" Variant	Conceptual Interior Model—"Chaotic and Overwhelming" Variant
1. Lighting	Cold light (color temperature 5000–6500 K, white with a bluish tint). Concealed light sources (linear fixtures integrated into suspended ceilings, wall recesses, and floor-embedded strips). Light reflections enhanced by smooth, reflective surfaces.	Warm light (2700–3000 K, yellowish tones). The play of chiaroscuro-strong contrasts of light and dark zones. Spot lighting, local lighting, bringing out sharp forms.

Table 8. *Cont.*

Aspekt	Interior Model—"Cold and Ethereal" Variant	Conceptual Interior Model—"Chaotic and Overwhelming" Variant
2. Form and Geometry	Fluid lines, waves, and curves expressed in ceilings, furniture, and details. A minimalist layout characterized by a small number of elements, emphasizing emptiness and airiness. Open space—absence of divisions and unnecessary partition walls.	Chaotic, disorderly arrangement-lack of alignment and compositional order. Geometric, sharp forms and angles-aggressive, irregular shapes. Density of elements-a feeling of limited, stuffy space. Layering-overlapping of forms, multiplication of shapes, impaired legibility of space.
3. Materials and Textures	Smooth surfaces (lacquered finishes, glass, polished stone, metal). Light, airy materials—semi-transparent textiles (voile, organza, mesh fabrics). Reflectivity—glass and polished surfaces reinforcing the sense of coolness.	Rough, rough surfaces (concrete, raw wood, brick, unpolished stone). No gloss, light absorbing materials. Variety of textures creating anxiety and tactile contrasts, textures soft, warm to the touch
4. Spatial Scale	A high ceiling conveying monumentality and spaciousness. Selected elements emphasizing verticality and lightness.	The space is getting tighter-walls, furniture and details are densely arranged, the feeling of being overwhelmed. A feeling of weight and confinement.
5. Sensory Impressions	Touch: cold, smooth, and hard surfaces (glass, stone). Acoustics: silence, with reverberation moderated by discreet acoustic treatments, absence of disturbances. Smell: freshness—e.g., ozonic, marine, or citrus notes. Thermal sensation: low temperature and a perception of coolness enhanced by light color, chromatic palette (white, blues, silver).	The space is getting tighter-walls, furniture and details are densely arranged, the feeling of being overwhelmed. A feeling of weight and confinement. Touch: roughness, unevenness. Hearing: dynamic sound effects-echo, reverberation, sudden acoustic contrasts (e.g., sounds reflected from hard surfaces). Smell: sweet smell (association of intensity, heaviness of the atmosphere). Thermal perception: feeling of warmth enhanced by the color of light and materials.

Operational guidelines for design practice can be summarized as follows:

Material selection: warm colors (e.g., red) should be paired with tactilely warm materials (wood, textured ceramics, textiles), while cool colors (e.g., blue) align better with smooth, glossy, and cold-to-touch materials (glass, metal, stone).

Lighting parameters: red-coded spaces benefit from warm light (2700–3000 K) and strong light–shadow contrasts, whereas blue-coded spaces should employ cool light (5000–6500 K) and diffuse illumination.

Acoustic design: multisensory coding can be reinforced by acoustic cues—red environments with sharper, more dynamic sounds; blue with low-pitched, muted, and calming soundscapes.

Olfactory integration: scents may be used to complement color associations (sweet or spicy notes for red, fresh or herbal notes for blue).

Study limitations include the relatively small sample size of the exploratory survey, as well as individual differences in sensory interaction (e.g., personal or cultural variability in crossmodal associations). These constraints highlight the need for further empirical validation, particularly across diverse cultural and age groups, and with larger participant samples.

Contemporary interior architecture design is increasingly moving away from treating color solely as a visual attribute of space, shifting instead toward a multisensory and inclusive approach. A comparative analysis of two contrasting hues—red and blue—illustrates how diverse effects they can evoke through sight, hearing, smell, touch, and thermal perception. Red, an energetic and "warm" color, stimulates and attracts attention but can also overwhelm and amplify noise; blue, on the other hand, soothes the senses, creates an impression of spaciousness and coolness, yet in excess may be perceived as detached or suppressing certain stimuli (e.g., appetite). Understanding these mechanisms allows designers to consciously shape the user experience—through an appropriate composition of colors and accompanying stimuli, mood, behavior, and comfort within interior spaces can be effectively modeled. The phenomenon of synesthesia serves as an inspiration to pursue sensory coherence in spatial design. Stimuli for different senses should not conflict but rather complement each other, creating a rich yet harmonious landscape of impressions. This translates into practical guidelines: when designing with color, one should simultaneously consider texture, sound.

At the same time, inclusivity in color design is indispensable for sustainable development. The right to fully experience space—including its aesthetic and sensory qualities—should belong to everyone, regardless of perceptual limitations. Solutions enabling blind, visually impaired, or otherwise disabled individuals to perceive color-related cues (through tactile markers, auditory or olfactory signals, etc.) are becoming a new standard of accessibility. Guided by the principle of "Nothing about us without us," designers should collaborate with users with special needs, testing and refining multi-channel information systems in space. Such initiatives are reinforced by international guidelines and standards, which means that inclusive color design is no longer a matter of goodwill, but a quality requirement in many contemporary projects.

In this context, it is worth including research that demonstrates the extraordinary plasticity of the human brain—the nervous system's ability to reorganize synaptic connections in response to environmental changes, damage, or functional loss—forming the basis of sensory compensation processes. One of the most fascinating examples is the adaptation observed in blind or deaf-blind individuals, where the visual cortex—deprived of typical visual stimuli—can be recruited for processing information from other modalities. This phenomenon may hold particular significance for the design of multisensory coding systems of information—including color—for blind and visually impaired individuals. Since the brain is capable of adaptation and can transfer functions from one modality to another, there is a real possibility of developing tools that translate visual stimuli into auditory, tactile, or thermal ones, which users could eventually learn to interpret as intuitively as sighted people perceive colors. In this sense, for blind and visually impaired individuals, this property of the brain paves the way for multisensory color coding. If the brain can "learn" to hear or feel by touch information that was originally processed visually, it is possible to create systems in which colors—such as red or blue—are encoded through sounds, tactile cues, temperature, or scents. Thanks to neural plasticity, users of such solutions could naturally "read" colors, much like implant users learn to perceive speech and sound. This means that appropriately designed multisensory technologies can become practical tools for restoring access to color-related information for people who have lost their sight, with effectiveness increasing over time as neural adaptation progresses.

For blind and visually impaired individuals, this property of the brain paves the way for multisensory color coding. If the brain can "learn" to hear or feel by touch information that was originally processed visually, it is possible to create systems in which colors—such as red or blue—are encoded through sounds, tactile cues, temperature, or scents. Thanks to neural plasticity, users of such solutions could naturally "read" colors, much like implant users learn to perceive speech and sound. This means that appropriately designed multisensory technologies can become practical tools for restoring access to color-related information for people who have lost their sight, with effectiveness increasing over time as neural adaptation progresses.

In conclusion, multisensory color experience represents an intersection of science, art, and technology in interior architecture. It deepens perception, making spaces more expressive and emotionally impactful. At the same time, it places responsibility on designers to shape environments that are socially sustainable—accessible and welcoming to all. An interdisciplinary approach—combining knowledge of perception with design practice and principles of universal design—enables the creation of interiors that engage the full spectrum of human senses, fostering stronger bonds between people and their surroundings. In such spaces, color ceases to be merely a background element and becomes a vivid experience, shared by both sighted and blind individuals, the young and the elderly, the able-bodied and those with disabilities—in other words, by the entire diverse community of users. In this way, interior architecture can fully realize its social role, creating environments that are simultaneously beautiful, functional and inclusive.

Author Contributions: Conceptualization, A.R.-L.; Software, A.K.; Validation, A.R.-L.; Formal analysis, A.R.-L. and A.K.; Investigation, A.R.-L.; Data curation, A.R.-L.; Writing—original draft, A.R.-L.; Supervision, A.R.-L. All authors have read and agreed to the published version of the manuscript.

Funding: This research received no external funding.

Institutional Review Board Statement: Not applicable.

Informed Consent Statement: Not applicable.

Data Availability Statement: The original work presented in this study can be found in the article. If you have any further questions, please contact the corresponding author.

Acknowledgments: While preparing this manuscript, the author used the Chat GPT 5 tool for data collection, generating a diagram and language translation. The author reviewed and edited the results and takes full responsibility for the content of this publication.

Conflicts of Interest: The author declares no conflicts of interest.

References

1. Xie, C.; Zhang, Z.; Guo, Y. Multisensory Experience and Emotional Response in Architectural Space: A Comparative Analysis of a Coffee Roasting Factory. *Buildings* **2023**, *15*, 1393.
2. Ziat, M.; Balcer, C.A.; Shirtz, A.; Rolison, T. A Century Later, the Hue-Heat Hypothesis: Does Color Truly Affect Temperature Perception? In *Haptics: Perception, Devices, Control, and Applications*; Springer: Berlin/Heidelberg, Germany, 2016; pp. 273–280. [CrossRef]
3. Jonauskaite, D.; Mohr, C. Do we feel colours? A systematic review of 128 years of psychological research linking colours and emotions. *Psychon. Bull. Rev.* **2025**, *32*, 1457–1486. [CrossRef]
4. Jonauskaite, D.; Abu-Akel, A.; Dael, N.; Oberfeld, D.; Abdel-Khalek, A.M.; Al-Rasheed, A.S.; Antonietti, J.-P.; Bogushevskaya, V.; Chamseddine, A.; Chkonia, E.; et al. Universal patterns in color–emotion associations are further shaped by linguistic and geographic proximity. *Psychol. Sci.* **2020**, *31*, 1245–1260. [CrossRef]
5. Takei, S.; Imaizumi, S. Effect of background color on facial emotion recognition: Influence of color–emotion associations. *i-Perception* **2022**, *13*, 1–16.
6. Alvarado, J. Text-to-image models reveal specific color-emotion associations. *Front. Psychol.* **2025**, *16*, 1593928. [CrossRef] [PubMed]
7. Bower, T.; Kim, J.; Kim, H. Influence of immersive room color on physiological and psychological responses in virtual reality. *Front. Psychol.* **2022**, *13*, 879325.
8. Wang, M.-Y.; Ku, Y.-C.; Lin, Y.-H. Changes in color preference and emotional associations during the COVID-19 pandemic. *Front. Psychol.* **2024**, *15*, 1365332.
9. Rui, L.; Firzan, M. Emotional Design of Interior Spaces: Exploring Challenges and Opportunities. *Buildings* **2025**, *15*, 153. [CrossRef]
10. Spence, C. Senses of place: Architectural design for the multisensory mind. *Cogn. Res. Princ. Implic.* **2020**, *5*, 46. [CrossRef]
11. Sandmann, P.; Dillier, N.; Eichele, T.; Meyer, M.; Kegel, A.; Pascual-Marqui, R.D.; Marcar, V.L.; Jäncke, L.; Debener, S. Visual activation of auditory cortex reflects maladaptive plasticity in cochlear implant users. *Brain* **2012**, *135*, 555–568. [CrossRef]

12. Miyamoto, K.; Taniyama, Y.; Hine, K.; Nakauchi, S. Congruency of color–sound crossmodal correspondence interacts with color and sound discrimination depending on color category. *i-Perception* **2023**, *14*, 1–18. [CrossRef]
13. Bartulienė, R.; Saudargienė, A.; Reinytė, K.; Davidavičius, G.; Davidavičienė, R.; Ašmantas, Š.; Raškinis, G.; Šatkauskas, S. Voice-Evoked Color Prediction Using Deep Neural Networks in Sound–Color Synesthesia. *Brain Sci.* **2025**, *15*, 520. [CrossRef] [PubMed]
14. Cho, J.Y.; Suh, J. Spatial Color Efficacy in Perceived Luxury and Preference to Stay: An Eye-Tracking Study of Retail Interior Environment. *Front. Psychol.* **2020**, *11*, 296. [CrossRef] [PubMed]
15. Ward, R.J.; Ashraf, M.; Wuerger, S.; Marshall, A. Odors modulate color appearance. *Front. Psychol.* **2023**, *14*, 1175703. [CrossRef]
16. Spence, C.; Levitan, C.A. Exploring the Role of Color in Flavor Perception: Recent Advances and Future Directions. *Food Qual. Prefer.* **2021**, *92*, 104124.
17. Reinoso Carvalho, F.; Wang, Q.J.; Spence, C. Assessing the Influence of Beer Color, Bottle Color, and Label Design on Consumer Expectations and Perceptions. *Food Res. Int.* **2021**, *141*, 110074.
18. Tu, Y.; Yang, Y.; Ma, M. Effect of Ambient Color on Taste Perception and Emotional Response. *Appetite* **2020**, *149*, 104630.
19. Wang, Z.-Y.; Cho, J.-Y. Older Adults' Response to Color Visibility in Indoor Residential Environment Using Eye-Tracking Technology. *Sensors* **2022**, *22*, 8766. [CrossRef]
20. Chen, Y.C.; Spence, C. Crossmodal Correspondences between Basic Tastes and Colors: A Replication and Extension Study. *Multisens. Res.* **2023**, *36*, 1–20.
21. Ho, H.-N.; Iwai, D.; Yoshikawa, Y.; Watanabe, J.; Nishida, S. Combining colour and temperature: A blue object is more likely to be judged as warm than a red object. *Sci. Rep.* **2014**, *4*, 5527. [CrossRef]
22. Ludden, G.D.; Schifferstein, H.N.; Hekkert, P. The multisensory perception of products: The influence of color–texture congruence. *Des. Stud.* **2020**, *68*, 100943.
23. Wastiels, L.; Schifferstein, H.N.J.; Wouters, I. Materiality matters: The influence of color and gloss on perceived material properties and touch expectations. *Front. Psychol.* **2021**, *12*, 624303.
24. Hagtvedt, H.; Brasel, S.A. Crossmodal correspondences between visual and tactile textures: The role of color. *J. Consum. Psychol.* **2022**, *32*, 482–495.
25. Ward, J.; Simner, J. The color of touch: A case of tactile–visual synaesthesia. *Neurocase* **2012**, *18*, 167–180. [CrossRef] [PubMed]
26. Iranzo Bartolomé, J.; Cho, G.; Cho, J.-D. Multi-Sensory Color Expression with Sound and Temperature in Visual Arts Appreciation for People with Visual Impairment. *Electronics* **2021**, *10*, 1336. [CrossRef]
27. Xu, H.; Zhao, J.; Jin, C.; Zhu, N.; Chai, Y. Research on the Multi-Sensory Experience. Design of Interior Spaces from the Perspective of Spatial Perception: A Case Study of Suzhou Coffee Roasting Factory. *Buildings* **2025**, *15*, 1393. [CrossRef]
28. Jaglarz, A. Color as a Key Factor in Creating Sustainable Living Spaces for Seniors. *Sustainability* **2024**, *16*, 10251. [CrossRef]
29. Jiang, A.; Fang, W.; Liu, J.; Foing, B.; Yao, X.; Westland, S.; Hemingray, C. The effect of colour environments on visual tracking and visual strain during short-term simulation of three gravity states. *Appl. Ergon.* **2023**, *110*, 103994. [CrossRef]
30. Wang, Q.J.; Spence, C. "That's What I Call Umami!" Color–Taste Associations in Children and Adults. *J. Sens. Stud.* **2022**, *37*, e12736.
31. Anstis, S.M. The Purkinje rod–cone shift as a function of luminance and retinal eccentricity. *Vis. Res.* **2002**, *42*, 2485–2491. [CrossRef]

sustainability

MDPI

Brief Report

Positive Associations between Women's Employment at the Community-Level and Opportunities and Interpersonal Sharing within Recovery Homes

Leonard A. Jason [1,*], Justin S. Bell [1] and Ted J. Bobak [2]

[1] Center for Community Research, DePaul University, Chicago, IL 60614, USA; jbell35@depaul.edu
[2] Center for the Study of Health & Risk Behaviors, University of Washington, Seattle, WA 98195, USA; tbobak@uw.edu
* Correspondence: ljason@depaul.edu

Abstract: The background of this study was an investigation of a sustainability issue involving how to deal with increasing deaths due to substance use as well as housing instability among those with substance use problems. Our study investigated the relationship between community-level women's employment and two key variables within Oxford House recovery homes: levels of employment and the sharing of resources among residents. These variables are crucial for the sustainability of recovery homes as they contribute to maintaining an ecological balance in communities. For our methods, we gathered data from recovery homes located in three different regions of the United States. The focus was on examining how women's employment rates in these communities correlate with the employment statuses of house residents and their resource-sharing behaviors. Regarding the results, we found a positive association between rates of employment among women in these communities and employment within the recovery housesand resource lending among residents. We conclude that women's employment at the community-level appears to be a significant macro-level factor that promotes supportive social and environmental policies. These community-level policies enhance employment opportunities and encourage interpersonal sharing within recovery homes, contributing to ecological balance and sustainability in terms of environmental, economic, and social dimensions.

Keywords: women's employment; zip codes; recovery homes; Oxford House

1. Introduction

Recovery homes for those with substance use disorders are complex and ubiquitous ecosystems, and residents' integration into these settings may be critical for facilitating the sustainability of substance use recovery resources in communities [1]. There is a need to develop more socio-economic and integrated approaches to understand sustainable processes for those with substance use disorders, particularly given the increasing problems involving the mortality of those with substance use disorders in part due to housing instability. Recovery homes might be part of a sustainable solution, as they provide communal living environments for individuals seeking a space to gain stability, structure, and social connection while attempting to stay abstinent from alcohol and/or substance use [2].

According to recent estimates, there are nearly 18,000 recovery homes in the United States [3], making them one of the largest shared living programs, and they play a vital role in addressing the community impact of alcohol and/or substance use issues. This influence has been found to be bidirectional, with the community environment impacting the success of recovery homes. As an example, residents' willingness to share resources is associated with outcomes such as increased wages, social support, and self-esteem [4], and it is also associated with lower rates of relapses [5]. These types of social network dynamics may contribute to understanding how these homes create desirable and sustainable recovery outcomes. It is possible and one of our hypotheses that economic conditions within the environment from which these houses are located could influence interpersonal dynamics in the recovery houses, such as with regard to loaning resources.

Therefore, it is important to identify exogenous conditions that predict outcomes among residents of recovery homes. For example, Jason et al. [6] evaluated a model that used a latent recovery factor to predict relapse, and a resident's probability of relapse was related to the "recoveries" of home peers rather than a member's own "recovery" status. In other words, this points to the social environment, as the resident relapse rate is related to the total recovery capital available in the homes. These house characteristics might make it more likely that improvements will be sustained after the residents leave these houses. Another of our hypotheses is that it is possible that economic conditions in the surrounding communities, particularly regarding employment, might make it more likely for residents to succeed in a house, as measured by their ability to find and retain employment.

At the house level, Jason et al. [7] examined friendship, advice seeking, and willingness to loan, as well as house president ratings, in relation to several variables such as house members' ability to pay monthly rent, their involvement in chapter activities, and how selective they were in accepting new residents. Using latent analyses, it was determined that the houses that had the lowest eviction rates were most likely to have residents who loaned others money, were more involved in chapter activities, and were better able to pay monthly house rent. This study suggests recovery homes have different ecological features that may influence and sustain recovery outcomes, and this finding is also connected with our hypotheses that what occurs within the surrounding community has sustainability impacts on the conditions influencing residents' ability to be successful.

Oxford House (OH) is the largest single network of recovery homes in the United States, serving approximately 28,000 individuals nationwide [8–10]. OH is distinct from other recovery home models in that it is entirely self-run; residents hold leadership positions, and there are no professional staff on site. Decisions, including the entrance of new residents, are made democratically, and all residents contribute to the upkeep and rent of a house [2]. Ferrari et al. [11] conducted a study to further assess the ecological characteristics of self-run OHs in a national sample. The OHs examined were in four distinct areas, such as urban, suburban upper middle class, and working-class environments. Neighborhood factors such as socioeconomic status were not significantly related to relapse rates. Kassanits et al. [12], however, found higher relapse rates among residents of OHs in neighborhoods with lower education and income levels. This finding suggests that recovery homes may house residents who have better outcomes when situated in ecological areas populated with people with higher incomes and greater education. However, the contrasting findings of [11,12] point to system-level characteristics that might affect relationships and behaviors among individuals in recovery. Environmental variables such as employment levels and social network measures may have important associations that have implications for understanding how to enhance successful and sustainable outcomes in these post-treatment settings.

The current study investigated the associations between women's employment at the community level and two variables within Oxford House recovery homes: levels of employment and sharing of resources. The data used in this study were collected from recovery homes in three regions of the US. The data were obtained from the 2015 American Community Survey (ACS), collected by the US Census Bureau. We hypothesized that female employment rates might suggest facilitating economic conditions in a community, especially as females in recovery confront more barriers to entering the labor force, including discrimination [13]. We hypothesized that when there is higher female employment in a community and therefore possibly more employment opportunities, rates of overall employment in recovery homes would be higher. In addition, we hypothesized that when rates of women's employment are higher within a community, residents of recovery homes might be more willing to share resources with peers because of having more access to resources due to being employed. If more macro-factors are related to these types of dynamics, they might point to mechanisms that help to sustain recovery resources in communities.

2. Method

The current study was conducted in 42 OHs in the US [3]. There are over 3000 OH recovery homes in the US, providing safe housing to about 28,000 individuals in recovery. Each house is self-run without any professional staff. They are gender-specific, usually including 6 to 12 residents. The OHs are democratically run, and requirements for remaining in the homes include not using illegal drugs or alcohol, paying monthly rent, and contributing to house maintenance. Residents enter OH from a variety of acute care and criminal justice settings, including in-patient treatment centers, incarceration, and drug court programs. Some enter OH directly from homelessness or other precarious living situations [14,15]. Individuals looking to live in OH must be voted in by current residents. An 80% majority is required to accept a new resident [2].

Data were collected from OHs located in North Carolina, Texas, and Oregon in the US. Including residences from different geographical regions provided some ability to address the generalizability of findings to recovery homes in other locations. Communication with residences about possible participation helped field staff assemble lists of residences to approach, and recruitment attempts were made in approximately the order that residence contact information became available. Member-elected house presidents were asked to introduce the study to residents by reading a project-provided script about the study; houses were accepted into the study if the house president and all or all but one member agreed to participate. The first thirteen consenting houses from each state were accepted, and three more houses were added, making a total of forty-two. One house dropped out completely, but another was added after wave 1, bringing the total to forty-three houses. However, only 42 houses had 2 or more waves of information available on their residents, constituting the data used in a prior study [6].

Participants were part of a longitudinal study that collected information every four months over a 2-year period for a total of 7 waves from 2015 to 2018. Participants were recruited and interviewed by field research staff in face-to-face meetings. If a participant left the home, their departure was classified as voluntary or involuntary. Involuntary exits included relapse, disruptive behavior, or failure to pay rent. Voluntary exits included leaving for other reasons such as desiring to live independently or with a spouse or friends. Participants were compensated USD 20 for completing each assessment. Permission was obtained through the DePaul University Institutional Review Board.

3. Measures

Resident demographic information included age, race/ethnicity, and sex. Employment status and education were obtained and classified as working part-time or full time (coded 1) or not working (unemployed, student, retired, on disability; coded as 0) and high school degree or less (0) vs. any college or other form of higher education (1), respectively.

In total, 93% of 714 eligible participants participated. Of these 666 residents, 602 were included in the current study. The sample was 51% male and 49% female. Their average age was 37.0 (SD = 10.5). Regarding race, 78.8% were White, 10% were Latinx, 8.5% were Black, and 2.7% were either Asian American, Alaskan Native, American Indian, or Pacific Islanders.

House Employment within the OH

The employment rate in the OHs was calculated by dividing the proportion of residents employed by the total number of residents living in the house. The median house employment rate was 87.5% and ranged from 25 to 100% ($M = 77.81$, $SD = 0.26$).

Community Employment Rate

Female employment rates were examined within 30 zip codes where OH recovery homes were located. The data were obtained from the 2015 American Community Survey (ACS) conducted by the US Census Bureau. The ACS collects detailed yearly demographic, social, economic, and housing information from communities across the United States from the U.S Census Bureau. The female community employment rate was measured as the percentage of females aged 16 years or older who were in the labor force. The median female employment rate was 56.3% (suggesting that about half of all females surveyed were employed). Female employment percentages ranged from 38.7 to 65.6 ($M = 55.17$, $SD = 5.98$) in the zip codes studied.

4. Loaning

The *Social Network Instrument* (SNI; [16]) was used to assess the social dynamics within each OH. All residents rated each other via what is referred to as a whole network. Money loaning, signifying willingness to lend resources, was counted as a connection with another resident if the person loaning money was willing to lend USD 500 or USD 100 but not considered a connection if he/she was only willing to loan USD 50, USD 10, or USD 0. This study examined patterns of bidirectional (or simply "directed") relationships, with each resident (ego) rating every other person in the house (alters). *Reciprocity* measures the tendency for mutual (bidirectional) connection in willingness to loan. Reciprocity is the proportion of all the edges for which a reciprocal edge is present, and this was the measurement used in our study.

5. Data Analysis

Linear mixed-effects models were utilized to evaluate the hypotheses due to the data set's structure. The nature of the data indicates a nested structure, meaning that observations within houses are likely to be correlated over time [17,18]. Statistical analysis was conducted using R 4.3.2, with the lme4 package [19] employed for multilevel analyses. Female community employment served as a fixed effect. House identification number (HID), zip code, and state were included as random effects for house employment, but state and zip code were removed in the model predicting willingness to loan as the variance component was estimated to be near zero.

6. Results

House Employment

The effect of the female community employment rate was positive and significant for house employment ($\beta = 0.79$, $SE = 0.23$, $z(1874) = 3.54$, $p < 0.001$). As the employment rate of women in the OH community increased, the overall employment rate of the residents in the OH increased. The fixed effect of community female employment accounted for 3.8% of the variance in this model (R^2 Marginal = 0.038), with random effects (State, Zip, HID) accounting for another 69.2% of the variance.

Willingness to Loan Money

In a separate model, the community-level female employment rate positively predicted willingness to loan reciprocity within the house ($\beta = 0.35$, $SE = 0.10$, $z(1874) = 3.56$, $p < 0.001$). As the employment rate of women in an OH's community increased, the willingness of residents to loan money increased. The fixed effect of female employment in a community accounted for 0.9% of the variance in the model (R^2 Marginal = 0.009), and the random effect of HID accounted for 0.7%.

7. Discussion

The rates of women's employment in a community were related to higher house employment and lending of resources. These findings suggest that higher female employment rates might reflect facilitating conditions in the environment that allow both higher employment within recovery homes and more reciprocity of loaning. This suggests an ecological balance involving environmental, economic, and social sustainability. However, this association does not indicate causality. Still, it is not surprising that when the surrounding community has enhanced employment opportunities for females, there would also be a greater likelihood of employment among the residents of the recovery houses.

The relationship between the macro-environmental levels of female employment and the sharing of resources was an association that was possibly due to higher resources within the house, which could have influenced the greater willingness to be generous and trusting with respect to loaning peers USD 100 or USD 500.

It is possible that higher female employment rates might signify a more facilitating and positive economic condition for everyone or might relate to other policies that positively affect sustainability, economic opportunities, the social services available, or public service transportation [20]. Higher community levels of female employment might be a marker of either the types of jobs available, the cost of living in such areas, or affordable transportation that enables residents of recovery homes to achieve high levels of employment and have access to more opportunities to work through role models, job contacts, or norms [21]. This is a complex set of associations, as there are multiple cultural and social norms that affect women in the workforce, such as having children [22].

In addition, when there were higher female community employment rates, levels of reciprocal sharing were higher. This finding suggests that with higher employment rates in the community, which is connected to higher house levels of employment, there could be more resources available, such as trust of the house residents, which might be the reason for the increased reciprocal loaning exchanges. Still, the relationship to external employment rates found in the current study and what occurs in terms of employment, as well as sharing resources, will need further investigation.

There are several limitations in the current study. We examined only a few of the possible variables of interest, and dropout and relapse rates could be examined in future studies. There are also multiple other variables that might have impacted the findings, including the available resources, both economic and social, that contributed to the findings, and future research will need to examine these other factors [23,24]. We also did not examine the influence of rates of male community employment on the house and social network variables or the differential rates of house employment among men and women. Finally, while there is a strong association between environmental/community-level factors when it comes to employment, for loaning, the percentage of variance explained is relatively small. Loaning behavior probably has a lot of individual and house-level factors that corroborate what has been found in previous research (e.g., social networking features like trust [4].

In summary, it does appear that the macroeconomic conditions of a community influence the following economic and social sustainability factors: rates of employment and the sharing of resources in Oxford House recovery homes. Kassanitis et al. [12] found higher relapse rates occurring in neighborhoods with lower income and education levels, whereas Ferrari et al. [11] found that neighborhood socioeconomic status did not predict relapse rates. These studies occurred during different decades, and the explanation for these differences may be changing economic or other factors during these time periods. Although we are not certain of how to explain the findings, it is important to continue to explore such macro sustainability factors that influence the potential outcomes of residents in recovery homes [25]. One practical implication is that locating recovery homes in settings with higher overall employment rates for women might be a way of enhancing the ability of these homes to potentially have beneficial employment and sharing outcomes for residents. Our study contributes to the literature on sustainability, particularly as it promotes an understanding of how environments and economic and social domains influence recovery and sustainability for those with substance use disorders.

Author Contributions: Conceptualization, L.A.J., J.S.B., T.J.B.; methodology, L.A.J., J.S.B., T.J.B.; writing—review and editing, L.A.J., J.S.B., T.J.B.; project administer; funding acquisition, L.A.J. All authors have read and agreed to the published version of the manuscript.

Funding: This work was supported by the National Institute on Alcohol Abuse and Alcoholism [AA022763].

Institutional Review Board Statement: The study was conducted in accordance with the Declaration of Helsinki, and approved by the Institutional Review Board of DePaul University (protocol code # LJ072314PSY and date of approval August 8, 2017).

Informed Consent Statement: Informed consent was obtained from all subjects involved in the study.

Data Availability Statement: The original contributions presented in the study are included in the article, further inquiries can be directed to the corresponding author.

Conflicts of Interest: The authors declare no conflict of interest.

References

1. Mericle, A.A.; Polcin, D.L.; Hemberg, J.; Miles, J. Recovery housing: Evolving models to address resident needs. *J. Psychoact. Drugs* **2017**, *49*, 352–361. [CrossRef] [PubMed]
2. Jason, L.A.; Ferrari, J.R. Oxford House recovery homes: Characteristics and effectiveness. *Psychol. Serv.* **2010**, *7*, 92–102. [CrossRef] [PubMed] [PubMed Central]
3. Jason, L.A.; Wiedbusch, E.; Bobak, T.; Taullahu, D. Estimating the number of substance use disorder recovery homes in the United States. *Alcohol. Treat. Q.* **2020**, *38*, 506–514. [CrossRef] [PubMed] [PubMed Central]
4. Jason, L.A.; Guerrero, M.; Salomon-Amend, M.; Lynch, G.; Stevens, E.B.; Light, J.N.; Stoolmiller, M.; Doogan, N.J. Network measures of advice-seeking and resource sharing are related to well-being in recovery homes. *Int. J. Drug Policy* **2021**, *92*, 102970. [CrossRef] [PubMed]
5. Jason, L.A.; Guerrero, M.; Salomon-Amend, M.; Lynch, G.; Stevens, E.; Light, J.M.; Stoolmiller, M. Advice seeking and loaning of money related to relapse in recovery homes. *J. Community Appl. Soc. Psychol.* **2021**, *31*, 39–52. [CrossRef] [PubMed]
6. Jason, L.A.; Guerrero, M.; Salomon-Amend, M.; Stevens, E.; Light, J.N.; Stoolmiller, M. Context matters: Home-level but not individual-level recovery social capital predict residents' relapse. *Am. J. Community Psychol.* **2021**, *67*, 392–404. [CrossRef] [PubMed]
7. Jason, L.A.; Stoolmiller, M.; Light, J.N.; Bobak, T.J. House level latent classes as predictors of recovery and evictions. *Soc. Work Public Health* **2023**, *38*, 58–71. [CrossRef] [PubMed]
8. Oxford House. Annual Report, Fiscal Year 2022. Available online: https://oxfordhouse.org/doc/ar2022.pdf (accessed on 15 May 2024).
9. Oxford House. Annual Report, Fiscal Year 2021. Available online: https://oxfordhouse.org/doc/ar2021.pdf (accessed on 15 May 2024).
10. Jason, L.A.; Guerrero, M.; Lynch, G.; Stevens, E.; Salomon-Amend, M.; Light, J.N. Recovery home networks as social capital. *J. Community Psychol.* **2020**, *48*, 645–657. [CrossRef] [PubMed]
11. Ferrari, J.R.; Groh, D.R.; Jason, L.A. The neighborhood environments of mutual-help recovery houses: Comparisons by perceived socio-economic status. Recovery from addiction in communal living settings: The Oxford House model. *J. Groups Addict. Recovery* **2009**, *4*, 100–109. [CrossRef] [PubMed]
12. Kassanits, J.; Bobak, T.J.; Stevens, E.; Guerrero, M.; Light, J.; Jason, L.A. The relationship of Oxford Houses across heterogeneous house and setting characteristics. *Am. J. Orthopsychiatry* **2020**, *90*, 324–327. [CrossRef] [PubMed]
13. Ostry, M.J.D.; Alvarez, J.; Espinoza, M.R.A.; Papageorgiou, M.C. Economic Gains from Gender Inclusion: New Mechanisms, New Evidence. International Monetary Fund. Available online: https://www.imf.org/en/Publications/Staff-Discussion-Notes/Issues/2018/10/09/Economic-Gains-From-Gender-Inclusion-New-Mechanisms-New-Evidence-45543 (accessed on 15 May 2024).
14. Jason, L.A.; Olson, B.D.; Ferrari, J.R.; Majer, J.M.; Alvarez, J.; Stout, J. An examination of main and interactive effects of substance abuse recovery. *Addiction* **2007**, *102*, 1114–1121. [CrossRef] [PubMed] [PubMed Central]
15. Whipple, C.R.; Jason, L.A.; Robinson, W.L. Housing and abstinence self-efficacy in formerly incarcerated individuals. *J. Offender Rehabil.* **2016**, *55*, 548–563. [CrossRef] [PubMed]
16. Jason, L.A.; Stevens, E. The reliability and reciprocity of a social network measure. *Alcohol. Treat. Q.* **2017**, *35*, 317–327. [CrossRef] [PubMed]
17. Pusponegoro, N.H.; Rachmawati, R.N.; Notodiputro, K.A.; Sartono, B. Linear mixed model for analyzing longitudinal data: A simulation study of children growth differences. *Procedia Comput. Sci.* **2017**, *116*, 284–291. [CrossRef]
18. U.S. Census Bureau. 2015 American Community Survey [Excel Data File]. Available online: https://factfinder.census.gov/faces/nav/jsf/pages/searchresults.xhtml?refresh=t (accessed on 15 May 2024).
19. Bates, M.E.; Voelbel, G.T.; Buckman, J.F.; Labouvie, E.W.; Barry, D. Short-term neuropsychological recovery in clients with substance use disorders. *Alcohol. Clin. Exp. Res.* **2006**, *29*, 367–377. [CrossRef] [PubMed]
20. Baxter, J.; Kane, E.W. Dependence and independence: A cross-national analysis of gender inequality and gender attitudes. *Gend. Soc.* **1995**, *9*, 193–215. [CrossRef]
21. Vartanian, T.P.; McNamara, J.M. Older women in poverty: The impact of midlife factors. *J. Marriage Fam.* **2004**, *64*, 532–548. [CrossRef]

22. Connelly, R. The effect of childcare costs on married women's labor force participation. *Rev. Econ. Stat.* **1992**, *74*, 83–90. [CrossRef]
23. Vartanian, T.P. Neighborhood effects on AFDC Exits: Examining the social isolation, relative deprivation, and epidemic theories. *Soc. Serv. Rev.* **1997**, *71*, 548–573. [CrossRef]
24. Sudha, T.; Reshi, I.A. Unleashing the Power: Empowering Women for a Stronger Economy. *Int. J. Educ. Rev.* **2023**, *3*, 826–833.
25. Bobak, T.; Majer, J.; Jason, L.A. Complex contexts within Oxford Houses: Psychiatric comorbid social networks. *Alcohol. Treat. Q.* **2023**, *41*, 237–249. [CrossRef] [PubMed Central]

buildings

MDPI

Review

The Impact of the Physical Environment on Staff in Healthcare Facilities: A Systematic Literature Review

Roshan S. Shetty [1], Giridhar B. Kamath [2,*], Lewlyn L. R. Rodrigues [2], Rama Devi Nandineni [1] and Sham Ranjan Shetty [3]

[1] Manipal School of Architecture and Planning, Manipal Academy of Higher Education, Manipal 576104, Karnataka, India; roshan.shetty@manipal.edu (R.S.S.); nandineni.rd@manipal.edu (R.D.N.)
[2] Department of Humanities and Management, Manipal Institute of Technology, Manipal Academy of Higher Education, Manipal 576104, Karnataka, India; l.rodrigues@manipal.edu
[3] T A Pai Management Institute, Manipal Academy of Higher Education, Manipal 576104, Karnataka, India; shamranjan.shetty@manipal.edu
* Correspondence: giridhar.kamath@manipal.edu

Abstract: This article investigates the staff outcomes that are influenced by physical environment features in a healthcare facility. Over the past few decades, research on the physical environment and its effects on hospital staff has advanced steadily. Focusing on the physical environment has become a research area with great potential because it is crucial to provide a favorable work environment for health professionals, which directly impacts the quality of care. A systematic literature review was conducted to identify research articles from two scientific databases and a healthcare journal. Articles that fulfilled the inclusion and exclusion criteria for relevant articles were included in this review. Relevant data were derived from the final shortlisted articles. The features identified in the articles were divided into three dimensions of the physical environment. The results demonstrate different outcomes for staff. This review shows the need to develop a comprehensive study involving the most important features of the three dimensions of the physical setting and its effects on staff. This study contributes invaluable insights for hospital administrators, architects, and designers by informing them about the critical factors shaping the healthcare workplace environment.

Keywords: healthcare facility; staff outcomes; physical environment

1. Introduction

The healthcare industry has more dynamic and complicated environments than other service industries [1]. It is a complex environment with continually functioning organizational systems, technological infrastructure, and users, including patients, families, and staff [2]. These interactions are dynamic since patient needs, service care, and technology are unpredictable over time and are likely to change, which places a heavy burden on how the built environment is designed. The physical environment is critical for enhancing patient well-being and providing a supportive workplace for healthcare workers [3]. Every aspect of the physical environment influences staff outcomes in a healthcare environment [4]. The physical environment is the physical features of the environment where users carry out tasks [5]. According to [6,7], the physical environment encompasses architectural elements, interior design, and ambient features. Within healthcare settings, staff outcomes are influenced by these physical factors and include a range of indicators, such as well-being, satisfaction, performance, productivity, stress and burnout, commitment, turnover, and other impacts on healthcare professionals in their workplace. The healthcare setting can create a stressful environment, as physical environmental factors act as environmental stressors [8]. In addition to providing treatment, healthcare settings face new challenges, including protecting users from infectious diseases [9]. According to [10], healthcare settings need to be designed to be resilient for continual development by integrating additional information about the influence of various design strategies on the users and other outcomes related to the organizations.

Studies have shown that variables like long work hours, variable schedules, patient distress, high patient volumes, workplace violence, and illness risk can lead to stressful work circumstances [11,12]. According to reports from Mental Health America (2020), 93% of healthcare personnel are stressed, 82% have mental and physical fatigue, 45% of nurses claim that they do not receive appropriate emotional care, and 22% of health professionals experience depression and post-traumatic stress [13]. In addition, 69% of doctors reported depression, 13% had suicidal thoughts, 64% of female doctors reported burnout [14], and frontline staff, nurses, and young staff had major psychological indications other than depression [15].

Many scientific studies [12,16–25] have gathered evidence demonstrating relationships between the physical healthcare environment and outcomes for patients, families, staff, and organizational outcomes. Consequently, there is an increasing consensus that an effectively designed environment can improve patient outcomes, promote a supportive work environment for users, and meet organizational and business goals. Although research on the effects of the work environment on healthcare workers has significantly increased, with more than 1.1 million hits in Google Scholar and 50,000 articles in the last five years [26], these studies have focused mainly on the task context, social context, organizational culture context, and physical context.

The healthcare environment has immense opportunities to be an effective intervention for reducing and preventing mental exhaustion [27] and can lessen stress and provide restorative conditions [28] for healthcare professionals. Research has also demonstrated the significance of physical environmental design on performance and efficiency among staff in healthcare facilities [5].

The majority of studies have proven that the physical environment in healthcare facilities has an impact on patient and staff outcomes. Healthcare professionals are increasingly conscious of the need for a well-designed physical workplace environment, as this environment is crucial for fostering positive and supportive work environments, and improved designs can make the jobs of staff much easier [12,16,20,26,29]. A great deal of potential exists for healthcare physical environment design to be an efficient approach for minimizing and eliminating negative impacts on staff [27]. Additionally, understanding the complex interplay between the physical environment and staff outcomes is essential for healthcare administrators, architects, designers, and policymakers. A supportive and conducive workplace environment enhances staff well-being and improves patient care quality, safety, and satisfaction.

Although the previously published research provides relevant information, those studies dealt with specific factors, as opposed to a more comprehensive assessment of physical environmental factors. It is vital to understand the extent to which physical environmental factors are studied and their impact on staff in healthcare settings. This article provides a systematic review of peer-reviewed articles to comprehensively understand the physical environmental factors that impact staff in healthcare facilities. This study aimed to address the following questions. Research question (RQ) 1: Which physical environmental factors contribute to staff outcomes in healthcare settings? RQ 2: What staff outcomes concerning the physical environment in healthcare settings have been studied? RQ 3: Which theories pertaining to the physical environment have been discussed in the literature?

2. Review Methods

A systematic review must employ an established method, as outlined by [30] for searching, filtering, reviewing, interpreting, synthesizing, and reporting findings obtained from various publications. We selected relevant studies for review in six steps: (i) identifying the concepts to be included in this study; (ii) selecting relevant databases; (iii) identifying keywords and search strategies; (iv) selecting the inclusion and exclusion criteria; (v) screening the data; and (vi) obtaining the search results.

In step (i), we identified the following concepts of interest: the population, setting(s), and outcome of the study. The population of interest was healthcare staff, the setting(s) of interest was the physical healthcare environment, and the outcome was the staff outcome. In step (ii), we selected two databases: PubMed and Scopus. In addition, manual searching was performed in the *Health Environments Research and Design (HERD) Journal*. PubMed was chosen because it is a healthcare-based database, and Scopus is one of the largest databases of peer-reviewed literature. The search included articles published between January 2010 and September 2022 to capture a decade of research given the surge in healthcare environment research starting from 2010 onward. In step (iii), we identified keywords and the search strategy. Using Boolean operators, the literature was searched for "physical environment", "staff outcomes", and "hospital" and their related terms. Table 1 lists the keywords used. The searches used standard search filters. The titles and abstracts were examined for relevant information. The approach involved search terms related to the hospital, physical environment, and staff outcomes and included several search strings for every database. Repeated searches were carried out using various terms and search string combinations.

Table 1. Keywords used in the search.

All Words Related to the Hospital	All the Words Related to The Physical Environment	All Words Related to Staff Outcomes
Healthcare Hospital Healthcare building Healthcare facility Healthcare setting Healthcare sector	Physical environment Healthcare design Built environment Building design Hospital environment Hospital design Environmental factors Healthcare environment	Employee outcomes Staff safety Staff stress Staff satisfaction Staff efficiency Staff performance Staff burnout Turnover intent Job satisfaction Workplace satisfaction perception

In step (iv), the inclusion and exclusion criteria were defined, and the criteria for selecting articles are given in Table 2.

Table 2. Inclusion and exclusion criteria.

Inclusion Criteria	Rationale
Only peer-reviewed journal publications	Ensuring that articles are peer-reviewed maintains a high standard of research quality and credibility.
Available in electronic databases	Limiting the search to electronic databases ensures the accessibility and verifiability of the sources, facilitating the replication of the review process and ensuring that the articles are easily retrievable.
Limited to the healthcare context only	Focusing on the healthcare context ensures that the findings are directly relevant to healthcare environments, where the impact of design features on staff outcomes is critically important.
All healthcare professionals at any level working in a healthcare facility	Including all healthcare professionals ensures a comprehensive understanding of how design features impact various roles within healthcare settings, providing a holistic view of staff outcomes across different professional levels.
Articles published in the English language only	Limiting the review to English-language publications ensures that the reviewers can accurately interpret and analyze the content, maintaining the integrity of the review process.
Full-text articles	Including only full-text articles ensures that all relevant details of the studies can be thoroughly examined and assessed, allowing for a complete understanding of the research findings and methodologies.
Studies employing robust research methodologies, including quantitative, qualitative, or mixed-method approaches	Including methodologically rigorous studies ensures that the findings are based on sound research practices, enhancing the validity and reliability of the review.

Table 2. *Cont.*

Exclusion Criteria	Rationale
Book chapters, conference proceedings, reports, and review papers	Excluding these articles ensures that the review focuses on original research studies that provide direct evidence of the impact of design features on staff outcomes.
Sectors other than healthcare	Excluding articles from other sectors ensures that the findings are specific to healthcare environments, where the specific challenges and dynamics may differ significantly from other settings.
Articles with only abstracts	Excluding articles that provide only abstracts ensures that the review is based on comprehensive data and detailed methodologies, which are necessary for a thorough analysis and accurate interpretation of the study results.

In step (v), data screening, this review was conducted using article titles within the database to screen out irrelevant articles, and duplicates were removed. The abstract of each article was screened for relevance. The full articles were critically reviewed, and all relevant information required for this review was documented.

In step (vi), for the search results, a total of 4869 articles were initially retrieved and subsequently checked. Based on the analysis of the titles and abstracts, 4172 articles (books, conference proceedings, dissertations, editorial commentaries, and review articles) were excluded. By employing the full-text inclusion and exclusion criteria, 439 articles were excluded. Consequently, 258 articles were assessed for eligibility. A total of 191 articles failed quality checks and were removed. Finally, 67 articles were selected for this review. These articles were thoroughly reviewed and verified by the authors. Figure 1 shows the flowchart of the article selection process.

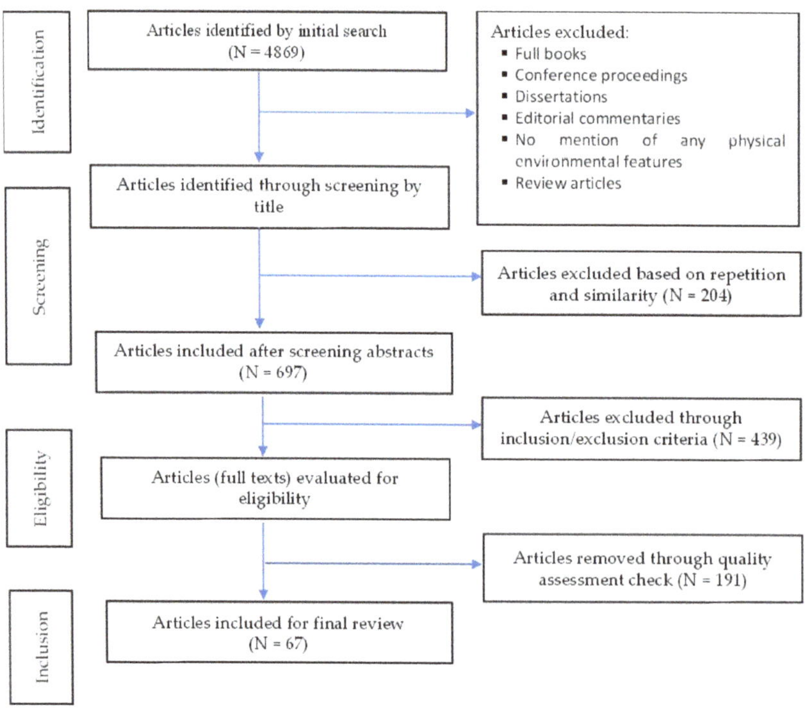

Figure 1. Flowchart of the study selection process for review.

3. Data Extraction and Synthesis

After the quality assessment, 67 articles were selected for this review. These articles were thoroughly reviewed and verified by the two authors. All 67 articles were analyzed and summarized. The first author gathered information from the included studies.

3.1. Analysis of the Included Articles

This section addresses RQ 1 and RQ 2. The information obtained included information such as (i) the articles published per year, (ii) the countries in which the studies were performed, (iii) the research techniques, (iv) the sampling types, (v) the sampling frames, (vi) the study settings, and (vii) the methods of data collection.

3.2. Identification of Variables

This section provides insights into the physical environmental variables considered to study the impacts on various staff outcomes. The physical environment was classified into three dimensions based on the study by [6,7,31]. The three dimensions included architectural design features, which represent relatively permanent aspects; interior design features, which include less-permanent elements of the environment; and ambient features, which encompass environmental conditions and sensory stimuli. Clusters were formed based on these three dimensions of the physical environment. Similarly, the staff outcome measures were also assessed for this study. This was crucial because it mapped the independent and dependent variables.

3.3. Theories Identified

Several theories have been established in the domains of environmental psychology, the environment and behavior, and organizational behavior. These theories provide valuable frameworks for analyzing the interaction between the environment and human behavior in an organizational context. Environmental psychology theories focus on how individuals perceive and interact with their immediate physical surroundings, while environment and behavior theory investigates how environmental factors influence individuals' behavior. Organizational behavior theory investigates individuals' behavior in organizations and how it influences organizational success. These theories provide support for the research undertaken to better understand the complex interplay between the environment and its impact on the individual. This understanding can guide the design of more supportive and conducive environments. This section presents the answer to RQ 3 by identifying the theories used in the review articles in the fields of environmental psychology, the environment and behavior, and organizational behavior.

3.4. Keyword Analysis

Figure 2 depicts a weighted cloud theme generated by Jason Davies's word cloud generator. It is a comprehensible technique for visualizing text data. The commonly used keywords in the articles were design, environment, satisfaction, and nurse. It can be concluded that most articles discussed environmental design and its impact on staff satisfaction.

Figure 2. Weighted focus cloud generated by a word cloud generator developed by Jason Davies.

4. Results and Discussions

4.1. General Characteristics

An analysis of the published articles led us to the conclusion that more articles were published in PubMed than in the Scopus database. Regarding the distribution of articles reported in the systematic review between 2010 and 2022 (Figure 3), there was a peak in the related research area during 2015, with twelve studies, and in 2018, with nine studies.

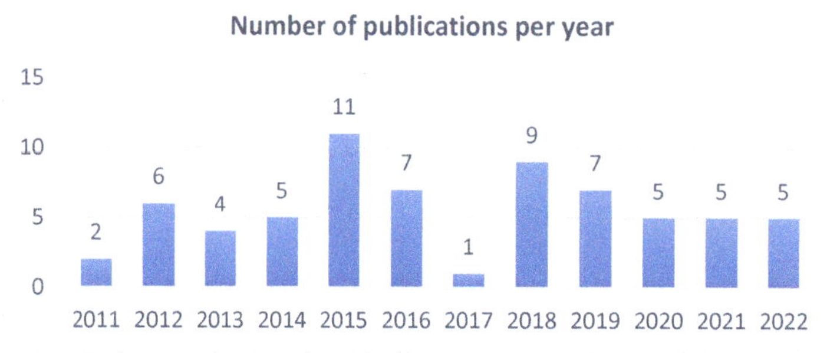

Figure 3. Number of publications per year.

Relevant research originated from multiple countries, including 30 (n = 30) from the USA; 5 from China; 3 from Australia; 2 each from Canada, the UK, and Iran; and 1 each from Denmark, Jordan, Uganda, Portugal, Brazil, Taiwan, Finland, Hong Kong, the Netherlands, and Japan (Figure 4).

4.2. Research Technique

Of the total articles reported for this review (Table 3), 45% used mixed methods, followed by quantitative methods (40%) and qualitative methods (15%). The quantitative methods used questionnaire survey methods for data collection, and the qualitative methods included focus group discussions, structured or semi-structured interviews, and observational studies. Both qualitative and quantitative mixed methods were used.

Table 3. Research methods reported in this review.

Sl. No.	Research Methods	Number of Articles Reported in this Review (n = 67)
1	Quantitative methods	27 [17,22,23,31–54]
2	Qualitative methods	10 [55–63]
3	Mixed methods	30 [8,18,19,56,64–89]

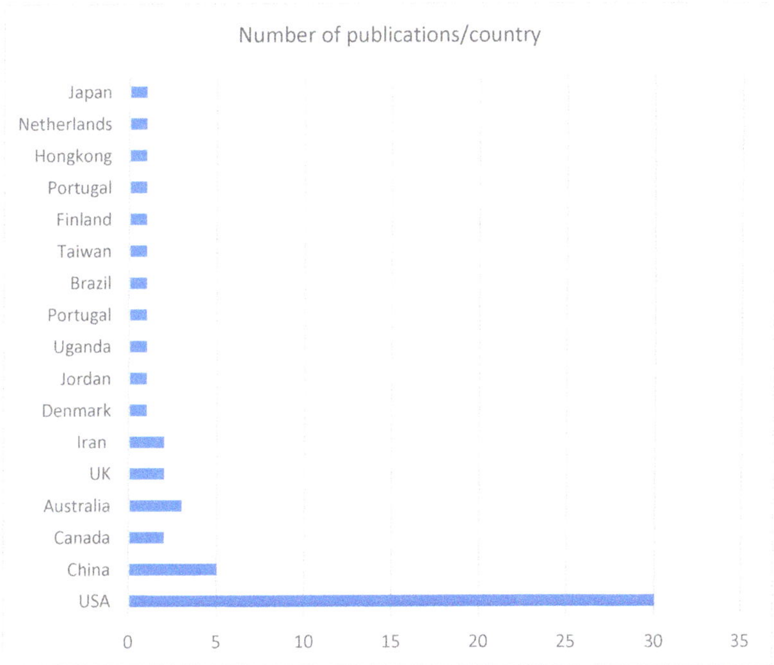

Figure 4. Number of publications per country.

4.3. Sampling Technique

Nonprobability sampling was adopted by 31 articles; 7 articles used the probability-sampling technique, and the remaining articles did not report the sampling technique adopted. The sampling techniques adopted in this review are reported in Table 4.

Table 4. Sampling techniques reported in this review.

Sampling Techniques	Types of Sampling Techniques	Number of Articles Reported in this Review (n = 67)
Probability sampling	■ A random sample (n = 4) [23,71,85] ■ Cluster random sample (n = 1) [33]	7
Nonprobability sampling	■ Convenience sample (n = 19) [5,8,34,37,38,40,48,49,51,56,60,64,70,76,77,85–87,89] ■ Volunteer/self-selected sample (n = 5) [59,72,78,83,88] ■ Purposive sample (n = 3) [57,58,74] ■ Sequential sample (n = 1) [18] ■ Convenience and snowball sample (n = 2) [67,81] ■ Convenience and purposive sample (n = 1) [55] ■ Non-random sample (n = 1) [45]	31
Not mentioned	■ [17,19,22,31,32,35,36,39,41–44,46,47,50,52,53,61,62,65,66,68,69,73,75,79,80,82,90]	29

4.4. Sampling Frame

There was a wide variety of participants in the studies. Nurses and doctors were the respondents in most of the studies. The studies on nurses were considered the highest sampling frame. The nurses included registered nurses, senior nurses, charge nurses, associate chief nurses, chief nurses, critical care nurses, circulating nurses, intensive care nurses, and nurse managers. The second most studied healthcare professionals were the care team staff members, comprising physicians, therapists, advanced practice specialists, physician supporters, and nurse practitioners. The clinical and non-clinical staff sample frame was the third most investigated sample frame. A detailed summary of the health professionals (study participants) included in this review is given in Table 5.

Table 5. Study participants included in this systematic review.

Healthcare Professionals (Study Participants)	Number of Articles Reported in this Review
Nurses	[8,17,19,22,39,40,42,47,55,56,61,64,65,70,75,78,79,82,84,89]
Care team staff members	[23,32,36,51,71]
Clinical and non-clinical Staff	[38,53,57–59,77,80]
Staff members	[63,66,67]
Physicians/medical doctors, nurses, and therapists	[5,18,31,33,37,41,44,46,67,71,73,81,86,91]
Nurses and technicians	[41]
Diagnostic, pharmacy, sterilization, rehabilitation, and health science department staff	[43]
Midwives and supporters	[60]
Medical directors	[49]
Nursing, social work, and therapy staff	[54]
Healthcare providers/employees	[50,62,88]
Technicians	[52]

4.5. Study Settings

Although the context of the studies varied, most of the studies were performed at general hospitals (n = 15), followed by acute care hospitals (n = 7). The detailed settings of the included studies in this review are provided in Table 6. The designs of the listed studies varied. The sample sizes varied considerably, from as low as 9 to 1561 staff participants.

Table 6. Settings of the selected studies.

Study Setting	Selected Studies Carried Out in This Setting
Hospital	15 [34,47,49,50,52,53,58,67,69,72–74,76,80,85]
Public nonprofit hospital	2 [64,81]
Somatic hospital	1 [55]
University teaching hospital/academic medical center	6 [17,35,43,51,61,87]
Maternity facility	2 [60,65]
Primary care facility	2 [32,77]
General hospital	2 [46,66]
Ambulatory care	1 [18]
Community hospital	3 [5,44,56]

Table 6. *Cont.*

Study Setting	Selected Studies Carried Out in This Setting
Acute care hospital	7 [22,31,39,57,77,78,89]
Mental healthcare facility	1 [77]
Tertiary care facility	3 [38,54]
Medical care center	1 [40]
Care and rehabilitation facility	1 [42]
Veterans' affairs healthcare facility	1 [62,63]
Children's hospital	1 [90]
No contextual information	18 [8,19,23,33,36,37,41,45,48,59,70,75,79,82–84,86,88]

4.6. Methods of Data Collection

Eleven studies used experimental and quasi-experimental designs [22,42,43,54,65,75,78,81,83,90], thirteen studies used interview methods [55–57,64,72,73,83], and six studies used focus group discussions [8,59,61,74,79,86]. One ethnographic study [60] and two thematic content analyses were used [18,59]. The remaining 53 studies [17,19,23,31–36,39,40,44–50,52,53,65–68,70,71,77,80,82,84,85,87–89,92] used survey questionnaire methods along with other methods. Few other studies used methods such as behavior mapping [48], participatory guided tours [62], visual assessment, simulation, and observational studies [41,51].

This review identified a wide range of research methods employed to investigate the impact of the physical environment on staff, which significantly helped us to understand the topic in depth. These studies investigated various aspects of environmental factors while gaining insights from various perspectives by employing a variety of research methods. The quantitative methods employed in the review articles included experimental and quasi-experimental research designs and survey research to acquire quantifiable data on staff outcome measures. The articles that employed experimental research allowed the researchers to manipulate specific environmental variables and observe their impacts on staff. These methods also provided useful information about the prevalence and extent of the impact of the environment on staff. The qualitative research methods, such as observational studies, interviews, and focus group discussions, enabled the researchers to delve deeper into the subjective experiences and perceptions of the physical environment. Furthermore, the mixed-method approach integrates both qualitative and quantitative methods, providing a more comprehensive understanding of complex phenomena. Overall, the diversity of the research methods allowed a more holistic approach to investigating various factors of the relationship between the physical environment and staff outcomes.

4.7. Identification of Physical Environmental Variables and Their Impacts on Staff Outcomes

4.7.1. Architectural Design Features and Their Impacts

Architectural design features were grouped into restorative elements, spatial layout and connections, design characteristics, workspace elements, and room and spatial dimensions. Table 7 shows the staff outcomes for architectural design features studied by various authors, and Figure 5 shows the number of articles on the effects of architectural design features on staff outcomes.

Table 7. Architectural design features and their impacts on staff outcomes.

Architectural Design Features	Staff Outcomes																								
	Satisfaction	Performance	Productivity	Burnout	Perception	Behavior	Communication	Experience	Efficiency	Collaborations	Teamwork	Privacy	Stress	Mood	Security/Safety	Work Pattern and Flow Disruption	Wellness	Activity	Well-Being	Attrition	Absenteeism	Medical Errors	Commitment	Comfort	Environmental Demand
Restorative elements	7	2	1	1	3		1		1			1	4	2			1		1	1	1	1			
Author/year	[8,23,34,42,43,50,53,56,64,66,78,80,85,93]																								
Spatial layout and connections	16	5	1	1	10	1	9	2	7	1	1	2	5			1		1	1				1	1	1
Author/year	[17,18,31–33,35,36,38,39,41,43,47,48,50,53,54,59,61,67,69–76,78,82–85,90]																								
Design characteristics	6	1			8		3	1	10		4	3													
Author/year	[5,37,41,57,59,67,69,73,78,82,86]																								
Workspace elements	4	3			3		1		3			1	1												
Author/year	[41,54,57,70,72,82,86,90]																								
Room and spatial dimensions	4	1			2		1	1	2			1	2		2										
Author/year	[43,49,53,57,58,73,74,78,82,89]																								

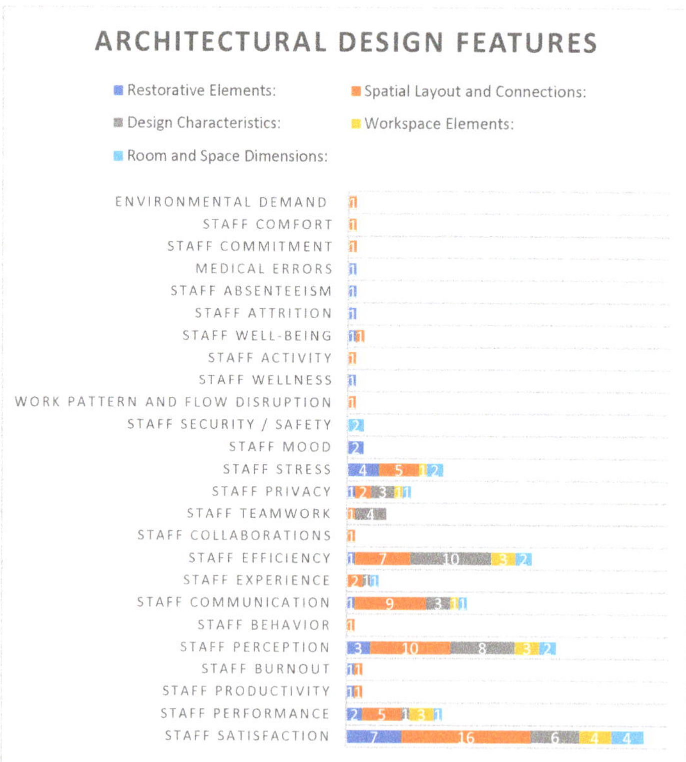

Figure 5. Total number of articles on impacts of architectural design features on staff outcomes.

Restorative Elements

Restorative elements in healthcare settings are designed to provide staff with environments that support relaxation, reduce stress, and promote overall well-being. The key restorative elements identified in the literature included break spaces [8,64], windows with views [23,34,42,50,53,80,85], balconies [64], and outdoor spaces [66]. Restorative break spaces are designated areas where staff can relax and recharge, often featuring comfortable seating and calming décor, significantly improving staff satisfaction [8,64], performance [64], and productivity [64] and reducing staff stress [8,64]. Windows that offer views of nature significantly enhance staff satisfaction [50,56,85], performance [22], communication [22], efficiency [56], and mood [5,22] and reduce stress [50,56]. Balconies, as outdoor extensions of interior spaces, provide fresh air, natural light, and views, offering a retreat from the high-pressure working environment. Outdoor spaces, such as gardens, are designed for relaxation and social interaction, allowing staff to connect with nature and take breaks, which are important for maintaining both mental and physical health.

Spatial Layout and Connections

Spatial layout refers to the organization and arrangement of various spaces within a healthcare facility. An effective layout ensures efficient workflows [48], easy access to essential areas, and minimal unnecessary movement, thereby enhancing staff satisfaction [17,18,37,41,43,47,50,54,71,73,75,78,84,85], efficiency [41,67,69,70,73,75,83], performance [36,37,43,70,91], productivity [83], behavior [17], communication [18,32,39,48,69,71,74], experience [71,73], privacy [69,85], collaboration [33], and teamwork [83]; improving perception [33,38,48,50,53,67,69,70,94], staff activity [38], well-being [83], commitment [43], and comfort [47]; and reducing burnout [35] and stress [35,50,61,83]. Visual connection, or the ability of staff to see and visually connect different areas within the facility, supports supervision, enhances communication [32], and improves the sense of safety and community among staff. The effective arrangement of spaces considers the need for privacy, ease of movement, and proximity to related functions, further enhancing workflow and reducing stress.

Design Characteristics

In healthcare settings, the design configuration plays a crucial role in determining how efficiently staff can perform their tasks, access necessary equipment, and interact with patients and coworkers. Flexibility refers to the ability of a space to adapt to different uses and needs over time. Flexible designs in healthcare can accommodate changing medical practices, emergencies, and evolving staff requirements without significant alteration and can help to improve staff perceptions [67,82], efficiency, and teamwork [67]. Accessibility ensures that all areas are easily accessible to staff by incorporating features such as wide doorways, ramps, and clear signage, which facilitate efficient and safe movement throughout the facility. Improving accessibility in healthcare settings enhances staff perceptions [57,67], efficiency [67], and privacy [57]. Good visibility supports patient supervision and enhances staff satisfaction [37,41,73], performance [71], communication [5,69], experience [73], efficiency [41,67,69,73], teamwork [5,67], and privacy [69] and promotes a collaborative environment among staff. Proximity is essential in healthcare settings for maintaining efficient workflows, ensuring quick response times, reducing physical strain on staff, and improving satisfaction [78], perception [67,82], communication [69], efficiency [67,69], teamwork [67], and efficiency.

Workspace Elements

An adequate workspace is essential for staff to perform their duties effectively and comfortably. Well-designed functional spaces promote staff performance [91]. Additionally, effective storage solutions are vital for organizations, reducing clutter, and ensuring that necessary items are easily accessible, all of which enhance staff satisfaction [41,43,54,63,86], performance [43,70,91], and efficiency [41,70].

Room and Spatial Dimensions

An appropriate room size is crucial for accommodating staff activities and equipment, ensuring a comfortable working environment, enhancing staff satisfaction [73,78], efficiency [73], and overall experiences [73], and reducing stress [89] in healthcare settings. Corridors should be wide enough to facilitate the easy movement of staff, patients, and equipment, as well-designed corridors can improve staff communication, privacy, and safety [74]. Personal spaces such as changing rooms and private offices are essential for staff well-being, offering personal comfort, and reducing stress [49]. Additionally, a well-designed layout of entrances, including doors and lobbies, provides easy access, security, and a welcoming atmosphere, benefiting staff by improving satisfaction [78], efficiency, and security [58]. The perception of space within a room or area is influenced by factors such as the ceiling height, lighting, and furniture arrangement. These elements contribute to a sense of spaciousness, enhancing comfort and reducing feelings of overcrowding and stress among staff.

While analyzing the distribution of research articles related to architectural design features, several key findings emerged. Staff satisfaction gained significant attention, with 23% of the research articles focusing on this aspect. Within staff satisfaction, spatial layout emerged as a prominent theme, constituting 43% of the discourse, followed by restorative elements, at 19%. Staff perceptions were researched in 16% of the articles, with a notable emphasis on design characteristics (31%) and spatial layout (38%). Staff efficiency, another vital aspect, was addressed in 14% of the articles, with a predominant focus on design characteristics (43%) and spatial layout (30%). These proportions reflect a nuanced exploration of how architectural design impacts staff, highlighting spatial layout as a common thread across satisfaction, perception, and efficiency considerations while also underscoring the importance of design characteristics in shaping staff efficiency and perceptions.

Considering the overall staff outcomes within architectural design features studies, spatial layout and connections emerged as the most explored aspects, encompassing 67% of the research, followed by design characteristics (36%) and restorative elements (28%). These findings underscore the central role of spatial design considerations in shaping staff outcomes, emphasizing the need for a comprehensive approach that integrates spatial layout, design characteristics, and elements conducive to restoration within built environments.

4.7.2. Interior Design Features and Their Impacts

Interior design features group elements such as natural elements, ergonomics and comfort, aesthetic and visual appeal, and navigation and orientation. Table 8 summarizes the staff outcomes for interior design features studied by various authors, and Figure 6 shows the number of articles on the effects of interior design features on staff outcomes.

Table 8. Interior design features and their impacts on staff outcomes.

Interior Design Features	Staff Outcomes											
	Satisfaction	**Performance**	**Burnout**	**Perception**	**Efficiency**	**Privacy**	**Stress**	**Wellness**	**Activity**	**Well-Being**	**Comfort**	**Environmental Environmental**
Natural elements	3		2	2			4	3	1			
Author/year	[34,40,49,52,53,64,80,81]											
Ergonomics and comfort	3	1	1	5	1	1	1					
Author/year	[31,34,43,53,70,78,82,84,86,89]											
Aesthetic and visual appeal	7	2		4	1		1					1
Author/year	[44,53,54,59,62,70,78,82,84,91]											
Navigation and orientation	3	1		1						1	1	1
Author/year	[37,42,47,54,59,82]											

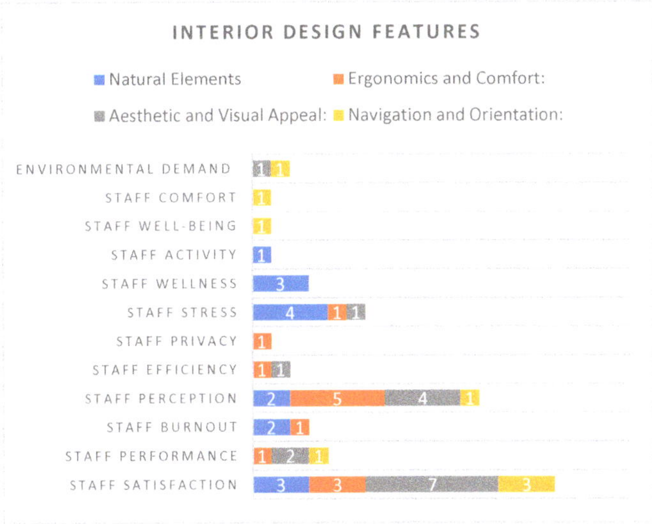

Figure 6. Total number of articles on the effects of interior design features on staff outcomes.

Natural Elements

In the healthcare context, incorporating natural elements such as gardens and landscaping can significantly impact staff well-being by reducing stress and providing a calming environment. Features such as aquariums also contribute to a soothing atmosphere, promoting staff wellness [80] and helping to alleviate anxiety and stress levels among staff [49]. Additionally, integrating artwork into healthcare spaces can boost morale and create a positive and inspiring atmosphere, ultimately reducing stress [49] and burnout [34] and improving satisfaction [64], wellness [80], activity [81], and perception [53] among staff. Furthermore, the projection of nature scenes (animations) on walls and ceilings improves staff satisfaction [52], and the presence of water features reduces stress [49].

Ergonomics and Comfort

In healthcare settings, prioritizing ergonomics and comfort for staff is essential for creating conducive work environments. Designing workspaces with ergonomic principles, such as adjustable desks and chairs, can significantly improve performance, efficiency, and perception [70] and reduce burnout [34] among staff members. Additionally, selecting comfortable and adjustable furniture supports good posture, reduces the risk of physical discomfort or injuries during long shifts, and improves satisfaction [84] and perception [31,53]. The incorporation of soft furnishings such as cushions and curtains further increases comfort levels, creating a warm and inviting atmosphere in staff areas and improving staff satisfaction [86] and perception [31,82].

Aesthetic and Visual Appeal

The aesthetic and visual appeal of healthcare settings are significantly influenced by elements such as flooring, carpeting, color schemes, and overall aesthetics. Comfortable and supportive flooring materials help reduce fatigue and minimize the risk of slips and falls, ensuring a safer environment for staff and improving satisfaction [44,78]. The use of calming colors and appealing design elements creates a welcoming and soothing atmosphere, which can positively impact staff satisfaction [78,84] and perception [53,82]. A visually appealing environment enhances the overall experience, making the workspace more enjoyable and uplifting for healthcare staff by improving satisfaction [54,84], performance [70,91], perception [62,70], and efficiency [70].

Navigation and Orientation

Effective wayfinding and signage are essential for navigation and orientation within healthcare facilities. Clear and strategically placed signs enable staff, patients, and visitors to easily find their way around, reducing confusion and stress. Well-designed wayfinding systems help minimize the time spent searching for specific locations, thereby enhancing satisfaction [37,47,54], performance [37], perception [82], and the overall workflow and safety within the healthcare environment.

While exploring research articles concerning interior design features, significant findings were obtained. Staff satisfaction emerged as a focal point, accounting for 30% of the research discourse. Within this realm, a significant proportion (44%) delved into the aesthetic and visual appeal of interior spaces, underscoring the importance placed on creating visually pleasing environments for staff. Staff perceptions, another critical aspect, were addressed in 23% of the articles, with a notable focus on spatial layout (38%) and design characteristics (31%). The efficiency of staff operations was also a key concern, as reflected in 42% of the articles focusing on ergonomics and comfort, followed closely by 33% on aesthetics and visual appeal. These findings highlight a multifaceted approach to understanding how interior design impacts staff experiences; emphasize the intersections between satisfaction, perceptions, and operational efficiency; and highlight the significance of aesthetics, spatial design, and ergonomic considerations in shaping these experiences within interior environments.

Delving deeper into interior design features studies, 31% of the articles examined aesthetics and visual appeal, indicating a continued interest in creating visually pleasing environments. Natural elements garnered attention in 29% of articles, highlighting the role of nature in enhancing staff experiences, while ergonomics remains significant at 25%, underlining the importance of designing spaces for comfort and well-being. These findings underscore the multifaceted nature of interior design features research, emphasizing the interplay between aesthetics, spatial design, natural elements, and ergonomic considerations in shaping staff satisfaction, perceptions, efficiency, and other outcomes within interior environments.

4.7.3. Ambient Features and Their Impacts

The ambient features were grouped into auditory environment, thermal comfort, lighting, air quality, and ventilation. Table 9 shows the staff outcomes for ambient features studied by various authors, and Figure 7 shows the number of articles on the effects of ambient features on staff outcomes.

Table 9. Ambient design features and their impacts on staff outcomes.

Ambient Features	Staff Outcomes																	
	Satisfaction	**Performance**	**Burnout**	**Perception**	**Communication**	**Efficiency**	**Privacy**	**Stress**	**Mood**	**Anxiety**	**Vigilance and Workplace Alertness**	**Wellness**	**Activity**	**Attrition**	**Absenteeism**	**Medical Errors**	**Comfort**	**Physiological and Psychological Response**
Auditory environment	9	1	2	6	1	1		4		1			1					
Author/year	[19,31,34,38,43,47,50,51,53,62,64,70,78,87,90]																	
Thermal comfort	4	3	1	4		2	1				1						2	
Author/year	[17,31,34,43,47,53,59,70,77,84,85]																	
Lighting	13	6	1	9	2	2	1	4	2		1	1	1	1	1	1	1	1
Author/year	[17,22,23,31,34,38,43,45,47,50,52,53,56,59,65,70,77,79,80,82,84,85,88,90,95]																	
Air quality and ventilation	3	1		3			1										1	
Author/year	[17,31,43,46,47,53,84,85]																	

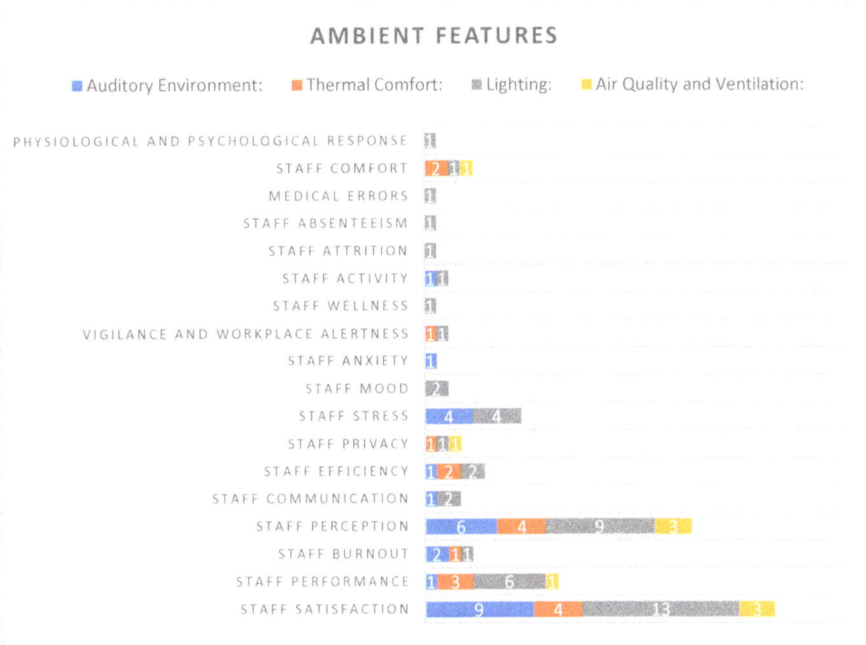

Figure 7. Total number of articles on the effects of ambient features on staff outcomes.

Auditory Environment

Ambient features are essential for creating a comfortable and efficient work environment. Integrating nature sounds or calming music into the auditory environment can help improve satisfaction [64], reduce stress, and improve focus, creating a more serene atmosphere for staff. Effective acoustic design in the healthcare setting impacts satisfaction [19,47,50,51,78,87,90], performance [70], perception [31,38,53,62,70,90], communication [90], efficiency [70], activity [38], comfort [47], burnout [19,34], stress [50,87,90], and anxiety [19].

Thermal Comfort

Thermal comfort is crucial for maintaining a comfortable work environment. Ensuring an optimal temperature is essential, as it prevents discomfort from excessive heat or cold, thereby enhancing staff satisfaction [47,53,84,85], performance [17,70], perception [31,53,70], efficiency [70], privacy [85], alertness [77], burnout [34], comfort [47], and productivity. Additionally, maintaining proper humidity levels is important for creating a comfortable atmosphere [47,70] as it reduces the risk of respiratory issues, contributes to a pleasant working environment, and improves efficiency and comfort.

Lighting

Lighting plays a pivotal role in creating an effective and pleasant work environment [96]. Access to natural daylight can significantly boost mood, energy levels, and overall well-being, which, in turn, leads to improved staff satisfaction [56,65,85,88], performance [17,22,70,88], communication [22], efficiency [56,70], privacy [85], mood [22,56], alertness [77], wellness [80], and activity [38] and reduced burnout [34], stress [56], attrition [23], absenteeism [23], and medical errors [23]. Additionally, adequate artificial lighting is essential for tasks that require precision, ensuring that staff can work efficiently even in areas with limited natural light, which can have an impact on satisfaction [45,47,50,52,65,68,79,84,90], performance [79,95], stress [50,52,90], and comfort [47]. Together, these lighting elements significantly contribute to making the workplace more satisfying.

Air Quality and Ventilation

In the healthcare context, air quality and ventilation are vital for maintaining a healthy and comfortable work environment. Good ventilation systems ensure a constant supply of fresh air, which reduces the risk of airborne infections and helps maintain the overall health of users. Additionally, maintaining high air quality is essential for creating a pleasant working atmosphere, which significantly contributes to staff satisfaction, performance [43,47,84,85], and comfort [47]. Together, these elements ensure a safe, healthy, and comfortable environment for healthcare staff.

While examining the impacts of ambient features on staff experiences, the research articles revealed a strong focus on various outcomes. Staff satisfaction was a major theme, representing 29% of the research, with a significant emphasis on lighting (45%) and the auditory environment (31%). Staff perceptions were addressed in 22% of the articles, with lighting, again, being a primary focus (41%), followed by the auditory environment (27%). Staff performance, although less extensive, at 11%, highlights the critical role of lighting (55%) and thermal comfort (27%). Overall, the ambient feature studies indicated that lighting was the most frequently explored aspect, accounting for 48% of the research, followed by the auditory environment (26%) and thermal comfort (18%). These findings underscore the paramount importance of lighting in shaping staff satisfaction, perceptions, and performance while also highlighting the significant roles of auditory and thermal environments in creating optimal working conditions.

4.8. Overall Impact of the Physical Environment on Staff Outcomes

The findings of research articles shortlisted for this study highlight the varying emphasis placed on various physical environmental features concerning staff outcomes. Architectural design features received the most attention, with 52% of studies focusing on how these features impact staff experiences and outcomes. Ambient features, accounting for 32% of the research, highlight the importance of elements such as lighting, the auditory environment, and thermal comfort in shaping staff satisfaction, perceptions, and performance. Interior design features were the least studied in this context, comprising 16% of the research, indicating a smaller but still significant interest in aspects such as aesthetics, visual appeal, and ergonomics. These proportions reflect the critical role of spatial and environmental design in influencing staff well-being and efficiency, highlighting areas of greater research concentration and potential gaps for future exploration.

4.9. Theories, Models, and Frameworks

There were a few requirements that needed to be met to qualify the use of relevant theories. Theories used to emphasize the findings of a study have to be supported by appropriate references. Seven existing theories (attention restoration theory, stress reduction theory, space syntax theory, systems theory, person–environment fit theory, emotional congruence theory, and organizational theory), two new theories through inductive analysis (grounded theories), and three conceptual frameworks (organizational ecology, the environmental comfort model, the Donabedian model, the naturalistic model, and the new care model) were relevant and were used in the review articles. Most of the theories used are from the domain of environmental psychology and examine the relationship between humans and their environment. The studies also included multidisciplinary approaches to examine individual behavioral responses as well as motives. Only a few of the review studies focused on developing theoretical propositions by inductive analysis of participant data, as opposed to using pre-existing theoretical frameworks. Using "structure", "process", and "outcomes", the conceptual Donabedian model offered a framework for reviewing health facilities and assessing the quality of healthcare. Table 10 lists the authors of the review articles and the theories or models they employed for the studies.

Table 10. Theories identified in this review.

Authors	Theory/Model
[64]	Attention restoration theory (ART)
[34]	Attention restoration theory and stress reduction theory
[74]	Becker and Parsons's (2007) organizational ecology Vischer's (2007, 2008) environmental comfort model of workspace quality
[39]	Space syntax theory
[75]	Systems theory
[57]	Grounded theory
[59,89]	Person–environment fit theory (Person–environment (P-E) congruence model)
[61,83]	New care model
[81]	Emotional congruence theory
[82]	Organization theory

5. Study Implications

This article consolidated findings from diverse geographical locations encompassing a wide range of physical environment factors, staff outcomes, theoretical applications, data collection techniques, etc. Although the researchers shared a common goal, i.e., to investigate the impact of physical environmental features on staff, most studies failed to identify the important features and levels at which these features have favorable and unfavorable effects on staff.

The majority of studies focused exclusively on outcomes among nurses (26 studies) in hospitals. The sample sizes ranged from 12 to 1561 staff. Few studies with multisite cross-sectional designs included large sample sizes. However, some studies with good study designs had relatively small sample sizes, limiting their generalizability. Most of the studies adopted non-random sampling (n = 31), while 29 did not report the sampling techniques adopted.

Our study revealed that only a few theories and conceptual frameworks (n = 10) were explicitly used in the review articles. The identified theories were from the field of environmental psychology and organizational behavior. Most of the studies did not mention or use any of the theories. It is crucial to employ theories when formulating research questions because they might affect the environmental characteristics that are analyzed and the connections between these characteristics and behavioral responses that are investigated. Theories are frequently employed to formulate research questions, devise research methodologies, and/or attempt to interpret results.

This review revealed the need to develop a comprehensive study involving the most important features of the three dimensions of the physical setting and to study the effects of these features on staff. The use of theories that explain the relationships between various constructs is of paramount importance in research. Table 11 shows the shortages in the reviewed articles and the possible directions for future research.

Table 11. Research gaps and directions for future research.

Shortages in Reviewed Articles	Research Gap Characteristics	Future Directions
The studies having a complex relationship between the environment and humans lacked empirical testing. Few studies used a dichotomous scale (yes versus no) while investigating the impacts of physical environmental features.	Evaluation gap	More empirical investigations by combining different techniques are needed. Multidimensional scaling techniques to understand the complex relationship between environmental variables and their impacts must be included in the research.

Table 11. *Cont.*

Shortages in Reviewed Articles	Research Gap Characteristics	Future Directions
The studies conducted to date provided no holistic solution for how the healthcare physical environment favorably affects staff. There were fewer studies from developing countries, as the majority of the previous studies were conducted in the United States. Generalizability was lacking. The majority of the studies also lacked how much the physical environmental features affected the various outcomes in healthcare professionals (efficiency, satisfaction, performance, productivity, etc.) The existing studies failed to include social, cultural, and behavioral considerations while studying the impact of the healthcare environment on its employees.	Knowledge void	Rigorous studies need to be conducted to identify the most important environmental features in the healthcare setting that have positive and negative outcomes among staff. This will help to strengthen evaluation methods by developing measurement instruments that describe different features of the physical environment. Social, cultural, and demographic variables should be included while investigating the impact of the environment on staff.
Theoretical application was absent in the majority of the studies. Most of the researchers did not show concern in applying environmental psychology theories, such as social learning theory (which promotes awareness of ingrained society standards); control theory (which discusses behavioral restrictions and how much control a person feels they have over their actions and behaviors); the theory of affordances (which provides an instantly recognizable function of environmental features); the preference model (positing that humans prefer engaging scenes to boring scenes); the pleasure–arousal–dominance hypothesis (positing that these three emotional responses translate into positive feelings); the behavior setting theory (positing that settings evoke particular patterns of behavior); and the biophilia hypothesis (positing that humans prefer being surrounded by nature).	Theoretical application void	The application of theories in studies about the healthcare environment has to be given more attention. It is crucial to use theories when posing research questions because theories can influence which environmental variables are addressed and which relationships between these elements and the behavioral responses of humans are examined. Researchers should describe theoretical constructs or conceptual structures and the relationships between physical environmental features based on these constructs or structures and link them to staff outcomes.
Even though there were a large number of mixed-method studies, the quantitative studies were equally large in number, and there were fewer qualitative studies. This implies that most studies did not consider both qualitative and quantitative methods, as the mixed method is a good approach to fully explore the research question. Most studies utilized only one research technique to study the effects.	Methodological void	Research problems and complicated phenomena can be better understood by combining qualitative and quantitative methodologies than by using just one of them. Future researchers should focus on utilizing both qualitative and quantitative tools.
The majority of earlier studies had study participants as nurses, implying a lack of studies on other healthcare professionals like physicians, technicians, and non-clinical staff.	Population void	A study on all healthcare professionals is required as they are exposed to a similar environment in healthcare settings.

6. Conclusions

To obtain insights into the impact of the physical environment on staff in healthcare settings, the research questions were defined, gaps were identified in the literature, and systematic directions were provided for prospective researchers. The results also suggest the need to include the social, cultural, and demographic information of staff while examining the impact of the physical setting on staff. The authors believe that this article offers helpful information about the state of the current literature on healthcare staff outcomes, which is critical because healthcare professionals are crucial in the healthcare service industry, as satisfied employees are more loyal to organizations and experience greater patient satisfaction. This review contributes to the literature by identifying the physical environmental variables in a healthcare setting. These variables were categorized based on the three dimensions of the physical environment, i.e., architectural design, interior design, and ambient elements. This study highlights that a comprehensive model needs to be developed to include important physical and environmental features to study their impacts on staff. The study also revealed important outcomes among staff in healthcare settings. These included a wide range of outcomes among staff. Among all the outcomes, staff satisfaction was the most investigated.

Future research should focus on several key areas to gain comprehensive insights into their impacts on staff outcomes. In architectural design, the effects of biophilic elements, flexible spaces, and integrating design characteristics into spatial layouts and connections need further exploration to better understand their impacts on staff and the interactions with patients and coworkers. Longitudinal studies should assess the long-term benefits of natural integration, while simulation models can be used to replicate and analyze real-world systems to understand the behavior and predict outcomes. In interior design, the influence of ergonomic furniture, color schemes, and material choices requires more investigation. Ergonomic assessments using tools like posture analysis and user feedback; experimental designs on color psychology to explore how various color schemes and psychometric testing determine beneficial colors; and comparative studies on materials to guide material selection will provide valuable insights into enhancing the physical environment. Ambient design research should examine the effects of lighting, acoustics, and thermal comfort. Controlled experiments on lighting systems can explore the effects of various lighting conditions, such as natural versus artificial light. Sound mapping, which involves creating acoustic profiles of different environments, should be conducted to help understand the impact of noise levels. Cross-sectional studies in different climatic conditions can provide insights into how temperature, humidity, and airflow affect occupant comfort. By addressing these areas with clear and actionable recommendations, future research can enhance evidence-based design practices and optimize built environments for improved staff outcomes.

The findings presented encourage the development of future investigations that can aid healthcare designers, hospital administrators, and designers in providing better design recommendations for healthcare settings. Furthermore, this study provides insights into the administrative aspects of healthcare facilities, with an emphasis on the well-being of workplace staff. It examined factors influencing satisfaction, productivity, performance, etc., and offers avenues for improvement in healthcare facility administration. The results serve as a basis for establishing policies and procedures to increase stakeholder satisfaction, optimize operational efficiency, and improve overall care.

This study investigated the theoretical implications of the review articles by providing insights into theoretical frameworks that strengthen the understanding of how the physical environment influences staff outcomes. Only a few articles used theories when presenting the rationale for research on the physical environment in healthcare. This review emphasizes that future researchers should explicitly use theories as they can powerfully influence how evidence is collected, analyzed, understood, and used.

This research focused on addressing specific research questions. Despite our best efforts, the sheer volume of articles may have prevented the literature search from capturing all pertinent studies. Nevertheless, this review contributes new insights to the knowledge on healthcare design. The gaps identified in this study will permit future researchers to undertake additional studies and provide design recommendations for staff workplaces in healthcare settings for overall well-being.

A few limitations of this study are that this review did not report contextual factors, such as organizational culture and management practices, which could significantly influence staff outcomes. Another limitation of this study is the lack of quantitative synthesis. While qualitative synthesis provides valuable insights, it does not offer the same level of statistical rigor that a meta-analysis would.

Author Contributions: This manuscript is a result of collaboration among all the authors. Conceptualization, R.S.S. and G.B.K.; methodology, R.S.S. and G.B.K.; validation, R.S.S., G.B.K. and L.L.R.R.; formal analysis, R.S.S. and G.B.K.; writing—original draft preparation, R.S.S., G.B.K. and S.R.S.; writing—review and editing, R.S.S. and S.R.S.; supervision, G.B.K. and R.D.N. All authors have read and agreed to the published version of the manuscript.

Funding: This research received no external funding.

Conflicts of Interest: The authors declare no conflicts of interest.

References

1. Plsek, P.E.; Greenhalgh, T. The Challenge of Complexity in Health Care. *Bmj* **2001**, *323*, 625–628. [CrossRef] [PubMed]
2. Dash, S.; Shakyawar, S.K.; Sharma, M.; Kaushik, S. Big Data in Healthcare: Management, Analysis and Future Prospects. *J. Big Data* **2019**, *6*, 54. [CrossRef]
3. Huisman, E.R.C.M.; Morales, E.; van Hoof, J.; Kort, H.S.M. Healing Environment: A Review of the Impact of Physical Environmental Factors on Users. *Build. Environ.* **2012**, *58*, 70–80. [CrossRef]
4. Joseph, A. *The Role of the Physical and Social Environment in Promoting Health, Safety, and Effectiveness in the Healthcare Workplace*; Center for Health Design: Concord, CA, USA, 2006.
5. Gharaveis, A.; Hamilton, D.K.; Pati, D. The Impact of Environmental Design on Teamwork and Communication in Healthcare Facilities: A Systematic Literature Review. *Health Environ. Res. Des. J.* **2018**, *11*, 119–137. [CrossRef] [PubMed]
6. Harris, P.B.; McBride, G.; Ross, C.; Curtis, L. A Place to Heal: Environmental Sources of Satisfaction among Hospital Patients. *J. Appl. Soc. Psychol.* **2002**, *32*, 1276–1299. [CrossRef]
7. Dijkstra, K.; Pieterse, M.; Pruyn, A. Physical Environmental Stimuli That Turn Healthcare Facilities into Healing Environments through Psychologically Mediated Effects: Systematic Review. *J. Adv. Nurs.* **2006**, *56*, 166–181. [CrossRef] [PubMed]
8. Nejati, A.; Shepley, M.; Rodiek, S.; Lee, C.; Varni, J. Restorative Design Features for Hospital Staff Break Areas: A Multi-Method Study. *Health Environ. Res. Des. J.* **2016**, *9*, 16–35. [CrossRef] [PubMed]
9. Mustafa, F.A.; Ahmed, S.S. The Role of Waiting Area Typology in Limiting the Spread of COVID-19: Outpatient Clinics of Erbil Hospitals as a Case Study. *Indoor Built Environ.* **2023**, *32*, 1914–1928. [CrossRef]
10. Brambilla, A.; Rebecchi, A.; Capolongo, S. Evidence Based Hospital Design. A Literature Review of the Recent Publications about the EBD Impact of Built Environment on Hospital Occupants' and Organizational Outcomes. *Ann. Ig.* **2019**, *31*, 165–180. [CrossRef]
11. Schooley, B.L.; Tarcan, M. Comparing Burnout Across Emergency Physicians, Nurses, Technicians, and Health Information Technicians Working for the Same Organization. *Medicine* **2016**, *95*, e2856. [CrossRef]
12. Joseph, A.; Choi, Y.S.; Quan, X. Impact of the Physical Environment of Residential Health, Care, and Support Facilities (RHCSF) on Staff and Residents: A Systematic Review of the Literature. *Environ. Behav.* **2016**, *48*, 1203–1241. [CrossRef]
13. Li, Y.; Scherer, N.; Felix, L.; Kuper, H. Prevalence of Depression, Anxiety and Posttraumatic Stress Disorder in Health Care Workers during the COVID-19 Pandemic: A Systematic Review and Meta-Analysis. *PLoS ONE* **2021**, *16*, e0246454. [CrossRef]
14. Lacy, B.E.; Chan, J.L. Physician Burnout: The Hidden Health Care Crisis. *Clin. Gastroenterol. Hepatol.* **2018**, *16*, 311–317. [CrossRef] [PubMed]
15. Vizheh, M.; Qorbani, M.; Arzaghi, S.M.; Muhidin, S.; Javanmard, Z.; Esmaeili, M. The Mental Health of Healthcare Workers in the COVID-19 Pandemic: A Systematic Review. *J. Diabetes Metab. Disord.* **2020**, *19*, 1967–1978. [CrossRef] [PubMed]
16. Ulrich, R.S.; Zimring, C.; Quan, X.; Joseph, A.; Choudhary, R. *The Role of the Physical Environment in the Hospital of the 21st Century: A Once-in-a-Lifetime Opportunity By*; Center for Health Design: Concord, CA, USA, 2004; pp. 1–69.
17. Obeidat, B.; Bani, M.; Al-shlool, E. Heliyon Investigations into the Impact of Nursing Unit Layout on Critical Care Nurses. *Heliyon* **2022**, *8*, e08929. [CrossRef] [PubMed]
18. Fauer, A.; Wright, N.; Lafferty, M.; Harrod, M.; Manojlovich, M.; Friese, C.R. Influences of Physical Layout and Space on Patient Safety and Communication in Ambulatory Oncology Practices: A Multisite, Mixed Method Investigation. *Health Environ. Res. Des. J.* **2021**, *14*, 270–286. [CrossRef] [PubMed]
19. Terzi, B.; Azizoğlu, F.; Polat, Ş.; Kaya, N.; İşsever, H. The Effects of Noise Levels on Nurses in Intensive Care Units. *Nurs. Crit. Care* **2019**, *24*, 299–305. [CrossRef] [PubMed]
20. Ulrich, R.S.; Zimring, C.; Zhu, X.; DuBose, J.; Seo, H.-B.; Choi, Y.-S.; Quan, X.; Joseph, A. A Review of the Research Literature on Evidence-Based Healthcare Design (Part I) Abstract Introduction Background. *Health Environ. Res. Des.* **2008**, *1*, 61–125.
21. Laursen, J.; Danielsen, A.; Rosenberg, J. Effects of Environmental Design on Patient Outcome: A Systematic Review. *HERD Health Environ. Res. Des. J.* **2014**, *7*, 108–119. [CrossRef]
22. Zadeh, R.S.; Shepley, M.M.; Ba, D.; Williams, G.; Sung, S.; Chung, E. The Impact of Windows and Daylight on Acute-Care Nurses' Physiological, Psychological, and Behavioral Health. *HERD Health Environ. Res. Des. J.* **2014**, *7*, 35–61. [CrossRef]
23. Shepley, M.M.; Gerbi, R.P.; Watson, A.E.; Imgrund, S.; Sagah-Zadeh, R. The Impact of Daylight and Views on ICU Patients and Staff. *Health Environ. Res. Des. J.* **2011**, *5*, 46–60. [CrossRef] [PubMed]
24. Rafeeq, D.A.; Mustafa, F.A. Evidence-Based Design: The Role of Inpatient Typology in Creating Healing Environment, Hospitals in Erbil City as a Case Study. *Ain Shams Eng. J.* **2021**, *12*, 1073–1087. [CrossRef]
25. Mustafa, F.A.; Ali, J.S. Active Design: Architectural Interventions for Improving Occupational Health Through Reducing Sedentary Behavior—A Systematic Review. *Am. J. Health Promot.* **2023**, *37*, 93–102. [CrossRef] [PubMed]
26. Maassen, S.M.; Van Oostveen, C.; Vermeulen, H.; Weggelaar, A.M. Defining a Positive Work Environment for Hospital Healthcare Professionals: A Delphi Study. *PLoS ONE* **2021**, *16*, e0247530. [CrossRef]

27. Jin, H.; Gold, C.; Cho, J.; Marzban, F.; Lim, L. The Role of Healthcare Facility Design on the Mental Health of Healthcare Professionals: A Literature Review. *HERD Health Environ. Res. Des. J.* **2023**, *16*, 270–286. [CrossRef] [PubMed]
28. Valipoor, S. In the Moment: Fostering Mindfulness and Reducing Stressors in the Healthcare Workplace. *HERD Health Environ. Res. Des. J.* **2021**, *14*, 386–398. [CrossRef] [PubMed]
29. Nordin, S.; Swall, A.; Anåker, A.; Koch, L. Von Does the Physical Environment Matter ?—A Qualitative Study of Healthcare Professionals' Experiences of Newly Built Stroke Units. *Int. J. Qual. Stud. Health Well-Being* **2021**, *16*, 1917880. [CrossRef] [PubMed]
30. Pati, D.; Lorusso, L.N. How to Write a Systematic Review of the Literature. *Health Environ. Res. Des. J.* **2018**, *11*, 15–30. [CrossRef]
31. Sadatsafavi, H.; Walewski, J.; Shepley, M.M. Factors Influencing Evaluation of Patient Areas, Work Spaces, and Staff Areas by Healthcare Professionals. *Indoor Built Environ.* **2015**, *24*, 439–456. [CrossRef]
32. Lim, L.; Kanfer, R.; Stroebel, R.J.; Zimring, C.M. Beyond Co-Location: Visual Connections of Staff Workstations and Staff Communication in Primary Care Clinics. *Environ. Behav.* **2020**, *54*, 3–35. [CrossRef]
33. Joshi, R.; Joseph, A.; Ossmann, M.; Taaffe, K.; Pirrallo, R.; Allison, D.; Perino, L.C. Emergency Physicians' Workstation Design: An Observational Study of Interruptions and Perception of Collaboration During Shift-End Handoffs. *Health Environ. Res. Des. J.* **2021**, *14*, 174–193. [CrossRef] [PubMed]
34. Mihandoust, S.; Pati, D.; Lee, J.; Roney, J. Exploring the Relationship Between Perceived Visual Access to Nature and Nurse Burnout. *Health Environ. Res. Des. J.* **2021**, *14*, 258–273. [CrossRef] [PubMed]
35. de Matos, L.B.N.; Fumis, R.R.L.; Nassar Junior, A.P.; Lacerda, F.H.; Caruso, P. Single-Bed or Multibed Room Designs Influence ICU Staff Stress and Family Satisfaction, But Do Not Influence ICU Staff Burnout. *Health Environ. Res. Des. J.* **2020**, *13*, 234–242. [CrossRef] [PubMed]
36. Valipoor, S.; Hakimjavadi, H.; De Portu, G. Design Strategies to Improve Emergency Departments' Performance During Mass Casualty Incidents: A Survey of Caregivers. *Health Environ. Res. Des. J.* **2020**, *13*, 206–220. [CrossRef] [PubMed]
37. Zamani, Z. Effects of Emergency Department Physical Design Elements on Security, Wayfinding, Visibility, Privacy, and Efficiency and Its Implications on Staff Satisfaction and Performance. *Health Environ. Res. Des. J.* **2019**, *12*, 72–88. [CrossRef] [PubMed]
38. Broom, M.; Kecskes, Z.; Kildea, S.; Gardner, A. Exploring the Impact of a Dual Occupancy Neonatal Intensive Care Unit on Staff Workflow, Activity, and Their Perceptions. *Health Environ. Res. Des. J.* **2019**, *12*, 44–54. [CrossRef] [PubMed]
39. Brewer, B.B.; Carley, K.M.; Benham-Hutchins, M.; Effken, J.A.; Reminga, J. Nursing Unit Design, Nursing Staff Communication Networks, and Patient Falls: Are They Related? *Health Environ. Res. Des. J.* **2018**, *11*, 82–94. [CrossRef] [PubMed]
40. Cordoza, M.; Ulrich, R.S.; Manulik, B.J.; Gardiner, S.K.; Fitzpatrick, P.S.; Hazen, T.M.; Mirka, A.; Perkins, R.S. Impact of Nurses Taking Daily Work Breaks in a Hospital Garden on Burnout. *Am. J. Crit. Care* **2018**, *27*, 508–512. [CrossRef] [PubMed]
41. Fay, L.; Carll-White, A.; Real, K. Emergency Nurses' Perceptions of Efficiency and Design: Examining ED Structure, Process, and Outcomes. *J. Emerg. Nurs.* **2018**, *44*, 274–279. [CrossRef]
42. Alvaro, C.; Wilkinson, A.J.; Gallant, S.N.; Kostovski, D.; Gardner, P. Evaluating Intention and Effect: The Impact of Healthcare Facility Design on Patient and Staff Well-Being. *Health Environ. Res. Des. J.* **2016**, *9*, 82–104. [CrossRef]
43. Schreuder, E.; van Heel, L.; Goedhart, R.; Dusseldorp, E.; Schraagen, J.M.; Burdorf, A. Effects of Newly Designed Hospital Buildings on Staff Perceptions: A Pre-Post Study to Validate Design Decisions. *Health Environ. Res. Des. J.* **2015**, *8*, 77–97. [CrossRef] [PubMed]
44. Harris, D.D. The Influence of Flooring on Environmental Stressors: A Study of Three Flooring Materials in a Hospital. *Health Environ. Res. Des. J.* **2015**, *8*, 9–29. [CrossRef] [PubMed]
45. Robinson, P.S.; Green, J. Ambient versus Traditional Environment in Pediatric Emergency Department. *Health Environ. Res. Des. J.* **2015**, *8*, 71–80. [CrossRef] [PubMed]
46. Horiguchi, M.; Shudo, E.; Sato, K.; Nakamura, M.; Sai, W.; Ohinata, T. Nurse Odor Perception in Various Japanese Hospital Settings. *Int. J. Nurs. Sci.* **2015**, *2*, 355–360. [CrossRef]
47. Kim, S.K.; Hwang, Y.; Lee, Y.S.; Corser, W. Occupant Comfort and Satisfaction in Green Healthcare Environments: A Survey Study Focusing on Healthcare Staff. *J. Sustain. Dev.* **2015**, *8*, 156–173. [CrossRef]
48. Bayramzadeh, S.; Associate, G.; Alkazemi, M.F. Centralized vs. Decentralized Nursing Stations: An Evaluation of the Implications of Communication Technologies in Healthcare. *HERD Health Environ. Res. Des. J.* **2014**, *7*, 62–80. [CrossRef] [PubMed]
49. Yen-Ju Lin, B.; Lin, Y.-K.; Wen Juan, C.; Lee, S.; Lin, C.-C. Moderating Role of Interior Amenities on Hospital Medical Directors' Patient-Related Work Stresses. *HERD Health Environ. Res. Des. J.* **2013**, *91*, 77–92.
50. Campos-Andrade, C.; Hernándezfernaud, E.; Lima, M.L. A Better Physical Environment in the Workplace Means Higher Well-Being A Study with Healthcare Professionals. *Psyecology* **2013**, *4*, 89–110. [CrossRef]
51. Trickey, A.W.; Arnold, C.C.; Parmar, A.; Lasky, R.E. Sound Levels, Staff Perceptions, and Patient Outcomes During Renovation Near the Neonatal Intensive Care Unit. *HERD Health Environ. Res. Des. J.* **2012**, *5*, 76–87. [CrossRef]
52. Quan, X.; Joseph, A.; Ensign, J.C. Impact of Imaging Room Environment: Staff Job Stress and Satisfaction, Patient Satisfaction, and Willingness to Recommend. *HERD Health Environ. Res. Des. J.* **2012**, *5*, 61–79. [CrossRef]
53. Mourshed, M.; Zhao, Y. Healthcare Providers' Perception of Design Factors Related to Physical Environments in Hospitals. *J. Environ. Psychol.* **2012**, *32*, 362–370. [CrossRef]

54. Marie Kotzer, A.; Koch Zacharakis, S.; Raynolds, M.; Aia, A.; Buenning, F. Evaluation of the Built Environment: Staff and Family Satisfaction Pre-and Post-Occupancy of the Children's Hospital. *HERD Health Environ. Res. Des. J.* **2011**, *4*, 60–78. [CrossRef] [PubMed]
55. Hjorhöy, L.G.; Thomsen, T.G.; Beck, M. Physical Environment as a Tool in Caring for the Hospitalized Patient: A Qualitative Study of Nurses' Experiences in Hospitals. *Health Environ. Res. Des. J.* **2022**, *15*, 143–153. [CrossRef] [PubMed]
56. Gharaveis, A.; Yekita, H.; Shamloo, G. The Perceptions of Nurses About the Behavioral Needs for Daylighting and View to the Outside in Inpatient Facilities. *Health Environ. Res. Des. J.* **2020**, *13*, 191–205. [CrossRef]
57. Haywood, A.; Barnes, S.; Marsh, H.; Parker, S.G. Does the Design of Settings Where Acute Care Is Delivered Meet the Needs of Older People? Perspectives of Patients, Family Carers, and Staff. *Health Environ. Res. Des. J.* **2018**, *11*, 177–188. [CrossRef] [PubMed]
58. Pati, D.; Pati, S.; Harvey, T.E. Security Implications of Physical Design Attributes in the Emergency Department. *Health Environ. Res. Des. J.* **2016**, *9*, 50–63. [CrossRef] [PubMed]
59. Durosaiye, I.O.; Hadjri, K.; Liyanage, C.L. Identifying Challenging Job and Environmental Demands of Older Nurses Within the National Health Service. *Health Environ. Res. Des. J.* **2016**, *9*, 82–105. [CrossRef] [PubMed]
60. Harte, J.D.; Sheehan, A.; Stewart, S.C.; Foureur, M. Childbirth Supporters' Experiences in a Built Hospital Birth Environment: Exploring Inhibiting and Facilitating Factors in Negotiating the Supporter Role. *Health Environ. Res. Des. J.* **2016**, *9*, 135–161. [CrossRef] [PubMed]
61. Rippin, A.S.; Zimring, C.; Samuels, O.; Denham, M.E. Finding a Middle Ground: Exploring the Impact of Patient- and Family-Centered Design on Nurse–Family Interactions in the Neuro ICU. *Health Environ. Res. Des. J.* **2015**, *9*, 80–98. [CrossRef]
62. Locatelli, S.M.; Turcios, S.; LaVela, S.L. Optimizing the Patient-Centered Environment Results of Guided Tours with Health Care Providers and Employees. *Health Environ. Res. Des. J.* **2015**, *8*, 18–30. [CrossRef]
63. Bosch, S.J.; Lorusso, L.N. Promoting Patient and Family Engagement through Healthcare Facility Design: A Systematic Literature Review. *J. Environ. Psychol.* **2019**, *62*, 74–83. [CrossRef]
64. Zhu, X. Assessing Preferences and Perceived Restorative Qualities of Break Spaces for Nurses in China. *HERD Health Environ. Res. Des. J.* **2022**, *15*, 126–142. [CrossRef] [PubMed]
65. Chang, W.; Cohen, J.; Mwesigwa, B.; Waiswa, P.; Rokicki, S. Impact of Reliable Light and Electricity on Job Satisfaction among Maternity Health Workers in Uganda: A Cluster Randomized Trial. *Hum. Resour. Health* **2022**, *20*, 30. [CrossRef] [PubMed]
66. Ma, M.; Adeney, M.; Long, H. Functional Settings of Hospital Outdoor Spaces and the Perceptions from Public and Hospital Occupant during COVID-19. *Healthcare* **2021**, *9*, 1534. [CrossRef] [PubMed]
67. Ahmadpour, S.; Bayramzadeh, S.; Aghaei, P. Efficiency and Teamwork in Emergency Departments: Perception of Staff on Design Interventions. *Health Environ. Res. Des. J.* **2021**, *14*, 310–323. [CrossRef] [PubMed]
68. Davis, R.G.; McCunn, L.J.; Wilkerson, A.; Safranek, S. Nurses' Satisfaction with Patient Room Lighting Conditions: A Study of Nurses in Four Hospitals with Differences in the Environment of Care. *Health Environ. Res. Des. J.* **2020**, *13*, 110–124. [CrossRef] [PubMed]
69. Xuan, X.; Chen, X.; Li, Z. Impacts of Nursing Unit Design on Visibility and Proximity and Its Influences on Communication, Privacy, and Efficiency. *Health Environ. Res. Des. J.* **2020**, *13*, 200–217. [CrossRef]
70. Freihoefer, K.; Lindval, S.; Bayramzadeh, S.; Hanson, L. Implications of a Decentralized Postpartum Unit Design and Clinical Operations. *Health Environ. Res. Des. J.* **2019**, *12*, 39–52. [CrossRef]
71. Zamani, Z.; Harper, E.C. Exploring the Effects of Clinical Exam Room Design on Communication, Technology Interaction, and Satisfaction. *Health Environ. Res. Des. J.* **2019**, *12*, 99–115. [CrossRef]
72. Wingler, D.; Machry, H.; Bayramzadeh, S.; Joseph, A.; Allison, D. Comparing the Effectiveness of Four Different Design Media in Communicating Desired Performance Outcomes with Clinical End Users. *Health Environ. Res. Des. J.* **2019**, *12*, 87–99. [CrossRef]
73. Xuan, X.; Li, Z.; Chen, X. An Empirical Examination of Nursing Units in China Based on Nurse Experience. *Health Environ. Res. Des. J.* **2019**, *12*, 108–123. [CrossRef] [PubMed]
74. Naccarella, L.; Raggatt, M.; Redley, B. The Influence of Spatial Design on Team Communication in Hospital Emergency Departments. *Health Environ. Res. Des. J.* **2019**, *12*, 100–115. [CrossRef] [PubMed]
75. Real, K.; Fay, L.; Isaacs, K.; Carll-White, A.; Schadler, A. Using Systems Theory to Examine Patient and Nurse Structures, Processes, and Outcomes in Centralized and Decentralized Units. *Health Environ. Res. Des. J.* **2018**, *11*, 22–37. [CrossRef]
76. Bayramzadeh, S.; Joseph, A.; San, D.; Khoshkenar, A.; Taaffe, K.; Jafarifiroozabadi, R.; Neyens, D.M. The Impact of Operating Room Layout on Circulating Nurse's Work Patterns and Flow Disruptions: A Behavioral Mapping Study. *Health Environ. Res. Des. J.* **2018**, *11*, 124–138. [CrossRef] [PubMed]
77. Sagah Zadeh, R.; Shepley, M.; Sadatsafavi, H.; Owora, A.H.; Krieger, A.C. Alert Workplace from Healthcare Workers' Perspective: Behavioral and Environmental Strategies to Improve Vigilance and Alertness in Healthcare Settings. *Health Environ. Res. Des. J.* **2018**, *11*, 72–88. [CrossRef] [PubMed]
78. Copeland, D.; Chambers, M. Effects of Unit Design on Acute Care Nurses' Walking Distances, Energy Expenditure, and Job Satisfaction: A Pre–Post Relocation Study. *Health Environ. Res. Des. J.* **2017**, *10*, 22–36. [CrossRef]
79. Hadi, K.; DuBose, J.R.; Ryherd, E. Lighting and Nurses at Medical–Surgical Units: Impact of Lighting Conditions on Nurses' Performance and Satisfaction. *Health Environ. Res. Des. J.* **2016**, *9*, 17–30. [CrossRef] [PubMed]

80. Trau, D.; Keenan, K.A.; Goforth, M.; Large, V. Nature Contacts: Employee Wellness in Healthcare. *Health Environ. Res. Des. J.* **2016**, *9*, 47–62. [CrossRef]
81. Ho, R.T.H.; Potash, J.S.; Fang, F.; Rollins, J. Art Viewing Directives in Hospital Settings Effect on Mood. *Health Environ. Res. Des. J.* **2015**, *8*, 30–43. [CrossRef]
82. Steinke, C. Assessing the Physical Service Setting: A Look at Emergency Departments. *Health Environ. Res. Des. J.* **2015**, *8*, 31–42. [CrossRef]
83. Pati, D.; Harvey, T.E.; Redden, P.; Summers, B.; Pati, S. An Empirical Examination of the Impacts of Decentralized Nursing Unit Design. *Health Environ. Res. Des. J.* **2015**, *8*, 56–70. [CrossRef] [PubMed]
84. Djukic, M.; Kovner, C.T.; Brewer, C.S.; Fatehi, F.; Greene, W.H. Exploring Direct and Indirect Influences of Physical Work Environment on Job Satisfaction for Early-Career Registered Nurses Employed in Hospitals. *Res. Nurs. Health* **2014**, *37*, 312–325. [CrossRef] [PubMed]
85. Rashid, M.; Boyle, D.K.; Crosser, M. Developing Nurse and Physician Questionnaires to Assess Primary Work Areas in Intensive Care Units. *Crit. Care Nurs. Q.* **2014**, *37*, 317–335. [CrossRef] [PubMed]
86. Guinther, L.; Carll-White, A.; Real, K.; Zephyr, D.; Bathke, A.; Smith, T.; Glass, B. One Size Does Not Fit All: A Diagnostic Post-Occupancy Evaluation Model for an Emergency Department. *HERD Health Environ. Res. Des. J.* **2014**, *7*, 15–37. [CrossRef] [PubMed]
87. Wang, Z. Role of a Service Corridor in ICU Noise Control, Staff Stress, and Staff Satisfaction: Environmental Research of an Academic Medical Center. *HERD Health Environ. Res. Des. J.* **2013**, *6*, 80–94. [CrossRef]
88. Dianat, I.; Sedghi, A.; Bagherzade, J.; Jafarabadi, M.A.; Stedmon, A.W. Objective and Subjective Assessments of Lighting in a Hospital Setting: Implications for Health, Safety and Performance. *Ergonomics* **2013**, *56*, 1535–1545. [CrossRef]
89. Dendaas, N. Environmental Congruence and Work-Related Stress in Acute Care Hospital Medical/Surgical Units: A Descriptive, Correlational Study. *HERD Health Environ. Res. Des. J.* **2011**, *5*, 23–42. [CrossRef] [PubMed]
90. Bosch, S.; Bledsoe, T.; Arnp-Bc, N.; Jenzarli, A.; Sykes, J.H. Staff Perceptions Before and After Adding Single-Family Rooms in the NICU. *HERD Health Environ. Res. Des. J.* **2012**, *5*, 64–75. [CrossRef] [PubMed]
91. Wingler, D.; Hector, R. Demonstrating the Effect of the Built Environment on Staff Health-Related Quality of Life in Ambulatory Care Environments. *Health Environ. Res. Des. J.* **2015**, *8*, 25–40. [CrossRef]
92. Gharaveis, A.; Hamilton, D.K.; Pati, D.; Shepley, M. The Impact of Visibility on Teamwork, Collaborative Communication, and Security in Emergency Departments: An Exploratory Study. *Health Environ. Res. Des. J.* **2018**, *11*, 37–49. [CrossRef]
93. Zadeh, R.S.; Shepley, M.M.; Waggener, L.T.; Sagha Zadeh, R. Rethinking Efficiency in Acute Care Nursing Units: Analyzing Nursing Unit Layouts for Improved Spatial Flow. *HERD Health Environ. Res. Des. J.* **2012**, *6*, 39–65. [CrossRef] [PubMed]
94. Sadatsafavi, H.; Walewski, J. Corporate Sustainability: The Environmental Design and Human Resource Management Interface in Healthcare Settings. *HERD Health Environ. Res. Des. J.* **2013**, *6*, 98–118. [CrossRef] [PubMed]
95. Davis, B.E. Rooftop Hospital Gardens for Physical Therapy: A Post-Occupancy Evaluation. *HERD Health Environ. Res. Des. J.* **2011**, *4*, 14–43. [CrossRef] [PubMed]
96. Joseph, A. *The Impact of Light on Outcomes in Healthcare Settings*; Center for Health Design: Concord, CA, USA, 2006.

buildings

MDPI

Article

Spatial Constraints and User Adaptation: A Mixed-Methods Study of High-Density Students Dormitories

Xinyu Song * and Demet Dincer

School of Built Environment, University of New South Wales, Kensington, NSW 2052, Australia; d.dincer@unsw.edu.au
* Correspondence: yukis0181@gmail.com

Abstract: This study employs a mixed-methods approach to investigate the current use and evolving needs of dormitory environments through a cross-analysis of existing research and user experience. A two-phased PRISMA-guided systematic literature review for both global and China-based content was conducted, which was then linked to an examination of eight high-density dormitory spaces in Shanghai. The case studies were analyzed using qualitative content analysis methods, incorporating both spatial analysis and netnography to provide a comprehensive understanding of dormitory use. The cross-analysis highlights a persistent design–use gap: efficiency-driven layouts underdeliver on privacy and comfort, prompting compensatory user modifications. Findings also reveal that most rooms fell below suggested standards; circulation was mainly limited to a single central aisle with no entrance buffer, constraining flexibility and privacy. Students responded to these limitations with small-scale personalization, such as bed curtains, mattress toppers, chair swaps, lighting, and storage hacks, to mitigate acoustic/visual exposure and comfort deficits. With the rising demand for dormitories and based on these insights, we identified three critical dimensions for future dormitory design: privacy, equity, and personalization. These findings underscore that current designs fail to fully satisfy even basic functional requirements, making it essential to develop strategies that respond to evolving needs and integrate considerations of well-being, comfort, and user autonomy.

Keywords: dormitory design; student accommodation; room occupancy; privacy; personalization; user experience; user adaptation

1. Introduction

The design of student dormitories plays a crucial role in shaping the lives of young people. Research indicates that designing youth-friendly, mentally supportive environments is essential for strengthening "the social fabric", especially by fostering social connectedness and a sense of belonging [1]. Considering that by 2050, 70% of youth will live in cities for educational and employment opportunities [2], the question of where they will reside has become an increasingly pressing topic. This applies to *dormitory* (the term *dormitory*, often abbreviated as *dorm* [3], is also referred to as *resident hall* [4]) design, which is defined as a large room or building containing many beds [5].

A *student dormitory* is a type of residential environment designed specifically for particular groups, such as university students. They serve as crucial spaces for living and engaging in exchange activities [6]. They are key spaces for promoting independence and are the first platforms for socialization within a community [7], identified as crucial factors for mental health [8]. For instance, students in high-density dormitories reported less trust, cooperativeness, and responsibility, as well as less socially responsible behavior compared to those in low-density dormitories [9].

Studies highlight the importance of designing spaces that consider individuals' diverse needs and preferences to prevent conflicts, which will affect their behavior and academic performance [10]. Amole (2011) mentions the importance of equitable design, arguing that all students should have access to fundamental facilities regardless of differences in gender or socioeconomic background [11]. The role of dormitories in students' social lives was also indicated to promote better social interactions [12] and higher levels of interaction with the faculty, as well as higher participation in student organizations [13,14]. This results in improved academic performance [15–17] and higher college satisfaction [8]. While these research outcomes provide valuable guidance on factors that benefit youth, each addressing different aspects of dormitories, the demand for purpose-built student accommodation far exceeds the supply of available beds in many countries such as the UK, Canada, and Australia [18].

China has a rapidly growing population with limited affordable accommodation availability in major cities. This pressure is further amplified by the increasing number of students relocating to cities for educational opportunities, creating a critical need for accessible and efficient living solutions. The number of universities in China has increased from 1731 to 2868 over the past 20 years. This growth of new universities and the expansion of university enrolments have put significant pressure on student accommodation [3]. Most universities in China offer on-campus living arrangements, where students typically share spaces with multiple peers. According to statistics from the Ministry of Education of the People's Republic of China, 82 percent of students reside in university dormitories, and 80 percent of them share a room with four to six individuals [4]. The city of Shanghai, which not only serves as China's economic hub but is also a major center of higher education, hosts 68 universities and colleges, which is the highest concentration of higher education institutions in East China [19].

With the rapid expansion of Chinese universities, the demand for student dormitories has surged, creating a growing imbalance between supply and demand [20]. Standard dormitory design in China has long emphasized functional efficiency, while largely overlooking how spatial layout and configuration influence students' day-to-day experiences [21]. While some research has briefly mentioned room layout or basic furnishings as part of broader investigations into student satisfaction [22], these aspects are rarely examined as independent spatial variables. Empirical understanding of how students interact with, adapt to, or personalize shared spatial environments remains limited, especially in the high-density contexts that characterize most Chinese dormitories.

Existing studies highlight the importance of the dormitory environment in shaping students' well-being and social dynamics. However, their insights are restricted by the temporal and geographical contexts their represent. Thus, there remains a lack of research on how students actively use these spaces today in response to their evolving needs. As improving the environment of university dormitories is suggested to be one of the main priorities of universities [23], a comprehensive synthesis of existing findings is needed to inform design strategies that address these changing requirements.

Within this framework, this study investigates the contemporary use and evolving needs of student dormitories, with a focus on one-room living arrangements shared by multiple users. It examines not only the spatial characteristics of these high-density environments, but also how students adapt to and interact with these conditions as their needs evolve in time, with the following aims:

- Aim 1: Evaluate the existing literature to identify key findings and gaps regarding the design and use of dormitory environments.

- Aim 2: Conduct a comprehensive analysis of the spatial characteristics of current high-density student dormitories as a comparative analysis.
- Aim 3: Examine the relationship between spatial layout and evolving needs of users to inform the future design of such spaces.

In the context of these aims, *spatial layout* refers to the arrangement and configuration of the physical spaces within a dormitory, including the positioning of furniture such as beds. *User experience* refers to the ways in which occupants interact with and utilize these spaces for their daily activities. In this case, the focus is how students engage with the dormitory physical space to meet their evolving needs.

The next section details the methodology, which is structured around the three aims and builds progressively from a systematic literature review with a general overview, then narrows down to studies in the Chinese context and informing the examination of a high-density dormitory environment in Shanghai, China (Figure 1). Following the explanation of the methodology, the final sections present the research findings (Aims 1 and 2) and discuss how current spatial characteristics and user adaptations can inform future design directions (Aim 3).

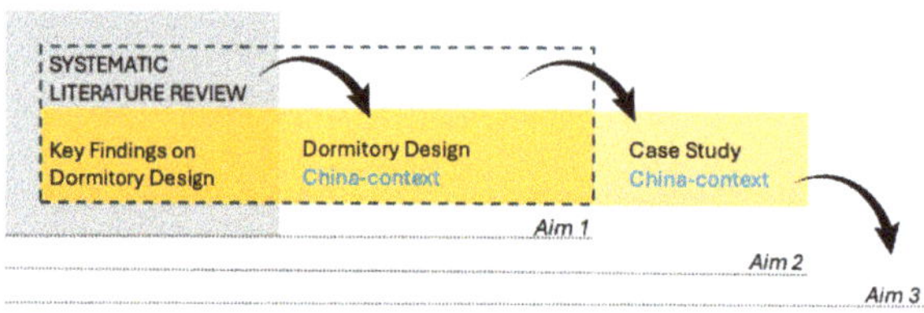

Figure 1. Diagram introducing the methodology structured around the three research aims, with arrows specifying the links between them.

2. Methodology

This study employs a mixed-method research approach to investigate the relationship between dormitory design and user experience, aiming to integrate diverse methods for a comprehensive analysis [24,25]. This method is preferred for its ability to offer a multidimensional perspective by providing richer insight into the research phenomena [26], conducted in two phases: 1. a systematic literature review and 2. a case study analysis, including a supporting examination of social media content to understand user perspectives. The details of each method, along with supporting analysis approaches, are presented in the following subsections.

2.1. Systematic Literature Review: Dormitory Design in Academic Research

The purpose of the systematic literature review (SLR) was to identify key academic insights regarding the design and use of dormitory environments to provide a foundation for analyzing the current usage of dormitory environments. The Preferred Reporting Items for Systematic Reviews and Meta-Analyses (PRISMA 2020) method with the suggested diagram template [27] was applied for this analysis, which is significant for identifying the relevant literature with the defined research objectives and is preferred for minimizing systematic errors during the selection, evaluation, and synthesis of studies [28]. SLR reduces such biases by employing a clear methodology to identify, select, and critically assess relevant studies. Also, this method is considered more transparent than traditional narrative literature reviews, resulting in a broader range of articles that allow for the mapping of specific trends and identification of gaps [28].

2.1.1. Database Selection

Given the interdisciplinary nature of dormitory-related research, a cross-database approach was adopted. We conducted a comprehensive search using three major academic platforms in June 2025: EBSCOhost, ProQuest, and Web of Science. Within EBSCOhost, the search included Art & Architecture Source, Avery Index to Architectural Periodicals, and Environment Complete. For ProQuest, relevant sub-databases were accessed, including Humanities, Design & the Arts, Health & Medicine, Social Sciences, and Education databases). Web of Science was searched without field limitation. The search retrieved all records published between 1 January 1998 and 5 June 2025, aligning with the temporal scope defined by the study's eligibility criteria.

2.1.2. Strategy and Analytical Approach

This study employed a two-phase systematic review approach to address Aims 1 and 2 (Figure 1). The first phase involved a global scan of the relevant literature aimed at identifying international research trends, methodologies, and thematic priorities. The second phase focused specifically on student dormitories in China, where demand has been rising significantly [3,4,19].

The global search included the following five keywords: dormitory, dormitory layout, dormitory use, and dormitory organization, along with user experience and user performance in dormitories. After removing duplicates from results obtained from the EBSCOhost (Arts and Architecture) database, titles and abstracts were screened in three consecutive steps, excluding articles that did not address university dormitory spaces, lacked full-text access, or omitted reference lists. When uncertainty arose, the decision was made in favor of inclusion, with all exclusions verified by both reviewers.

The second phase of the search included the keywords "China" and "Chinese", combined with "dormitory". To ensure the relevance and contextual specificity of the literature included in this second phase of the systematic review, the following four eligibility criteria were applied. These criteria were designed to focus the review on studies that directly address the spatial design and user experience of student dormitories in China.

i. Geographic and topical relevance: Since the geographic focus was justified by the high density of dormitories and growing demand, it was essential to maintain contextual consistency, reflecting the context-based factors influencing dormitory design in China. Studies conducted outside of China were excluded.

ii. Publication type: Articles were published in academic journals, though not strictly peer-reviewed articles. This decision was made due to the relatively niche scope of research on Chinese student dormitory spatial design; excluding non-peer-reviewed studies would have significantly reduced the available literature and valuable insights that might inform further research.

iii. Temporal scope: Only articles published in 1998 and later were included. This cutoff was defined based on a pivotal policy shift: in 1998, the Ministry of Education of the People's Republic of China proposed "The Action Plan to Vitalize Education in the Twenty-first Century" [29], which marked the beginning of a rapid expansion in higher education enrolment [30]. This expansion had a profound impact on student housing demand and design, making studies prior to this period less relevant to the current context.

iv. Language: To ensure accessibility and consistency in analysis, only articles published in English were included.

All retrieved records were imported into EndNote 21 for duplicate removal, followed by title, abstract, and full-text screening in Covidence according to the agreed inclusion criteria.

2.2. Case Study Analysis: Dormitory Environments in Use

The second methodology we employed was a case study analysis, focusing on a single location in China. This method was chosen to provide insights into how leading institutions address student dormitory needs by analyzing how dormitory spaces are currently organized, inhabited, and adapted by students in their daily use. Shanghai's diverse institutional types contribute to a wide range of dormitory configurations, making it an ideal site for conducting comparative analyses of student housing [19].

Eight dormitories across two highly ranked universities were selected to represent a range of spatial conditions as high-density case studies. Both universities are in high demand and provide a strong basis for examining current student dormitory design and use in China, as shown in Table 1 [31]. Shanghai Jiao Tong University, with 18,582 undergraduate students, accommodates approximately 30,000 students in its dormitories, including both undergraduate and postgraduate residents. Fudan University reported an undergraduate population of 15,523 in 2024, although dormitory-specific data is not publicly available.

Table 1. Selected universities for case study context.

University	Establishment	Student Population	Key Characteristics
Fudan University (FDU)	1905	15,523	One of China's most prestigious comprehensive universities, with strengths in the humanities, sciences, and medicine.
Shanghai Jiaotong University (SJTU)	1896	18,582	Renowned for engineering and technology programs, with a significant research output.

The case studies were analyzed using qualitative content analysis methods, incorporating both spatial analysis and netnography. Publicly accessible data in various formats was collected to capture informal, user-generated perspectives on dormitory rooms. Including these digital sources allowed a comprehensive understanding of how students perceive and navigate their dormitory environments.

Spatial analysis is essential for understanding the spatial characteristics and social interactions that the space hosts. For instance, Chrysikou [32] highlights the importance of spatial analysis in healthcare research as an architectural methodology, as the outcomes can reveal insights into institutional behaviors that may not be evident through traditional qualitative methods. In this study, spatial analysis focused on dormitory room occupancy, spatial characteristics, and user-led personalization. Personalization is defined as the adjustment of design elements and/or spatial configurations to reflect individual preferences and/or needs. In this context, it describes students' spatial modification behaviors aimed at optimizing their living environments, including layout adjustments. The eight dormitory spaces in these two universities were analyzed accordingly, visualizing their spatial characteristics and examining the user experience. Overlay diagrams were used as tools to superimpose multiple layers of information, including architectural plans combined with furniture layouts and usage data gathered from public social media posts shared by users of these specific dormitories.

Netnography is employed as a qualitative methodology and is considered a form of online ethnographic research involving the systematic observation of groups [33]. As this method is identified as an adaptation of the traditional ethnographic techniques to the online context [34], it serves as a significant tool for gaining insights into user-generated content and interactions, particularly content shared by younger generations. A digital platform specific to Chinese users, Rednote, was selected to analyze users' viewpoints due to its user group characteristics being highly compatible with this study. This platform is considered the largest lifestyle-sharing community in China, also functioning as an e-commerce platform driven by user-generated content (UGC) social networking [35]. As of March 2025, 44 percent of active users are specified as under 24 years old [36]. This user structure ensures that a large portion of the feedback collected directly reflects the experiences and perspectives of university students [37], who mainly live in student dormitories. This data collection method provides a user-focused analysis of spatial elements across different dormitories, offering insights into their variations.

3. Results

The results are organized in three sections. Section 3.1 presents findings from the systematic literature review, including both global and China-based research outcomes. Section 3.2 follows with the analysis of case studies, and Section 3.3 concludes the results with a comparative discussion that synthesizes findings across all methods.

3.1. Systematic Literature Review

3.1.1. SLR Findings: Global Review

As the first phase of the literature review, a global overview offered insights into international research on dormitory design, including methodologies, thematic priorities, and spatial considerations, which informed the second phase of the China-based literature framework of this study. The preliminary search yielded 189 articles published between 1980 and December 2024. In total, 42 articles were assessed as full texts after the screening process, as shown in Figure 2.

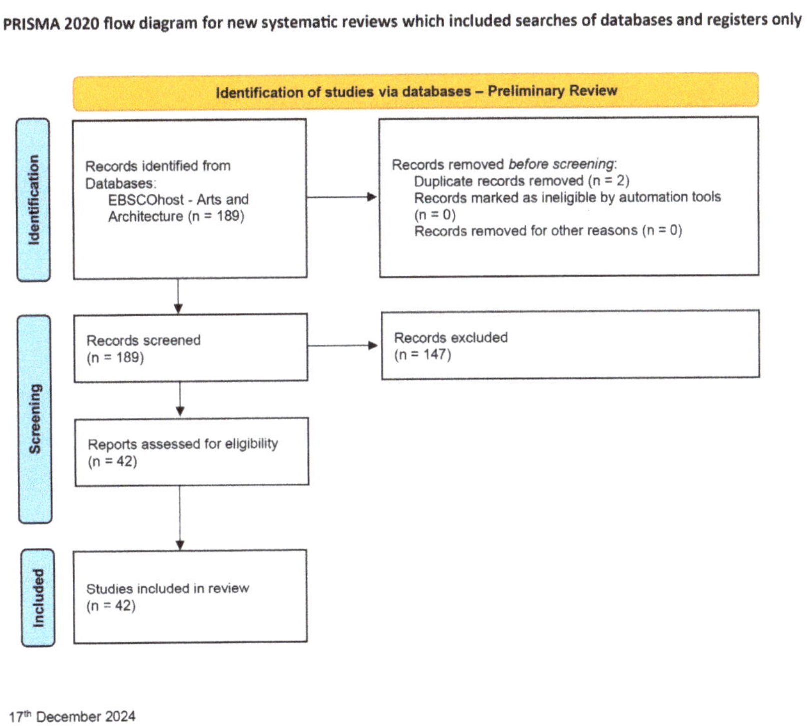

Figure 2. Flow diagram showing the preliminary literature review process.

In total, 42 articles were analyzed to provide a general overview, and the results were categorized into two dimensions to identify key findings regarding the dormitory environments (Aim 1). Research outcomes categorized under physical characteristics emphasize environmental quality and spatial configuration, highlighting factors such as acoustic and air quality, access to natural light, effective thermal control, and architectural suggestions. The second dimension, experiential characteristics, addresses aspects not covered by physical characteristics, focusing on user-related factors such as well-being needs, privacy, and social adaptability. Table 2 summarizes the outcomes, with one key reference highlighted from the literature review.

Table 2. Literature review findings.

Content	Key Outcomes from the 1st Phase of the SLR	Key Reference
Physical Characteristics	Indoor environment quality: acoustic performance, air quality, and natural light	[38]
	Adequate daylighting and effective thermal control	
	Low-rise building typologies with balconies and/or semi-open spaces	[11,39]
	Preference for single and double rooms where possible	[11]
	Small clusters for shared rooms with localized facilities	
	Bathrooms inside clusters, limit users per shared facility to reduce crowding	
	Cluster-plex layouts instead of corridor-style arrangements	[40]
	Biophilic design helps improve dormitory environments and enhances students' physical and mental recovery	[41]
	Dormitory air quality (CO_2, temperature, humidity, and ventilation) impacts students' health, comfort, and satisfaction with their living environment	[42]
Experiential Characteristics	Cultural and gender-sensitive spatial design	[7]
	Visual and acoustic privacy through barriers and buffer zones at entries	
	Balanced public–private spaces, spatial hierarchy with semi-private social zones	
	Modular design for adaptability to social	[43]
	Opportunities for personalization to enhance user satisfaction	[44]
	The in-between space most directly influences daily communication and the use of personal space within dormitories	[45]

Our objective was to undertake a more rigorous and in-depth analysis to strengthen the case study and inform future design strategies (Aims 2 and 3). Thus, after systematically examining and thematically categorizing the research outcomes, we identified eight studies based on "spatial characteristics", including spatial design, spatial organization, and spatial use. The details and outcomes of each study are presented in Table 3.

Table 3. Summary of nine studies focusing on "spatial design" resulted from the preliminary literature review.

Focus and Country	Dormitory Type and Occupancy	Method	Key Outcomes	Reference
Gender differences and housing preferences Nigeria	Resident Hall/NA	Multi-stage sampling technique and questionnaire	• Building height: Not exceeding three stories • Room density: Each apartment or floor has a maximum of four bedrooms • Room type: Prioritization single and double rooms	[39]
Students' performance and privacy Nigeria	Resident Hall/2–4 persons per room	Questionnaire (stratified sampling by course, length of stay, and gender)	• Building height: Support low-rise buildings with balconies • Room type: Prioritization of single and double rooms • Common facilities: Limit users in shared facilities	[11]
Crowding perception and dormitory satisfaction Turkiye	Corridor-style dormitory/2 persons per room	Quota sampling and paper questionnaire administered	• Privacy: Crowding reduces privacy; higher floors perceived as quieter and more private	[40]
Spatial organization, social interaction, sense of community and privacy United States of America	Traditional; corridor-plex; cluster-plex; specialty housing/2 persons per room	Online questionnaire	• Privacy: Cluster-plex dorms enhance privacy; corridor-plex dorms least preferred	[46]
Privacy aspects (visual, acoustic, olfactory) of female dormitory Indonesia	Boarding-school dormitory/4–8 persons per room	Qualitative case study analysis: participatory observation, semi-structured interviews, documentation, and spatial/photographic analysis	• Privacy: Recommended to establish entry buffer zones, visual barriers, gender-specific facilities, and culturally sensitive spaces within dormitories	[7]
Spatial characteristics, layout analysis, space efficiency Slovakia	Prefabricated modular dormitory cells/single rooms 2 to 4 persons sharing cells with common facilities	Mixed-method case study analysis: documentation review of 35 European modular dorm projects	• Privacy: Considered as a key objective in designing modular dorms but one that has to be balanced with social flexibility	[43]
Dormitory satisfaction, density, and privacy Turkiye	Dormitory/2–4 persons per room	Questionnaire and case study	• Privacy: Enhance privacy with layout and acoustic separation • Density: Reduce room resident density • Layout: Use clustered mini-halls of 3–4 rooms; place a shared bath inside each cluster	[23]
Dormitory personalization, satisfaction, and privacy United States of America	Corridor-style dormitory/2 persons per room	Questionnaire and photographic content analysis	• Privacy: Explicitly measured via 18 privacy-regulation mechanisms (contact-seeking and contact-avoiding) plus effectiveness ratings	[47]
Adaptation of biophilic design in university dormitories Iraq	University dormitory buildings/NA	Field experiment	• Focus on how biophilic elements positively impact student well-being indicators • Indicates a correlation between biophilic design features and improvements in students' mental health and comfort levels	[45]

From an architectural perspective, two studies recommend limiting the building height to a maximum of three stories (including the ground floor) to enhance accessibility and encourage interaction to support a human-scaled living environment [11,39]. Where feasible, the inclusion of a balcony is recommended to improve ventilation and natural lighting and provide a semi-private transitional zone that enhances residents' connection to the outdoor environment [11]. In terms of room density, same studies also highlighted that single and double rooms should be prioritized, while acknowledging that economic considerations remained an important constraint. In addition, communal facilities were identified as critical, proposing a limit of five users per kitchen and ten per bathroom to reduce overcrowding and associated stressors [11].

The critical role of privacy emerges as a central concern in dormitory design, particularly under conditions of high resident density. Previous outcomes have shown that overcrowding can diminish trust, cooperation, and friendliness among residents [9], while also contributing to psychological stress, interpersonal conflict, and reduced academic performance [10]. Across genders, students express heightened sensitivity to privacy concerns when selecting housing, with preferences shaped by factors such as shared facilities, number of occupants, bedroom type, and study location. Crowding negatively impacts privacy satisfaction, with higher floors perceived as more private due to reduced noise and expansive views. Quantitative assessments using a 26-item architectural rating scale revealed that cluster-plex dorms scored highest on the "quiet" factor (mean: 6.20), indicating superior privacy and study suitability, while corridor-plex dorms scored lowest (mean: 4.23) [46]. One study specifically focused on the impact of privacy in dormitory environments using qualitative indicators such as window height, indoor bathroom configurations, and student interviews [7]. The results pointed to factors like noise and odor disturbances caused by bathroom locations, as well as the effects of window placement and size on residents' visual privacy, underscoring the importance of multi-sensory privacy. Privacy preservation is thus recognized as a key objective in dormitory design, regardless of gender.

Awal [45] highlights flexibility needs by pointing out that dormitory spaces are neither entirely public nor entirely private as their nature dynamically shifts with usage and time. The in-between zones within dormitories, in fact, are important in providing this flexibility of use. Through forms such as buffer zones, adaptable spaces, and semi-open areas, the balance between public and private can be adjusted, providing users with more control over their privacy while fostering a stronger sense of belonging.

3.1.2. SLR Findings: China-Based Review

In the second phase of the literature review, we focused on the Chinese context and retrieved a total of 189 articles, including 36 from the EBSCOhost database, 37 from the ProQuest database, and 116 from the Web of Science database. To ensure focus on dormitory design, we excluded studies that did not investigate university student housing and lacked full-text access. Based on this process, 48 studies were included in the Chinese context review.

Table 4 presents three themes that emerged from the analysis of China-specific studies, spanning topics such as thermal comfort, energy performance, ventilation, and occupancy. Due to the extensive number of references, one key resource was identified for each theme (Table 4).

The majority of the analyzed studies (n = 35) focused on physical environmental conditions in Chinese dormitories, with a focus on building performance, energy consumption (n = 12), and indoor environmental parameters (n = 12). Although environmental factors play a significant role in shaping user experience, most of the studies overlooked the spatial conditions of the dormitory spaces and disregarded user-specific needs. Two studies considered room types, offering insights into 1. spatial layout and user satisfaction and 2. spatial adaptability and user-driven personalization. The necessity of balancing space usage and comfort was emphasized by Dong et al. [56], suggesting that dormitory design should consider both physical layout and psychological needs. The study found that dormitory space layout design—including factors such as floor area per capita, spatial organization, and furniture design—has the greatest impact on overall indoor environmental quality satisfaction among all indoor environmental factors, even exceeding traditional indoor environment quality (IEQ) factors such as thermal environment, air quality, lighting environment, and acoustic environment. Single rooms and two-person rooms with balconies are most capable of improving student satisfaction. As the average living space per person increases, student satisfaction improves significantly, but once the average space per person reaches approximately 13.5 square meters, further improvements in satisfaction are limited [56]. Part of Wang et al.'s [57] research covered the impact of modular dormitory layouts on user satisfaction. The results showed that the shift from traditional crowded dormitory designs to more personalized, independent living spaces significantly improved thermal comfort, privacy, and the overall living environment, leading to higher resident satisfaction.

Table 4. Categorical review of the analyzed studies specific to the Chinese context.

Theme Category	Research Focus	Number of Studies	Key Reference
Physical Environmental Conditions	Building performance and energy consumption	12	[48]
	Outdoor/ventilation studies	3	[49]
	Thermal comfort and heat adaptation	8	[50]
	Indoor air quality	12	[51]
Functional and Safety Considerations	Functional optimization of dormitory interiors	7	[52]
	Evacuation and safety design	2	[53]
	Occupancy and usage patterns	2	[54]
User Behavior and Design Adaptation	Renovation and student preferences	2	[55]
Total		48	

Research by Dong et al. [56] noted that universities should adopt space designs incorporating sufficient private and public areas to enhance student well-being and satisfaction. However, their focus was primarily on the impact of university dormitory space design on residents' IEQ satisfaction, the relationship between space design and other IEQ factors, and the halo effect within space design. Wang et al. [57] focused on the thermal comfort and health impacts of unit-type dormitories, comparing unit-type layouts with traditional layouts. The study found that unit-type layouts can improve thermal environmental conditions and enhance user satisfaction but did not explore how students interact with or modify these spatial layouts [57].

3.2. *Case Study Analysis*

This case study examines eight dormitories across two leading universities in Shanghai, China, analyzing how student dormitory environments function in practice and comparing these findings with the SLR review through cross-analysis. Table 5 presents the characteristics of the eight dormitory spaces, detailing specifications such as room size and occupancy density.

The case study analysis employed a multi-step approach to capture user-specific dimensions of dormitory use. First, a spatial analysis was conducted to examine the physical characteristics of the dormitory rooms, including floor area, layout, and patterns of use, and to assess their compliance with recommendations identified in the systematic literature review (SLR). Second, user-driven adaptations were investigated to understand modifications made beyond the standard room design. This stage incorporated netnographic analysis of user-generated content. Visual analysis was performed using publicly available photographs, which were mapped and overlaid onto architectural drawings from the initial spatial analysis. Additionally, discourse analysis of online posts was undertaken to capture user perspectives and contextualize observed adaptations. The case study analysis process and its integration with the SLR is illustrated in Figure 3.

Table 5. Key characteristics of the dormitories analyzed as case studies.

Case ID	Room Size (Excluding Balcony and Bathroom)	Number of Residents	Bed Arrangement	Storage Type	Bathroom Use	Other Features
D-01	15.84 m^2	4	Loft Bed with Desk	Personal Locker	Shared	Enclosed Balcony
D-02	15.84 m^2	3	Loft Bed with Desk	Personal and Shared Storage	Shared	Enclosed Balcony
D-03	14.55 m^2	3	Mixed Bunk and Loft Beds	Personal and Shared Storage	Private	Enclosed Balcony
D-04	15.52 m^2	2	Standard Single Bed	Shared Storage	Shared	Standard Balcony
D-05	34.20 m^2	5	Standard Single Bed	Shared Storage	Shared	Enclosed Balcony
D-06	15.76 m^2	4	Loft Bed with Desk	Personal and Shared Storage	Shared	Bay Window
D-07	17.60 m^2	4	Loft Bed with Desk	Personal and Shared Storage	Shared	None
D-08	18.02 m^2	4	Loft Bed with Desk	Personal and Shared Storage	Shared	None

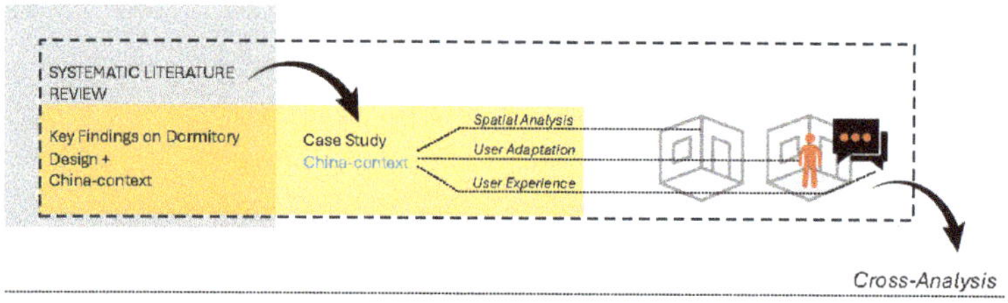

Figure 3. Illustration presenting the case study analysis workflow for the eight selected dormitories and its integration with the SLR for cross-analysis, specified with the arrows.

3.2.1. Spatial Analysis

Spatial characteristics of each dormitory room were documented using publicly available resources, such as floor plans and photographs illustrating overall room use. When official floor plans or measurements were unavailable, layouts were reconstructed based on standard dormitory bed dimensions, photographic evidence, and precise furniture measurements provided on the university's official website, with necessary spatial approximations. Each dormitory type was then drawn in detail, highlighting key spatial features, including circulation zones, furniture arrangements, storage areas, and window placement. This process enabled a comparative spatial analysis across all cases, focusing on layout and functional use. The following visuals and captions present a detailed description of each dormitory type, emphasizing layout, occupancy, and spatial characteristics.

Room occupancy, defined as the number of residents accommodated within each dormitory space, was analyzed across eight selected cases representing four distinct occupancy types, ranging from two-person to five-person rooms. Figures 4–11 illustrate each dormitory layout, with captions and legends detailing the specific spatial characteristics of the rooms. In all figure labels, '*' indicates the dimension format (length × width).

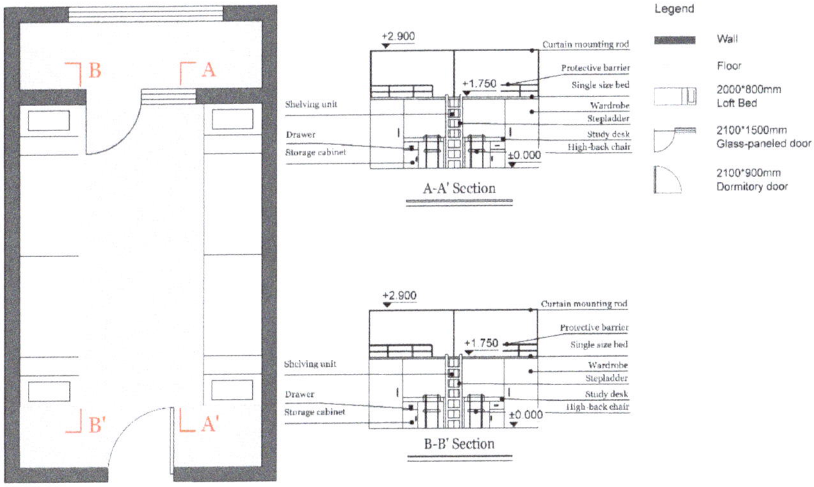

Figure 4. D-01 SJTU 4-person dormitory floor plan: Accommodates up to four residents, providing a per capita area of 3.96 m^2. Features include a balcony, loft beds with integrated desks for each resident.

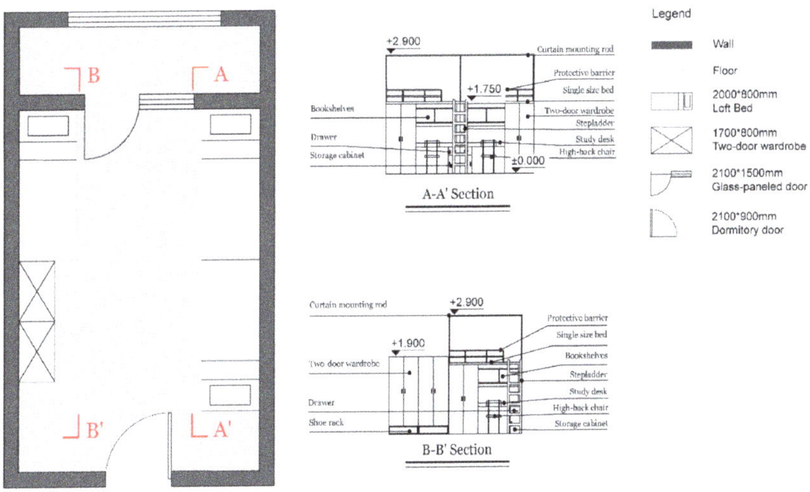

Figure 5. D-02 SJTU 3-person dormitory floor plan (type 1): Designed for three residents with a per capita area of 5.28 m^2, including a balcony and loft beds with integrated desks. Additional storage is provided via two shared cabinets.

In the studied dormitories, circulation is primarily organised along a single centralized axis, creating straightforward yet constrained pathways between furniture elements. The pass is the space between two sets of beds or tables, with a clear width ranging from 1.4 m to 1.7 m, which meets basic circulation needs. The layout of Dormitory D-04 is an exception, with its L-shaped circulation route providing slight flexibility and differentiation in the use of space.

All analyzed dormitories feature a symmetrical furniture layout arranged along the walls. Each furniture unit within the same dormitory (including beds, desks, and wardrobes) has a modular design with consistent alignment, and the number of furniture items directly corresponds to the number of occupants, creating a predictable and unified spatial arrangement. In

Chinese dormitory design, there are two major furniture layout strategies: integrated loft beds and traditional floor bed arrangements. Integrated loft bed modules (such as D-01, D-02, D-03, D-06, D-07, and D-08) enhance spatial compactness, while floor bed and desk layouts (such as D-04 and D-05) offer greater visual openness but require larger room areas. Storage facilities within the case can generally be divided into two categories: personal storage space (such as ladder-integrated storage units, personal wardrobes connected to loft beds, bookshelves, and desk drawers) and shared cabinets. These storage components are usually arranged on the entrance wall or near the sleeping area.

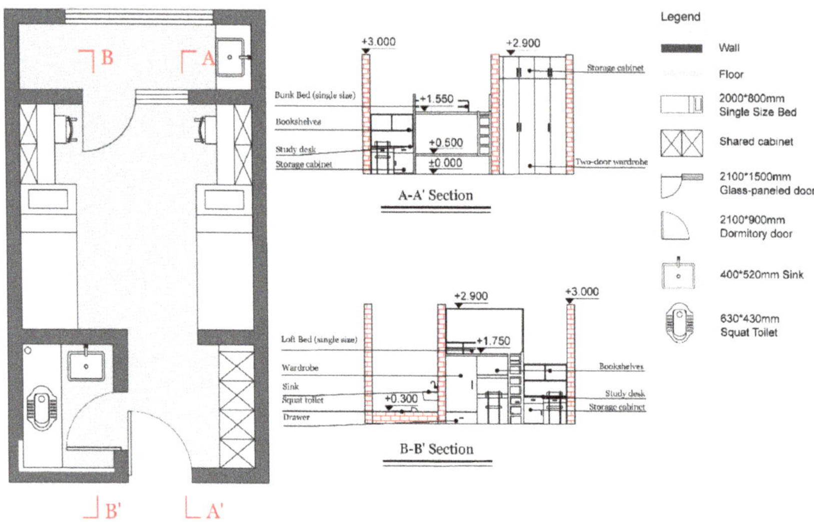

Figure 6. D-03 SJTU 3-person dormitory floor plan (type 2): Originally intended for two residents but currently housing three, offering a per capita area of 4.85 m^2. Includes a balcony, toilet, and a mix of loft and bunk beds with desks, supplemented by four communal storage cabinets.

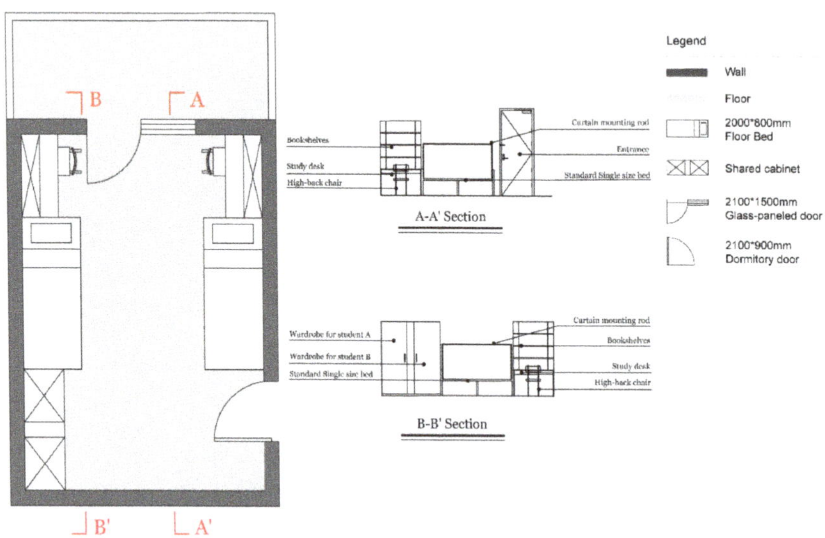

Figure 7. D-04 SJTU 2-person dormitory floor plan: Two-person occupancy, with the largest per capita area of 7.76 m^2, featuring individual floor beds and desks, a balcony, and shared storage cabinet.

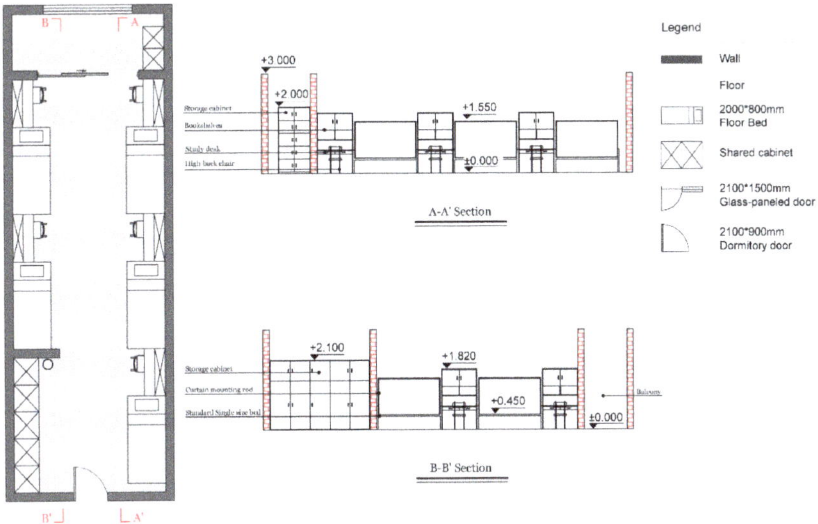

Figure 8. D-05 FDU 5-person dormitory floor plan: Five-person dormitory providing 6.84 m^2 per capita. Facilities include floor beds, individual desks, a balcony, and seven shared storage cabinets.

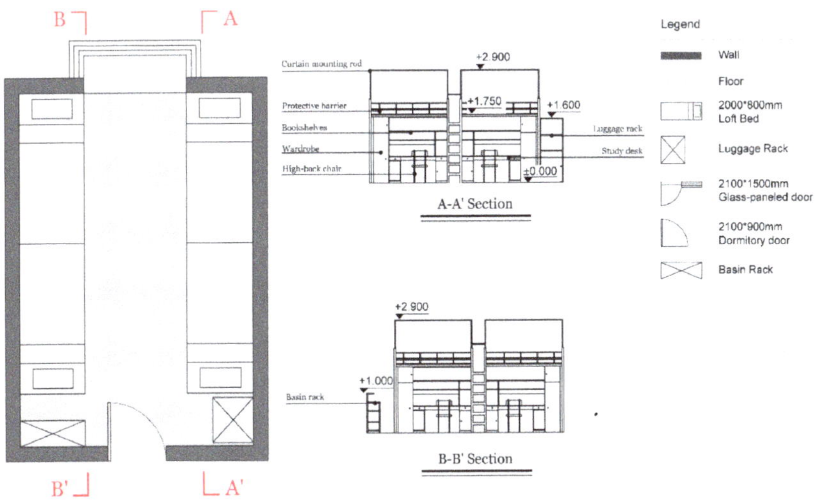

Figure 9. D-06 FDU 4-person dormitory floor plan (type 1): Four-person occupancy, the smallest per capita area at 3.94 m^2. Furnished with loft beds integrated with desks, bay windows, and communal storage for luggage and shoes.

According to the Chinese national standards titled "Code for Design of Dormitory Buildings" [58], dormitories are classified into four types based on bed configuration and number of occupants (Table 6).

Excluding balcony and bathroom areas, each category has a minimum recommended per-person space standard. Analysis revealed that six out of the eight dormitory cases (D01, D03, D04, D06, D07, and D08) studied did not meet the national standards, indicating that per capita space allocation generally falls short of the recommended guidelines (Table 7).

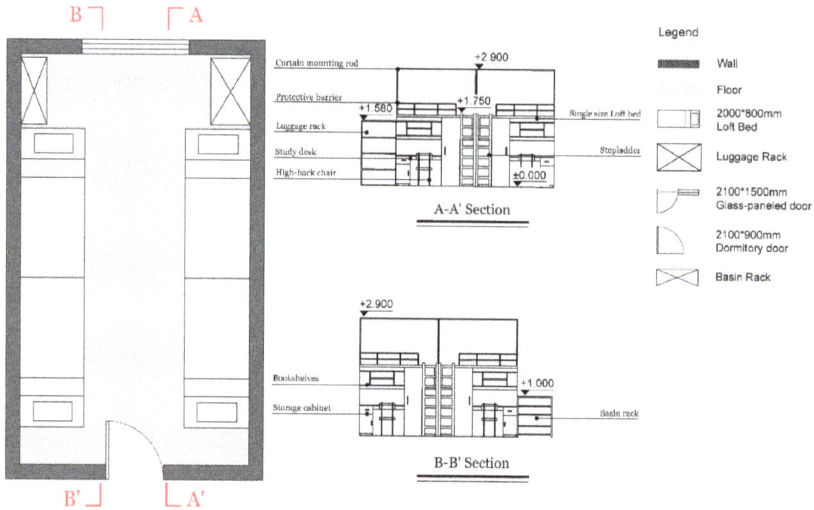

Figure 10. D-07 FDU 4-person dormitory floor plan (type 2): Accommodates four residents with 4.4 m^2 per capita. Features loft beds with integrated desks, sliding windows, and shared storage racks for luggage, shoes, and basins.

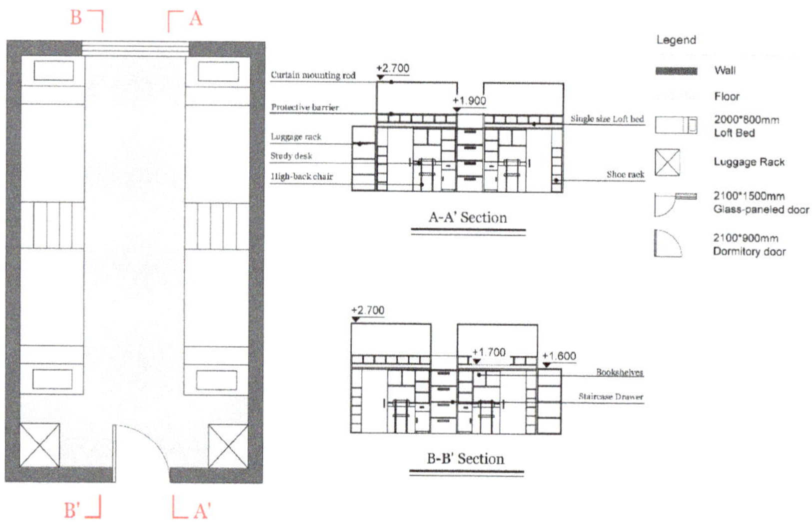

Figure 11. D-08 FDU 4-person dormitory floor plan (type 3): Another four-person room providing 4.5 m^2 per capita. Equipped similarly to D-07 with loft beds and integrated desks, shared stair drawers between beds, and communal storage racks.

Table 6. Chinese national standards, excluding toilets, bathrooms, and balconies [58].

Standards		Type I	Type II	Type III	Type IV
Number of Occupants per Room		1	2	3-4	6-8
Minimum Usable Area (m^2/person)	Floor Bed/Loft Bed	16	8	5	-
	Bunk Bed	-	-	-	4-3
Storage Space	Wall Cabinets, Hanging Cabinets, Bookshelves				

Table 7. Measured per capita usage area evaluation across eight case studies.

Case Number	Measured Area (m^2/Person)	Standard Requirement (m^2/Person)	Compliance
D01	3.96 m^2	≥5	Below
D02	5.28 m^2	≥5	Meets
D03	4.85 m^2	≥5	Below
D04	7.76 m^2	≥8	Below
D05	6.84 m^2 (5 person)	≥5	Meets
D06	3.94 m^2	≥5	Below
D07	4.4 m^2	≥5	Below
D08	4.5 m^2	≥5	Below

3.2.2. User Adaptation

Drawing on publicly available photographs and social media content, student-led modifications and personalization of dormitory spaces can be broadly categorized into three key areas: (i) bed zone upgrades, (ii) seating replacements, and (iii) personalized study zone. These categories reflect common strategies students use to adapt their living environments to better suit individual needs and preferences. Each category is discussed in detail below (Figure 12).

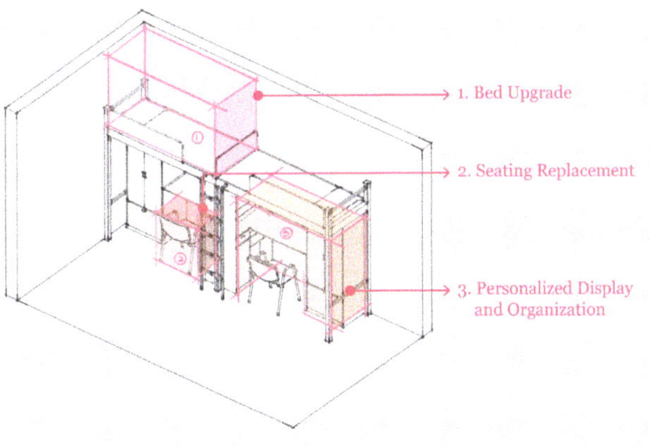

Figure 12. Representation of the three categories defining how students adapt their living environment to suit their needs.

i. Bed zone upgrades

To enhance sleep quality, comfort, and privacy, students frequently modify their bed spaces through targeted upgrades. Common adaptations include the addition of blackout bed curtains, supplementary mattresses or mattress toppers, bedside storage baskets, and small environment-related items such as hanging night lights and portable fans. These modifications aim to create a more comfortable and personalized sleeping environment within constraints of limited space, contributing to both physical comfort and a sense of security (Figure 13).

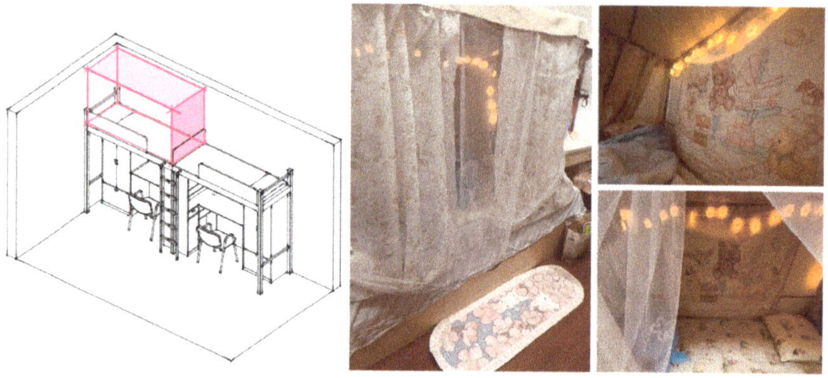

Figure 13. Bed zone modifications for improved comfort and privacy. (Illustration by the authors; photographs sourced from Rednote users with permission.).

Previous research findings have already highlighted the importance of privacy and noise control in bedroom use: 89% of the respondents considered bedroom privacy important, and 62% stated that their bedroom reflected their personal identity [59]. Such findings resonate with the personalization needs observed in these dormitories, where students actively reshape their spaces to assert control and express individuality.

Beyond functional improvements, students also engage in symbolic and aesthetic modifications, such as decorating vertical surfaces with posters, adding plush toys and decorative pillows, and using bed-top study desks. These practices further illustrate the diverse ways students negotiate spatial limitations and personalize their environments to support both emotional well-being and daily routines.

ii. Seating replacement

The default seating provided in Chinese dormitories is a chair with a wooden seat surface and a metal frame. Some students choose to replace the default chairs with more ergonomic office chairs. Others place cushions or pillows on the default seats to reduce discomfort from sitting on hard wooden chairs for extended periods (Figure 14).

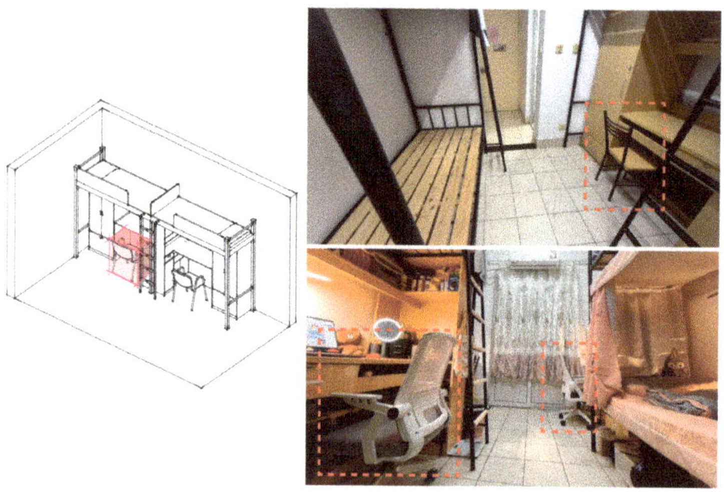

Figure 14. Standard seating and user modifications. (Illustration by the authors; photographs sourced from Rednote users with permission.).

iii. Personalized study zone

Visual analysis of student desk areas reveals students' personalization and adaptation of their study zones. It can be observed that students incorporate a variety of personal items into their desk setups, including lighting fixtures, decorative objects, storage boxes, posters, and photographs. Some students apply wallpaper to alter the original color of the furniture, install LED light strips for ambiance, or repurpose open bookshelves into curated display zones (Figure 15). These modifications reflect both their functional needs, such as improved lighting and organization, as well as expressive intentions, allowing them to assert identity and create a sense of ownership within shared living environments.

3.2.3. User Experience

In the dormitories analyzed, all eight allow basic personalization, with limitations on the use of high-power appliances (for example, D-01, D-02, D-03, and D-04 strictly prohibit students from using such appliances, while the remaining dormitories only allow the use of small appliances with a power rating of 450 watts or less). While personal storage space is generally available, some dormitories (D-04 and D-05) rely entirely on shared wardrobes. Regarding sanitary facilities, all dormitories except D-03 rely on communal toilets and bathhouses. D-03 is unique in offering private toilets, though students must still use the public bathhouse for showers. The analysis of students' posts which were shared on social media platforms provides additional confirmation of this finding.

Figure 15. Personalization of study desks and shelves. (Illustration by the authors; photographs sourced from Rednote users with permission.).

Student feedback shared on the Rednote social media platform details their lived spatial experiences. An analysis of posts related to dormitory design, students' daily use of their living spaces, and room modifications reveal recurring themes and concerns. The posts between 2022 and 2025 were analyzed and some anonymized excerpts from representative posts and comments are quoted to show how students perceive, adapt to, and critique their living environments, all of which have been anonymized.

At one selected university, user discussion intensity significantly increased starting in 2024, with dissatisfaction expressed over the university's decision to convert dorm rooms originally designed for two people into three-person rooms. Other complaints primarily focused on issues such as aging dormitory buildings, poor living conditions, shared bathroom facilities, and unfair dormitory allocation:

> *"This dormitory is hard to accept. The ventilation is very poor, the walls are moldy, and the toilet pipes are old. The ceiling of a dormitory in the same building fell off in half."*
>
> *"The water pressure in the bathrooms is very low, and the environment is poor. I even saw on the university discussion board that a student was bitten on the foot by a centipede while showering."*
>
> *"The room is too small. With narrow corridors, and extremely low ceiling that will hit your head when you sit up on the bed."*
>
> *"The one who came up with the idea of turning two-people dormitory into three-people must be a genius. Moving in feels like my life is over."*

In contrast, another university recently completed a major renovation of its undergraduate dormitories, receiving mostly positive feedback. Online posts generally praise the improved living experience and updated facilities. However, some students raised concerns about inequities between residents living in renovated versus old dormitory spaces, although these comments lack visual evidence:

> *"Overall, the new dormitory is good, with plenty of storage space and luxurious decoration."*
>
> *"Why don't you post about main dormitory, only choosing to post the good-looking ones? You don't mention a thing about mosquitoes, dampness, and years-old stains."*
>
> *"The conditions are really good. I live in the East District, in an old building, and the dormitory isn't this new."*

The primary complaints about this university, as reflected in students' posts, center on insufficient personal space, inadequate storage solutions for personal belongings, and the additional cost required for air conditioning. Practical concerns such as limited areas for drying clothes are also frequently mentioned.

> *"The space is too small to fit student's things, so we can only be messy and happy."*
>
> *"The loft bed are really bad. The space utilization is too poor. It is difficult to put clothes in the closet. The cabinet below is very deep and can't be used at all. The dormitory is small, and I don't even have a place to put a clothes hanger."*
>
> *"The air conditioning is utterly unreasonable; the money each students pays to rent an air conditioner is enough to buy a much better new one. Moreover, most students never get their deposit back."*

To examine which spatial elements students consider most important and to understand their overall perceptions of current dormitory conditions, we conducted a word-frequency analysis of the 61 collected comments and visualized the results using a word cloud using the website Word Cloud Generator (Figure 16). The analysis included synonyms, excluded stop words and irrelevant verbs, and applied a minimum word length of three letters to ensure clarity in interpretation. The words "dormitory" and "dorm" were removed from the analysis since they appeared in all entries and added no meaningful variation.

As shown in Figure 16, the term "small" appears most frequently (n = 9) in students' posts, highlighting the constrained layouts that shape daily activities, often described as "limited space" (n = 7). Storage-related concerns are also prominent, with references to "excessive belongings" (n = 7) and "cabinet" (n = 6). Additionally, issues related to room maintenance and hygiene emerge through words such as "dirty" (n = 4), "messy" (n = 4), and "bad conditions" (n = 4).

Figure 16. Word cloud mapping.

Table 8 summarizes a thematic analysis of 61 posts describing students' dormitory experiences at the two case study universities. Six themes emerged, with comments on spatial layout being the most prevalent (n = 21). Among these, concerns about bed height, desk design, and overall narrowness were mentioned most frequently. Beyond layout issues, recurring problems included hygiene, noise, humidity, and psychological stress.

Table 8. Rednote user comment categories.

Comment Topic	Count of Topic
Uncomfortable spatial layout (bed height, desk design, narrowness)	21
Poor maintenance and hygiene issues	14
Improved conditions after renovation	3
Emotional stress	3
Inequity experience	11
Environmental conditions (noise, humidity, and indoor air quality)	9

3.3. Cross-Analysis of Findings (Comparative Insights)

This section integrates findings from the systemic literature review and case study analysis to provide comparative insights into dormitory design. Literature screening results without location specification indicated that dormitory design research primarily addresses indoor environmental quality (IEQ) topics such as thermal comfort, acoustics, and air quality, which are factors that significantly influence students' experience within their rooms. However, there is a clear need for further investigation given the increasing demand and density of these rooms, as existing studies on user experience are limited (n = 9). A recurring theme among these resources identifies privacy as one of the fundamental needs. Based on a single study, recent research examining users' experience with biophilic design adaptations in student housing indicates a positive impact on students' well-being [41]. Although not specific to dormitory settings, evidence from the broader educational context on implementing biophilic design strategies demonstrates benefits for student health and academic success, highlighting a promising direction for further research [60].

Dormitory research within the Chinese context emphasizes building performance and indoor air quality (n = 24), while spatial characteristics are largely overlooked as key factors shaping user experience. Only two studies address user needs, which is critical for high-density dormitories in cities such as Shanghai. Dong et al. [56] proposed a reference threshold for student satisfaction of approximately 13.5 m^2 per person. Accordingly, most Chinese dormitories fall short of even the national minimum standard of less than 4 m^2 per person. While research generally agrees that two-person or private unit-style dormitories enhance student well-being, the trend in population density is moving in the opposite direction, which indicates an emerging need for further research.

The SLR emphasizes that privacy is a key element of user satisfaction but lacks in-depth empirical analysis. Spatial characteristics observed in case studies reveal that dormitory layouts lack transition zones or buffer areas, resulting in the main living area being directly exposed upon entry. This design severely compromises residents' privacy, a conclusion further supported by social media user feedback, where users explicitly expressed psychological discomfort due to insufficient privacy. Figure 17 presents the two dormitory layouts analyzed as examples of visual exposure based on entrance visibility. Beds located within these zones, particularly near the entrance or in direct line of sight when doors are unexpectedly opened, are more vulnerable to visual privacy concerns. In contrast, beds positioned outside these zones benefit from visual shielding. Such spatial inequalities may influence how students personalize their areas, negotiate boundaries, or experience daily disturbances.

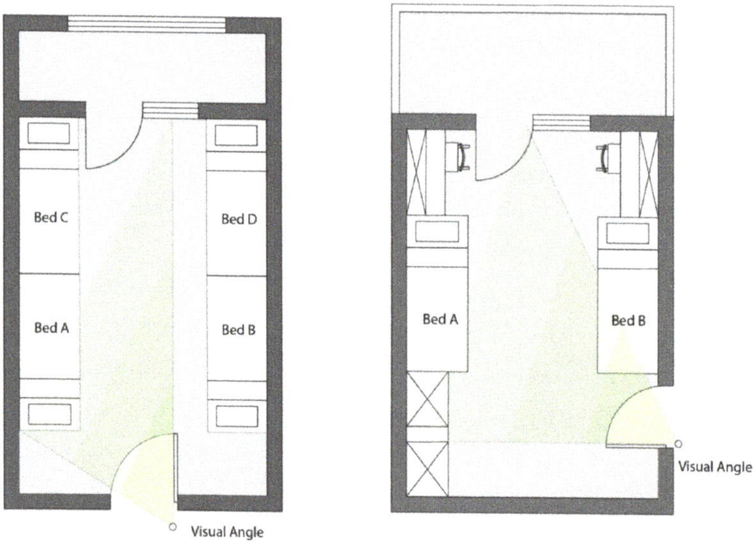

Figure 17. Plan-based analysis of potential visual exposure.

Complementary analyses of social media content further confirm widespread user dissatisfaction, particularly in relation to overcrowding, lack of personal space, and inadequate storage solutions. The growing number of social media posts about dormitory renovations indicate emerging user needs and students' proactive efforts to compensate for functional shortcomings. Common interventions include blackout curtains, custom seating, enhanced storage, decorative elements, and wallpaper applied through personal renovation initiatives. Students' posts also reveal discomfort due to spatial layouts (n = 21) and concerns about poor maintenance and hygiene (n = 14). Cross-analysis results review the correlation between dormitory space layout (including factors such as area, furniture arrangement, and room layout) and student satisfaction [57,58]. The results indicate that satisfaction increases with more spacious, private, and personalized layouts.

4. Conclusions

Existing research on dormitories acknowledges physical environmental needs as well as the importance of spatial adaptability and user-centered design. However, with rising demand for dormitory spaces, current layouts and organizational structures prioritize structural efficiency, which can lead to student discomfort and negatively affect their well-being. In examining the current use of shared dormitory spaces, insights are revealed regarding evolving user needs, as students actively reconfigure their spaces by using bed curtains, additional storage solutions, and personalized accessories. These behavioral patterns indicate that beyond meeting functional requirements, particularly in response to growing demand, dormitory design must also address the emotional, social, and sensory dimensions that shape how students interact with and inhabit their spaces. Through cross-analysis of the systematic literature review results and case studies, three dimensions emerged as critical in shaping students' experience beyond what the current literature addresses:

1. Privacy needs should be embedded within spatial organization, particularly through strategies that enable visual and acoustic separation. Although privacy is widely acknowledged as essential, its physical translation into practice requires more empirical study. The existing literature recognizes insufficient dormitory space and lack of privacy as major concerns; however, practical guidance on addressing these issues through design strategies remains limited. Potential solutions include layouts incorporating entrance buffer zones, visual screening, and acoustic isolation. In most cases, a single circulation route and the absence of transitional buffer zones between the entrance and activity areas result in immediate exposure to the main living space upon entry. While this configuration offers visual transparency, it provides minimal privacy.

2. Equity is an emerging need in dormitory design, particularly in the context of increasing standardization. In shared rooms, this issue becomes more pronounced, as students do not have equal opportunities to personalize or control their immediate space, leading to disparities in comfort and experience. In addition, standardized layouts often lack inclusivity and adaptability, failing to accommodate diverse student needs, such as physical accessibility or cultural preferences, which remain largely unaddressed in current design practices.

3. Personalization should be considered a fundamental need, rather than a discretionary feature, as is reflected in students' active efforts to adapt and customize their living spaces. However, there is a lack of empirical research capturing how students personalize and use these spaces—an important area for understanding evolving user needs. This study provides an initial foundation by illustrating diverse strategies students employ to modify their environments.

This study argues that future dormitory designs must balance structural efficiency with spatial flexibility, enabling students to express their personal identity, control their sensory experience, and establish micro-boundaries within shared environments. Providing flexibility for spatial adaptation, such as designing a bed area that allows small modifications for separation, which is suggested to be part of standardized layouts, is a fundamental need to support students' sense of comfort and privacy. While standardization remains essential for managing density and ensuring affordability, incorporating opportunities for personalization can enhance adaptability and comfort. Simple design interventions, such as pin-up boards behind desks, optional curtains for separation, or varied lighting options, offer flexibility and support students' well-being. Our findings also highlight that bed placement and proximity to entrance doors often create unequal conditions among occupants in shared rooms. Including a transitional space, such as an entrance hallway or spatial separation between the entry and sleeping areas, could provide effective solutions.

Overall, the lack of research on design solutions, combined with the tendency of most dormitory studies to prioritize physical parameters over user experience, underscores the need for further investigation, particularly as demand for dormitories continues to rise. Empirical studies linking quantifiable spatial characteristics (e.g., room type, density, per capita area, furniture modules, circulation areas, and buffer zones) to student experience and performance remain limited. Furthermore, the contemporary needs in student dormitories, conceived as residential environments for university students, offers insights beyond their target group. These spaces function as multi-purpose environments optimized for usability, revealing not only physical efficiency, but also evolving social and psychological needs. A notable trend is the increasing number of studio apartments or micro-flats in densely populated cities. Research indicates that the number of micro-flats in Hong Kong has doubled in recent years [61], reflecting shifting patterns of urban living and an increasing demand for compact spaces. In fact, future research on dormitory environments will provide more insights into how spatial layout and organizational strategies can positively influence people's well-being, particularly within high-density environments.

5. Limitations

There may be omissions in the selected literature due to the inherent limitations of the search strategy; English-language publications were considered, potentially excluding relevant studies published in other languages. While the scope of dormitory-related research is extensive, the review does not account for the diverse needs of specific user groups, such as students with disabilities or mental health conditions.

Regarding empirical data, social media content was sourced primarily from Rednote. While such user-generated content offers valuable first-hand accounts, it remains limited in scope and lacks the depth that an interview-based study could provide. Future research is needed to gain a more comprehensive understanding of users' perspectives. The online data was carefully considered, and only the evidence-based content that aligned with the broader findings and/or supported the other posts was included.

The two case study universities provide a wide range of dormitory types, which vary significantly across campuses, academic disciplines, and student cohorts. Due to the dynamic nature of annual housing allocations, it is difficult to ensure that the dormitory layouts included in this study fully represent the full diversity of student living environments. The spatial documentation was also constrained by the absence of official architectural drawings or publicly available floor plans. As a result, room layouts and dimensions were inferred using standard floor tile measurements. While care was taken to ensure consistency, minor deviations between measured and actual room sizes are likely.

Author Contributions: Conceptualization, X.S. and D.D.; methodology, X.S. and D.D.; software, X.S.; validation, D.D.; formal analysis, X.S.; investigation, X.S. and D.D.; resources, X.S. and D.D.; data curation, X.S. and D.D.; writing—original draft preparation, X.S. and D.D.; writing—review and editing, X.S. and D.D.; visualization, X.S.; supervision, D.D.; project administration, D.D. All authors have read and agreed to the published version of the manuscript.

Funding: This research received no external funding.

Data Availability Statement: The original contributions presented in the study are included in the article, further inquiries can be directed to the corresponding author.

Conflicts of Interest: The authors declare no conflicts of interest.

References

1. Collins, P.Y.; Sinha, M.; Concepcion, T.; Patton, G.; Way, T.; McCay, L.; Mensa-Kwao, A.; Herrman, H.; de Leeuw, E.; Anand, N.; et al. Making cities mental health friendly for adolescents and young adults. *Nature* **2024**, *627*, 137–148. [CrossRef]
2. UNICEF Innovation & ARM. *Innovation for Children in an Urbanizing World: A Use-Case Handbook*; UNICEF: New York, NY, USA, 2017. Available online: https://www.unicef.org/innovation/reports/innovation-children-urbanizing-world (accessed on 22 November 2025).

3. Ministry of Education of the People's Republic of China. List of National Colleges and Universities. 2024. Available online: http://www.moe.gov.cn/jyb_xxgk/s5743/s5744/A03/202406/t20240621_1136990.html (accessed on 14 January 2025).
4. Ministry of Education of the People's Republic of China. National Statistical Communiqué on the Development of Education. 2019. Available online: http://www.moe.gov.cn/jyb_sjzl/sjzl_fztjgb/202005/t20200520_456751.html (accessed on 14 January 2025).
5. Cambridge Academic Content. Dormitory. Available online: https://dictionary.cambridge.org/fr/dictionnaire/anglais/dormitory (accessed on 16 January 2025).
6. Li, Y.; Cheng, F.; Wei, R. Optimization evaluation model of student dormitory design scheme based on multiple evaluation principles. *Expert Syst. Appl.* **2023**, *219*, 119605. [CrossRef]
7. Firmansyah, R.; Shaari, N.; Ismail, S.; Utaberta, N.; Usman, I.M.S. Observation of female dorm privacy in islamic boarding schools in west java, indonesia. *J. Islam. Archit.* **2021**, *6*, 360–368. [CrossRef]
8. Rodger, S.; Johnson, A.M. The impact of residence design on freshman outcomes: Dormitories versus suite-style residences. *Can. J. High. Educ.* **2005**, *35*, 83–99. [CrossRef]
9. Bickman, L.; Teger, A.; Gabriele, T.; McLaughlin, C.; Berger, M.; Sunaday, E. Dormitory Density and Helping Behavior. *Environ. Behav.* **1973**, *5*, 465–490. [CrossRef]
10. Daliri Dizaj, M.; Hatami Khanghahi, T. Students' residential preferences: A case study is dormitories of University of Mohaghegh Ardabili. *J. Asian Archit. Build. Eng.* **2022**, *21*, 1348–1363. [CrossRef]
11. Amole, D. Students housing preferences in Southwestern Nigeria. *J. Archit. Plan. Res.* **2011**, *28*, 44–57. Available online: http://www.jstor.org/stable/43030926 (accessed on 23 December 2024).
12. Kim, W.P. Effects of Physical Living Environment on the Social Interaction and Perceived Sense of Community among Students in University Dormitory. *J. Korean Inst. Educ. Facil.* **2015**, *22*, 3–10. [CrossRef]
13. Astin, A. Impact of dormitory living on students. *Educ. Rec.* **1973**, *54*, 204–210.
14. Astin, A. Student involvement: A developmental theory for higher education. *J. Coll. Stud. Dev.* **1999**, *40*, 518–529.
15. Erden, G.; Özdoğru, A.A.; Çoksan, S.; Ögel-Balaban, H.; Azak, Y.; Altınoğlu-Dikmeer, İ.; Ergül-Topçu, A.; Yasak, Y.; Kıral-Uçar, G.; Oktay, S.; et al. Social contact, academic satisfaction, COVID-19 knowledge, and subjective well-being among students at Turkish universities: A nine-university sample. *Appl. Res. Qual. Life* **2021**, *17*, 2017–2039. [CrossRef]
16. Hountras, P.T.; Brandt, K.R. Relation of student residence to academic performance in college. *J. Educ. Res.* **1970**, *63*, 351–354. [CrossRef]
17. Nowack, K.M.; Hanson, A.L. Academic achievement of freshmen as a function of residence hall housing. *NASPA J.* **1985**, *22*, 22–28. [CrossRef]
18. ICEF Monitor Global Report Finds That Demand for Student Housing Is Still Far Greater than Supply. Available online: https://monitor.icef.com/2025/02/global-report-finds-that-demand-for-student-housing-is-still-far-greater-than-supply/ (accessed on 21 November 2025).
19. Government Portal of the Ministry of Education of the People's Republic of China. List of General Higher Education Institutions in the Country. 2023. Available online: https://hudong.moe.gov.cn/qggxmd/ (accessed on 14 January 2025).
20. China International Engineering Consulting Corporation. Reflections and Recommendations on University Student Dormitory Construction. Available online: https://www.ciecc.com.cn/art/2025/2/27/art_2218_111889.html?utm_source (accessed on 14 January 2025).
21. He, W.; Zeng, N. The Relationship Between University Dormitory Environmental Factors and Students' Informal Learning Experiences: A Case Study of Three Universities in Guangdong Province. *Buildings* **2025**, *15*, 2518. [CrossRef]
22. Ning, Y.; Chen, J. Improving Residential Satisfaction of University Dormitories through Post-Occupancy Evaluation in China: A Socio-Technical System Approach. *Sustainability* **2016**, *8*, 1050. [CrossRef]
23. Beder, D.; Imamoğlu, Ç. Correlates of dormitory satisfaction and differences involving social density and room locations. *J. Hous. Built Environ.* **2023**, *38*, 2307–2323. [CrossRef]
24. Bryman, A. *Social Research Methods*; Oxford University Press: Toronto, ON, Canada, 2016.
25. Creswell, J.W. *A Concise Introduction to Mixed Methods Research*; SAGE publications: Thousand Oaks, CA, USA, 2021.
26. Dawadi, S.; Shrestha, S.; Giri, R.A. Mixed-Methods Research: A Discussion on its types, challenges, and criticisms. *J. Stud. Educ.* **2021**, *2*, 25–36. [CrossRef]
27. PRISMA 2020. Available online: https://www.prisma-statement.org/prisma-2020 (accessed on 14 January 2025).
28. O'Brien, A.M.; Mc Guckin, C. *The Systematic Literature Review Method: Trials and Tribulations of Electronic Database Searching at Doctoral Level*; SAGE Publications: London, UK, 2016. [CrossRef]
29. Ministry of Education of the People's Republic of China. The Action Plan to Education in the 21st Century. Available online: http://www.moe.gov.cn/jyb_xwfb/xw_zt/moe_357/s3580/moe_2448/moe_2470/tnull_41377.html (accessed on 6 May 2025).
30. Mok, K.H.; Jiang, J. Massification of higher education and challenges for graduate employment and social mobility: East Asian experiences and sociological reflections. *Int. J. Educ. Dev.* **2018**, *63*, 44–51. [CrossRef]

31. Times Higher Education University Rankings. Available online: https://www.timeshighereducation.com/world-university-rankings/2024/world-ranking (accessed on 18 April 2025).
32. Chrysikou, E. Psychiatric institutions and the physical environment: Combining medical architecture methodologies and architectural morphology to increase our understanding. *J. Healthc. Eng.* **2019**, *2019*, 4076259. [CrossRef]
33. Wallace, R.; Costello, L.; Devine, A. Netnographic slog. *Int. J. Qual. Methods* **2018**, *17*, 1609406918797796. [CrossRef]
34. Sandlin, J.A. Netnography as a consumer education research tool. *Int. J. Consum. Stud.* **2006**, *31*, 288–294. [CrossRef]
35. Xu, Y.; Wang, Y. The Influence of Chinese Social Media "Xiaohongshu". In Proceedings of the Appearance Anxiety Among University Students in Shanghai International Conference on Science Education and Art Appreciation, Chengdu, China, 24–26 June 2022.
36. Ma, Y. Active User Age Distribution of Xiaohongshu (Rednote) as of March 2025. Available online: https://www.statista.com/statistics/1053545/china-xiaohongshu-user-age-distribution/ (accessed on 23 May 2025).
37. Zou, Y.; Sharpe, A.; Demant, D. The effectiveness of podcasts in promoting health among young people: A scoping review. *Health Promot. Int.* **2025**, *40*, daaf102. [CrossRef] [PubMed]
38. Abdollahzadeh, N.; Farahani, A.V.; Soleimani, K.; Zomorodian, Z.S. Indoor Environmental Quality Improvement of Student Dormitories in Tehran, Iran. *Int. J. Build. Pathol. Adapt.* **2022**, *41*, 258–278. [CrossRef]
39. Adebayo, J.; Amole, D. Gender and Students' Housing Preferences. *Int. J. Constr. Environ.* **2019**, *10*, 27–40. [CrossRef]
40. Kaya, N.; Erkip, F. Satisfaction in a Dormitory Building. *Environ. Behav.* **2001**, *33*, 35–53. [CrossRef]
41. Hasa, A.; Husein, H. Promoting Students' Well-Being Indicators through Adapting Biophilic Design Attributes in Salahaddin University Dormitories. *Al-Qadisiyah J. Eng. Sci.* **2023**, *16*, 180–186. [CrossRef]
42. Nazarpour, M.-T.; Norouzian-Maleki, S. A Comparative Study of Satisfaction Evaluation between Students of Mid-Rise and High-Rise Student Housing. *Facilities* **2020**, *39*, 508–524. [CrossRef]
43. Vráblová, E.; Czafík, M.; Puskár, B. Spatial Characteristics of Contemporary Prefabricated Modular Dormitory Cells. *Spatium* **2022**, 64–74. [CrossRef]
44. Gosling, S.D.; Craik, K.H.; Martin, N.R.; Pryor, M.R. Material Attributes of Personal Living Spaces. *Home Cult.* **2005**, *2*, 51–87. [CrossRef]
45. Awal, S. Public, Private & In-Between Spaces in Student Housing. Master's Thesis, Chulalongkorn University, Bangkok, Thailand, 2020. [CrossRef]
46. Devlin, A.S.; Donovan, S.; Nicolov, A.; Nold, O.; Zandan, G. Residence Hall Architecture and Sense of Community Everything Old Is New Again. *Environ. Behav.* **2008**, *40*, 487–521. [CrossRef]
47. Vinsel, A.; Brown, B.B.; Altman, I.; Foss, C. Privacy Regulation, Territorial Displays, and Effectiveness of Individual Functioning. *J. Personal. Soc. Psychol.* **1980**, *39*, 1104–1115. [CrossRef]
48. Hou, H.C.; Law, K.K. Examining the Relationship between Student Residents' Self-Reported Energy Consumption Behavior and Their Energy-Saving Intention. *J. Build. Eng.* **2024**, *98*, 111490. [CrossRef]
49. Chen, X.; Kang, H.; Zhao, J.; Liu, Q. Optimization Design Research of Architectural Layout and Morphology in Multi-Story Dormitory Areas Based on Wind Environment Analysis. *Buildings* **2025**, *15*, 1747. [CrossRef]
50. Cheng, M.-J.; Hwang, R.-L.; Lin, T.-P. Field Experiments on Thermal Comfort Requirements for Campus Dormitories in Taiwan. *Indoor Built Environ.* **2008**, *17*, 191–202. [CrossRef]
51. Jian, W.; He, H.; Wang, B.; Liu, Z. Investigating the Impact of Green Space Ratio and Layout on Bioaerosol Concentrations in Urban High-Density Areas: A Simulation Study in Beijing, China. *Sustainability* **2024**, *16*, 3688. [CrossRef]
52. Kong, D.; Han, D.; Zhai, R.; Wang, C.; Zhang, J.; Xia, Y.; Nian, X.; Liu, C.; He, Y.; Wang, D. A Case Study on Tropical Bed Bug, Cimex Hemipterus (Hemiptera: Cimicidae) Infestation and Management in Dormitories. *J. Econ. Entomol.* **2022**, *116*, 5–11. [CrossRef] [PubMed]
53. Huang, Y.; Guo, Z.; Chu, H.; Sengupta, R. Evacuation Simulation Implemented by ABM-BIM of Unity in Students' Dormitory Based on Delay Time. *ISPRS Int. J. Geo-Inf.* **2023**, *12*, 160. [CrossRef]
54. Wen, K.; Fang, Y. Daily Information Management System for Postgraduates to Check In and Out of the Dormitory Based on Mobile Edge Computing. *Mob. Inf. Syst.* **2021**, *2021*, 5167395. [CrossRef]
55. Ren, J.; Esangbedo, M.O. Grey Preferences Selection Index with Trimmed Group Preference for Evaluating University Dormitory Renovation Design. *Sci. Rep.* **2024**, *14*, 28218. [CrossRef]
56. Dong, Z.; Zhao, K.; Ren, M.; Ge, J.; Chan, I.Y.S. The impact of space design on occupants' satisfaction with indoor environment in university dormitories. *Build. Environ.* **2022**, *218*, 109143. [CrossRef]
57. Wang, Z.; Wu, Y.; Jia, Z.; Gao, Q.; Gu, Z. Research on health and thermal comfort of unit-type student apartments in the Western China Science and Technology Innovation Harbor. *Front. Public Health* **2022**, *10*, 850107. [CrossRef] [PubMed]
58. *JGJ 36-2005/J408-2005*; Code for Design of Dormitory Building. Ministry of Construction of the People's Republic of China: Beijing, China, 2005. Available online: https://aimg8.dlssyht.cn/u/816266/ueditor/file/409/816266/1723686368312784.pdf (accessed on 22 November 2025).
59. Dincer, D.; Tietz, C.; Dalci, K. Beyond Sleep: Investigating User Needs in Today's Bedrooms. *Buildings* **2024**, *14*, 1061. [CrossRef]

60. Peters, T.; D'Penna, K. Biophilic Design for Restorative University Learning Environments: A Critical Review of Literature and Design Recommendations. *Sustainability* **2020**, *12*, 7064. [CrossRef]
61. Lau, M.H.M.; Wei, X. Housing size and housing market dynamics: The case of micro-flats in Hong Kong. *Land Use Policy* **2018**, *78*, 278–286. [CrossRef]

sustainability

MDPI

Article

Active and Healthy Case Della Comunità: Model Research for Spatial Requirements of Waiting Spaces

Elena Bellini [1,*], Nicoletta Setola [1], Lorena Rossi [2] and Vera Stara [2]

1 TESIS Centre, Department of Architecture, University of Florence, Via San Niccolò 93, 50125 Firenze, Italy; nicoletta.setola@unifi.it
2 Innovative Models for Ageing Care and Technology, IRCCS INRCA, Via S. Margherita 5, 60124 Ancona, Italy; l.rossi@inrca.it (L.R.); v.stara@inrca.it (V.S.)
* Correspondence: elena.bellini@unifi.it

Abstract: The built environment is one of the determinants of health as it acts as a promoter of healthy lifestyles. This research deals with design solutions to promote healthy and active ageing in socio–healthcare facilities for primary care in Italy. This three-year research study aims to develop Design Guidelines for waiting spaces in Casa della Comunità (CdC: House of the Community, a new model of primary care facility in Italy) to promote good health and well-being in sustainable cities and communities. In accordance with these goals, the study applied different research methods in three main phases: Background research, starting from three fundamental groups of theories derived from the scientific literature to define a Theoretical Framework; data collection and field research, dealing with technical analysis of international best practices, as well as perceptive analysis through interviews and questionnaires conducted with the staff, patients, and caregivers of the socio–healthcare facilities, in order to define spatial requirements for waiting spaces; and finally, the results phase, involving the development of tools and design solutions of health-promoting waiting spaces according to the Sustainable Development Goals (the Design Guidelines were applied in two experimental pilots: a VR-based pilot and a physical pilot conducted at an existing CdC in Florence, Italy). In this contribution, we focus on the background and field research phases, describing the process leading to the outcomes of the second phase: spatial requirements for CdC waiting spaces.

Keywords: health promotion; human behaviours; healthy and active ageing; sensory design; social sustainability; primary care

1. Introduction

The Italian health reform DM 77/2022 [1] is promoting the spread of local socio–healthcare facilities on the national territory to support well-being and health promotion in communities. The House of the Community (Case della Comunità, or CdC) is a new model of primary care facility [2] with the aim of treatment, prevention, and health promotion, focusing on chronic diseases and continuous care for the elderly. In this facility, there is an integration between the social and health field, involving social assistants, general practitioners, and nurses and medical specialists who work together to address the needs of complex chronic patients. CdCs are physical places where all people should be equally welcomed and have access to health and social knowledge and assistance. For this reason, the configuration of spaces and the characteristics of the environment are very important and should aim to promote comfort and well-being [3]. CdCs are public healthcare buildings that should be completely "accessible" to represent a "collective resource" [4–6] and promote equity and inclusion.

It is now consolidated in the scientific literature that the built environment influences public health, as it acts as a promoter of healthy lifestyles [7,8] that are represented by performing physical activity [9], having a healthy diet, and being involved in positive social interactions. Healthy lifestyles help to prevent chronic diseases (such as cardiovascular diseases, diabetes, and some types of cancer) and physical and cognitive decline [10]. Thus, actions aimed at supporting fragility and combating loneliness also benefit from the design of a health-promoting and inclusive built environment.

In this process of favouring comfort, well-being, and inclusion, waiting spaces can have a specific and crucial role, using the time of waiting to encourage people to interact with the space to promote health and the social community. Despite a growing understanding of the importance of the built environment for health, there remains limited research on how specific architectural features of waiting spaces can impact well-being in the context of primary care, and there is no evidence in relation to this about the design of Case della Comunità. Moreover, our research offers a contribution of spatial models for both new construction and existing facilities, which represent the majority of Italian healthcare buildings, while also increasing sustainability by promoting the idea of reuse.

For this reason, this research, as part of the Age-It project (Next Generation EU—"Age-It—Ageing well in an ageing society" project (PE0000015), National Recovery and Resilience Plan (NRRP)—PE8—Mission 4, C2, Intervention 1.3) for active and healthy ageing [11], deals with the study of waiting spaces of local socio–healthcare facilities for primary care (CdCs), proposing sensory-based waiting environments which promote health according to the Sustainable Development Goals, as follows: reducing stress before healthcare interventions and favouring relations with professionals; supporting physical and mental well-being and rehabilitation by providing psychological support and promoting physical activity; favouring the prevention of chronic diseases and the promotion of healthy lifestyles. Moreover, in this period of rapid demographic change and increasing demands on healthcare systems, rethinking the role of waiting environments is not only a matter of comfort but also a question of public health and social sustainability.

This research has been conducted for three years in three main phases, as summarised in the diagram in Figure 1: background research, data collection and field research, and results and application (tools and solutions).

Each phase has generated the next one: the background research was the basis for the definition of a Theoretical Framework, a tool to lead the data collection and the field research phase (technical and perceptive analysis), while the field research phase generated case study sheets and space and perceptive requirements, which represent the basis for the development of the Design Guidelines.

The Design Guidelines present strategies and solutions to design CdC waiting spaces. These have then been applied in two pilots:

- A virtual model of a sensory waiting space, built in collaboration with STIIMA CNR (Intelligent Industrial Technologies and Systems for Advanced Manufacturing, Italian national research council), IRCCS INRCA (Innovative Models for Ageing Care and Technology), and UNICATT (Catholic University of the Sacred Heart);
- A physical pilot was implemented by the renovation of waiting spaces of an existing CdC in Florence.

These two pilots are going to be validated within the next year, and feedback will be used for a final revision of the Design Guidelines, to verify the efficacy of this design tool to support professionals.

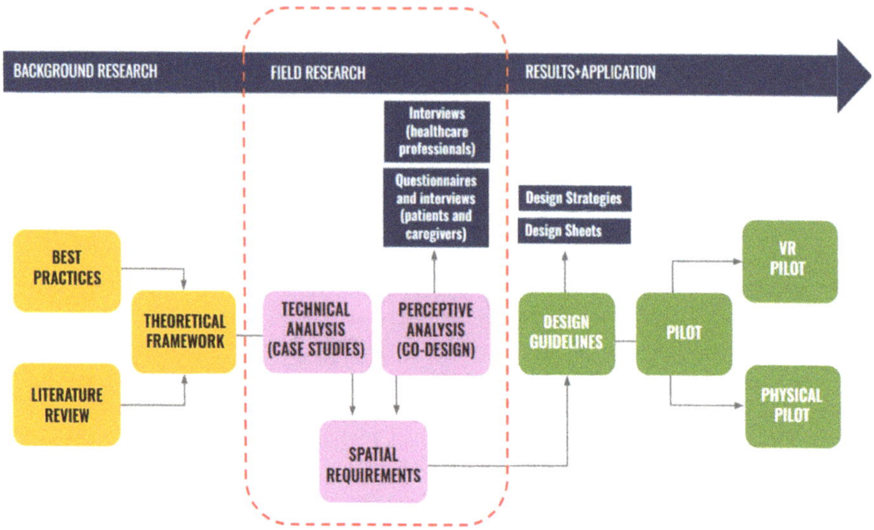

Figure 1. Diagram of the three-year research process. The main object of this paper is represented by the research phase highlighted in dashed red.

In this contribution, we focus on the background and field research phases, going into depth about the processes leading to the outcomes of the second phase: space requirements for CdC waiting spaces. A future publication will focus on the third phase, presenting the Design Guidelines for healthy and active waiting spaces in CdCs.

2. Background: Fundamentals of Research in the Specific Context of CdC

As presented in the introduction, the research aims to support the design process of waiting spaces to achieve the following: (i) reduce patients and caregivers/visitors' stress, in particular during the waiting time, in the phase preceding the healthcare service; (ii) support people's psychological comfort; (iii) promote healthy and active ageing and prevention of chronic diseases (Figure 2).

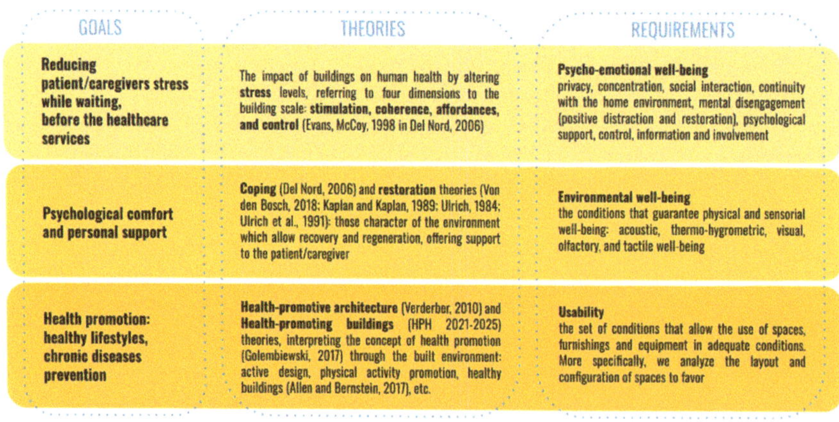

Figure 2. Diagram of the fundamentals of the research [12–21].

These main goals refer to three fundamental groups of theories (Figure 2) we identified from the scientific literature, which represent the main principles of the research.

The first group of theories refers to the concept that the built environment has an impact on people's health by altering stress levels [22–24]. Del Nord 2006 [12] dedicates one chapter to "the stress and the stressors in the hospital" [25], referring to the factors of the physical dimension of the space which contribute to altering people's stress to improve the healthcare experience and affect their health. It refers to four specific dimensions of the environment at the building scale, which can have an impact on stress/health:

- Stimulation refers to creating a balance between sensory stimuli. Overwhelming (e.g., noises, strong lights, etc.) causes stress in people; at the same time, a low stimulation causes sensory deprivation and boredom.
- Coherence refers to configuring a comprehensible and clear building layout to avoid stress;
- Affordance refers to the ability of an environment to induce the right behaviours and functions/activities;
- Control refers to the ability of a person to modify the environment to adapt it to the person's specific needs.

All physical elements such as light, temperature, smells, sounds, colours, textures, etc., act as sensory stimuli for users in the space, engaging their body and senses in interacting with the environment [26]. The information received through our senses shapes our perceptions of the environment and influences our emotions, thoughts, and behaviours [27].

Multi-sensory rooms and equipment deal with stimulation and self-regulation of senses (especially for individuals with vulnerabilities) to generate positive sensations and emotions, reduce stress, promote relaxation and sense of choice, and control acquisition or recovery [28–30]. There are many examples of sensory rooms in healthcare settings, hospitals and care centres, as presented in Bellini, Setola, 2024 [31]. Del Nord, in the chapter "the environment and the perceptive-sensory factors" [32], suggests operative guidelines for the design project to ensure the environment acts as a healing resource that satisfies the physical and psychosocial needs of all the users and contributes positively to healthcare outcomes and caregiving.

Among these salutogenic properties of the environment, the second group of theories refers to those environmental characters that allow patients, caregivers/visitors and the staff to have "restoration" [13–16] (by colours, nature, art, etc.) and to employ "coping" [17] to manage stress conditions. "Restoration" refers to the process of recovering cognitive and emotional energies by interaction with the environment, achieving physical and mental balance [33]. The restoration process can occur when a person is observing or experiencing nature or art to generate well-being and influence human health [34]. The role of art in healthcare settings is widely recognised [35], particularly in reducing stress, pain, suffering and feelings of isolation. Relating to waiting spaces, art can reduce the perception of time passing and the anxiety of waiting, as well as encouraging interaction with the staff [36,37]. The concept of "coping" concerns what a person does (from a cognitive, emotional and behavioural point of view) to face a critical condition. Research has well demonstrated that the environment has an important role in the process of stress/coping. The ability to control the environment and the development of social relations are two main resources of coping [17].

Finally, the third group of theories refers to the concept that the Casa della Comunità has health promotion as its main mission. For this reason, the design of this facility should be based on theories of health-promoting architecture and health promoting buildings [18,19], interpreting the concept of health promotion [20] through the built environment [38], e.g., active design [39], promotion of physical activity, Healthy Buildings approach [21].

Starting from these groups of theories and following the Requirement Classes of the Building System (UNI 8289:1981) [40], this research focuses on the Well-being and Usability classes, also according to the Guidelines for the Humanization of Healthcare Spaces [41] (Figure 2):

- Psycho-emotional well-being and the needs for privacy, concentration, social interaction, continuity with the home environment, mental disengagement (positive distraction and restoration), psychological support, control, information and involvement;
- Environmental well-being, i.e., the conditions that guarantee physical and sensorial well-being: acoustic, thermo-hygrometric, visual, olfactory, and tactile well-being;
- Usability, i.e., the set of conditions that allow the use of spaces, furnishings and equipment in adequate conditions. More specifically, we analyse the layout and configuration of spaces to favour.

These fundamentals (theories and requirements) have been guiding our research and its application in the specific context of Casa della Comunità (CdC), to be able to define Design Guidelines for health-promoting waiting spaces in these facilities. In fact, in Italy, the layout and architectural features of CdC are still not widely explored because it represents a new model of socio-healthcare facility. Therefore, these spaces should be very different from typical waiting spaces of other healthcare buildings, such as hospitals or clinics. There is a lack of specific guidelines about waiting spaces in this new architectural model. In Italy, architects in the healthcare field commonly refer to UK Building Notes as one of the most complete design guideline tools. Considering that CdC model of facility can be assimilated to primary care facilities, we can in part refer to UK Health Building Notes 11-01, 2013 [42] which deals with "Facilities for primary and community care services", giving "best practice" guidance on the design and planning of new healthcare buildings and on the adaptation/extension of existing facilities. However, the notes regarding waiting spaces are quite essential and synthetic. Health Building Notes 00-01, 2014 [43] and 00-03, 2013 [44] provide more details about waiting spaces, but they refer to "General design guidance for healthcare buildings" and "Clinical and clinical support spaces", so not specifically to primary and community care services. Moreover, these documents do not explore the main idea we propose about health promotion through the built environment and during waiting time. There are some studies regarding the idea of health promotion in these spaces, which are more based on health literacy [45] and the divulgation of health information [46–48], not specifically on the primary care context (such as dental clinics) [49,50], and mainly in a passive way [8]. Some studies are promoting the importance of information to promote physical activity in primary care contexts [9,51,52] and using waiting areas and sensory design principles to encourage physical and cognitive training [53], but it is still a field that should be explored in the architectural context and in the related integrated technologies.

In Italy, we could not identify complete and effective tools for professionals to drive the design of CdC buildings and, more specifically, CdC waiting spaces. There are some Italian documents which aim to describe the context of the CdC, defining this new model and giving some indications to design CdC facilities, as follows. However, as mentioned before, the promotion of health through these spaces is not explored.

Brambilla and Maciocco, 2022 [2] introduce the process of transformation of Case della Salute in Casa della Comunità, after the Italian health reform DM 77/2022 [1], but this does not represent an operational tool for professionals to use in designing CdC buildings (nor waiting spaces). Prima la Comunità, 2021 [54] published a document which is a sort of identity card of the CdCs, expressing aims and mission, stakeholders involved, and governance, roles and organisational systems/management, but it does not refer to the architectural features of CdC spaces.

Agenas, 2022 [55] presents the "Directional document for the meta-design project of the Casa della Comunità", referring to the aims and strategies of the project of a CdC, presenting its functional areas and some general references to the architectural and technological features. Although waiting spaces are mentioned in the document, it does not present specific requirements for these spaces. Peretti and Torricelli, 2023 [56] is a document about "Houses and Hospitals of the Community", also presenting in that case the context and then some operative design strategies for the CdC, identifying five main architectural topics to be provided in a CdC, which should be "a place with meaning, flexible and adaptable, integrated into the local area, which respects the environment and puts people and quality care at the centre". However, waiting spaces are not mentioned in the document. Finally, Quintelli et al., 2025 [57] proposes a new architectural and urban model called a "Community Health Center" as an evolution of the Casa della Comunità and identifies waiting spaces as one of the "components of the typological device" to be designed in the compositional framework. Therefore, this publication proposes some layouts of waiting spaces but it does not go into detail about all the aspects dealing with environmental characters of the spaces and does not refer to the idea of health promotion.

In conclusion, this research could make a contribution in this field, aiming to fill this gap in design guideline tools. To be able to develop Design Guidelines, it is necessary to define spatial requirements for CdC waiting space, which are the object of this paper.

3. Methodology

The next sections will describe in detail the methodology and tools that have supported the development of space requirements: (i) Theoretical Framework; (ii) Technical Analysis; (iii) Perceptive Analysis; (iv) Integrated Requirements.

These methods and the related tools can be applied to different contexts, starting from background theories that are valid for every typology of healthcare buildings, aiming at people's well-being, physical and psychological comfort, and health promotion. Emphasising that the research starts from these universal principles (e.g., the impact of the built environment on stress levels or the use of Sensory Design) makes the approach robust and transferable at a conceptual level. When the context changes, it will be necessary to adapt the technical analysis by selecting best practices to be aligned with the specific context of the new research (see Section 3.2.1), but the method used for the analysis (Case-based design/reasoning—CBR, see Section 3.2) will still be valid. Regarding the perceptive analysis, it will be necessary to adapt questions in the case of a specific target group of users in the healthcare facility being analysed.

3.1. The Theoretical Framework

As anticipated in the background section (see Section 2), the Background research identified three fundamental groups of theories derived from the scientific literature. These theories represent the core principles of the research and were used to establish and define a Theoretical Framework (Figure 3). This framework serves as an operative tool to guide the data collection and subsequent analysis of best practices, enabling the definition of spatial requirements for health-promoting waiting spaces. This Theoretical Framework is based on three main areas of investigation: (i) Territorial primary care and CdC, (ii) Waiting Areas, (iii) Sensory Design.

First of all, we focused on local primary care facilities, such as Case della Comunità/Case della Salute. As anticipated in the first section, it was not possible to find operative documents, such as Design Guidelines, in this field, as it represents a new model of healthcare facility. Moreover, there is little research in the area of health promotion in these spaces, particularly in the waiting areas. For this reason, it was necessary to broaden our research, also examining Primary Care facilities and Community Centres to identify existing design solutions in waiting spaces for health promotion, consistent with the WHO (World Health Organisation)'s idea of age-friendly and healthy cities [58]. Moreover, we decided to expand this research, looking at good examples (best practices) of waiting rooms in healthcare facilities for outpatient departments, medical and healthcare centres. Finally, Sensory Design was identified as a suitable approach to solve both the goals of comfort and health promotion by obtaining relaxing, homely and customisable environments for different types of users [31]. Therefore, we decided to explore this field, collecting sensory solutions found in welcome and waiting spaces of both outpatient departments and primary care facilities. The use of Sensory Design has been investigated in the literature across a wide range of settings, such as health and education [31], especially concerning mental health, managing distress and anxiety. There are also many examples of sensory rooms in healthcare settings, hospitals and care centres [59]. Conversely, there is a lack of similar experiences in the area of primary care, so we decided to spread the research in other contexts, in which sensory environments are commonly used, such as children's hospitals, geriatric facilities, physiotherapy, etc., to collect effective solutions to be applied to CdC settings.

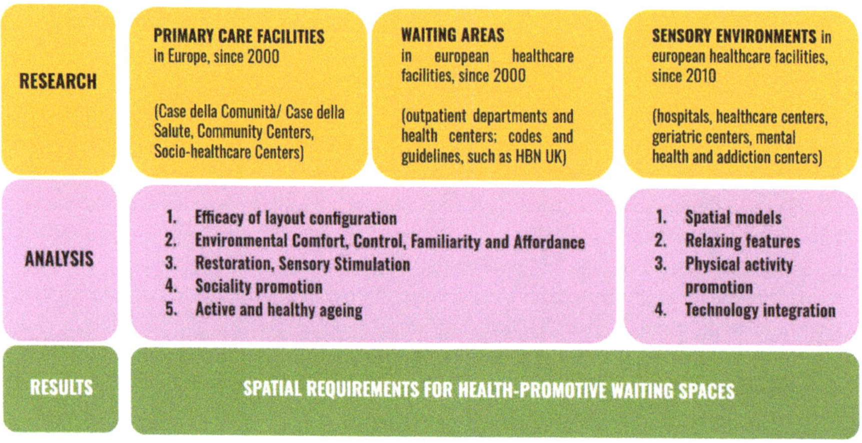

Figure 3. Diagram of the Theoretical Framework.

In these areas of investigation, starting from the background research, we were able to identify different categories of collection and analysis (see Section 3.2.2 for description). For the first two areas:

- Efficacy of Layout Configuration [12,22–25,60];
- Environmental Comfort, Control, Familiarity and Affordance [12,22–25];
- Restoration, Sensory Stimulation [13–17,26–37];
- Sociality Promotion [18–20,38];
- Active and Healthy Ageing [18–21,38,39].

For the third area of investigation (Sensory Design), as described in Bellini, Setola, 2024 [31]:

- Space Models;
- Relaxation Strategies;
- Movement Promotion;
- Technology Integration.

3.2. Technical Analysis

We identified a gap in the literature regarding guidelines for the design of CdC and the promotion of health through waiting spaces. Starting from case-based design [61] theory, the authors analysed relevant European case studies to derive innovative solutions to be applied in the design of waiting spaces in the primary care context and to address these gaps through the application of significant technical-design solutions in this specific context:

- Identification of invariants that are common in case studies (welcome and waiting spaces of various types of social and healthcare facilities, e.g., hospital, clinic, health centre, etc.) and that can be translated into input for the project;
- Innovative ideas related to other contexts or different users, to be reinterpreted and applied to the project. To explore innovation, Sensory Design was identified as a suitable approach to both provide comfort and promote health in waiting spaces, by creating relaxing, welcoming and customizable environments for different users through the integration of digital technologies.

Case-based reasoning (CBR) is a cognitive model developed in the fields of artificial intelligence and cognitive psychology, which has given rise to a theory of how we reason in problem solving [61]. In accordance with this model, to solve or understand a new problem, we can learn from existing good solutions to find answers to a similar problem. The goal is to capture what is common among a variety of similar solutions and formulate rules that are generally applicable [62]. Case-based design is the application of CBR to the design process, using elements of previous design solutions in developing new design solutions [63]. "Cases" are represented by existing architectures or design solutions used as sources of knowledge.

3.2.1. Selection Criteria

For the first and the second area of investigation, best practices were chosen among healthcare buildings with innovative design solutions and characteristics in at least two or three categories described in paragraph Section 3.2.2, so much so that they could be defined as "Best Practices" (Figure 4). In particular, they were selected among hospitals with an outpatient department, outpatient clinics, health houses, or local primary care facilities that presented interesting or innovative solutions for welcome and waiting spaces. European buildings were selected in order to evaluate examples close in size and social and cultural context, designed and built since the 2000s. Specialised healthcare facilities were excluded, such as emergency departments, cancer centres, paediatric hospitals, dental clinics, etc., as they would represent different solutions for requirements and context. The research was carried out through literature review and online search engines on dedicated portals (e.g., google, archdaily, healthcare design, pinterest, architect magazine, architecture for health, dezeen, etc.) using some keywords: hospital AND outpatient AND waiting spaces; european AND hospital AND outpatient AND waiting spaces; (different countries) hospital AND outpatient AND waiting spaces; outpatient department AND elderly; house of community; community hospital; community centre; primary care; medical centre; healthcare architecture AND waiting spaces; healthcare architecture AND elderly.

For the third area, the authors collected Sensory Design examples in European healthcare facilities, built since 2010, according to the criteria presented in Bellini and Setola, 2024 [31].

In total, 37 case studies were selected: 20 for the first and second areas, as summarised in Table 1, and 17 for the sensory area [31].

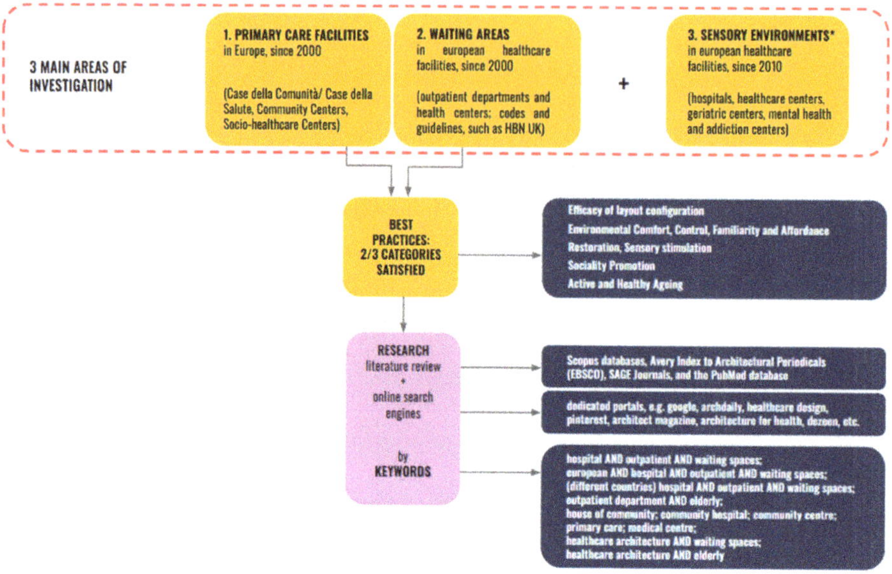

Figure 4. Chart of case studies investigation. * Results of sensory environments investigation have been illustrated in Bellini, Setola, 2024 [31].

Table 1. Case studies analysed in the technical analysis phase: waiting spaces of European healthcare facilities, built since the 2000s.

	Case Study (Web Sites Accessed on 30 July 2025)	Place	Year	Designer	Area (mq)
a1	Tergooi Medical Centre https://www.archdaily.com/1001687/tergooi-medical-center-wiegerinck	Hilversum, The Netherlands	2023	Wiegerinck, Arnhem, The Netherlands	55.000
a2	Dijklander Hospital https://www.archdaily.com/995948/dijlander-hospital-revitalization-bureau-ira-koers-and-plus-studio-roelof-mulder?ad_medium=gallery	Purmerend, The Netherlands	2022	Studio Roelof Mulder and Ira Koers, Amsterdam, The Netherlands	1.780
a3	Bijgaardehof Co-Housing and Healthcare Centre https://www.archdaily.com/988277/bijgaardehof-co-housing-and-healthcare-center-bogdan-and-van-broeck	Ghent, Belgium	2022	Bogdan & Van Broeck, Brussels, Belgium	9.375
a4	Health Municipal Clinic in Liffol-Le-Grand https://www.archdaily.com/1006257/health-municipal-clinic-in-liffol-le-grand-studiolada	Liffol-Le-Grand, France	2021	Studiolada, Nancy, France	615
a5	CAP Riells i Viabrea https://archello.com/project/cap-riells-i-viabrea	Riells i Viabrea, Spain	2021	Comas-pont Arquitectes, Barcelona, Spain	566
a6	Steno Diabetes Center Copenhagen https://www.archdaily.com/1006085/steno-diabetes-center-copenhagen-vilhelm-lauritzen-architects-plus-mikkelsen-architects-plus-sted	Copenhagen, Denmark	2021	Vilhelm Lauritzen Arch. + Mikkelsen Arch. + STED, Copenhagen, Denmark	18.200

Table 1. *Cont.*

	Case Study (Web Sites Accessed on 30 July 2025)	Place	Year	Designer	Area (mq)
a7	Taverny Medical Centre https://www.archdaily.com/932080/taverny-medical-center-maaj-architectes	Taverny, France	2020	Maaj Architectes, Paris, France	1.095
a8	Hospital Nova https://www.theplan.it/architettura/hospital-nova-l-ospedale-del-futuro-che-pensa-fuori-dagli-schemi	Jyvaskyla, Finland	2020	JKMM Architects, Helsinki, Finland	116.000
a9	Health Centre at Gibraleón https://www.archdaily.com/956493/health-center-at-gibraleon-javier-terrados-estudio-de-arquitectura	Gibraleon, Spain	2020	Javier Terrados Estudio de Arquitectura, Sevilla, Spain	
a10	Lindesberg health centre https://whitearkitekter.com/project/lindesberg-health-centre/	Lindesberg, Sweden	2020	White Architects, Gothenburg, Sweden	
a11	Health Municipal Clinic In Audun-Le-Roman https://www.archdaily.com/1000868/health-municipal-clinic-in-audun-le-roman-studiolada	Audun-Le-Roman, France	2019	Studiolada, Nancy, France	
a12	The Jean Bishop Care Centre https://www.archdaily.com/909641/the-jean-bishop-integrated-care-centre-medical-architecture	Kingston upon Hull, UK	2018	Medical Architecture, London, UK	2.761
a13	Luz Saúde Vila Real Hospital https://www.archdaily.com.br/br/945349/hospital-luz-saude-vila-real-openbook-architecture?ad_medium=gallery	Villa Real, Portugal	2018	OPENBOOK Architecture, Lisboa, Portugal	6.800
a14	Zaans Medical Centre https://www.archdaily.com/874330/zaans-medical-centre-mecanoo	Zaandam, The Netherlands	2016	Mecanoo, Delft, The Netherlands	38.500
a15	Närsjukhus in Angered https://www.sweco.se/projekt/angereds-narsjukhus/	Gothenburg, Sweden	2015	Sweco, Stockholm, Sweden	
a16	Porreres Medical Centre https://www.archdaily.com/189622/porreres-medical-center-maca-estudio	Porreres, Majorca, Spain	2011	MACA Estudio, Guadalajara, Jalisco, Mexico	900
a17	Orbis (now Zuyderland) Medical Centre https://dezwartehond.nl/en/projecten/orbis-zuyderland-medisch-centrum/	Sittard-Gele, The Netherlands	2010	De Zwarte Hond, Groningen, Rotterdam, Cologne and Berlin	73.850
a18	Kentish Town Health Centre https://modulo.net/it/realizzazioni/kentish-town-health-centre	London, UK	2009	Allford Hall Monaghan Morris, Lndon, UK	
a19	Centros Municipales de Salud https://www.archdaily.cl/cl/02-208005/3-centros-municipales-de-salud-en-madrid-san-blas-usera-villaverde-estudio-entresitio	Madrid, Spain	2008–2010	Estudio Entresitio, Madrid, Spain	5.430
a20	Heart of Hounslow Polyclinic https://commercialpropertyphotography.com/health-care/heart-of-hounslow-medical-practice/	London, UK	2007	Penoyre & Prasad (now Perkins&Will), London, UK	8.800

Through the technical analysis, case studies were summarised in 37 Case Study Sheets, which illustrated the contribution of each best practice in the categories defined by the Theoretical Framework in order to highlight the best solutions in designing waiting spaces for application to the CdC context.

3.2.2. Observation Criteria

Aiming at the research objectives, case studies in the first and second areas of investigation were analysed on the basis of the five categories:

- Efficacy of Layout Configuration: solutions to configure welcome and waiting areas to optimise the use and quality of healthcare environments. In this category, the authors evaluated the type of spaces, their position and relation with the healthcare services, their integration and connection with the entrance, the configuration of the distribution areas and public spaces, and the clarity and comprehensibility of the routes;
- Environmental Comfort, Control, Familiarity and Affordance: the effective use of colours and materials to promote comfort and a sense of familiarity, recognizability, and welcoming. The authors analysed design solutions for light (natural and artificial), ventilation and all the aspects that impact the environmental well-being; furnishing solutions to promote the sense of welcoming and homeliness; the ability of control by the patients/caregivers/visitors; the affordance capabilities of the proposed furnishings;
- Restoration, Sensory Stimulation: strategies for support and recovery through art, nature or other regenerative elements; the variety of sensory stimulations and the possible regulation of stimuli;
- Sociality Promotion: promotion of a functional mix and design solutions to foster relationships and a sense of community;
- Active and Healthy Ageing: strategies and design solutions to promote active design (e.g., through the use of stairs), healthy food, Healthy Buildings construction, etc.

Figure 5 represents an example of one of the Case Study Sheets in the first and second areas of investigation.

As regards the Sensory Design area of investigation, the following analysis criteria were followed:

- Space Models: Designing the building according to a multisensory approach; sensory equipment diffused in public spaces; sensory atrium or sensory waiting spaces; snoezelen rooms; portable sensory equipment;
- Relaxation Strategies: specific elements of the project which represent targeted solutions to reduce stress, promote psychological comfort and support restoration;
- Movement Promotion: solutions to favour the interaction between the person and the space (or the object) and encourage movement and physical activity, in accordance with the idea of active and healthy waiting time;
- Technology Integration: devices and automation technologies which promote the control of space by the person, the regulation of sensory stimuli, and the interaction between the person and the environment. The authors also analysed the integration of these elements into the environment through the architectural project.

Figure 6 represents an example of one of the Case Study Sheets in the Sensory Design area of investigation.

Steno Diabetes Center Copenhagen

a06

Place: Copenhagen, Denmark
Year: 2021
Designers: Vilhelm Lauritzen Architects + Mikkelsen Architects + STED
Area: 18.200 mq

The Steno Diabetes Center Copenhagen (SDCC) is the largest center in Northern Europe for the prevention and treatment of diabetes. It is located at Herlev Hospital and specializes in the treatment, research, and prevention of diabetes, as well as the training of healthcare professionals in this field. It treats 13,000 diabetic patients annually. It redefines the way in which we perceive health, as it is founded upon the concept that architecture and nature can contribute not only to treatment, but also to prevention and education.
Designed in collaboration with users, their involvement was crucial in creating a pleasant experience across all phases, from arrival and waiting to the consultation.

Sources: https://architectures.jidipi.com/j00091342/en/steno-diabetes-center-copenhagen, https://www.archdaily.com/1006085/steno-diabetes-center-copenhagen-vilhelm-lauritzen-architects-plus-mikkelsen-architects-plus-sted, https://architizer.com/projects/steno-diabetes-center-copenhagen/, https://www.arch2o.com/steno-diabetes-center-copenhagen-vilhelm-lauritzen-architects-mikkelsen-architects/#google_vignette, https://vilhelmlauritzen.com/project/steno-diabetes-center-copenhagen, https://hudevad.com/en/cases/steno-diabetes-center-copenhagen/, https://www.dezeen.com/awards/2023/longlists/steno-diabetes-center-copenhagen-sdcc/, https://en.mikkelsengroup.dk/nyheder/sdcc

Photos: © Sjavit Maestro, Rasmus Hjortshøj

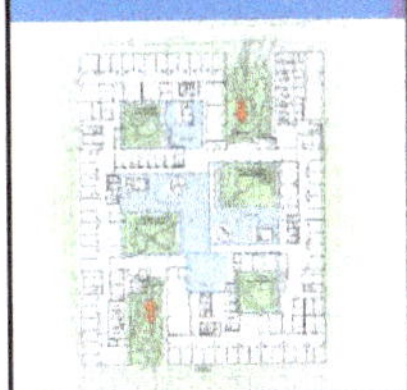

The main entrance is south-facing, thereby welcoming visitors into a warm, illuminated space that is integrated with the surrounding landscape and leads into the interior.
The common areas and treatment facilities are situated on the ground floor, establishing a simple and intuitive connection between arrival at the center and the primary patient areas. Conversely, the research areas, along with associated laboratories, and the inpatient hospitalization areas are located on the first floor, where the access is provided via an open staircase. From the common areas, movement proceeds gradually toward the treatment areas, which are organized around dedicated spaces and courtyards. These areas are interspersed with a series of waiting areas, silence rooms, libraries, and desks. The wards on the first floor are surrounded by spacious common areas, all oriented toward the large central garden, where patients and relatives may relax, dine, and spend time together. Access to the staff areas is provided by open staircases that lead to spacious social zones. These zones can be used daily for informal meetings, social gatherings, and coffee breaks. From these areas, there is easy access to the office facilities. The offices are arranged in enclosed spaces along the building's exterior façade, as well as in open-plan areas designated for teamwork or informal meetings. All these areas are oriented toward the central garden.

Efficacy of layout configuration

It is a hospital that, by virtue of its size, the materials used, and the atmosphere it creates at all levels, places the human being at the center. It seeks a familiar atmosphere, as far removed as possible from the perception of a healthcare facility, also by prioritizing natural materials: wooden floors, wooden ceilings, and green courtyards.
The area is designed to arouse curiosity: from the outset, patients and visitors should feel welcome and encouraged to explore. In-situ poured concrete cladding creates a beautiful path leading to the center, while a large landscaped staircase invites access to the public rooftop garden. The main entrance faces south to ensure the best conditions for a bright and sunny welcome.
The common areas possess a typically Nordic atmosphere, featuring light wooden slat floors and ceilings.
Nature is perceived throughout the building as a central element of the structure. The change of seasons creates a beautiful variation throughout the year. In winter, the snow contrasts with the warm and welcoming character of the building; in summer, the changing interplay of colors in the vegetation lends the building a rich and varied character.

Env. Comfort, Control, Familiarity, Affordance

Drawing upon the biophilic approach, the omnipresence of nature constitutes a key point of the SDCC architecture. This omnipresence serves as an integral component of treatment and the promotion of a healthy lifestyle.
The SDCC building and its garden are closely linked, establishing a symbiosis between the interior and the exterior. This structure is organized around an expansive two-story garden, which incorporates six smaller, lush courtyards and a public rooftop garden.
This architectural environment is further enhanced by the variety of the seasons, which ensures that the landscape remains consistently diverse and vital. The garden functions as a center for biodiversity, maintaining accessibility to all individuals twenty-four hours a day, seven days a week.
Ultimately, this symbiosis between the landscape and the building facilitates the creation of an architectural environment designed to stimulate and nourish both patients and visitors, thereby encouraging exploration and movement.

Restoration, Sensory Stimulation

The communal areas are designed for patients, relatives, and staff.
The activities within the zone are organized around smaller squares, each of which frames the center's core themes. For instance, a food laboratory and cafeteria are located at the "Nutrition Square". A library and exhibitions are situated at the "Knowledge Square". Exercise and training rooms are found at the "Fitness Square", and workshops are held at the "Exhibition Square". Additionally, these areas feature public computer stations, comfortable lounges and reading niches, and designated children's areas positioned near the main entrance.

Sociality Promotion

The project re-envisions the function of communal spaces, converting waiting time into active time and supporting a flow of activities centered around the themes of nutrition, physical activity, and new knowledge. Activities within the zone are organized around smaller squares, each of which frames the central themes of the center: the "Nutrition Square," the "Knowledge Square," the "Fitness Square," and the "Exhibition Square".
In accordance with the biophilic approach, the omnipresence of nature constitutes a key point of the SDCC's architecture and serves as an integral component of treatment and the promotion of a healthy lifestyle. The landscape guides visitors inward and, via a pathway constructed with cast-in-situ concrete flooring, naturally directs them toward the grand staircase, which invites visitors to ascend to the public rooftop garden. This symbiosis between the landscape and the building facilitates the creation of an architectural environment that stimulates and nourishes patients and visitors, thereby encouraging exploration and movement.

Active and Healthy Ageing

Figure 5. Case Study Sheet a06.

Sight&Sound Centre, Great Ormond Street Hospital (GOSH)

b07

Place: London, UK
Year: 2021
Designers: Art in Site

Inaugurated in June 2021, the new GOSH Sight and Sound Centre is transforming the experience and care of children and young people from across the United Kingdom, through a bespoke multidisciplinary centre specifically designed for the sensory needs of these children. As the first of its kind for children in the United Kingdom, the centre features state-of-the-art soundproofed booths for hearing tests, an ophthalmic imaging suite, an optical centre, and other testing facilities.
The Sight and Sound Centre has been designed down to the smallest detail, from the artwork to the colour schemes, ensuring that it does not resemble a hospital clinic but rather a welcoming place of well-being. Children with sight and hearing problems and their families were involved in the design brief for this specialized facility, thereby helping to create an environment that is welcoming, fully accessible, and easy to navigate.

Sources: https://www.gosh.r hs.uk/news/welcome-to-our-new-sight-and-sound-centre/, http://www. artinsite.co.uk/projects/sight-and-sound-centre, https://www.youtube.com/watc h?v=Z5ufNZ_3XI8&t=115s

Space Models

The Sight and Sound Centre was designed so that multisensory design is an intrinsic part of the spaces, addressing children with sight and hearing difficulties and their inclusion, starting with collaborative design alongside their families.
The new Sight & Sound Centre at Great Ormond Street incorporates illustrations, signage, branding, and interactive multisensory artwork, thereby assisting patients with hearing and sight difficulties to navigate and move through the space via play.
Furthermore, artists collaborated with the hospital's art programme, GOSH Arts, placing specially commissioned artworks in the centre, designed with all the senses in mind, so that children can see, touch, and smell them.
The centre also features a sensory garden with plants that children can see, touch, smell, and listen to, and with which they can talk, in addition to interacting with the artworks there.

Relaxation Strategies

At the new Great Ormond Street Sight & Sound Centre, illustrations, signage, branding, and interactive multisensory artworks have been integrated, primarily to facilitate orientation and movement through playful engagement for patients experiencing hearing and vision impairments. In this context, "dolls' houses" were specifically designed to welcome families within the main waiting area. Children are able to "meet" an illustrated cohort of clinicians and patients in engaging, animated scenarios by interacting with the house's various windows. Upon activation of a button, a window illuminates, and recorded voices deliver reassuring statements aloud. For instance, a recording states: "I was scared of eye drops, but they aren't so bad once you've had them!". Significantly, all 27 windows feature recordings derived from real patients, who were interviewed via Zoom during the Covid period (2020).
Interactive play allows children to virtually explore various parts of the hospital, thereby enhancing enjoyment, providing insight into the centre's operations, and simultaneously mitigating patient anxiety. A further benefit is that the dolls' house instills a vital sense of occupation and human presence within a large facility where patients might not ordinarily interact. Subsequently, throughout other areas of the centre, the same cohort of illustrated characters appears on large vinyl wall surfaces, providing words of encouragement and directional guidance presented in speech bubbles.
The illustrated cohort is intentionally characterized by gender neutrality and encompasses a diverse range of ages and abilities. Representations include individuals with cochlear implants, glasses, varied skin tones, and fluctuating heights. This deliberate design ensures that any patient entering the facility feels included, as they are able to observe representations of themselves celebrated as integral components of the community. Furthermore, the artists collaborated with the hospital's arts programme, GOSH Arts, to install specially commissioned artworks designed with all senses in mind, allowing children the opportunity to see, touch, and smell the pieces.
The centre also incorporates a sensory garden featuring plants that children can perceive visually, tactually, and olfactorily. Children are also encouraged to interact aurally (listening to and talking to the plants), alongside further opportunities for interaction with artwork situated within the garden.

Movement Promotion

The orientation and navigation through the building's spaces are facilitated by the presence of the illustrated gang and various interactive artworks, which prompt movement through play and discovery. Lighting elements are also integrated to highlight the graphical elements and the pathways.
Furthermore, in addition to the graphics, multisensory tactile panels are positioned along the corridors to promote play and interaction.
The staircases are additionally defined by a sound artistic installation, which encourages their utilization by stimulating users to traverse them.
Finally, the wall installation of the dollhouse motivates individuals to rise during periods of waiting and interact playfully with the gang characters, whom they may subsequently encounter during their "journey" throughout the facility.

Technology Integration

The "dollhouse" has been designed to accommodate families in the main waiting area. Children can "meet" an illustrated group of doctors and patients in appealing animated scenarios, interacting with the various windows of the house. Upon pressing a button, a window illuminates, and voices sound audibly, offering reassuring words. All 27 windows feature recordings of actual patients, who were interviewed via Zoom calls during the Covid period (2020).

Figure 6. Case Study Sheet b07.

3.3. Perceptive Analysis

A semi-structured interview model was submitted to healthcare professionals (40 persons among doctors, nurses, management and administration) of the relevant healthcare facilities to collect data on the use and perception of the waiting spaces. The interviews were conducted in INRCA facilities in Ancona, Casa della Comunità Le Piagge (Florence), Casa della Comunità Morgagni (Florence), Department of Multidimensional Medicine of Florence.

Then, a semi-structured interview model was submitted to patients (41 persons) and caregivers (26 persons), at the same facilities where the interviews with healthcare professionals were carried out, obtaining qualitative results to be integrated into the Design Guidelines.

The interviews posed questions about three main topics: the spatial configuration; the presence of sensory elements that could improve relaxation or mental distraction (e.g., music, projections, coloured lights, etc.); and health promotion (e.g., having information on health, doing physical activity, etc.).

During the interviews, pictures of different waiting rooms (Table 2) selected in the technical analysis were shown to respondents to support questions about the space and facilitate the discussion. Pictures were selected to respond to the related question of the interview to support people's comprehension; the criteria of selection and the composition of the interview will be described in a future publication we are working on, focusing only on the definition of perceptive requirements.

Table 2. List of images used to discuss with patients and caregivers during the interviews.

	Case Study	Typology of Space	Web Sources (Accessed on 30 July 2025)
im01	Zaans Medical Centre	Waiting area near the medical rooms. This is an example of a waiting space where people can see the doors of the medical room during waiting time. Colours are vivid, artworks are present on the walls, and the furniture is coloured, modern and homey.	https://www.archdaily.com/874330/zaans-medical-centre-mecanoo/594ce485b22e38e9290004a9-zaans-medical-centre-mecanoo-photo?next_project=no
im02	Luz Saúde Vila Real Hospital	Waiting room filtered by a timber frame screen. The colours of the room are warm and the lights are soft.	https://www.inap2.com/kr-sq/gallery/gallery_e951861bd33b87c365b190e678d03048/
im03	Porreres Medical Centre	Waiting room on the first floor. It is an example of the model "waiting in a separate room". All the walls of the room are yellow, the seats are simple, and at the end of the room there is a big window looking outside in an urban context.	https://www.archdaily.com/189622/porreres-medical-center-maca-estudio/5016efad28ba0d235b00053b-porreres-medical-center-maca-estudio-photo
im04	Tergooi Medical Centre	Waiting room next to a green courtyard. It is a good example of favouring relations with comfortable and homey furniture. Sofas are oriented in different directions to let people look at each other or outside at nature. The walls are white and clean.	https://www.archdaily.com/1001687/tergooi-medical-center-wiegerinck/6475318820e8d30376dd6c9e-tergooi-medical-center-wiegerinck-photo

Table 2. *Cont.*

	Case Study	Typology of Space	Web Sources (Accessed on 30 July 2025)
im05	Tergooi Medical Centre	Waiting pod integrated into the wall of a corridor. Walls are white and the pod is highlighted by the orange colour. While sitting, it is possible to look outside through the windows on the opposite wall.	https://www.archdaily.com/1001687/tergooi-medical-center-wiegerinck/647531883d39be5bd4fa41e0-tergooi-medical-center-wiegerinck-photo
im06	Health Municipal Clinic in Liffol-Le-Grand	Waiting room on the second floor. The walls are white and the room is filtered by a timber screen, which filters the large amount of natural light. As seats, the typical "Liffol chairs" are used, which are a symbol of the local culture.	https://www.archdaily.com/1006257/health-municipal-clinic-in-liffol-le-grand-studiolada/64f0b142bbf563583c72ac19-health-municipal-clinic-in-liffol-le-grand-studiolada-photo
im07	Zaans Medical Centre	Waiting area on the first floor. It is an open space, connected with openings to different floors. All the walls are characterised by big art illustrations. Furniture is homey, coloured and oriented to make people interact.	https://www.oneclub.org/awards/adcawards/-award/27298/zaans-medical-centre-healing-environment/
im08	Steno Diabetes Centre Copenhagen	Waiting area along the glazing wall, looking at the green courtyard.	https://www.archdaily.com/1006085/steno-diabetes-center-copenhagen-vilhelm-lauritzen-architects-plus-mikkelsen-architects-plus-sted/64ed052452867b5c9aedc962-steno-diabetes-center-copenhagen-vilhelm-lauritzen-architects-plus-mikkelsen-architects-plus-sted-photo
im09	Dijklander Hospital	Waiting area in the street of the hospital. Plants and greenery are growing around the seats, creating some green rooms where people can wait.	https://www.archdaily.com/995948/dijlander-hospital-revitalization-bureau-ira-koers-and-plus-studio-roelof-mulder/63dd1e07e8dab046150a93c5-dijlander-hospital-revitalization-bureau-ira-koers-and-plus-studio-roelof-mulder-photo
im10	Zaans Medical Centre	Waiting area around the green courtyard. The big glazing wall creates a direct connection between the interior and the exterior.	https://www.archdaily.com/874330/zaans-medical-centre-mecanoo/594ce8a8b22e3898a7000814-zaans-medical-centre-mecanoo-image?next_project=no
im11	Taverny Medical Centre	Waiting area in the green courtyard. It explores the idea of waiting outside.	https://www.archdaily.com/932080/taverny-medical-center-maaj-architectes/5e211fc03312fd970e0003cf-taverny-medical-center-maaj-architectes-photo

After the interviews, a questionnaire model was also submitted to patients and caregivers, which provided more quantitative data on specific questions about the waiting spaces' configuration.

Perceptive requirements are the result of the co-design analysis and will be described in another publication related to this specific topic.

3.4. Integrated Requirements

The Theoretical Framework was used to select and analyse case studies. The results of the technical analysis were represented by spatial models and environmental characters identified as invariants in designing waiting rooms and as points of innovation to be applied in the primary care context of CdC. These aspects were discussed in the co-design phase, by showing pictures of best practices selected (Table 2) and discussing the results of the technical analysis with users. Moreover, both the results of technical and perceptive analysis were summarised by the five categories of the Theoretical Framework, defining spatial requirements for designing health-promoting waiting spaces (Figure 7) and then to develop Design Guidelines.

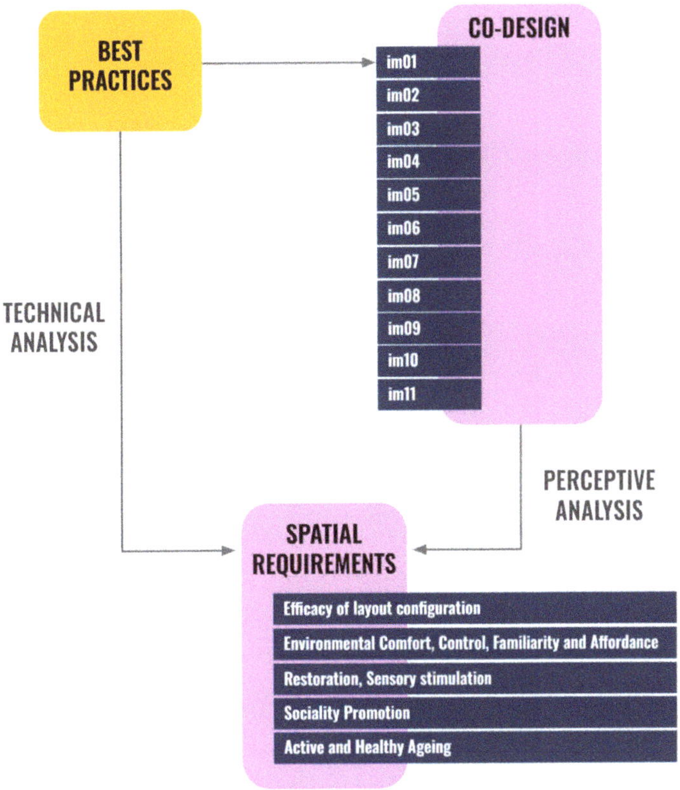

Figure 7. Diagram about the process of mixing research methods.

4. Results and Discussion: Space Requirements

In accordance with the aim of this paper, in this section we present the results of the definition of space requirements for active and healthy waiting spaces.

Space requirements represent the result of both technical and perceptive analysis, mixing indications derived from the study of best practices and qualitative responses with the interviews related to the point of view of patients, caregivers and staff regarding waiting spaces' models of the technical phase.

Space requirements form the basis for the development of Design Guidelines, which represent the final result of the research. Design Guidelines are briefly presented in this contribution to illustrate the impact of the previous research phases in reaching this operative tool.

4.1. Efficacy of Layout Configuration

4.1.1. Case Studies Results

According to the different layouts of the healthcare facilities analysed, we identified eight different models of waiting spaces based on the configuration layouts (Table 3). In the table, we also present the references to the case studies where the different layouts are applied to show the recurrence of the spatial models.

Table 3. Models of waiting space layouts derived from the analysis of case studies. The red represents the waiting spaces, the yellow the corridors/connections, and the blue the outpatient area. The arrows represent the users' paths. The proportion of spatial models is calculated on the total number of case studies analysed.

	Model	Diagram of the Spatial Model	Case Studies	Proportion Among Case Studies
M01	Waiting in the main hall	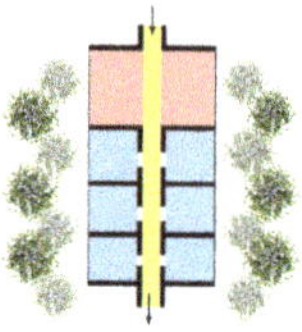	a11, a12, a13, a15, a16, a18	30%
M02	Waiting in the main street	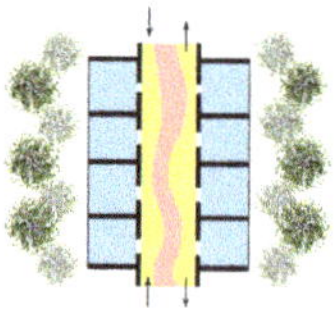	a02, a08, a10, a14, a20	35%
M03	Waiting around the courtyards	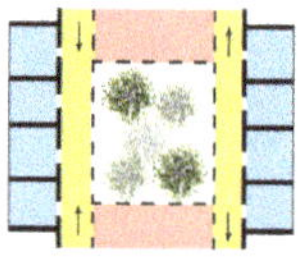	a01, a05, a06, a07, a09, a12, a14, a15, a19	45%
M04	Waiting in the courtyards	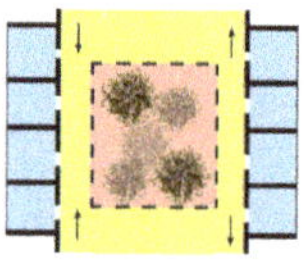	a06, a08, a12, a14, a15	35%
M05	Waiting in a separate room	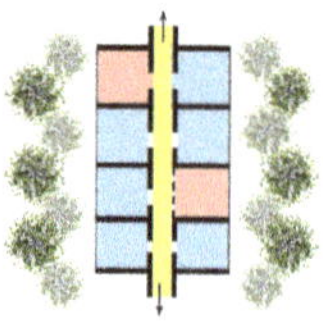	a04, a11, a13, a16, a19	35%

Table 3. *Cont.*

	Model	Diagram of the Spatial Model	Case Studies	Proportion Among Case Studies
M06	Waiting room in front of the outpatient area	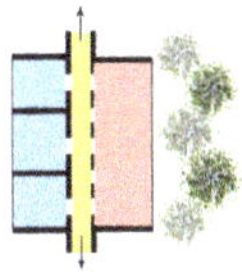	a01, a04, a06, a13, a14, a16, a18, a19	40%
M07	Waiting corners	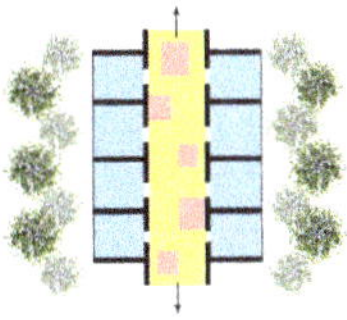	a02, a12, a14, a15, a17, a20	30%
M08	Waiting in the corridor		a01, a08, a12	15%

Waiting areas may be situated directly in the atrium (M01), articulated through various seating arrangements to foster social relations, while simultaneously guaranteeing privacy, contingent upon different users' needs (referring, for example, to case studies a12, a13 in Table 1). This area can also be characterised by colours (a12, a16), art or nature, to enhance comprehension and recognizability.

Alternatively, waiting areas may be developed along "the street" (M02), which runs longitudinally through the whole healthcare facility, establishing differentiated waiting corners along the path, such as small designated resting points. Nature assumes a crucial role in this area, such as creating "green rooms" through flower beds with tall vegetation that segments the spaces (a02), or using flower beds with low-spreading plants (a10), as well as art that can be spread along the path, such as artworks, sculptures, lighting systems (a08, a14) or large multi-level walls (a14, a18). Social interaction may be facilitated by furniture, such as through seating oriented to look at each other or positioned around a table to encourage relationships; similarly, games or art placed along the route (a10, a14, a17) promote shared experiences and dialogue. The street can be on a single level, double or triple volume, overlooking the other floors through galleries and raised connections (a08, a17, a18, a20), also creating larger areas of interaction such as a central big square (a18, a20) or smaller ones at various floors, creating a connection by stairs or slides that lead from one floor to another (a14).

Another recurring typology is represented by waiting spaces built around patios or courtyards (M03), which are highly amenable to green landscaping (a01, a05, a06, a07, a09, a14, a15), to maintain continuous contact with nature and encourage distraction during the waiting time. Courtyards can be accessible (M04) or non-accessible. They can have different shapes and dimensions: for instance, long and rectangular spaces facilitate the provision of natural yet indirect, non-dazzling light internally, and may offer seating along the openings (a5, a9), thereby preserving a connection with nature even when external access is not utilized. Smaller and more contained spaces, with the same dimensional relationship between waiting and green space, can project the exterior into the interior and expand the space up (a19), also thanks to big transparent glass openings. When accessible, these areas function as places for rest and relaxation, featuring seats in a structured greenery (a07, a14) or in a wild and more natural context (a01, a06); they may include covered spaces to host artworks (a08) or to carry out different activities, such as yoga, children's play, etc., (a15). Green courtyards can represent the core of the structure, also having a function linked to health promotion, physical and mental support activities and sociality (a06, a15).

A different model can be represented by a well-defined waiting space, such as a separate waiting room (M05), which operates autonomously from the other areas of the healthcare facility, while maintaining visual contact and access control of the relevant healthcare services/spaces. For example, this typology involves the placement of several rooms in different areas of the building, adjacent to the healthcare services/spaces, completely separated and oriented outwards to sustain the relationship with the surrounding context and landscape (a11, a16). These waiting areas can be highlighted by colour to facilitate recognisability, both from inside and outside (a16). They can be closed and contained, to favour privacy and relaxation, supported also by dim light (a11) or other sensory elements. Alternatively, the spatial separation may be less pronounced, establishing a filtering element (M06), such as glass walls (a13), wooden slats (a13), or a semi-transparent perforated surface (a04), which allow light to filter through and preserve the visual continuity with the different areas, thereby fostering interaction among users and ensuring the visibility of outpatient entrances, reducing stress.

The last typology is designated as "the corner" (M07). This model encompasses several configurations, serving as some waiting points situated within a larger space, positioned along circulation paths or around the courtyards, and identified by colour, specific seating elements (a2, a12, a17), or artworks (a15); niches/pods open to corridors or other public spaces, built into the walls or with an independent structure, identified by colours or materials (a12); additionally, furniture creates a point of rest, such as armchairs featuring higher backrests and sides, designed to envelop the user, thereby enhancing privacy (a12, a14, a17).

Model M08 is highly prevalent in the existing healthcare; however it should be avoided in new designs due to several critical issues, such as lack of privacy, high level of pedestrian traffic, noise pollution, lack of space, and challenges facilitating social interaction. Nevertheless, we decided to analyse this model within the Design Guidelines because we are also referring to the refurbishment of existing buildings.

The definition of the eight spatial models (Table 3) is a significant result of the technical analysis, which serves as the foundation for discussion during staff interviews, to understand their opinion regarding the impact of configuration on the perceived waiting experience across the different models. The staff themselves emphasised the importance of "the organisation of spaces to enhance the perceived quality of the facility and consequently of the healthcare service, as well as reducing stress and fostering relationships of patients/caregivers with the staff". A clear and comprehensible layout is paramount in the design process.

The most contentious element determining the varying layouts of waiting spaces is the visibility of the healthcare service entrance door from the waiting area: on the one hand, observing the door facilitates monitoring and can thus reassure the patient, for example, that they will not miss their turn; on the other hand, this constant visibility can induce anxiety, as it prevents patients from fully distracting themselves and relaxing, thereby maintaining a persistent state of stress. It is perhaps more important for the patient to be certain they are in the correct location (with the provision of appropriate information via both spatial layouts and signage), and to know that a member of the staff will call them, or at least that they can have access to call systems for the scheduled visit/exam. In this regard, the waiting room may be designed as a completely separate room, offering the distinct advantage of being a private place where patients/caregivers can relax and distract themselves.

It is highly recommended to have a secluded area away from the main corridors to promote calm and privacy during waiting, while avoiding waiting directly in the corridors, as "it is a transit point and, consequently, various health, pathological, and social situations may be observed," which can generate anxiety. This also promotes privacy, not looking at the doors during waiting and ensuring visual and acoustic privacy.

The configuration of waiting spaces should be as varied as possible to meet the different needs of users. For example, some people prefer seclusion (such as a seating area in an alcove), while others prefer a shared space where they can more easily pass the time, talk, and interact with others. The waiting area should foster relationships between patients or caregivers, for example, by having a connected coffee or tea area, which promotes greater conviviality. Open large space is also suitable for people with dementia or Parkinson's, where they can move freely and safely. In any case, it is important to be able to find privacy in an open space, to recreate a sort of intimacy between the caregiver and the patient, to have the opportunity to read, for example, and to distract oneself, reducing the feeling of waiting time.

Outdoor waiting is a very interesting model, but it is affected by weather conditions, so at least some coverage is required, and obviously it cannot be the only waiting area. It is beneficial to have space for walking while waiting, to "promote active ageing," or an area dedicated to "sensory stimulation"; "this could also be a corridor/greenhouse containing plants". In any case, it is important that the green area be adjacent to the waiting room to maintain a connection.

Greenery can also be imagined within waiting areas, to "mentally escape and thus reduce rumination regarding the procedures patients are about to undergo" and to create "privacy" between seats. However, it is important to have big windows to allow proper oxygenation, that there are no flowers that can cause allergies and infections, and that maintenance and cleaning are guaranteed. The sight of greenery already positively influences the waiting experience.

4.1.2. Interview Results

Regarding the layout, we first discussed the importance of seeing the door with users. This is a rather controversial topic, with the majority saying they are indifferent or, conversely, prefer not to see it, but with a small difference. Caregivers show a similar result, with the majority being indifferent or preferring not to see the door. The desire to see the door is often linked to anxiety of "not seeing what's happening," or "losing their turn," to check they "didn't make a mistake," or "to make the visit seem closer"; in general, a sense of control. Many say that if they cannot see the door, a display or call support is necessary. However, the situation is different for caregivers who prefer to maintain eye contact to ensure they can intervene if needed. Those who prefer not to see the door often experience a situation of anxiety or distress caused by constantly seeing the door and waiting for the other person to exit; furthermore, avoiding the door view allows for positive distraction, such as "looking at a beautiful view".

According to this idea, the use of a filter between the waiting area and the outpatient doors seems to be an effective solution. This allows for monitoring but also ensures a more private waiting experience. On the other hand, the filter used should be pleasant and not too restrictive, like "sticks" or a "barbed wire fence," which makes the area feel "like a prison".

Many, however, report the importance of waiting in a large, open space. For this reason, many users dislike a separate waiting room, as it often feels small and cramped. This can be addressed by a large window overlooking the outside, which is always appreciated for its view, natural light, and connection to the outside world. Those who prefer to see the door will obviously find a separate room unpopular, especially if it is far from the outpatient area, but this problem can be addressed for many with a display or call system. The problem of small spaces is they are often associated with the potential for crowding, which is considered very negative. Anyway, there are some who prefer this model because it is separate, quieter, and more secluded.

The connection with the outdoors, and in particular the view of greenery, is greatly appreciated by virtually all users. In this sense, many of them would gladly wait outside to "get some fresh air," especially if there are many people and it is hot, and to pass the time: "It's more distracting, you can look around," "you can completely disconnect from the inside". Displays and call devices are clearly necessary here as well. Many prefer a situation where the outdoor space is adjacent to the waiting room, to maintain visual control. Finally, many report the problem of waiting outside in adverse weather conditions, such as rain, cold in winter, or heat in summer. They suggest adding seating, activity equipment, children's games, fountains, social spaces, or connected services such as a cafeteria outside. The advantages of waiting outside for those with children were also highlighted, as it allows for easier entertainment during the wait.

4.2. *Environmental Comfort, Control, Familiarity and Affordance*

4.2.1. Case Studies Results

The design of patient comfort in healthcare facilities fundamentally relies on the quality of light (natural and artificial) and ventilation. Natural ventilation can be favoured by the use of courtyards and patios, but also through openings on the ceiling, for example, in pitched roofs (a7). Acoustics are also foundational to comfort in healthcare settings. For instance, implementing acoustic false ceilings is recommended in waiting areas to mitigate noise and reverberation (a5, a11).

Spaces, openings and facade design should vary according to solar orientation to maximise comfort and energy saving (a5, a6, a11, a16). The use of large openings or transparent glass surfaces, including on the roof, should be prioritised, particularly in public areas (a1, a2, a4, a10, a12, a13, a14, a15, a16, a17) to enhance natural light and connection with the landscape. Facade shading systems are necessary to regulate incoming light, thereby preventing glare and overheating (a5). An internal filter may also be incorporated to regulate natural light and allow visibility, favouring comfort and privacy (a4). Moreover, indirect light filtered through internal green courtyards ensures a more comfortable light distribution (a8, a9, a14).

Artificial lighting significantly impacts the quality and comfort of the internal spaces of healthcare facilities. Lighting fixtures that are not dazzling and give a warm, welcoming and familiar atmosphere are to be favoured, ensuring a bright, comfortable, and welcoming environment regardless of weather (a8, a13, a17). Lighting elements can also be integrated into the artworks, favouring interaction and navigation (a8).

The choice of colours and materials contributes to the warm and welcoming diffusion of light (a15), such as the use of wood, whereby the light is reflected and acquires a warm tone that offers softer contrasts (a15). The use of warm and neutral colours should be encouraged (a3, a5, a8, a12, a16); these, combined with the use of natural materials, can inspire a sense of familiarity (a8). Colours can liven up the environment and convey a bright and positive tone, particularly near the entrances and along the pathways (a12, a16), promoting the idea of a welcoming and friendly place. For instance, yellow can promote a sense of positivity (a12, a16), while green and other natural hues can foster a warm, bright and welcoming atmosphere (a17); in contrast, cold colours are deemed more functional for staff areas, supporting the creation of a stimulating and appealing work environment (a17). Then, the use of colour can also promote recognisability, comprehension, and navigation.

Wood is highly appropriate (a1, a4, a5, a6, a7, a13, a14) to promote warmth, welcoming, familiarity, and a sense of regional belonging, aligning with the use of local and natural materials (a5, a6, a7). It evokes external nature and provides a positive associative effect, generating feelings of calm and well-being, and positive memories (a15). Then, wood can be used in flooring to distinguish waiting areas, setting them apart from the paths, and imparting a more domestic and comfortable feel (a14). Moreover, wood can be applied to the facade, reflecting from the outside of the building the desire to provide the city with a healthy and environmentally friendly building (a7) and creating a warm character on the outside, also encouraging people to enter (a15). The principle of using local materials also extends to stone, enhancing the integration with the surrounding context and supporting identity and recognisability (a11). This contextual integration is also influenced by the size of the building and its volumes, integrating with the surrounding urban landscape, while making itself recognisable via materials and colours (a18, a19). Indeed, the use of different materials on the façade completely changes the exterior appearance and the interior relationship with the surrounding landscape (urban and natural), allowing designers to either contrast with or complement the context (a19).

The familiarity of spaces, as well as their usability, is also enhanced by furniture. Diverse seating arrangements and different types of seating are appropriate to foster relations and facilitate control by users, facing outwards to encourage engagement with nature. Options such as sofas, chairs with tables (low or high), soft seats, armchairs, and coloured or wooden chairs, promote familiarity, comfort and privacy (a1, a8, a11, a12, a14, a15, a18, a19, a20). Furniture can also reinforce identity and attachment with the community through the selection of iconic and local products (a4). Citizens' involvement in the design, including for furniture and iconic elements, such as patterns on the internal and external coverings of the building, is also encouraged (a15). Another important role of furniture is to reduce potential imbalance between patient and doctor, for example, through the use of curved elements, such as round tables, where all participants can sit at the same level (a13, a15). Accessories such as lockers (a1) further contribute to user comfort.

4.2.2. Interview Results

According to the staff, brightness in a waiting room positively impacts the waiting experience, particularly mental health. Natural light is preferred where feasible, or artificial lighting that is well-diffused and non-overwhelming (for example, avoiding flickering neon lights). In addition to lighting, openings are considered crucial, both to ensure natural ventilation—especially in crowded waiting rooms—and to provide a view of the exterior (see next section).

Colours greatly help enliven the environment and enhance its aesthetic appearance. Light, non-vibrant colours are favoured over colours that might induce anxiety, such as light blue, which is ideal for relaxation. Colours traditionally associated with healthcare and institutional environments, such as light green or white, should be avoided. It is deemed appropriate to use the same colours for walls and doors.

Materials are very important; for example, wood is a natural, warm, and welcoming material that can be very beneficial in waiting areas, and "Curved lines are also transmitting the same feelings of calm and welcoming".

These spaces should also convey a sense of cleanliness and order, with linear furnishings. Comfortable sofas or armchairs are preferred for both seating comfort and conviviality. They evoke the idea of a more informal and familiar environment, where, for example, "you could also have a table with magazines to distract yourself while waiting." The ergonomics of the seats should also be considered to facilitate standing, especially for elderly patients or those with mobility issues. As an example, it is important to have seats with armrests.

However, hygiene, health, and safety standards must always be respected.

According to patients and caregivers, a significant portion of users prefer a familiar environment, through the use of natural materials such as wood and a homier furnishing style. Wood, in particular, is appreciated by most people because it "provides warmth," is "more welcoming, and less cold." On the other hand, we did not expect a significant percentage (more than half) of people to be so concerned about hygiene and, for this reason, less inclined to use more domestic fabrics or furnishings. They prefer a functional, easy-to-clean environment that "gives a sense of health," reflecting the function of this space.

Regarding colours, users prefer soft, pastel, bright, and vibrant colours, provided they are not too strong. Almost all users reject white, except for a few who prefer it above all else, because it "gives a sense of cleanliness" and "makes the space seem more open." In general, blue is highly appreciated because it "relaxes," "reminds one of the sky," "of water," or "of the sea." Green is also highly appreciated, and some prefer yellow, but only if it is not too strong, although many say they do not like it.

Regarding lighting, natural light is preferred by virtually all users, as it "reminds one of being outdoors," "is more vital and joyful," and allows one to "maintain a connection with the environment and the passing of time," except for a few who appreciate the use of coloured lights. As for artificial lighting, soft, not-too-bright, warm, "relaxing" lights are preferred, reproducing outdoor light or modulated at different times of the day.

They do not like bare spaces, but they also do not like messy or chaotic spaces that are overcrowded. A modern, linear furnishing style is often appreciated because it is clean and tidy, but it shouldn't be too cold. Some prefer single, unpadded chairs, as mentioned, but many instead point to the benefits of using sofas because they are more comfortable and, above all, more convivial, as they actively foster social interaction.

4.3. Restoration, Sensory Stimulation

4.3.1. Case Studies Results

Most studies place nature as the principal source of restoration, alongside art and sensory elements.

The relationship with nature can be promoted in healthcare facilities, both inside and outside. Internally, nature can permeate the spaces through indoor greenery, which makes the environment more vital, familiar, healthy and therapeutic (a2, a10), in addition to establishing a filter via taller vegetation, thereby functioning as authentic "green rooms" (a2). Furthermore, an indirect visual connection can be established with the surrounding landscape (a1, a6, a11, a14, a15), green patios and courtyards (a5, a9, a19), or even the sky (a19), facilitated by glass openings in the waiting and connecting spaces. When feasible, it is suitable for green courtyards to be accessible (a1, a5, a6, a7, a9, a14, a15), allowing for a direct relationship with nature, more or less structured, with rest and interaction areas, or spaces equipped for various activities, such as yoga or children's playing (a15). This approach can be extended across the entire structure, creating a symbiosis between interior and exterior, in a variety of essences and colours, based on the seasons (a6), through larger and smaller courtyards or green terraces hanging in the volumes (a3, a14, a15, a18), and roof gardens (a6, a18). The configuration of green courtyards can also be linked to specific functional areas of the structure (atrium, bar, staff, children, etc.), thereby extending their use outside, and may incorporate distinct sensory stimuli, such as those based on the four elements of earth, air, fire and water (a15). The arrangement of the external greenery is crucial from the perspective of both welcoming and involving citizens. Greenery situated at the entrance (a4, a12) or around the building (a17, a20) allows for creating both a link with and a buffer against the urban context, and it can be freely used by citizens (a15). It can be equipped with pathways and seating (a4), artworks or other elements to encourage interaction and sensory experiences (a12), or connections with other functions such as the café or the waiting area (a18).

On the other hand, nature may be conveyed through art, recreated in different artworks, colours and materials (a8), in graphics and illustrations (a14) or as a pattern that recurs throughout the building, also creating a link with the territory and promoting a sense of identity and belonging (a15).

The illustrations can become intrinsic to the architecture, where the space is fully permeated by art (a14, a17). Otherwise, the colours can liven up the environment, make it less institutional, and foster sensory stimulation (a19, a20). Art can be reproduced in artworks, paintings or sculptures, which invite interaction with the space and positive distraction; facilitate navigation, in focal points of the building such as the entrance or reception, in the waiting rooms or in the paths (a8, a10, a12, a15, a17); create a connection with the city (a3), with sculptures (a10) or street art outside. It is recommended to use artworks by local artists (a12), potentially through rotating exhibitions (a18), and open after closing the healthcare facility to favour visits and awareness of the structure by new groups of citizens (a15).

Sensory stimulations aim to activate the person and generate interest in the surrounding environment, thereby facilitating interaction with spaces and aiding navigation; on the contrary, sensory deprivation can lead to negative outcomes of stress, anxiety, depression and poor motivation, which affect the quality of life [31]. In this category, sensory features aim to relax and support restoration, to foster a positive effect on mental comfort and health. In Bellini and Setola, 2024 [31], case studies in the sensory design field illustrate the impact of sensory environments/features on health promotion. In the case studies of the first and second areas of investigation, sensory perception is encouraged by dim and coloured lighting, and music, for example, by placing a piano (a2) or disseminating sounds and music in the public area. Lastly, intimate spaces, such as sensory rooms, are appropriate for relaxation, containment, and sensory rebalancing (a11).

4.3.2. Interviews Results

Staff emphasised the importance of establishing a comfortable, welcoming, intimate, and safe waiting area where people can relax, allowing time to pass without worrying about when their turn will come. "If you feel more relaxed, you're also a little more cooperative, perhaps more willing to talk and able to clarify, for instance, symptoms." It should be a sort of "decompression room." "A quiet and serene environment, with comfortable seating, paintings or murals on the walls, succulents (which require no maintenance), and soft lighting." The idea of "being able to involve local young people in creating the mural and have it donated to seniors, to foster the idea of multigenerationalism and the feeling of belonging", is an interesting concept. Colour, art, and greenery are all good in all their forms. Green elements keep people active and promote mental health. Even the pathways could be imagined as "tree-lined paths, with a view of greenery in the background, which can convey a message of greater tranquillity and health promotion."

Projections of landscapes and natural scenery, video screens, or touch screens can distract, induce relaxation, and refocus attention, a function also served by background music. The music should be "low-volume, non-disturbing (e.g., loud sounds can create discomfort in people with dementia or make some people nervous), calm, and preferably instrumental, even better if you can choose based on your preferences," or "sounds of nature, especially water," "wind, rain, and animals." Even objects that can be played and interacted with manually, such as a large chess set, are desirable.

In any case, "whether it is a game, entertainment, music, or a connection... The primary objective is to ensure that patients do not perceive they are wasting their time while waiting."

"A sensory preparation room can be ideal," "a relaxing, meditative place where you can be alone, even with and for those accompanying you." "A place where you can stay for an hour and not feel the need to do anything else, a warm, natural, familiar environment, darkened with curtains," "illuminated by chromotherapy," with "video projections for distraction," "comfortable seating, waterfalls and sounds," with "water, which also provides a sense of coolness." Such a space "can also improve pain tolerance," for instance, in the case of invasive tests, "can feel better in case of bad news by the doctor", "in case of fragility" and for "people with cognitive or behavioural diseases". In general, a dedicated and "possibly isolated" space is preferred, as it is quieter and more intimate, offering greater privacy; however, some prefer the idea of having sensory elements in a larger waiting room, where they can continue to interact with others, or use a sensory alcove overlooking the waiting area, so as to still be able to isolate themselves somewhat.

As for the staff, art is appreciated by almost all patients/caregivers, as it "distracts"—especially if you're alone—"stimulates curiosity," "makes you stop and look," "makes time pass," "relaxes and calms," and "provides a sense of beauty." Only a few say it "isn't interesting" or "not relevant," but no one makes negative comments about it or disdains it.

Nature and the view outside are universally appreciated by all users, as they "are relaxing," "provides calm," "openness" and a "sense of respite," conveys "the idea of being outside," and allows you to "watch time pass." Some did, however, mention the importance of seeing the outside, but without having a completely transparent glass wall overlooking the waiting area. This is likely also linked to the cultural context, where it is uncommon to have completely open glass facades.

More controversial is the issue of having greenery within the waiting area; the majority of users appreciate it, but some are concerned about the potential for allergies, the presence of insects, and the potential for debris caused by plants. Above all, they are concerned about the possibility of inadequate maintenance, which could result in the plants becoming damaged or drying out. This can be addressed by choosing plants that are hypoallergenic, highly durable, and require little maintenance, or by using preserved plants.

For the same reason, although most people prefer real natural elements, some prefer the artificial view of nature, such as through screens or video projections, which "allow your imagination to travel."

This raises the topic of sensory spaces, which are appreciated by almost all patients, and the majority of caregivers because they "relax," "calm," "relieve anxiety," "help you sleep," "distract you, and stop you from thinking." Many have never experienced a sensory space, so the results may be partially affected by a lack of knowledge. Some refer to the use of this space not only during the wait, but also after the appointment, or for activities such as those of psychologists, or for specific pathologies and vulnerable users.

They also question whether they can truly relax in such a context. For this reason, the majority prefer a separate space, since it is isolated and "makes you feel more at peace," and more than one person can use it at a time. For the same reason, the sensory pod is less appreciated. Some also raise the issue of hygiene, having to use it collectively. Those who generally prefer to isolate themselves appreciate it, as they can remain in the waiting room, yet still be isolated and have an individual experience. In this sense, for instance, it can be useful for breastfeeding.

Screens and projections that reproduce natural scenarios are appreciated by almost all users, preferably with background music, as it "distracts and takes you away from your thoughts," or "travel scenarios that let your imagination soar."

Music is also appreciated by everyone, but it is important that it be soft and light, just a background. Some people prefer to be able to choose it; otherwise, relaxing music, classical or instrumental, is generally preferred, along with the sounds of nature, especially water, "like a waterfall, ideally real."

4.4. Sociality Promotion

4.4.1. Case Studies Results

The integration of diverse functions and a robust connection with the surrounding territorial and social context should be actively promoted. Consequently, the decision to situate healthcare facilities in buildings that also have a residential function (a3, a10), or more public functions such as an atelier/carpentry shop (a3), a gallery to exhibit the work of local artists, a space where local groups meet and a base for neighbourhood initiatives (a18), an area for firefighters (a12), a gym, physiotherapy services, a library (a18), spaces for children (a10) or family consultants (a18), outdoor spaces also used by citizens, green areas (a3, a4, a10, a12) or market areas (a4), or the urban square itself (a4, a13). Otherwise, another strategy is to connect to functions already existing in the area, such as cafes, or sports facilities (a5), a contemporary art museum, or an arena for outdoor events (a9). Also relevant is the choice to locate the structure in a central position in the city (a10) or well connected by public transport, for instance by placing itself near the train station (a10).

The public atrium, the street and the courtyards, constitute the principal nexus for connectivity, thereby fostering meeting and exchange among the facility's different users and establishing a real public space comparable to an urban street or a town square (a2, a8, a10, a14, a17, a18, a20), which are furthermore complemented by public services and functions such as café, hairdresser (a2), and similar provisions.

The establishment of permeability between internal spaces and through paths is crucial for guaranteeing visibility and facilitating relations. It can be provided by the use of transparent walls (a19); filters created with wooden slats (a11, a13) or grilled surfaces (a4); walkways, balconies and views of the double volume of the street (a2, a10, a11, a14, a17, a18, a20); corners or seats along the route that foster relationships during stops and passages (a12, a15). Furthermore, external permeable pathways that promote circulation in the form of tunnels, elevated roads where residents meet, children play, plants grow, etc., (a3) are favourable for developing sociality. Finally, the relationship between inside and outside can promote permeability, for instance by opening and making the central distribution axis transparent, as if it were an urban road (a4).

Likewise, patios and courtyards are a valid strategy not only to maintain the relationship with the greenery, but also to foster relationships between people, both if they are accessible, creating relaxation and leisure spaces where it is easier to meet and relate with other users (a6, a7, a15), and if they are not, placing waiting areas around the open spaces to promote continuous visibility and permeability (a1, a5, a9, a19). Internal courtyards can also be enriched by artworks to improve distraction and restoration (a8), and encourage socialisation (a2), also organising various activities, such as yoga or games for children (a15). This concept can be extended to the common spaces of the hospital by organising various information and awareness-raising activities, distracting and regenerative ones, reading, physical activity, or workshops (a6).

Furnishings constitute another social strategy, enabled through the use of different seating arrangements to maximise usability for various user types, oriented in multiple directions, including tables or low tables (a1, a2, a4, a6, a11, a12, a14, a17, a18, a20), allowing users the autonomy to choose whether to have privacy or social interaction. Tables, chairs and workstations with electrical sockets or for public use (a6) can also allow working while waiting (a2). Designated staff areas also allow for strengthening the relationship between professionals, especially if characterised by furnishings and materials that encourage familiarity and informality (a17, a18, a20).

4.4.2. Interviews Results

In general, the staff considers the presence of a convivial environment to be highly favourable, such as "a fertile intergenerational space," "useful for both patients and caregivers," since "talking with others allows them to share and discuss their problems and find comfort," "distraction, reduce tension," and "support, especially if they are without a caregiver." For instance, "it can be helpful to bring together young people with diabetes, with the aim of getting them to know each other and better accept the situation." Similarly, sharing is very important for pregnant women, as well as for the elderly, who "often suffer from loneliness and need to talk about their illnesses, grandchildren, etc." However, it is crucial that the waiting room is not overcrowded, mitigating the risk that "one patient's anxiety can be transmitted to another." A suggestion proposes implementing a "living room as a waiting space, where people can feel comfortable, looking outside and talking with other people".

The majority of patients and caregivers emphasise the importance of connecting with others while waiting, "especially if you're alone," because "seeing people reduces anxiety," "keeping others company," "allowing you to hear about other experiences," and "taking your mind off what you have to do." Open spaces are considered optimally suited for this purpose, as they allow individuals to choose whether or not to engage in this interaction and "not necessarily find yourself face-to-face." Nonetheless, ensuring acoustic comfort is essential, particularly by reducing noise from conversations, "which can be very annoying." Furthermore, the provision of "a space for privacy" is necessary, contingent upon the specific moment and the person's mood. A pod is a "cozy space" where one can "seclude and isolate yourself." It is appreciated by only a few patients and less than half of caregivers, because it is considered "too closed off," consequently inducing a feeling of being "excluded and isolated." Some people expressed their willingness to use the pod specifically for breastfeeding.

4.5. Active and Healthy Ageing

4.5.1. Case Studies Results

Health can be promoted by information panels (a15) or screens, as well as by activities, such as courses. The promotion of physical activity should be achieved through dedicated fitness spaces (a6), green courtyards (a6), and active and social spaces in which to encourage movement.

According to the active design approach, stairs should be prominently highlighted, potentially via scenic shaping or strategic central positioning (a4, a6, a14, a15, a20), or even through the colour (a8), materials (a4, a11) or lights (a4, a8), encouraging their use; connecting elements can also include games such as a slide (a14). Conversely, elevators can be relegated to a secondary area, contrasting with the intended centrality of the staircase (a4, a11).

Slow and green mobility should be prioritised (a12), the provision of slow traffic areas (a1), bicycle parking near the entrance (a20), guarded and covered (a3), reducing car parking spaces (a3), charging points for electric vehicles and foldable bicycles (a18), and access to public transport (a3).

The provision of healthy food and accessible, potable, free water should be suitable. To promote healthy behaviours, spaces dedicated to awareness-raising of healthy eating (a6) can be provided, also through an educational kitchen where families can learn to cook healthy meals (a15), or a city market with local products (a4).

4.5.2. Interview Results

The staff like very much the idea of promoting health during waiting; the main goal is defined as enhancing "improve patient/caregiver's health literacy and make them feel relaxed after waiting".

All staff expressed interest in the idea of "information points" where patients could "know where to go," obtain information on the "needs of seniors," or generally promote healthy lifestyles, in various forms: video seemed the most suitable solution, through a projection, a monitor, or a touchscreen, which could also be interacted with; posters or "something large on the walls"; and pamphlets and brochures, which could also be "picked up and taken home." Most, however, preferred something shared over individual tablets. Informing could "be useful for distracting, shifting attention, and passing the time." It is important to provide information "on nutrition, exercise, and social engagement"; on "prevention, such as smoking cessation, or the promotion of specific diets, which, for instance, reduce vascular risk, including by providing recipes"; "education for expectant mothers"; "on gender-sensitive medicine for young people, including by organizing peer training or meetings with the public"; on "other patients' experiences, through video stories, which can provide reassurance, particularly for young diabetics, to facilitate acceptance of the disease"; "on interventions to be performed or on specific pathologies, such as in-depth analysis, again through videos, of anatomy, since many things are often taken for granted when talking to the patient"; "on how to use the devices"; "on local itineraries that can be taken for patients with mobility difficulties to encourage walks, for instance in the woods, which reduce behavioural disturbances." "Disseminating this information within a quiet, peaceful, and intimate setting is certainly more beneficial." It would also be interesting if "they could watch these videos both before and after healthcare services, even at home, to prepare or, conversely, get suggestions for continuing the treatment." Language should be simple, and explanations should be "with cartoon-like illustrations" or presented "in a fun, entertaining, and interactive way."

Similarly, interaction and play can engage people and encourage them to exercise. The idea of "going outside and engaging in physical activity" is particularly positive, combining exercise with the opportunity to "see plants and stimulate the senses," such as "a walk along a tree-lined avenue," or a "wellness trail," even with covered areas. This would be ideal, especially for diabetic patients.

We also discussed the opportunity of using passive training equipment, such as seating elements which make people stay in a position in order to activate the muscles and restore their motor function. As an example, we were showing a picture of MyActiveBench by the Italian company (Metalco s.r.l.) [53], a "standing" seat for correct posture and active muscle tone, specifically designed and conceived for "fast recovery" moments. Anyway, passive training equipment has been less successful, as the staff appear concerned about the safety of using the devices independently, imagining elderly patients and/or patients with mobility issues. However, some are very interested in the idea and have proposed using these devices for caregivers, suggesting including an educational message on how to use them. They are considered more suitable for use in a dedicated physical activity area, especially outdoors, in the area adjacent to the waiting room.

Staff consensus regarding the implementation of physical and cognitive self-testing devices remained polarised: an interactive screen/projection aiming at making people do individual tests such as getting up from a chair, walking, lifting weights, bending and stretching, etc. [53] to understand their healthy and active ageing level and a dedicated cognitive training interface to experiment their neuro skills. The staff think self-testing can be very useful for some patients (e.g., diabetics), on the other hand there is concern about doing the tests independently, due to the patient's interpretation of the results and the worry and anxiety it may generate. Some of the tests considered useful, especially thinking about healthy and active ageing, are represented by "the test of balance, the analysis of movement, the speed of walking, the memory test".

The majority of patients and caregivers were interested in having access to information about health promotion. They preferred screens/video projections as a means of communication because they were more immediate and engaging, offering brief, "commercial-like" information, since people do not read and share, thus fostering greater relations. They also referred to multiple shared screens, but not individual tablets that would perform the same function as a phone.

Among patients, those who responded that they preferred not to have this information cited concerns and anxiety about the procedure, preferring to distract themselves while waiting and escape from their current state. Some also preferred to receive this information from the doctor, as it was targeted and "safe."

Those who proposed using paper-based support instead referred to the possibility of taking the information home for further study at a later time.

The majority of patients reported they would not use passive training equipment, while almost the half of caregivers said they would; this result may also be attributed to the age of the interviewees. In general, several patients stated "they want to be comfortable" while waiting and performing physical activity at home or in the gym, or at most outside, even being intimidated by the idea of doing so in front of other people. Some said they "prefer the idea of a walk rather than sitting in a position like this, as they need to release their anxiety." Many reported that they do not consider these devices suitable for the elderly.

Regarding movement, the majority of patients and caregivers said they would enjoy playing and interacting with various devices "to pass the time." Some also enjoyed light physical activity. This is certainly considered beneficial for children. Again, reference is made to the embarrassment of doing these activities in front of others, so they would be happy with a dedicated or screened area.

Finally, regarding the idea of being able to perform self-tests (physical and cognitive), more than half of patients and the majority of caregivers are in favour of using these devices. Many are concerned that they lack professional interpretation and may fail to constitute "serious and reliable information," or data "based on a doctor's directives."

4.6. From Space Requirements to Design Guidelines

Starting from these results, Design Guidelines for waiting spaces have been developed and organised through the same five categories, which also represent the drivers of a good design of health-promoting spaces. Design Guidelines are structured according to "Design Strategies", which aim at proposing general strategies to promote health through the built environment, and "Design Sheets", which present specific spatial requirements and solutions for the waiting spaces. The eight spatial models are used in a transversal way as a field of application of the Design Strategies, and they are detailed in the Design Sheets where specific spatial requirements are presented for every model, according to its morphology and environmental characteristics.

The objective of Design Strategies is to indicate the actions necessary to promote healthy and conscious lifestyles and behaviours through the built environment and to design waiting spaces able to improve people's health and well-being conditions and, more specifically, healthy and active ageing.

Strategies considered useful in promoting healthy and active ageing through waiting spaces are the following:

1. Create easily accessible places and routes;
2. Involve, welcome and encourage sociality;

3. Create environments for the person's psychological comfort and support;
4. Promote active waiting time;
5. Promote health and activate healthy behaviours.

The objective of the Design Sheets is to provide design indications regarding the waiting areas and the integration of digital devices in these spaces, as an operational tool of concrete support for architects and designers for both the refurbishment and the new construction of waiting spaces.

Design Sheets developed are the following:

1. Inclusion systems;
2. Nature as restoration;
3. Relax spaces and Sensory scenarios;
4. Sitting and spaces for relations;
5. Sitting and spaces for privacy;
6. Children's area;
7. Breastfeeding area;
8. Movement and soft physical activity;
9. Passive physical activity;
10. Auto-assessment for prevention of chronic diseases;
11. Cognitive training;
12. Knowledge and information on health.

Both Design Strategies and Design Sheets may serve as an opportunity to promote dialogue between different actors involved in the co-design process of a CdC, in supporting the definition of the framework of needs of the organisational system of CdC and developing the briefing phase of the design.

Design Guidelines will be presented in a future publication, presenting an answer to a gap in tools to support professionals in designing healthy-promoting waiting spaces and in the specific context of CdC, which represents a new model of healthcare facility, not yet comprehensively investigated.

4.7. Limits and Future Developments of the Research

This study is subject to some limitations inherent to its design and scope. First, data collection and analysis of this part of the research relied exclusively on qualitative methods. The research was conducted within the Italian context, focusing on case studies situated in Europe, as it was aligned with the Italian context of local primary care health facilities (CdC). This limits the generalizability of the findings to other national or non-European settings. A promising area for future research involves the evaluation of the application of spatial requirements and Design Guidelines to different sociocultural and economic contexts, such as different countries, considering the needs of different communities and the adaptability of this tool. Moreover, the selection of case studies and interviewed participants introduces potential sources of bias, as the process was not randomised but rather purposive, aimed at capturing contexts closely aligned with the objectives of the study. While this approach is appropriate for exploratory research and for gaining in-depth insights, it may not reflect the full diversity of experiences or practices present elsewhere. Interviews report people's perceptions of different spaces, which can differ a lot, especially regarding sensory features. In this research, the interviews were conducted with both female and male people of different ages (from 18 to 85 years old), but we did not consider specific groups of users, especially regarding disabilities. Interviews conducted with experts supported the consideration of different needs, but they cannot be completely representative for all people and their specific needs, especially in the case of frailty. In the future, Design Guidelines can be evaluated by specific users' needs groups to enhance the universality and effectiveness of the design of CdC waiting spaces.

These limitations have been considered in the interpretation of the findings, and future work in this broader research project will employ complementary quantitative methods and expanded sampling to enhance the transferability and rigour of the findings.

This study highlights how thoughtfully designed waiting spaces in healthcare facilities can actively support well-being and healthy ageing. As we move forward, such research will become even more relevant: our environments are not just backgrounds; indeed, they shape experiences, emotions, and health outcomes. Future studies in this field will likely explore new ways to blend technology, nature, and inclusivity, making everyday places healthier and more responsive to people's needs. The emphasis on inclusivity in healthcare design ensures that all individuals feel included and valued, contributing to the creation of welcoming, empowering, and health-promoting environments for all. The continued development of research on health-promoting built environments is crucial for creating spaces that promote active living, inclusivity, and well-being, thereby directly addressing contemporary health challenges and advancing sustainable healthcare models.

5. Conclusions

This contribution presents a portion of a three-year research project dealing with waiting spaces to promote good health for all people in the social community.

We were focusing on the background and field research phases (Figure 1), describing how we used these methods to define spatial requirements for active and healthy waiting spaces in Casa della Comunità, a new model of territorial socio-healthcare facility in Italy.

The Background research focused on the identification of three fundamental groups of theories derived from the scientific literature, which represent the core principles of the research and were used to establish and define a Theoretical Framework as an operative tool to guide the data collection in the subsequent phases. The Theoretical Framework was based on five categories: Efficacy of layout configuration; Environmental Comfort, Control, Familiarity and Affordance; Restoration, Sensory Stimulation; Sociality Promotion; Active and Healthy Ageing.

These categories represent the first result of our study and summarise the main principles for active and healthy waiting spaces in a Casa della Comunità.

A Mixing Method approach based on Technical and perceptive analysis was used to define space requirements for waiting spaces in CdC. The technical analysis was based on the Case-based design/reasoning (CBR) approach to identify innovative design inputs from existing solutions. International case studies of waiting spaces of primary care facilities were selected and analysed by the five categories of the Theoretical Framework. The technical analysis was useful to analyse best practices and identify invariants, which can be translated into input for the project according to these categories.

According to this approach, a result of the technical analysis has been the definition of 8 layout models for waiting spaces, based on the configuration layouts (Table 3) which differ regarding the place of the waiting area (e.g., main hall, corridor, courtyard, etc.), the typology (e.g., separate room or waiting corner), and the relation of the waiting spaces with the outpatient area.

The space requirements (and consequently the Design Guidelines) have been correlated to these spatial models of waiting areas, both in the existing structures and in the new ones, to detail the recommended interventions to get active and health-promoting waiting spaces.

These layout models and the case studies' invariants were also used as a basis of discussion in the perceptive analysis, conducting interviews with the staff, patients and caregivers by iconic pictures to improve the comprehension of their needs during waiting.

By mixing methods it has been possible to clarify people's preferences about different spatial models, such as waiting in an open space, in a separate room, or directly outside, expressing advantages and disadvantages of each layout; characters of the environment to be used as drivers for designing waiting spaces, such as colours, materials, lights, or furniture; activities people prefer to do during waiting time, such as relaxing, chatting, reading, playing, walking around or performing physical activity; interests people have about health and its promotion, such as improving knowledge in this field, testing their physical and mental level of ageing, etc.

As a result of these two phases (Figure 1), spatial requirements have been summarised according to the five categories, defining different spatial models and environmental characters to be developed in the Design Guidelines.

Design Guidelines have been developed and organised through the same five categories which also represent the drivers of a good design of health-promoting waiting spaces in Casa della Comunità and are structured according to "Design Strategies" and "Design Sheets". Design Guidelines will be presented in detail in a future publication as an effective tool for professionals in order to design healthy and active environments able to keep all people in good health in a new model of healthcare facility, which actually presents no evidence about optimal design.

Contribution and Originality

Going into depth with the application of the spatial requirements through the five categories, we can identify different contributions and elements of originality of this study.

In the first category (Efficacy of Layout Configuration), we focused on spatial models derived from the analysis of existing healthcare facilities, to be able to differentiate the application of spatial requirements according to the morphology and the environmental characteristics of each model. This system also considers the opportunity of enhancing existing buildings, which represent the majority of healthcare facilities in Italy, and increasing sustainability with the idea of reuse and refurbishment. The spatial models will also form the basis of Design Sheets, operative tools for professionals to design CdC waiting spaces, included in the Design Guidelines.

The second (Environmental Comfort, Control, Familiarity and Affordance) and the third (Restoration, Sensory Stimulation) categories are already thoroughly described in the scientific literature [12–17,22–37]; our contribution was to confirm and enhance these theories by the analysis of case studies and the application to the specific context of CdC waiting spaces. We also gave strength to these concepts by reporting people's opinions about the impact of the environment on comfort, restoration, familiarity and well-being.

The fourth category (Sociality Promotion) can be included in the idea of social cohesion, dealing with the general concept of health promotion, which also represents the main originality of this contribution, as expressed in the fifth category (Active and Healthy Ageing).

Therefore, one of the greatest originalities of this paper is represented by the idea of interpreting the concepts of active and healthy ageing [11] through the built environment. While the concept of health-promoting architecture and health-promoting buildings [18,19] are explored in the literature, showing the contribution of the built environment to prevent illnesses and chronic diseases and to support the healing process, limited research has focused on how healthcare buildings can interpret the concept of health promotion in terms of building design [20]. This study has made a contribution to this field and applied these concepts to the specific context of CdC, whose mission is strictly connected to health promotion.

Moreover, we have introduced the concept of "active waiting time" and "widespread health promotion in waiting spaces", focusing on the design of CdC waiting spaces and enhancing the activation of positive behaviours and healthy lifestyles by performing healthy activities during waiting time, favoured by the built environment. According to these concepts, the idea of transforming passive waiting time into an opportunity of improving knowledge about health [45–48] and healthy lifestyles [8], performing physical activity [9] and cognitive training [45], and increasing social cohesion was very much appreciated by all users (staff, patients, caregivers).

Therefore, with this project, we have enhanced the capacity of aesthetics of the built environment to influence health, moving away from the pathogenic model of health that is often dominant in the healthcare sector [20].

Author Contributions: Conceptualisation, E.B. and N.S.; methodology, E.B. and N.S.; investigation, E.B. and L.R.; data curation, E.B., V.S. and L.R.; writing—original draft preparation, E.B.; writing—review and editing, E.B. and N.S.; visualisation, E.B.; supervision, N.S.; funding acquisition, N.S. All authors have read and agreed to the published version of the manuscript.

Funding: This research was funded by Next Generation EU—"Age-It—Ageing well in an ageing society" project (PE0000015), National Recovery and Resilience Plan (NRRP)—PE8—Mission 4, C2, Intervention 1.3.

Institutional Review Board Statement: This study was conducted in accordance with the Declaration of Helsinki. Italian regulations require approval from an Ethics Committee only for studies involving clinical interventions, the use of drugs/medical devices, or diagnostic/therapeutic procedures on participants (Legislative Decree 14 May 2019, n. 52; Law 11 January 2018, n. 3). The present study is limited to qualitative interviews, without any clinical intervention or collection of biological data, and therefore does not fall within the types of studies subject to mandatory ethical review.

Informed Consent Statement: Informed consent was obtained from all subjects involved in the study.

Data Availability Statement: The raw data supporting the conclusions of this article will be made available by the authors on request.

Acknowledgments: This paper was developed within the project funded by Next Generation EU—"Age-It—Ageing well in an ageing society" project (PE0000015), National Recovery and Resilience Plan (NRRP)—PE8—Mission 4, C2, Intervention 1.3". The views and opinions expressed are only those of the authors and do not necessarily reflect those of the European Union or the European Commission. Neither the European Union nor the European Commission can be held responsible for them. This contribution represents one of the results of the research project "Sensory design for spaces in social and healthcare facilities for healthy and active ageing", led in Spoke 9—Advanced Gerontechnologies for active and healthy ageing, Task 1.2—Design strategies to improve active/healthy ageing in primary healthcare facilities.

Conflicts of Interest: The authors declare no conflicts of interest.

References

1. Ministero della Salute. DECRETO 23 Maggio 2022, n. 77. In *Regolamento Recante la Definizione di Modelli e Standard Per lo Sviluppo Dell'assistenza Territoriale nel Servizio Sanitario Nazionale*; GU Serie Generale n.144; Ministero della Salute: Rome, Italy, 22 June 2022.
2. Brambilla, M.; Maciocco, G. *Dalle Case della Salute alle Case Della Comunità. La Sfida del PNRR per la Sanità Territoriale*; Carocci Editore: Roma, Italy, 2022.
3. Setola, N.; Borgianni, S. Promoting Health and Well-Being—Tools for Designing Community Health Centres and Healthy Neighbourhoods. *AGATHÓN Int. J. Archit. Art Des.* **2025**, *17*, 226–241.
4. Lauria, A. Progettazione Ambientale & Accessibilità: Note sul Rapporto Persona-Ambiente e sulle Strategie di Design. *TECHNE J. Technol. Archit. Environ.* **2017**, *13*, 55–62.
5. Conti, C.; Tatano, V. Accessibilità, tra tecnologia e dimensione sociale. In *Progettare Resiliente*; Lucarelli, M.T., Mussinelli, E., Daglio, L., Eds.; Maggioli Editore: Santarcangelo di Romagna, Italy, 2018; pp. 41–48.
6. Baratta, A.; Conti, C.; Tatano, V. Abitare inclusivo. Studi, ricerche e sperimentazioni. In *Abitare Inclusivo*; Baratta, A., Conti, C., Tatano, V., Eds.; Il Progetto per una vita autonoma e indipendente; Anteferma: Conegliano, Italy, 2019; pp. 14–17.
7. Jackson, R.J.; Kochtitzky, C. *Creating a Healthy Environment: The Impact of the Built Environment on Public Health*; Monograph Series; Sprawl Watch Clearinghouse: Washington, DC, USA, 2001; pp. 1–19.

8. Cass, S.J.; Ball, L.E.; Leveritt, M.D. Passive interventions in primary healthcare waiting rooms are effective in promoting healthy lifestyle behaviours: An integrative review. *Aust. J. Prim. Health* **2016**, *22*, 198–210. [CrossRef]
9. Sanchez, I.; Bully, P.; Martinez, C.; Grandes, G. Effectiveness of physical activity promotion interventions in primary care: A review of reviews. *Prev. Med.* **2015**, *76*, S56–S67. [CrossRef] [PubMed]
10. Rao, M.; Prasad, S.; Tissera, H.; Adshead, F. The built environment and health. *Lancet* **2007**, *370*, 1111–1113. [CrossRef] [PubMed]
11. Engelen, L.; Rahmann, M.; de Jong, E. Design for healthy ageing—The relationship between design, well-being, and quality of life: A review. *Build. Res. Inf.* **2021**, *50*, 19–35. [CrossRef]
12. Del Nord, R. *Lo Stress Ambientale Nel Progetto Dell'ospedale Pediatrico. Indirizzi Tecnici e Suggestioni Architettoniche*; Motta Architettura: Milano, Italy, 2006.
13. Von den Bosch, M. *Oxford Textbook of Nature and Public Health: The Role of Nature in Improving the Health of a Population*; Oxford University Press: Oxford, UK, 2018.
14. Kaplan, R.; Kaplan, S. *The Experience of Nature: A Psychological Perspective*; Cambridge University Press: New York, NY, USA, 1989.
15. Ulrich, R. View through a Window May Influence Recovery from Surgery. *Science* **1984**, *224*, 420–421. [CrossRef]
16. Ulrich, R.; Simons, R.; Losito, B.; Fiorito, E.; Miles, M.; Zelson, M. Stress recovery during exposure to natural and urban environments. *J. Environ. Psychol.* **1991**, *11*, 201–230. [CrossRef]
17. Evans, G.W.; McCoy, J.M. When buildings don't work: The role of architecture in human health. *J. Environ. Psychol.* **1998**, *18*, 85–94. [CrossRef]
18. Verderber, S. *Innovations in Hospital Architecture*; Routledge: New York, NY, USA, 2010.
19. HPH. *Global HPH Strategy 2021–2025*; HPH: Hamburg, Germany, 2020.
20. Golembiewski, R.T. Salutogenic architecture in healthcare settings. In *The Handbook of Salutogenesis*; Mittelmark, M., Ed.; Springer International Publishing: Cham, Switzerland, 2017; pp. 267–276.
21. Allen, J.; Bernstein, A.; Cao, X.; Eitland, E.S.; Flanigan, S.; Gokhale, M.; Goodman, J.M.; Klager, S.; Klingensmith, L.; Laurent, J.G.C.; et al. *Building Evidence for Health. The 9 Foundations of a Healthy Building*; Harvard T.H. Chan School of Public Health: Boston, MA, USA, 2017; pp. 1–36.
22. Campos Andrade, C.; Sloan Devlin, A. Stress reduction in the hospital room: Applying Ulrich's theory of supportive design. *J. Environ. Psychol.* **2015**, *41*, 125–134. [CrossRef]
23. Beukeboom, C.J.; Langeveld, D.; Tanja-Dijkstra, K. Stress-Reducing Effects of Real and Artificial Nature in a Hospital Waiting Room. *J. Altern. Complement. Med.* **2012**, *18*, 329–333. [CrossRef]
24. Ulrich, R.S.; Cordoza, M.; Gardiner, S.K.; Manulik, B.J.; Fitzpatrick, P.S.; Hazen, T.M.; Perkins, R.S. ICU Patient Family Stress Recovery During Breaks in a Hospital Garden and Indoor Environments. *Health Environ. Res. Des. J.* **2020**, *13*, 83–102. [CrossRef]
25. Del Nord, R. Lo stress e gli stressori in ospedale. In *Lo Stress Ambientale Nel Progetto dell'Ospedale Pediatrico. Indirizzi Tecnici e Suggestioni Architettoniche*; Del Nord, R., Ed.; Motta Architettura: Milano, Italy, 2006; pp. 60–79.
26. Aroua, I.; Hussein, F. Ambiance as a Key for a Better Birth Experience. *Stud. Health Technol. Inf.* **2024**, *29*, 205–220.
27. Skouboe, E.B.; Højlund, M. Crafting Atmospheres for Healthcare Design. *J. Somaesthetics* **2022**, *8*, 8–29.
28. Berkheimer, S.D.; Qian, C.; Malmstrom, T.K. Snoezelen Therapy as an Intervention to Reduce Agitation in Nursing Home Patients with Dementia: A Pilot Study. *J. Am. Med. Dir. Assoc.* **2017**, *18*, 1089–1091. [CrossRef] [PubMed]
29. Cavanagh, B.; Haracz, K.; Lawry, M.; James, C. Receptive Arts Engagement for Health: A Holistic and Trans-Disciplinary Approach to Creating a Multisensory Environment. *Sage Open* **2020**, *10*, 2158244020978420. [CrossRef]
30. Unwin, K.L.; Powell, G.; Jones, C.R. The use of Multi-Sensory Environments with autistic children: Exploring the effect of having control of sensory changes. *Autism* **2021**, *26*, 1379–1394. [CrossRef]
31. Bellini, E.; Setola, N. Sensory Design in healthcare welcome spaces for active and healthy aging. In *Effects of Design on Health and Wellbeing*; Verma, I., Arpiainen, L., Eds.; Studies in Health and Information Technology Book (HTI) Series; IOS Press: Amsterdam, The Netherlands, 2024; pp. 165–182.
32. Del Nord, R. L'ambiente e i fattori percettivo-sensoriali. In *Lo Stress Ambientale Nel Progetto dell'Ospedale Pediatrico. Indirizzi Tecnici e Suggestioni Architettoniche*; Del Nord, R., Ed.; Motta Architettura: Milano, Italy, 2006; pp. 102–149.
33. Barbiero, G. Biophilic design reframed. The theoretical basis for experimental research. *Ri-Vista Ric. Progett. Paesaggio* **2024**, *21*, 80–91. [CrossRef]
34. Setola, N.; Bellini, E.; Marcheschi, E. Santa Maria Nuova as a case study in evidence-based research. In *Art, Identity and Care. The Public Spaces of Santa Maria Nuova Hospital*; Diana, E., Geddes, M., Setola, N., Eds.; Edizioni Polistampa: Firenze, Italy, 2019; pp. 67–100.
35. Ullán, A.M. Artes visuales en hospitales pediátricos. El papel del arte en el bienestar psicológico de los niños hospitalizados. *Arte Individuo Soc.* **2022**, *34*, 1479–1501. [CrossRef]
36. Pati, D.; Nanda, U. Influence of positive distractions on children in two clinic waiting areas. *HERD Health Environ. Res. Des. J.* **2011**, *4*, 124–140. [CrossRef]

37. Awtuch, A.; Gbczyska-Janowicz, A. Art and Healthcare—Healing Potential of Artistic Interventions in Medical Settings. *IOP Conf. Ser. Mater. Sci. Eng.* **2017**, *245*, 042037. [CrossRef]
38. Miedema, E.; Lindahl, G.; Elf, M. Conceptualizing Health Promotion in Relation to Outpatient Healthcare Building Design: A Scoping Review. *HERD Health Environ. Res. Des. J.* **2018**, *12*, 69–86. [CrossRef]
39. Sport England. Active Design: Planning for Health and Wellbeing Through Sport and Physical Activity. 2015. Available online: https://www.sportengland.org/guidance-and-support/facilities-and-planning/design-and-cost-guidance/active-design (accessed on 24 July 2025).
40. *UNI 8289:1981*; Building. Functional Requirements of Final Users. Classification. UNI: Milano, Italy, 1981.
41. Del Nord, R.; Peretti, G. Ministero della Salute. In *L'Umanizzazione Degli Spazi di Cura. Linee Guida*; Tesis: Firenze, Italy, 2012.
42. NHS England. Technical Standards and Guidance (Health Building Notes/Health Technical Memoranda Documents). HBN 11-01: Facilities for Primary and Community Care Services. 2013. Available online: https://www.england.nhs.uk/publication/facilities-for-primary-and-community-care-services-hbn-11-01/ (accessed on 10 October 2025).
43. NHS England. Technical Standards and Guidance (Health Building Notes/Health Technical Memoranda Documents). HBN 00-01: General Design Guidance for Healthcare Buildings. 2014. Available online: https://www.england.nhs.uk/publication/designing-health-and-community-care-buildings-hbn-00-01/ (accessed on 10 October 2025).
44. NHS England. Technical Standards and Guidance (Health Building Notes/Health Technical Memoranda Documents). HBN 00-03: Clinical and Clinical Support Spaces. 2013. Available online: https://www.england.nhs.uk/publication/designing-generic-clinical-and-clinical-support-spaces-hbn-00-03/ (accessed on 10 October 2025).
45. McDonald, C.E.; Voutier, C.; Govil, D.; D'Souza, A.N.; Truong, D.; Abo, S.; Remedios, L.J.; Granger, C.L. Do health service waiting areas contribute to the health literacy of consumers? A scoping review. *Health Promot. Int.* **2023**, *38*, daad046. [CrossRef] [PubMed]
46. Maskell, K.; McDonald, P.; Paudyal, P. Effectiveness of health education materials in general practice waiting rooms: A cross-sectional study. *Br. J. Gen. Pract.* **2018**, *68*, e869–e876. [CrossRef] [PubMed]
47. Gignon, M.; Idris, H.; Manaouil, C.; Ganry, O. The waiting room: Vector for health education? the general practitioner's point of view. *BMC Res. Notes* **2012**, *5*, 511. [CrossRef] [PubMed]
48. Tamsuri, A.; Widati, S. Factors Influencing Patient Attention toward Audiovisual-Health Education Media in the Waiting Room of a Public Health Center. *J. Public Health Res.* **2020**, *9*, 81–84. [CrossRef] [PubMed]
49. Jawad, M.; Ingram, S.; Choudhury, I.; Airebamen, A.; Christodoulou, K.; Sharma, A.W. Television-based health promotion in general practice waiting rooms in London: A cross-sectional study evaluating patients' knowledge and intentions to access dental services. *BMC Oral Health* **2017**, *17*, 24. [CrossRef]
50. Reddy, J.R.; Pravallika, T.S.; Maddela, J.; Akula, S. Health Information in Hospital Waiting Rooms, Can It Act as a Vector in Health Promotion? Survey among Patients Attending Medical and Dental Hospitals. *J. Dent. Res. Rev.* **2017**, *4*, 97–100. [CrossRef]
51. Wangler, J.; Jansky, M. The influence of GP advice on physical activity and health promotion in elderly patients—Findings from a quantitative waiting room survey in Germany. *J. Public Health* **2025**. [CrossRef]
52. Van der Wardt, V.; di Lorito, C.; Viniol, A. Promoting physical activity in primary care: A systematic review and meta-analysis. *Br. J. Gen. Pract.* **2021**, *71*, e399–e405. [CrossRef]
53. Bertiato, F.; Bellini, E.; Setola, N. KALI project: A new concept for health promotion within waiting rooms. In *Ambient Assisted Living. ForItAAL 2024. Lecture Notes in Bioengineering*; Fiorini, L., Sorrentino, A., Siciliano, P., Cavallo, F., Eds.; Springer: Cham, Switzerland, 2024; pp. 30–50.
54. Prima la Comunità; La Casa Della Comunità. 2021. Available online: https://www.primalacomunita.it/wp-content/uploads/2021/04/Casa-della-comunita_PROGETTO-COMPLETO.pdf (accessed on 10 October 2025).
55. Agenas. Documento di indirizzo per il metaprogetto della Casa della Comunità. In *Quaderno di Monitor*; Agenzia Nazionale per i Servizi Sanitari Regionali: Rome, Italy, 2022.
56. Peretti, G.; Torricelli, M.C. (Eds.) Documento CNETO per le Case della Comunità e Ospedali di Comunità. Centro Nazionale per l'Edilizia e la Tecnica Ospedaliera (C.N.E.T.O.). 2023. Available online: https://www.saluteinternazionale.info/wp-content/uploads/2023/12/Case-della-Comunita-e-Ospedali-di-Comunita.pdf (accessed on 10 October 2025).
57. Quintelli, C.; Prandi, E.; Vertrame, G.; Furlotti, G.; Simbari, A. *Dalla Casa Della Comunita' Al Centro Di Salute Comunitaria. Il Progetto Architettonico E Urbano*; Festival Architettura Edizioni: Parma, Italy, 2025.
58. World Health Organization. *National Programmes for Age-Friendly Cities and Communities: A Guide*; World Health Organization: Geneva, Switzerland, 2023.
59. Bellini, E.; Macchi, A. Design Flexibility: Starting from ASD to learn how senses and sensory preferences impact people's wellbeing. In *Design for Dementia, Mental Health and Wellbeing Co-Design, Interventions and Policy*; Niedderer, K., Ludden, G., Dening, T., Holthoff-Detto, V., Eds.; Routledge: London, UK, 2024; Volume 3, pp. 318–330.
60. Setola, N.; Borgianni, S. *Designing Public Spaces in Hospitals*; Routledge: New York, NY, USA, 2016.

61. Zambelli, M. *La Conoscenza per il Progetto. Il Case-Based Reasoning nell'Architettura e nel Design*; Firenze University Press: Firenze, Italy, 2022.
62. Kolodner, J. *Case-Based Reasoning*; Morgan Kaufmann Publishers: Mateo, CA, USA, 1993.
63. Watson, I.; Perera, S. Case-based design: A review and analysis of building design applications. *Artif. Intell. Eng. Des. Anal. Manuf.* **1997**, *11*, 59–87. [CrossRef]

MDPI AG
Grosspeteranlage 5
4052 Basel
Switzerland
Tel.: +41 61 683 77 34

MDPI Books Editorial Office
E-mail: books@mdpi.com
www.mdpi.com/books

www.ingramcontent.com/pod-product-compliance
Lightning Source LLC
Chambersburg PA
CBHW061544220926
51850CB00005B/247